Calculus Concepts

An Informal Approach to the Mathematics of Change

Brief First Edition

Calculus Concepts

An Informal Approach to the Mathematics of Change

Brief First Edition

Donald R. LaTorre
John W. Kenelly
Iris B. Fetta
Clemson University

Laurel L. Carpenter
Lansing, Michigan

Cynthia R. Harris
Reno, Nevada

HOUGHTON MIFFLIN COMPANY   **Boston**   **New York**

EDITOR-IN-CHIEF: *Charles Hartford*

ASSOCIATE EDITOR: *Elaine Page*

SENIOR PROJECT EDITOR: *Maria Morelli*

EDITORIAL ASSISTANT: *Christian Zabriskie*

SENIOR PRODUCTION/DESIGN COORDINATOR: *Carol Merrigan*

SENIOR MANUFACTURING COORDINATOR: *Sally Culler*

MARKETING MANAGER: *Ros Kane*

COVER DESIGN: Deborah Azerrad Savona

COVER IMAGE: New York, Midtown skyline from Queensboro Bridge. Photograph by Jon Ortner. © Tony Stone Images

Printed in the U.S.A.

International Standard Book Numbers:
  Brief Edition Text: 0-669-39859-4
  Brief Edition Exam Copy: 0-669-45197-5
  Brief Edition Text and Graphing Calculator Keystroke Guide: 0-669-46127-X
  Complete Edition Text: 0-669-45125-8
  Complete Edition Exam Copy: 0-669-45196-7
  Complete Edition Text and Graphing Calculator Keystroke Guide: 0-395-90604-0

  School Edition Text: 0-669-45126-6
  School Edition Text and Graphing Calculator Keystroke Guide: 0-669-46152-X

  Graphing Calculator Keystroke Guide: 0-669-39864-0

1 2 3 4 5 6 7 8 9–DC–02 01 00 99 98

CONTENTS

TO INSTRUCTORS

What This Book is About

This book presents a fresh, new approach to the concepts of calculus for students in fields such as business, economics, liberal arts, management, and the social and life sciences. It is appropriate for courses generally known as "brief calculus" or "applied calculus."

Philosophy

Our overall goal is to improve learning of basic calculus concepts by involving students with new material in a way that is significantly different from traditional practice. The development of conceptual understanding, not mastery of algebraic skill and technique, is our guiding force coupled with a commitment to make calculus meaningful to the student. Thus, the material in this book is data-driven and technology-based, with a unique modeling approach. It considers the ability to correctly interpret the mathematics of real-life situations of equal importance to the understanding of the concepts of calculus in the context of change.

Data-Driven

Many everyday, real-life situations involving change are discrete in nature and manifest themselves through data. Such situations often can be represented by continuous or piecewise continuous mathematical models so that the concepts, methods, and techniques of calculus can be brought to light. Thus we seek, when appropriate, to make real-life data a starting point for our investigations.

The use of real data and the search for appropriate models also exposes the students to the reality of uncertainty. We emphasize that sometimes there can be more than one appropriate model and that answers derived from models are only approximations. We believe that exposure to the reality that mathematics is not always right or wrong is valuable.

Technology-Based

Calculus has traditionally relied upon a high level of algebraic manipulation. However, many non-technical students are not strong in algebraic skills, and an algebra-based approach tends to overwhelm them and stifle their progress. Today's easy access to technology in the form of graphing calculators and microcomputers breaks down barriers to learning imposed by the traditional reliance on algebraic methods. It creates new opportunities for learning through graphical and numerical representations. We welcome these opportunities in this book by assuming continual and immediate access to technology.

This book requires that students use graphical representations (scatter plots of data and graphs of functions) freely, make numerical calculations routinely, and fit functions to data. Thus, continual and immediate access to technology is absolutely essential. Because of their low cost, portability, and ability to personalize the mathematics, the authors prefer graphing calculators. These materials have also been successfully taught using microcomputer software (such as Maple) and spreadsheets.

It is worth noting that different technologies may give different model coefficients than those given in this book. We used a TI-83 graphing calculator to generate the models in the text and the answer key. Other technologies may use different fit criteria for some models than the criteria used by the TI-83.

Modeling Approach

We believe that modeling is an important tool and introduce it at the outset. Both linear and nonlinear models of discrete data are used to obtain functional relationships between the variables of interest. The functions given by the models are the ones used by students to conduct their investigations of calculus concepts. It is the connection to real-life data that most students feel shows the relevance of the mathematics in this course to their lives and adds reality to the topics studied.

Interpretation Emphasis

This book is substantially different from traditional texts, not only in the philosophy but also in its overall focus, level of activities, development of topics and attention to details. Interpretation of results is a key feature of this text that allows students to make sense of the mathematical concepts and appreciate the usefulness of those concepts in their lives.

Informal Style

While we appreciate the formality and precision of mathematics, we also recognize that this alone can deter students from access to mathematics. Thus, we have sought to make our presentations as informal as possible by using non-technical terminology where appropriate and a conversational style of presentation.

Projects

Projects included after each chapter are intended to be group projects with oral and/or written presentations. We recognize the importance of helping students develop the ability to work in groups, as well as hone presentation skills. The projects also give opportunity for students to practice the kind of writing that they will likely have to do in their future careers.

Other Pedagogical Features

Chapter Opener   Each chapter opens with a real-life situation and several questions about the situation that relate to the key concepts in the chapter.

Concept Inventory   A Concept Inventory is listed at the end of each section, giving students a brief summary of the major ideas developed in that section.

Section Activities   The Section Activities begin by cementing concepts followed by explorations of topics using, for the most part, actual data in a variety of real-world settings. Questions and interpretations pertinent to the data and the concepts are always included in these activities. The activities do not mimic the examples in the chapter discussion and thus require more independent thinking on the part of the students. Possible answers to odd activities are given at the end of the book.

Chapter Summary A Chapter Summary connects the results of the chapter topics and further emphasizes the importance of knowing these results.

Chapter Review Test A Chapter Review Test at the end of each chapter provides practice with techniques and concepts. Answers to the Chapter Review Tests are included in the answer key.

Supplements

The Instructor's Guide gives practical suggestions for using the text in the manner intended by the authors. It contains sample tests, ideas for in-class group work, suggestions for implementing and grading projects and complete activity solutions.

The technology supplements provide technology-specific instructions ordered to match the organization of the text chapters. An open-book icon appears at places in the text where a new concept or skill is presented in the technology supplements.

A Student Solutions Guide is also available. A Test Item File is available PC and Macintosh formats.

Acknowledgments

We gratefully acknowledge the help and support of many people during the development of this book.

We express our gratitude to the Fund for the Improvement of Post-secondary Education (FIPSE), U.S. Department of Education, which provided substantial funding for this project. FIPSE, your dedication to innovation and improvement of American post-secondary education is genuinely appreciated.

We appreciate the thoughtful advice of several of our colleagues at Clemson University: P.M. Dearing, Matt Saltzman, Herman Senter, and Sherry Biggers, as well as that of Bruce Blackadar at University of Nevada Reno.

The many hours spent by the FIPSE project evaluators, Dr. Jim Wilson and Mary Beth Searcy of the University of Georgia, and Denise Johnson and Susan Kadles at Alma College, were extremely helpful in guiding the direction of the text. We sincerely appreciate all their advice.

The materials were used in hundreds of classes at Clemson since 1994. The authors appreciate the comments of the Clemson University students and graduate teaching assistants in these classes who provided valuable feedback as the material was being developed and tested.

Special thanks to April Haynes, Gloria Orr, Becky Singletary, and Karen McConkie who word processed the manuscript.

Class testing was also conducted by many different teachers at many different schools. Initial class testing was conducted by the teachers listed below. The authors would like to thank these instructors and their students for their many thoughtful comments and valuable contributions.

Ellen King, Kim Freeman, *Anderson College, SC*

John Haverhals, Mary Jane Sterling, Tom McKenzie, Mike McAsey, Chris Stewart, Libin Mou, *Bradley University, IL*

Linda Nash, Martha Wicker, Mary Stephens, Paul Myers, Catherine Aust, *Clayton State College, GA*

Nancy Mauldin, Gary Harrison, *College of Charleston, SC*

Daniel Alexander, Larry Naylor, *Drake University, IA*

Heidi Staebler, *Texas A & M at Commerce*

Joe Cieply, Gina Kietzmann, *Elmhurst College, IL*

Jeffrey Clark, Terri Johnson, *Elon College, NC*

Hugh Williams, Robin Baumgarner, Glenn Jacobs, *Greenville Technical College, SC*

Marlene Sims, *Kennesaw State College, GA*

Jerry Bolick, Vicki Schell, Lloyd Smith, Ron Butler, *Lenoir-Rhyne College, NC*

Kathleen Bavelas, Kathy Peters, *Manchester Community Technical College, XX*

Ann Preston, Cheryl Slayden, *Pellissippi State Technical Community College, TN*

Siham Alfred, Lance Hemlow, *Raritan Valley Community College, NJ*

Jacqueline Fernandez, *Santa Barbara City College, CA*

Richard Sauvageau, *Staples High School, CT*

Paul Ache III, Dan Lewis, Robert Main, Elton Lacey, Yvette Hester, *Texas A&M University, TX*

Carollyne Guidera, *University College of the Fraser Valley, Canada*

Mary Beth Searcy, *University of Georgia, GA*

Bruce Blackadar, *University of Nevada Reno, NV*

Suzanne Smith, John Thornton, *University of North Carolina Charlotte, NC*

Stephen King, Frank Townsend, Patricia Brown, David Jaspers, Jack Leifer, Mike May *University of South Carolina Aiken, SC*

Eddie Warren, Jim Harvey, *University of Texas Arlington, TX*

Helen Read, *University of Vermont, VT*

Audrey Borchardt, Robert DeVos, Bruce Pollack-Johnson, *Villanova University, PA*

Jim Snodgrass, Bernd Rossa, Sheila Doran, Martha Holland, Danny Otero, *Xavier University, OH*

Special thanks go to Emily Keaton for her careful work in checking the text and answer key for accuracy. The authors express their sincere appreciation to Charlie Hartford, Elaine Page, Maria Morelli, and their staffs at Houghton Mifflin Company for all their work in bringing this first edition into print.

TO STUDENTS

What this Book is About

This book is written to help you understand the inner workings of how things change and to help you build systematic ways to use this understanding in everyday real-life situations that involve change. Indeed, a primary focus of the material is on change, since calculus is the mathematics of change.

Even if you have studied calculus before, this book is probably different from any other mathematics textbook that you have used. It is based on three premises:

1. Understanding is as important as the mastery of mathematical manipulations. Algebraic skill and the ability to manipulate expressions must be regularly practiced, or they will fade away. If you understand concepts, you will be able to explain some things in your life forever.

2. Mathematics is present in all sorts of real-life situations. It is not just an abstract subject in textbooks. In real life, mathematics is often messy and not at all like the tidy, neat equations that you were taught to factor and solve. Speaking of equations, where do they come from? Nature seldom whispers an equation into our ears.

3. The new graphics technology in today's calculators and computers is a powerful tool that can help you understand important mathematical connections. Like many tools in various fields, technology frees you from tedious, unproductive work; enables you to engage situations more realistically; and lets you focus on what you do best . . . think and reason.

How to Use this Book

- Begin by throwing away any preconceived notions that you may have about what calculus is and any notion that you are "not good" in mathematics.

- Make a commitment to learn the material: not just a good intention, but a genuine commitment.

- Study this book. Notice that we said "study", not "read". Reading is a part of study, but study involves much more. You should not only read (and re-read) the discussions, but work through each example to understand its development.

- Use paper, pencil and your graphing calculator or computer when you study. These are your basic tools, and you cannot study effectively without them.

- Find a study partner, if at all possible. Each of you will be able to help the other learn. Communicating within mathematics, and about mathematics, is important to your overall development toward understanding mathematics.

- Write. Write your solutions clearly and legibly, being certain to interpret all of your answers with complete sentences using proper grammar. Careful writing will help you sort through your ideas and focus your learning.

- Make every effort not to fall behind. You know the dangers, of course, but we remind you nevertheless.
- Finally, remember that there is no substitute for effective study. You have your most valuable resource with you at all times—your mind. Use it.

Calculus Concepts

An Informal Approach to the Mathematics of Change

Brief First Edition

Ingredients of Change: Functions and Linear Models

The primary goal of this book is to help you understand the two fundamental concepts of calculus—the derivative and the integral—in the context of the mathematics of change. This first chapter is therefore devoted to a study of the key ingredients of change: functions and mathematical models.

Functions give us a way to analyze the mathematics of change, because they enable us to describe relationships between variable quantities. Examples are physical quantities such as time and distance traveled; business quantities such as sales and revenue; social quantities such as birth rates, populations, and airline travel; and economic quantities such as interest rates and investment yields. Because linear functions are central to many situations involving change, we examine them in detail.

This chapter introduces you to the process of building mathematical models. The models are often constructed from data, and an important and necessary part of our focus is mathematical decision making and the interpretation of results that accompanies the use of the models. In later chapters, we apply methods and techniques of calculus to the equations the models provide as we investigate rates of change and their implications in a variety of everyday situations.

A corn chopper cuts stalks of corn into silage that is used as feed for livestock. We could ask several questions about the harvesting of corn that can be answered mathematically using functions and calculus: (1) How many tons of silage can be chopped from 40 acres of corn? 80 acres? 120 acres? (2) How many acres of corn must be chopped in order to produce 2 tons of silage? 3 tons? 4 tons? (3) How many stocks are chopped per minute? (4) At what rate is silage being poured from the corn chopper?

Can you think of other questions that might have numerical answers that could be obtained from mathematical models?

1.1   Fundamentals of Modeling

Watch television, read a newspaper, look at any magazine, visit a manufacturing plant, or ask a business executive about information used to determine daily courses of action. Chances are that you will see a graph or table presenting data or a conclusion that was reached as a result of the collection of data. Important decisions are made each day using the results of data.

To analyze the information contained in real-world data, we often build a mathematical **model**. The model should describe the current situation as closely as possible and allow for modification in case unforeseen events affect the situation. No matter what career you choose, you will need to know how to apply the basic principles of mathematical modeling. Because calculus is often regarded as the mathematics of change between related variables, the modeling process will prove to be an important tool as we attempt to describe the relationship between the variables and their changing nature.

You have already encountered situations in which you used models. Consider the process of learning to drive a car. You may have taken a driver's education course in which you used a driving simulator to model actual highway conditions. If not, you learned the rules of driving from a pamphlet obtained from your state highway department. Did you expect to get in a car and immediately be an expert driver? Of course not! Along with committing to memory the rules for safe driving and knowing how to operate your vehicle, you found that expert driving requires understanding the relationships among many separate pieces of information and then skillfully putting into practice all that you learned. Even though learning to drive is not directly connected with mathematical formulas or expressions, you had to make important decisions on the basis of your information and experience, and you had to have a thorough understanding of the situation in order to deal with changes in circumstances.

In addition to developing your understanding of basic calculus concepts and providing you with some techniques for solving problems that you will encounter in the practical applications of calculus, the material in this book will also help you formulate mathematical models to describe real-life situations and interpret the results so they can be applied to solving the associated problems. The models that we study are fairly simple. In actual practice, many other factors that we do not consider are present. However, the techniques that we study are consistent with those used in more complex models and will suffice for the situations we shall encounter here.

What Is Mathematical Modeling?

In order to get a feeling for what the words *model* and *modeling* mean in mathematics, we start with two simple situations. How do insurance companies set the rates (premiums) for automobile insurance? First, data describing accidents and drivers are obtained. These data are likely to show that some of the factors affecting persons involved in an accident or the cost of repairs to automobiles are

- age of the driver
- sex of the driver
- physical condition of the driver
- past performance of the driver
- make and model of the vehicle
- speed of the vehicle
- county in which the vehicle is registered

Can you think of other factors that the insurance company might consider in constructing a model to determine premiums for automobile insurance? Once the factors are determined, the data are used to estimate the contribution of each factor to the model, and the premium is set by applying the model to specific situations. How can the insurance company determine whether the model is appropriate? Well, if the company generates enough income to be profitable and its rates are competitive with other companies, then the model is successful. The process of translating the problem from its real environment to a mathematical setting and then back again to the real-life situation is the very essense of **mathematical modeling**.

As a second example, suppose you wish to determine how many hours you should study for the final exam in this course. What factors might influence this decision? You should probably consider

- your number of class absences
- the level of difficulty of the course
- the time you spend on the course outside of class
- your grades on quizzes
- your grades on major tests
- your level of understanding of the material
- the number of final exams you must take that day
- your physical condition

What other factors should you consider? Can you assign a value to the contribution of each of these variables and formulate a mathematical model that produces the number of hours that you should study? It's not so easy, is it? One of the things that you will notice about the material in this book is that there is not always a unique approach or a single solution to a problem. This fact, in itself, models the real world! One of our goals is to help you develop the confidence to try your own approach and be able to support your conclusions. That is exactly what you will need to do when you begin to apply these techniques in your chosen profession.

The Role of Graphs

One of the first steps in modeling a situation is to understand clearly the facts that are known. This information is often presented in the form of a graph. Consider the graph in Figure 1.1, which shows the speed of two cars on an interstate highway.

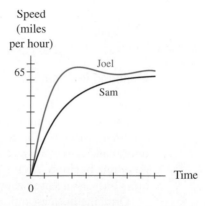

Speed of Two Cars

FIGURE 1.1

■ What general information is presented by this graph?

It appears reasonable to assume that Joel and Sam are the drivers of the cars. We also can probably assume that 65 miles per hour (mph) is the speed limit on this interstate highway.

■ Has either car exceeded the speed limit (assuming the speed limit is 65 mph)?

Yes. Imagine a horizontal line 65 units above the time axis. The graph representing Joel's speed is above this line after the second mark to the right of 0 on the time axis.

■ Which car appears to be accelerating faster during the first 3 units of time?

Joel reaches the speed limit earlier than Sam, so Joel's speed must be increasing faster than Sam's. In other words, the graph of Joel's speed is steeper than the graph of Sam's speed during the first 3 units of time. Remember: **Acceleration** is the rate of change of speed.

■ Note that no units (labels) have been placed on the time axis. On the basis of your own driving experience, give a numerical value and unit of measure for the first tick mark on the horizontal axis.

This value will, of course, differ for each person and type of vehicle. It appears to take Joel the time represented by approximately 2 tick marks to reach 65 mph, whereas Sam approaches that speed at a more leisurely pace. Because most vehicles can reach 65 mph in about 8 seconds, let us assume that the first tick mark on the horizontal axis represents 4 seconds.

■ How fast is each car traveling after 12 seconds?

At 3 tick marks past zero, Sam is traveling approximately 50 mph and Joel is traveling approximately 70 mph.

■ Does either car appear to be using cruise control?

The graph indicates that Sam's speed increases and then levels off, whereas Joel's speed fluctuates. Sam may be using cruise control after reaching the speed limit.

■ Will Sam pass Joel?

There is not enough information given in the graph to answer this question. We are not told whether these two vehicles are traveling in the same direction. We do not even know whether they are on the same portion of the interstate highway.

Modeling in Business

Modeling in the business world often involves terms that are easily understood. Examples of such business terms are *fixed costs* (also called *start-up costs* or *overhead*), *variable costs*, *total cost*, *revenue*, *profit*, and *break-even point*. To help you understand these terms, consider an enterprising college student who opens a dog-grooming business. An initial capital investment is needed to purchase equipment, buy a business license, set up a lease agreement for a shop, and so on.

Once the business officially opens, the owner sets aside money at the beginning of each week to pay rent, utilities, salaries, and the like. These costs remain the same regardless of the number of dogs that are groomed during the week and are called **start-up costs** or **fixed costs**. Other costs change with the number of customers. These **variable costs** include expenditures such as advertising, supplies, laundry and custodial fees, and overtime. The fixed costs plus the variable costs give the **total cost** of doing business.

Fixed costs + variable costs = total cost

When the commodity produced (dogs groomed in this example) can be measured in units, then the total production cost divided by the number of units produced is known as the **average cost**.

$$\frac{\text{Total cost}}{\text{number of units produced}} = \text{average cost}$$

Businesses receive **revenue** which is, in general, the quantity sold times the price. In the pet grooming business, revenue is the number of dogs groomed times the grooming price. **Profit** is total revenue minus total cost. The graph in Figure 1.2 shows the weekly profit for this business as function of the number of dogs groomed.

Profit = total revenue − total cost

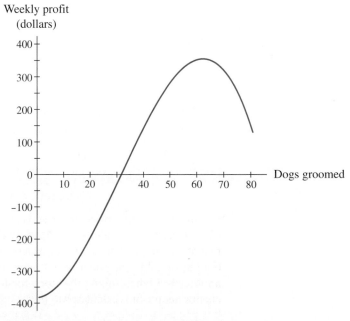

Pet Grooming Weekly Profit

FIGURE 1.2

The places where a graph crosses (or touches) the horizontal and vertical axes are called **intercepts**. The intercepts often have interpretations in the context of the situation that the graph represents. The horizontal axis intercept in Figure 1.2 is approximately 31, and the vertical axis intercept is about −380. Therefore, from this graph we can conclude that the fixed costs per week are approximately $380 (the cost associated with grooming zero dogs). Profit is negative until 31 dogs are groomed. This point on the graph is called the **break-even point**, the point at which revenue equals cost so that profit is zero.

> The **break-even point** occurs when revenue equals cost and profit equals 0.

Profit appears to peak at approximately $350. This occurs when about 62 dogs are groomed each week. Thus 62 is the optimum number of dogs for this business to groom each week. If more than 62 dogs were groomed each week, profit would actually decline, perhaps because of the need to pay overtime or hire another employee. Cost, profit, revenue, and break-even points are fundamental concepts that are important for business and nonbusiness students to understand. We will refer to them regularly throughout this text.

Using Models in Real Situations

As we look at real-life models, we will often ask you to evaluate numerical expressions, to solve equations, to estimate and check the reasonableness of answers, and to graph functions. We expect that in so doing, you will use technology appropriate to the setting: a graphing calculator or a computer.

EXAMPLE 1   *A Stockbroker Transaction*

A stockbroker's client purchases 250 shares of stock at $29.75 a share. Two months later, the client sells the stock at $43.50 a share. If the stockbroker's fee is $45 per transaction, what is the client's net profit?

Solution: The model that the stockbroker uses to determine the client's net profit is given by the equation

$$\text{Net profit} = \text{selling price} - (\text{purchase price} + \text{transaction fees})$$

If we use the letters s, p, and t to represent selling price, purchase price, and transaction fees respectively, we can write the equation for net profit as

$$\text{Net profit} = s - (p + t) \text{ dollars}$$

Calculate the selling price of the stock as $s = (250)(\$43.50) = \$10,875$ and the purchase price as $p = (250)(\$29.75) = \$7437.50$. The transaction fee is fixed at $45; that is, it remains the same regardless of the number of shares of stock that are bought or sold. The client must therefore pay $45 when purchasing the stock and another $45 when selling the stock for total transaction fees of $t = \$90$. Thus, the client's net profit is calculated to be

$$\text{Net profit} = s - (p + t) \text{ dollars}$$
$$= \$10,875 - (\$7437.50 + \$90)$$
$$= \$3347.50$$

EXAMPLE 2   *A Savings Account*

Table 1.1 lists the accumulated amount A in a savings account earning interest at a yearly rate of 5%. The variable t represents the number of years since the account was opened. Assume that the initial deposit is $1000 and that no other deposits or withdrawals are made from the account.

TABLE 1.1

t	0	1	2	3	4
A	$1000	$1050	$1102.50	$1157.63	$1215.51

The model giving the amount in the account at the end of each year is

$$A = 1000(1+0.05)^t \text{ dollars after } t \text{ years}$$

Graph this equation, and then predict the amounts in the account 5, 7, and 10 years after it was opened.

Solution: In order to draw an appropriate graph of the amount model, we should first consider which values of t and A we want to see. What values of t make sense in this problem? Because t represents time in years since the account was opened, and we want the accumulated amount after 10 years, we probably want to view the graph with the minimum value of t as 0 and the maximum value of t as 10. In order to determine the correct output range for these input values, we note that the amount in the account is increasing and calculate the minimum value of A as $1000 when $t = 0$ and the maximum value of A as $1000(1 + 0.05)^{10} \approx \$1628.89$ when $t = 10$. If you graph $A = 1000(1 + 0.05)^t$ with t ranging from 0 to 10 and A from 900 to 2000, then you should see the graph in Figure 1.3.

FIGURE 1.3

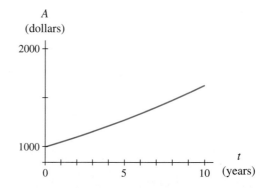

Accumulated Amount in a Savings Account

You can obtain the requested values by evaluating the function $A = 1000(1 + 0.05)^t$ at each of the input values $t = 5, 7,$ and 10. The exact[1] values are shown in Table 1.2.

TABLE 1.2

Year	5	7	10
Amount	$1276.28	$1407.10	$1628.89

---

1. We use the word *exact* in the sense that the values are calculated from the function and not obtained by estimation from viewing or tracing the graph.

In each of the foregoing examples, we were able to describe a real-life situation in terms of a general mathematical framework. This is the essence of mathematical modeling. In each case, we used the available information to produce a mathematical equation that described the relationship between the variables of interest. This, too, is typical of mathematical modeling. However, as the next example shows, it is not always possible to produce an equation.

EXAMPLE 3   *Bacteria Growth*

We all have bacteria on our skin, and they reproduce at a rapid rate. The graph in Figure 1.4 describes the growth of a colony of bacteria on the human body.

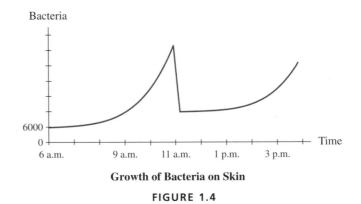

Growth of Bacteria on Skin

FIGURE 1.4

a. Describe a situation that might result in the graph in Figure 1.4 for the growth of bacteria on your skin.

b. Fill in values for the unlabeled tick marks on the vertical axis.

c. Use the graph to estimate the number of bacteria on your skin at 9 a.m. and at 3 p.m.

d. At what times is the concentration of bacteria on your skin the highest? Give an estimate of the maximum number of bacteria on your skin during the time interval indicated on the graph.

e. Estimate the intervals of time during which the number of bacteria is increasing.

f. Estimate the intervals of time during which the number of bacteria is decreasing.

Solution:

a. There are many possible descriptions you could give. Consider this one. While you are sleeping in your air-conditioned apartment, the rate of growth of the bacteria is slow. After you arise at 8 a.m., the colony grows faster, and its growth really accelerates when you play tennis between 9 a.m. and 11 a.m. You then take a 15-minute shower and wash off most of the pesky critters! You relax, eat lunch in your apartment, and go jogging at about 2 p.m. After jogging, you remain outside to talk with a friend until approximately 3:30 p.m. In the humid, moist atmosphere created by jogging, the bacteria grow rapidly.

b. The vertical axis on the graph indicates that each tick mark represents 6000 bacteria. Thus, reading upward, the remaining tick marks are labeled 12,000, 18,000, 24,000, 30,000, 36,000, and 42,000.

c. The number of bacteria on your skin at 9 a.m. is between 6000 and 12,000. One possible visual estimate is 11,500. The number at 3 p.m. is between 18,000 and 24,000. A possible answer is 22,000 bacteria.

FIGURE 1.5

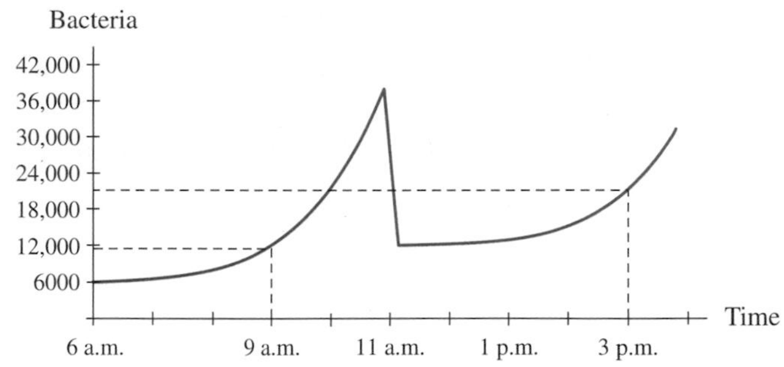

Bacteria

Growth of Bacteria on Skin

d. The times at which the concentration of bacteria on your skin appears the highest in relation to nearby times are 11 a.m. and around 4 p.m. The maximum number of bacteria is a value of the variable *Bacteria*. The value is approximately 38,000 bacteria at 11 a.m. and 36,000 bacteria at 4 p.m.

e. If a graph is rising as you move from left to right along the horizontal axis, then it is said to be **increasing**. This graph is increasing between approximately 6 a.m. and 11 a.m. It is also increasing from the time you finish your shower, say 11:15 a.m., until approximately 4 p.m.

f. A graph is said to be **decreasing** if it is falling as you move from left to right along the horizontal axis. This graph is decreasing between 11 a.m. and 11:15 a.m. A graph that neither rises nor falls is called a **constant** graph. This graph is never constant. ■

Our final example takes us straight to the heart of mathematical modeling in calculus.

EXAMPLE 4   *Building a Tunnel*

A civil engineer is planning to dig a tunnel through a mountain. The tunnel will begin 575 feet above sea level and will be constructed with a constant upward slope of 5%. The engineer uses this information and determines the equation for y, the elevation above sea level of the tunnel, to be $y = 0.05x + 575$ feet, where x is the horizontal distance in feet from where the tunnel begins at the base of the mountain.

FIGURE 1.6

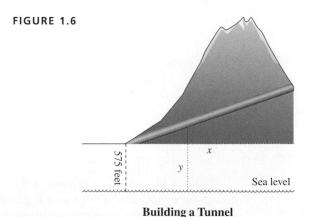

575 feet

Building a Tunnel

(figure not drawn to scale)

a. How much is the tunnel rising for each 1 foot of horizontal increase?

b. How much is the tunnel rising for each 100 feet of horizontal increase?

c. Find the elevation of the tunnel at a horizontal distance of 2500 feet from its starting point.

d. If the tunnel exits the mountain at a horizontal distance of 7000 feet from where it began, what is the elevation of the tunnel when it emerges from the mountain?

e. What is the length of the tunnel?

Solution:

a. Substituting $x = 1$ in the model $y = 0.05x + 575$, we obtain $y = 575.05$ feet. However, this is the elevation of the tunnel above sea level. Because the elevation is 575 feet when $x = 0$, we see that the tunnel has risen only 0.05 foot for the first (and each succeeding) 1 foot of horizontal increase.

b. Using a similar explanation, we see that the tunnel is rising 5 feet for each 100 feet of horizontal increase.

c. Any of the following three methods can be used to determine the elevation. *Algebraically*: Substitute $x = 2500$ in the model to obtain 700 feet above sea level. *Graphically*: Draw a graph of the function $y = 0.05x + 575$ with input values, x, from 0 to 5000, and a suitable output range. Read the y-value from the graph that corresponds to $x = 2500$. It is $y = 700$. *Numerically*: Create a table of values using the answers to parts a and b of this problem (see Table 1.3). Because the tunnel rises 0.05 feet for each 1 foot of horizontal increase and 5 feet for each 100 feet of horizontal increase, it should rise 25 feet for each 500 feet of horizontal increase.

TABLE 1.3

horizontal increase (feet)	500	1000	1500	2000	2500
vertical increase (feet)	25	50	75	100	125

Because the elevation at the starting point is 575 feet, the height of the tunnel 2500 feet from the starting point is $125 + 575 = 700$ feet above sea level.

d. Using any of the methods described in part c, we find that the elevation of the tunnel when it emerges from the mountain is 925 feet.

e. Using the Pythagorean Theorem, we find that

$$\text{Length of tunnel} = d = \sqrt{7000^2 + 350^2} = \sqrt{49,122,500} \approx 7009 \text{ feet}$$

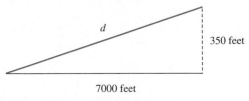

7000 feet

350 feet

d

FIGURE 1.7

Example 4 serves as a preview to our use of mathematical models in calculus. It uses the available data to produce a mathematical equation that describes the relationship between the variable quantities of interest (elevation of the tunnel above sea level and horizontal distance from where the tunnel enters the mountain). It then uses the model to investigate the rate of change of the tunnel's elevation (how rapidly the tunnel is rising) and to answer associated questions. Throughout the book, you should keep in mind this overall view of mathematical modeling. Modeling is an important tool that will ultimately help us bring the power of calculus to a variety of real-life situations.

1.1 Concept Inventory

- Mathematical modeling
- Business terms:
    fixed costs (start-up costs)
    variable costs
    total cost
    average cost
    revenue
    profit
    break-even point
- Graph terms:
    increasing
    decreasing
    constant
    axis intercept

1.1 Activities

1. The graph in Figure 1.1.1 shows the cost of buying compact disks at a music store that offers one free CD with the purchase of four CDs priced at $12 each (tax included).

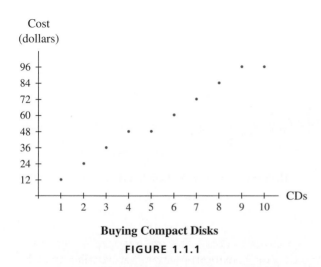

Cost (dollars)

Buying Compact Disks

FIGURE 1.1.1

a. What is the cost of 6 CDs?

b. How many CDs could you buy if you had $36?

c. How many CDs could you buy if you had $80?

d. What is the average price of each CD if you buy 3 CDs? 6 CDs?

2. A fraternity is selling T-shirts on the day of a football game. The shirts sell for $8 each.
   a. Complete the following revenue table (Table 1.4).

TABLE 1.4

Number of shirts sold	Revenue (dollars)
1	
2	
3	
4	
5	
6	

b. Construct a revenue graph by plotting the points in the table.

c. How many T-shirts can be purchased with $25?

d. If an 8% sales tax were added, how many T-shirts could be purchased with $25?

3. You are interested in buying a used car and in financing it for 60 months at 10% interest. As a special promotion, the dealer is offering to finance with no down payment. The graph in Figure 1.1.2 shows the value of the car as a function of the amount of the monthly payment.

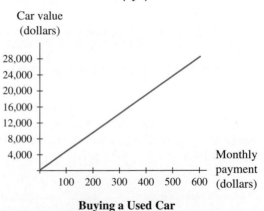

Car value (dollars)

Monthly payment (dollars)

Buying a Used Car

FIGURE 1.1.2

a. Estimate the value of the car you can buy if your monthly payment is $200.

b. Estimate the monthly payment for a car that costs $16,000.

c. Estimate the amount that your monthly payment will increase if you buy a $20,000 car rather than a $15,000 car.

d. How would the graph change if the interest rate were 12.5% instead of 10%?

4. You have decided to purchase a car for $18,750. You have 20% of the purchase price to use as a down payment, and the purchase will be financed at 10% interest. The graph in Figure 1.1.3 shows the monthly loan payment as a function of the number of months over which the loan is financed.

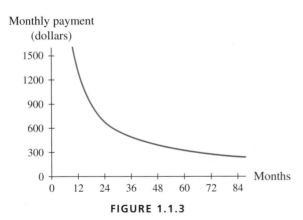

FIGURE 1.1.3

a. What is the actual amount being financed?

b. Estimate your monthly payment if you finance the purchase over 36 months.

c. Estimate the number of months you will have to pay on the loan if you can afford to pay only $300 a month.

d. If you decrease the time financed from 48 to 36 months, how much will your payment increase?

e. How would the graph change if you were buying a $20,000 car?

5. The graph[2] in Figure 1.1.4 shows the percentage by which social security checks have increased as a result of cost-of-living adjustments from 1989 through 1996.

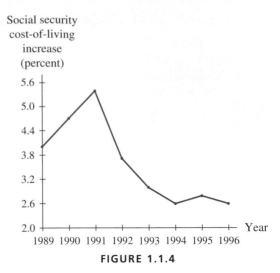

FIGURE 1.1.4

a. What was the cost-of-living increase in 1992?

b. When was the cost-of-living increase the greatest? Estimate the cost-of-living increase in that year.

c. When was the cost-of-living increase 2.6%?

d. Did social security benefits increase or decrease between 1991 and 1994? Explain.

6. The graph[3] in Figure 1.1.5 compares the minimum wage to its adjusted value in constant 1995 dollars.

Real and Adjusted (1995 dollars) Minimum Wage

FIGURE 1.1.5

---

2. Social Security Administration.

3. Bureau of Labor Statistics.

a. Estimate the minimum wage in 1970 and the adjusted minimum wage in that same year.

b. During the years when the minimum wage did not increase, what happened to the buying power of the minimum wage?

c. In what years was the adjusted minimum wage below $5.75?

7. Suppose gasoline costs $1.25 per gallon.

   a. How much does $5\frac{1}{2}$ gallons cost?

   b. How many gallons can you buy for $12?

   c. Write a model for the cost of g gallons of gas.

8. Suppose that a vending machine in your dorm dispenses cans of soda for $1 and bags of potato chips for 75 cents.

   a. How much would 4 sodas and 3 bags of chips cost?

   b. How many bags of chips could you buy if you wanted 2 cans of soda and had $5.25 to spend?

   c. Write a model giving the cost of s cans of soda and c bags of chips.

9. In a certain neighborhood, homes are valued at $40 per square foot, driveways at $1.25 per square foot, and decks at $2.85 per square foot. If you were an appraiser, what mathematical model would you use to determine the value of an 1850-square-foot home with a 350-square-foot deck and a driveway that is 8 feet wide and 50 feet long? Use your model to find the appraised value of this home.

10. Corporations normally reimburse their employees for business-related travel. The expenses that companies pay typically include lodging and meals. Many companies also pay a fixed amount for mileage used. Write a mathematical model that represents the reimbursement made by a company that pays its employees $55 per night for lodging, $30 per day for meals, and 27¢ per mile for gas.

11. A baby weighing 7 pounds at birth loses 7% of her weight in the 3 days after birth and then, over the next 4 days, returns to her birth weight. During the next month, she steadily gains 0.5 pound per week. Sketch a graph of the baby's weight from birth to 4 weeks. Accurately label both axes.

12. Harry Eagle is putting on the 18th hole of a golf course he has never played. His approach shot

has landed on the green, and his putts are described by the graph in Figure 1.1.6. (The dot on the vertical axis represents the distance from the hole that Harry's approach shot landed on the green.)

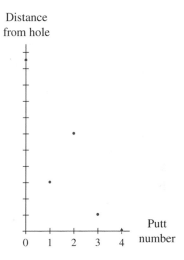

Harry Eagle's Putting Results

FIGURE 1.1.6

a. Assign appropriate units to the tick marks on the vertical scale.

b. What do you think happened on Harry's second putt?

c. If Harry's approach shot that landed on the green was his third shot, what was his score on the hole?

13. The plot in Figure 1.1.7 on page 14 depicts a certain girl's height plotted according to her age.

   a. What are the units along the vertical axis?

   b. How long was the girl when she was born?

   c. Approximately how much did she grow during the first year?

   d. At approximately what age did she reach her full height? How tall did she become?

   e. Did the girl grow faster during her first 3 years or during the last 3 years before she attained her full height?

   f. Would you expect the graph to increase or decrease after age 20? Why?

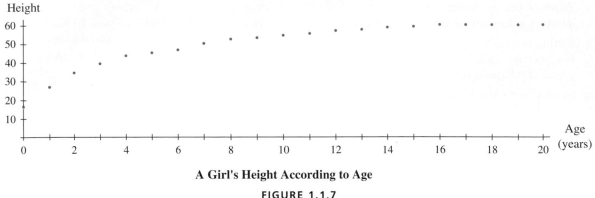

A Girl's Height According to Age

FIGURE 1.1.7

14. A graph depicting the average monthly profit for Slim's Used Car Sales for last year is shown in Figure 1.1.8. Slim needs your help in interpreting this graph.

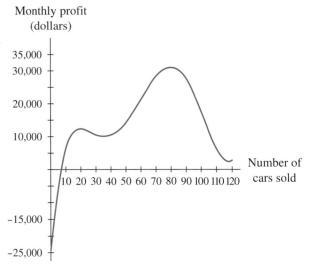

Slim's Used Car Sales

FIGURE 1.1.8

a. Slim pays rent on the lot where the cars to be sold are displayed, and he pays himself and his two salespeople a fixed salary at the beginning of each month. There is also the cost of electricity, water and sewer, health insurance, and so on. These costs remain the same regardless of the number of cars sold during the month. Estimate Slim's fixed costs at the beginning of each month.

b. Slim also has variable costs, such as the prices he pays for the cars. Other variable costs include paying someone to wash and maintain the cars on the lot, minor repairs to the cars, and so on. Slim receives revenue from the customer for each car he sells. Where is Slim's total cost more than his revenue?

c. What is Slim's monthly break-even point?

d. Between what numbers of cars sold is Slim's profit increasing?

e. The monthly profit peaks at two places. Estimate the number of cars sold and the monthly profit generated from the sale of those cars at these two points.

f. Give plausible reasons why Slim's profit sometimes decreases.

g. On the basis of the information presented in the graph, how many cars should Slim try to sell each month to maximize his monthly profit?

15. Donner Pass in the Sierra Nevada of northern California is the most important transmontane route between Reno and San Francisco. Donner Pass is named after George and Jacob Donner, who in 1846 attempted to lead a party of more than 80 immigrants through the pass to the Sacramento Valley. They were unable to make it through the pass before they became snowbound. A little over half the party survived the winter and finished their journey.

The graph[4] in Figure 1.1.9 represents the depth of the snow cover at Donner Memorial State Park from December 1993 through February 1994.

4. Compiled from data in *Monthly Climatological Data: California*, vol. 97, no. 12 (December 1993) and vol. 98, Nos. 1 and 2 (January/February 1994).

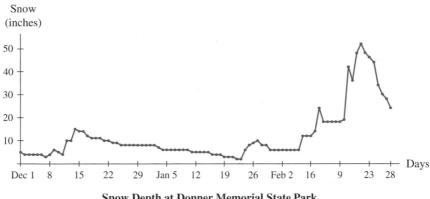

Snow Depth at Donner Memorial State Park
(December 1993 – February 1994)

FIGURE 1.1.9

a. How deep was the snow on December 1, 1993?

b. For approximately how many days after December 2 did the snow remain at the same depth?

c. What do you think happened on December 8–9, December 12–15, January 21–27, February 8–10, February 17, and February 19–20?

d. Why might there have been an immediate decrease in the depth of the snow cover after each one of the peaks?

e. On what day was the snow deepest? How deep (in feet and inches) was the snow at that time?

f. What event may have taken place to cause the rapid decrease of the graph after the last peak?

g. During what day(s) did the most snowfall probably occur?

16. You operate a small printing company. Today's jobs are three printings that use Elite Bond paper that you purchased for $6.50 per ream. One job requires $\frac{4}{5}$ of a ream, another requires $1\frac{3}{8}$ reams, and the third will use $\frac{3}{4}$ of a ream.

a. If you have 3 reams of this paper on hand, are you overstocked or understocked for the job?

b. The fixed costs of operating your business are $50 per day. What might these costs include?

c. You charge for each printing job on the basis of $30 per ream of Elite Bond used. That is, the first job costs the customer $\left(\frac{4}{5}\right)(30) = \$24$. What is your net profit for today's work?

17. A patient is instructed by her doctor to take one pill containing 500 milligrams of a drug. Assume that her body uses 20% of the original amount of the drug daily. Let y equal the number of milligrams of the drug remaining in the patient's body after x days.

a. Complete Table 1.5 to describe numerically the elimination of the drug from the patient's body.

TABLE 1.5

x	0	1	2	3	4	5
y	500	400				

b. Find a model that describes the elimination of the drug from the patient's body.

c. What are the smallest and largest values of x that makes sense in this problem? What are the smallest and largest values of y that make sense in this problem? Graph your model.

d. Calculate the x- and y-intercepts for the graph of your model. Interpret these values in the context of this problem.

e. For which values of x is y decreasing?

f. Use a graph of the model to estimate how much of the drug is left in the patient's body after 3.5 days. Use your model to find the exact value.

g. Use a graph of the model to estimate when the concentration of the drug will be 60 milligrams. Use your model to find the exact value.

18. Repeat Activity 17, assuming that each day the patient's body uses 20% of the amount of drug that was in her body *the previous day*.

■ 1.2 Functions and Graphs

Calculus is the study of change—how things change and how quickly they change. We begin our study of calculus by considering how we describe change. Let us start with something that many Americans desire to change, their weight. Suppose Joe weighs 165 pounds and begins to diet, losing 2 pounds a week. We can describe how Joe's weight changes in several ways.

With a table:

TABLE 1.6

Weeks after diet starts	0	1	2	3	4	5
Weight (pounds)	165	163	161	159	157	155

With a graph:

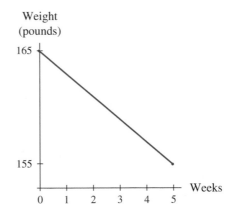

Joe's Weight versus Weeks on a Diet

FIGURE 1.8

With words:

> $W(t)$ is Joe's weight (in pounds) t weeks after he begins to diet if he weighs 165 pounds at the start and loses 2 pounds each week.

With an equation:

> $W(t) = 165 - 2t$ pounds, where $W(t)$ is Joe's weight after t weeks of dieting

How Joe's weight changes during the course of his diet is an example of a **function**. Our four different representations describe the same function. Informally, a function is a description of how one thing changes (Joe's weight) as something else changes (the number of weeks he's been on his diet). To understand functions thoroughly, however, we need a more precise definition.

Think of Joe's weight function as a *rule* that tells you what he weighs when you know how long he has been dieting. We visualize this with the **input/output diagram** shown in Figure 1.9.

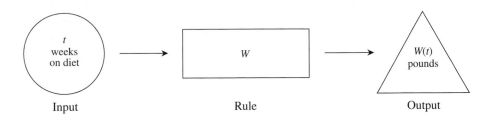

Input   *Rule*   *Output*

FIGURE 1.9

The rule is a function if each input produces exactly one output. If any particular input produces more than one output, then the rule is not a function.

> A **function** is a rule that assigns to each input exactly one output.

To verify that the rule for Joe's weight is a function, we must ask, "Can Joe have more than one weight after a certain number of weeks on his diet?" Assuming that Joe's weight is measured at a fixed time on the same scale once a week, there will be only one weight that corresponds to each week. Thus Joe's weight is a function of the number of weeks he has been on his diet.

You probably recall from previous math courses that the standard terms for the set of inputs and the set of outputs of a function are *domain* and *range*, respectively. Other terms for input and output include *independent variable* and *dependent variable* and *controlled variable* and *observed variable*. In this book, however, we use the terms *set of inputs* and *set of outputs*.

In the table representation of Joe's weight function, the set of inputs is {0, 1, 2, 3, 4, 5} and the set of outputs is {165, 163, 161, 159, 157, 155}. In the continuous graph representing Joe's weight, the set of inputs is all real numbers (not just integers) between 0 and 5 ($0 \leq$ weeks ≤ 5) and the set of outputs is all real numbers between 155 and 165 ($155 \leq$ weight ≤ 165).

EXAMPLE 1   *Student Grades*

Suppose $Q(s)$ is the grade (out of 30 points) that student s scored on the first quiz in a course attended by the five students listed in Table 1.7.

TABLE 1.7

Student, s	J. DeCarlo	S. Dyers	J. Lykin	E. Mills	G. Schmeltzer
Grade, $Q(s)$	28	12	25	21	25

a. Is Q a function of s?

b. Identify the set of inputs and the set of outputs.

Solution:

a. Q is a function of s because each name (input) corresponds to only one quiz grade (output). The input/output diagram for $Q(s)$ is shown in Figure 1.10.

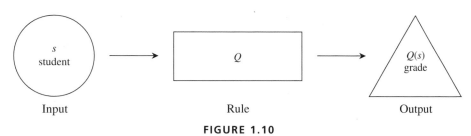

FIGURE 1.10

b. The set of inputs is simply the students' names, and the set of outputs is {12, 21, 25, 28}.   ∎

Determining Outputs

In the example of Joe's diet, t is the number of weeks on the diet (the input) and W is the name of the rule. The name of the rule can also be used to denote the output, but a more precise notation for the output is $W(t)$. The t is inside the parentheses to remind us that t is the input, and the W is outside to remind us that it is the rule that gives the output. Thus $W(4) = 157$ means that when Joe has been on his diet for 4 weeks (the input), he weighs 157 pounds (the output). It is also correct to say that $W = 157$ pounds when $t = 4$ weeks.

The way you find the output that corresponds to a known input depends on how the function is represented. In a table, you simply locate the desired input in the input row (or column). The output is the corresponding entry in the adjacent column or row. For example, the output corresponding to an input of 1991 is in color in Tables 1.8 and 1.9.

TABLE 1.8

Year	1989	1990	1991	1992	1993	1994
Amount	7.3	5.9	5.8	6.4	3.2	3.2

TABLE 1.9

Year	Amount
1990	496,321
1991	527,379
1992	546,108
1993	599,630
1994	654,127
1995	683,044

In a function represented by a graph, the input is traditionally on the horizontal axis. You locate the desired value of the input on the horizontal axis, move directly up (or down) until you reach the graph, and then move left (or right) until you encounter the vertical axis. The value at that point on the vertical axis is the output. In the graph[5] of faculty salaries shown in Figure 1.11, the output is between $55,000 and $56,000 when the input is 1991.

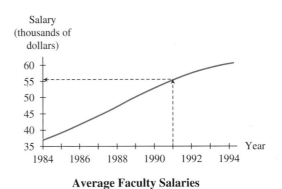

Average Faculty Salaries

FIGURE 1.11

Finally, if the function is represented by a formula, you simply substitute the value of the input everywhere that the variable appears in the formula and calculate the result. To use the formula $W(t) = 165 - 2t$ to find Joe's weight after 3 weeks, substitute 3 for t in the formula: $W(3) = 165 - 2(3) = 165 - 6 = 159$ pounds.

EXAMPLE 2:   *Land Value*

The value of a piece of property is given by the equation

$$v(t) = 3.622(1.093)^t \text{ dollars}$$

where t is the number of years since the end of 1900.

a. What was the value of the land in 1985?

b. When did the value reach $15,000?

Solution:

a. In 1985 $t = 85$, so the value of the land is $v(85) = 3.622(1.093)^{85} \approx \$6945$.

b. In this question we know the output, and we need to find the input. This requires us to solve for t in the equation $15,000 = 3.622(1.093)^t$. You can either solve the equation algebraically (using logarithms) or use a calculator or computer to solve it. In either case, you should find that $t \approx 93.7$ years. Because $t = 93$ corresponds to the end of 1993, the land reached a value of $15,000 during 1994.   ∎

---

5. Based on data from *1992 Information Please Almanac* and *Statistical Abstract of the United States,* 1994. Future references to the latter source will read simply *Statistical Abstract* followed by the year.

Recognizing Functions

Consider the following input/output tables:

TABLE 1.10

t	1	2	0	3	1	4
$F(t)$	7	8	9	10	11	12

TABLE 1.11

q	1	2	3	4	5	6
$C(q)$	10.6	8.3	10.6	8.3	10.6	8.3

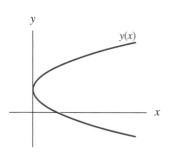

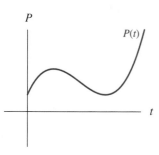

Are These Graphs Functions?

FIGURE 1.12

To determine whether these tables represent functions, we must decide whether each input produces exactly one output. Table 1.10 does not represent a function because when the input is $t = 1$, the output can be either 7 or 11. In Table 1.11, each input has only one associated output, so this table does represent a function. This is true even though different inputs ($q = 1, 3, 5$) correspond to the same output [$C(q) = 10.6$].

It is easy to determine whether a graph represents a function. Examine the two graphs in Figure 1.12.

The graph of $y(x)$ does not describe y as a function of x because each positive x-value produces two different y-values. The graph of $P(t)$ does show P as a function of t because every input produces only one output. The quickest way to find out whether a graph represents a function is to apply the **vertical line test**.

Vertical Line Test

Suppose that a graph has inputs located along the horizontal axis and outputs located along the vertical axis. If at any input you can draw a vertical line that crosses the graph in two or more places, then the graph does not represent a function.

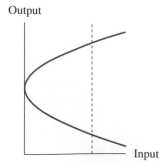

This Graph is Not a Function

FIGURE 1.13

Figure 1.13 illustrates the vertical line test failing for a graph.

To determine whether a formula represents a function, first graph the equation and then apply the vertical line test.

Discrete versus Continuous

As we have seen, functions can be represented with tables, graphs, words, or equations. Although each is a valid way to describe functions, there are important differences between these representations. Table 1.6 gives Joe's weight at six different moments in time: when he begins dieting, 1 week later, 2 weeks later, and so on. In contrast, the graph in Figure 1.8 shows Joe's weight at every time during the 5 weeks of his diet. We use the terms **discrete** and **continuous** to distinguish between the type of information given in Table 1.6 and that given in Figure 1.8. The table contains discrete information: Joe's weight at six different moments in time. In Figure 1.14, we plot these six data points and get a discrete graph called a **scatter plot**.

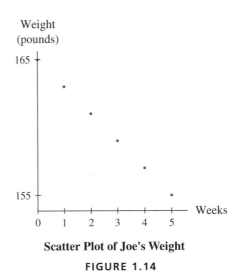

Scatter Plot of Joe's Weight

FIGURE 1.14

All the information we encounter in tables is discrete, but it is possible to record continuous information. Continuous information is recorded in some type of graph, such as an EKG strip chart or output from a seismograph.

Once we draw a line through the points on the scatter plot, we change the discrete graph to a continuous one showing Joe's weight constantly changing as he diets. See Figure 1.15. The continuous graph has a weight associated with every possible time during the 5 weeks of Joe's diet and can be drawn without lifting the writing instrument from the page. The continuous graph is a better description of how Joe's weight changes, because he doesn't lose weight all at once at the end of each week. An equation [in this case, $W(t) = 165 - 2t$] that represents a continuous graph is a continuous model.

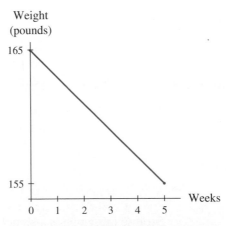

Continuous Graph of Joe's Weight

FIGURE 1.15

There are times when we use a continuous model, represented by a continuous graph, in a situation that is not actually continuous. Perhaps the best example of this is compounded interest. Suppose you invest $100 at a 7% annual percentage rate. We call the initial amount invested the **principal**. If interest is compounded annually, then the formula for the amount in your account t years after you made the investment is $A(t) = 100(1.07)^t$ dollars. Its graph is shown in Figure 1.16.

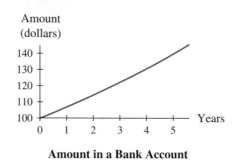

Amount in a Bank Account

FIGURE 1.16

Although the formula and graph are continuous (they are defined at all values of t, and the graph can be drawn without lifting the writing instrument off the page), they make sense only for integer values of $t \geq 0$ in the context of annually compounded interest.

Suppose that 6 months after you invest the $100, you have to withdraw the money. According to the formula, when $t = \frac{1}{2}$ year, the amount in the account should be $100(1.07)^{\frac{1}{2}} \approx \$103.44$. However, because interest is compounded once a year (at the end of the year), your money has earned no interest after 6 months, and the amount in the account is only $100. The only points on the continuous graph that have meaning in this investment situation are those shown in Figure 1.17.

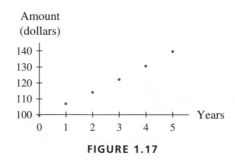

FIGURE 1.17

Because this graph is a scatter plot, we say that $A(t)$ is a continuous function with discrete interpretation.

A more accurate graph of the amount, which shows the balance changing only at integer input values, is given in Figure 1.18. (The open circles at the end of each year indicate that the amount is the value of the solid dot at the beginning of the next year.)

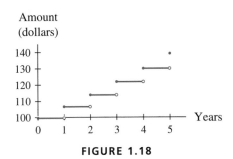

FIGURE 1.18

The equation for the graph in Figure 1.18 describes the balance after t years and can be written as

$$A(t) = \begin{cases} \$100 & \text{when } 0 \le t < 1 \\ \$107 & \text{when } 1 \le t < 2 \\ \$114.49 & \text{when } 2 \le t < 3 \\ \$122.50 & \text{when } 3 \le t < 4 \\ \$131.08 & \text{when } 4 \le t < 5 \\ \$140.26 & \text{when } t = 5 \end{cases}$$

This function is not a continuous function, and its graph is not a continuous graph; however, we do not use the term *discrete* to describe either the model or the graph. We reserve the word *discrete* for scatter plots, tables of data, and the interpretation of continuous functions in some situations.

As another example, consider that although sales of a product for a large corporation are made on a daily basis, the quarterly (or monthly or yearly) reports for that corporation give only totals for the period of time covered by the report.

Table 1.12 gives discrete information. If we were to find a continuous model for the data, it would have discrete interpretation; that is, it would have meaning for the corporation only at input values corresponding to the ends of quarters. The model could not be used to answer such questions as "What were the daily sales on the tenth day of the second quarter?" and "What were the total sales for the month of June?"

TABLE 1.12

Quarter	1	2	3	4
Quarterly sales (millions of dollars)	1.8	1.4	1.1	1.9

There are situations that are not technically continuous but for which we use a continuous model, not discretely interpreted. For example, consider the cumulative sales of the same corporation (see Table 1.13).

TABLE 1.13

Quarter	1	2	3	4
Total sales (millions of dollars)	1.8	3.2	4.3	6.2

Table 1.13 contains discrete information, and a scatter plot of the data is a discrete graph. We also know that if it were possible actually to record cumulative sales of the corporation at every moment in time, the graph would not be smooth. It would "jump" up as sales were made (or down in the case of refunds) because sales is always a number that has two decimal places. For this reason, a graph of sales at every moment in time is not continuous. However, because the "jumps" in the graph are insignificant in the context of total sales of the corporation, we use a continuous model, interpreted as a continuous model, to describe total sales. Situations similar to total sales include cumulative totals of anything and growth of populations.

The question to ask yourself in every modeling situation is "For this model, do inputs of any value make sense in context?" If the answer is no, then the model must have discrete interpretation. To illustrate, consider the four function situations we have discussed:

1. *Joe's weight:* Because Joe has a weight at every possible moment, and because the model gives an approximation of his weight at all times, we use the continuous model without restriction.

2. *Annually compounded interest:* Although there is an account balance at every possible time, and although the continuous model exists at every possible time, the output from the model does not describe the account balance for input other than integers. For this reason, if we use the continuous model, it must be discretely interpreted at integer values only.

3. *Quarterly sales:* Because quarterly sales exist only at the ends of quarters, a continuous model would not have meaning for any inputs other than those associated with ends of quarters. For this reason, a continuous model of quarterly sales would have discrete interpretation.

4. *Cumulative sales:* Because cumulative sales exist at all times and because a continuous model is an approximation of those sales, we use the continuous model without restriction.

Examples of situations that can be represented by continuous models but have discrete interpretation include interest that is not compounded continuously and anything that is a yearly total or average, such as sales, profit, revenue, number of live births, number of overseas telephone calls, and average price of gasoline. Because calculus is primarily the study of continuous change rather than discrete change, the ability to transform discrete data into a continuous function (a model) is a vital part of our study of calculus. It is important, however, to understand for what input values the model makes sense in context.

EXAMPLE 3   *Higher Education*[6]

Determine whether each of the following function representations is discrete, continuous, or continuous with discrete interpretation.

a. **TABLE 1.14**

Year	1987	1988	1989	1990	1991	1992
Higher education enrollment each fall (millions)	12.8	13.1	13.5	13.8	14.4	14.5

b. Fall enrollment in
    higher education
       (millions)

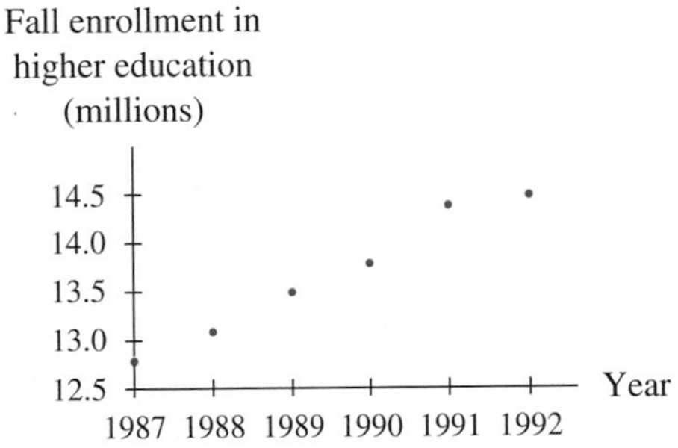

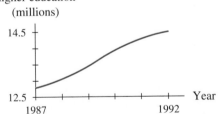

FIGURE 1.19

c. Fall enrollment in
    higher education
       (millions)

FIGURE 1.20

d. $C(t) = 0.0167t^3 + 0.121t^2 + 0.152t + 12.814$ million students enrolled each fall, where t is the number of years since 1987.

e.   Expenditure
    (billions of dollars)

283.7

0 — Month
  Mar   Jun   Sep   Dec

**Total Consumer Expenditure on
Recreation in 1992**

FIGURE 1.21

6. Based on data from *Statistical Abstract*, 1994.

Solution: The table of data in part *a* and the scatter plot in part *b* are discrete. The graph in part *c* and the equation in part *d* are both continuous functions with discrete interpretations, because they are defined only at values of *t* that correspond to the fall of each year. The graph in part *e* is a continuous function. This function can give us year-to-date expenditure at any time throughout 1992. ∎

1.2   Concept Inventory

- Function
- Inputs and outputs
- Input/output diagram
- Discrete and continuous
- Scatter plot

1.2   Activities

Draw an input/output diagram for each of the rules in Activities 1 through 10, and determine whether the rules are functions.

1. $R(w)$ = the first-class postal rate (in cents) of a letter weighing *w* ounces

2. $G(t)$ = your score out of 100 on the first test in this course when you studied *t* hours the week before the test

3. $B(x)$ = the number of students in this class whose birthday is on the *x*th day of the year (assuming it is not a leap year)

4. $D(p)$ = the distance from the origin to a point *p* on the circle $x^2 + y^2 = 1$

5. $A(m)$ = the amount (in dollars) you pay for lunch on the *m*th day of any week

6. $C(m)$ = the amount of credit (in dollars) that Citibank Visa will allow a 20-year-old with a yearly income of *m* dollars

7. a. $B(t)$ = the amount in an investment account (in dollars) after *t* years, assuming that no deposits or withdrawals are made during the *t* years

   b. $A(t)$ = the amount in an investment account (in dollars) after *t* years, assuming that deposits and withdrawals are permissible

8. $D(r)$ = the number of years it takes for an investment to double if the annual percentage rate (APR) is *r*%

9. $P(x)$ = the hair color of movie starlet *x* during 1997

10. $H(a)$ = your height in inches when you were age *a* years old

11–20. State the set of inputs and the set of outputs for each of the functions you found in Activities 1 through 10.

Determine whether the tables in Activities 21 through 24 represent functions. Assume that the input is in the left column.

21. **TABLE 1.15**

Age[7]	Percent with flex schedules at work
16–19	10.6
20–24	12.0
25–34	15.7
35–44	16.5
45–54	15.3
55–64	12.2
over 64	16.4

7. *Statistical Abstract*, 1994.

22.

TABLE 1.16

Military rank[8]	Basic monthly pay in 1993 (dollars)
Second Lieutenant	1697
First Lieutenant	2311
Captain	2924
Major	3588
Lt. Colonel	4359
Colonel	5371
Brigadier General	6469
Major General	7330
Lt. General	8075
General	9017

23. **TABLE 1.17**

Person's height	Person's weight (pounds)
5′ 3″	139
6′ 1″	196
5′ 4″	115
6′ 0″	203
5′ 10″	165
5′ 3″	127
5′ 8″	154
6′ 0″	189
5′ 6″	143

24. **TABLE 1.18**

Year[9]	Millions of workers with taxable earnings
1985	120
1986	123
1987	126
1988	130
1989	132
1990	133
1991	132
1992	132

25. $P(m)$ is the median sale price (in thousands of dollars) of existing one-family homes in metropolitan area m in 1993. Write the following statements in function notation.

a. The median sale price in Honolulu was $358,500.

b. In San Antonio the median sale price was $77,000.

c. $106,000 was the median sale price in Portland, OR.

26. $E(t)$ is the value of cotton exports[10] in millions of dollars in year t. Write sentences interpreting the following mathematical statements.

a. $E(1988) = 1975$

b. $E = 1999$ when $t = 1992$

27. Which of the following graphs represent functions? (The input axis is horizontal.)

a.

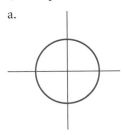

FIGURE 1.2.1

b.

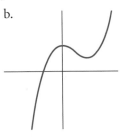

FIGURE 1.2.2

c.

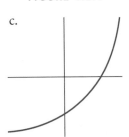

FIGURE 1.2.3

8. *Statistical Abstract*, 1994.
9. *Statistical Abstract*, 1994.
10. *Statistical Abstract*, 1994.

28. Which of the following graphs represent functions? (The input axis is horizontal.)

a.

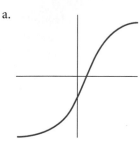

FIGURE 1.2.4

b.

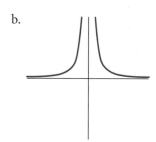

FIGURE 1.2.5

c.

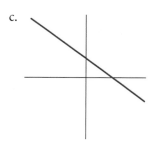

FIGURE 1.2.6

29. Which of the following statements describe continuous functions, and which describe discrete functions?

a. $H(a)$ is your height at age a years.

b. $W(t)$ is the percentage of the popular vote that the winning candidate for president received in year t.

c. $A(t)$ is the amount in a savings account after the principal has been invested for t years at 7% compounded continuously.

d. $P(x)$ is the number of passengers traveling on an airline during the xth month of 1997.

In Activities 30 and 31, categorize each function representation as discrete, continuous, or continuous with discrete interpretation.

30. a. **TABLE 1.19**

Year	Average basic cable rate (dollars per month)
1970	5.50
1980	7.69
1985	9.73
1990	16.78
1991	18.10
1992	19.08
1993	19.39

b.

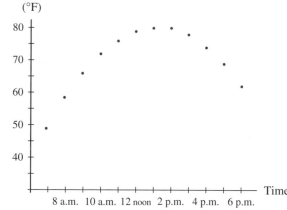

FIGURE 1.2.7

c. $M(t) = 0.0067t^3 - 39.7900t^2 + 79{,}043.5958t - 52{,}340{,}433.71$ thousand medical students[11] in the fall of year t.

11. Based on data from *Statistical Abstract*, 1994.

d.

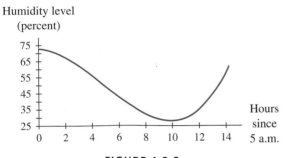

Humidity level
(percent)

FIGURE 1.2.8

31. a.

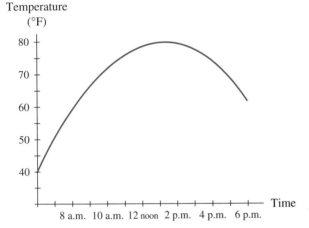

Temperature
(°F)

FIGURE 1.2.9

b.

U.S. medical students
(thousands)

FIGURE 1.2.10[12]

c. $G(m) = (2.798 \cdot 10^{-4})m^2 - 0.041m - 9.956$ gallons of gas in a car's gas tank after m minutes of driving.

12. *Statistical Abstract*, 1994.

d.

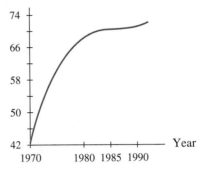

U.S. medical students
(thousands)

FIGURE 1.2.11[13]

For Activities 32 through 35, find the output of the function corresponding to each input value given.

32. $y = 7x + 3$; $x = 20, x = -2$

33. $R(w) = 39.4\,(1.998)^w$; $w = 3, w = 0$

34. $S(t) = \dfrac{120}{1 + 3.5e^{-2t}}$; $t = 10, t = 2$

35. $Q(x) = 0.32x^3 - 7.9x^2 + 100x - 15$; $x = 12, x = 3$

For Activities 36 through 39, find the input of the function corresponding to each output value given.

36. $y = 7x + 3$; $y = 24, y = 13.5$

37. $R(w) = 39.4\,(1.998)^w$; $R(w) = 78.8, R(w) = 394$

38. $S(t) = \dfrac{120}{1 + 3.5e^{-2t}}$; $S(t) = 60, S(t) = 90$

39. $Q(x) = 0.32x^3 - 7.9x^2 + 100x - 15$;
$Q(x) = 515, Q(x) = 33.045$

For each of the functions in Activities 40 through 43, determine whether an input or an output value is given, and find its corresponding output or input.

40. $g(x) = 49x^2 + 32x - 134$; $g(x) = 5086$

41. $A = 3200e^{0.492t}$; $t = 15$

42. $p = \dfrac{100}{1 + 25e^{-0.37m}}$; $p = 95$

43. $y = (39.4)(\ln 1.998)(1.998^x)$; $y = 2.97$

13. Based on data from *Statistical Abstract*, 1994.

1.3   Constructed Functions

In the previous section, we examined what a function is and how functions can be represented. Now we consider constructing more complicated functions.

Inverse Functions

We begin by examining the quiz grade function from Table 1.7 in the previous section.

Student, s	J. DeCarlo	S. Dyers	J. Lykin	E. Mills	G. Schmeltzer
Grade, Q(s)	28	12	25	21	25

Let us turn the relationship around, and let the input be the grade on the quiz and the output be the student who made that grade. The input/output diagram is shown in Figure 1.22.

FIGURE 1.22

The set of inputs is now {12, 21, 25, 28} and the set of outputs is {J. DeCarlo, S. Dyers, J. Lykin, E. Mills, G. Schmeltzer}. Is s a function of Q? This is the same as asking, "Does each grade correspond to only one student?". The answer is no. If the input grade is 25, the output is both J. Lykin and G. Schmeltzer. This relationship is not a function. If, however, the quiz grades had been as shown in Table 1.20,

TABLE 1.20

Student, s	J. DeCarlo	S. Dyers	J. Lykin	E. Mills	G. Schmeltzer
Grade, Q(s)	28	12	30	21	25

then swapping the inputs and outputs would have resulted in a function because no two students had the same grade. When the input and output of a function are reversed, we obtain a new rule. If this rule is also a function, we call it the **inverse** of the original function.

EXAMPLE 1 *New-Home Prices*[14]

TABLE 1.21

Year	1970	1975	1980	1985	1990
Average sale price of a new home	$23,400	$39,300	$64,600	$84,300	$122,900

a. Does Table 1.21 show new-home prices as a function of the year?

b. If inputs and outputs are reversed, is the result an inverse function?

Solution:

a. Because each year (input) corresponds to exactly one average price (output), Table 1.21 shows the average new-home price as a function of the year.

b. If we swap the inputs and outputs, we get

TABLE 1.22

Average sale price of a new home	$23,400	$39,300	$64,600	$84,300	$122,900
Year	1970	1975	1980	1985	1990

Each average price (input) corresponds to only one year (output), so Table 1.22 also represents a function, the inverse of the original function.   ■

EXAMPLE 2 *Office Space*[15]

The scatter plot in Figure 1.23 shows the percentage of vacant office space in Chicago from 1987 through 1993.

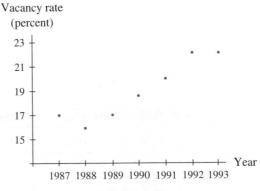

FIGURE 1.23

Does this scatter plot have an inverse function?

14. *Statistical Abstract,* 1994.
15. *Statistical Abstract,* 1994.

Solution   This scatter plot represents a function because there is only one vacancy rate (output) for each year (input). Swapping the inputs and outputs for this graph requires that we put years on the vertical axis and vacancy rates on the horizontal axis.

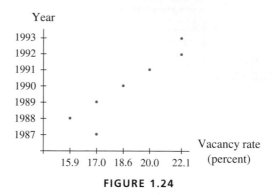

FIGURE 1.24

Note in Figure 1.24 that a rate of 17.0% corresponds to two different years (a vertical line through 17.0% passes through two points). This also occurs at 22.1%. The year is not a function of the vacancy rate, so this scatter plot does not represent an inverse function.   ■

Combining Functions

Swapping inputs and outputs is one way in which we can transform one function into another function (if, indeed, the result is a function). We can also combine two or more functions to create a new function. Adding or subtracting is the simplest way to do this.

Consider Tables 1.23 and 1.24, which show sales and costs for a corporation from 1990 through 1995.

TABLE 1.23

Year	1990	1991	1992	1993	1994	1995
Sales (millions)	$3.570	$3.945	$4.359	$4.817	$5.323	$5.881

TABLE 1.24

Year	Costs (thousands)	Year	Costs (thousands)
1990	$1061	1993	$2329
1991	$1562	1994	$2594
1992	$1984	1995	$2782

Both tables represent functions, and by subtracting costs from sales, we can get a new function representing profit. Before doing so, however, we need to ask ourselves two questions: "Is the set of inputs the same for both functions?" and "Are the outputs given in the same units?" The answer to the first question is yes, but the answer to the second is no. However, we can easily change one of the tables to achieve the same output units. The resulting profit function is shown in Table 1.25.

TABLE 1.25

Year	1990	1991	1992	1993	1994	1995
Sales (millions)	$3.570	$3.945	$4.359	$4.817	$5.323	$5.881
Cost (millions)	$1.061	$1.562	$1.984	$2.329	$2.594	$2.782
Profit (millions)	$2.509	$2.383	$2.375	$2.488	$2.729	$3.099

The equation $S(t) = 3.570(1.105)^t$ gives the sales in millions of dollars t years after 1990, and the equation $C(t) = -39.2t^2 + 540.1t + 1061.0$ gives the costs in thousands of dollars t years after 1990. As we did with the tables, we can find the profit equation by subtracting costs from sales as long as the output units correspond.

To construct an equation $P(t)$ for profit, we must first multiply $C(t)$ by 0.001 to convert the output units from thousands of dollars to millions of dollars. Then profit can be written as

$$P(t) = S(t) - 0.001C(t)$$
$$= 3.570(1.105)^t - 0.001(-39.2t^2 + 540.1t + 1061.0) \text{ million dollars}$$

where t represents the number of years since 1990.

Further, we can now use $P(t)$ to find the profit in any given year. For instance, the profit in 1992 was $P(2) = 2.375$ million dollars.

Product functions are created by multiplying two functions. Again, this makes sense only if both functions have the same set of inputs and if the output units are compatible so that, when multiplied, they give a meaningful result.

Suppose that $S(x)$ is the selling price (in dollars) of a gallon of milk on the xth day of last month, and $G(x)$ is the number of gallons of milk sold. The input/output diagrams are shown in Figure 1.25.

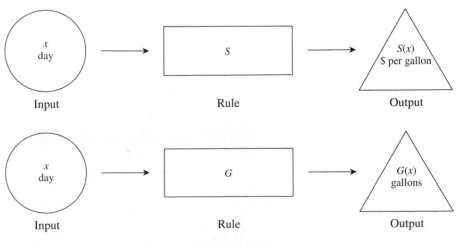

FIGURE 1.25

The inputs of both functions are the same, and when the outputs are multiplied ($ per gallon)(gallons), the result is the total sales in dollars on the xth day of last month. The input/output diagram for our new function $T(x) = S(x) \cdot G(x)$ is given in Figure 1.26.

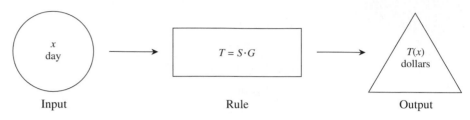

| Input | Rule | Output |

FIGURE 1.26

$S(x) = 0.007x + 1.492$ dollars per gallon and $G(x) = 31 - 6.332(0.921)^x$ gallons are functions modeling the selling price and the number of gallons of milk sold on the xth day of last month. Thus, total sales from milk can be modeled by the function

$$T(x) = [S(x)] \cdot [G(x)]$$
$$= [0.007x + 1.492] \cdot [31 - 6.332(0.921)^x] \text{ dollars}$$

on the xth day of last month.

Another way in which we combine functions is called **function composition.** Consider Tables 1.26 and 1.27, which give the altitude of an airplane as a function of time and give air temperature as a function of altitude.

TABLE 1.26

t = time into flight (minutes)	$F(t)$ = feet above sea level
0	4500
1	7500
2	13,000
3	19,000
4	26,000
5	28,000
6	30,000

TABLE 1.27

F = feet above sea level	$A(F)$ = air temperature (degrees Fahrenheit)
4500	72
7500	17
13,000	-34
19,000	-55
26,000	-62
28,000	-63
30,000	-64

It is possible to combine these two data tables into one table showing air temperature as a function of time. This process requires using the output from Table 1.26 as the input for Table 1.27. The only restriction is that both the numerical values and the units in the output of one function match those in the input of the other function. The resulting new function is shown in Table 1.28.

TABLE 1.28

t = time into flight (minutes)	0	1	2	3	4	5	6
$A(F)$ = air temperature (degrees Fahrenheit)	72	17	-34	-55	-62	-63	-64

Because we used the output of Table 1.26, $F(t)$, as the input for Table 1.27, the new function output is commonly written as $A(F(t))$. The A outside the parentheses is to remind us that the output is air temperature, and putting $F(t)$ inside the parentheses reminds us that altitude from Table 1.26 is the input. It is common to refer to A as the *outside function* and to F as the *inside function*. The mathematical symbol for the composition of an inside function F and an outside function A is $A \circ F$, so

$$(A \circ F)(t) = A(F(t))$$

The altitude data can be modeled as $F(t) = -222.22t^3 + 1755.95t^2 + 1680.56t + 4416.67$ feet above sea level, where t is the time into flight in minutes. The air temperature can be modeled as $A(F) = 277.897(0.99984)^F - 66$ degrees Fahrenheit, where F is the number of feet above sea level. The composition of these two functions is

$$(A \circ F)(t) = A(F(t))$$
$$= 277.897(0.99984)^{(-222.22t^3 + 1755.95t^2 + 1680.56t + 4416.67)} - 66 \, °F$$

where t is the time in minutes into the flight.

Let us consider another example to help make function composition clearer.

EXAMPLE 3   *Lake Contamination*

Consider the word descriptions of the following two functions and their input/output diagrams shown in Figure 1.27.

$C(p)$ = parts per million of contamination in a lake when the population of the surrounding community is p people

$p(t)$ = the population in thousands of people of the lakeside community in year t

Note that the output of the second function is almost the same as the input of the first function. If we multiply $p(t)$ by 1000 to convert the output from thousands of people to people, we have the function $P(t) = 1000p(t)$ people in year t. Now we can compose the two functions $C(p)$ and $P(t)$ to create the new function whose input/output diagram is shown in Figure 1.28.

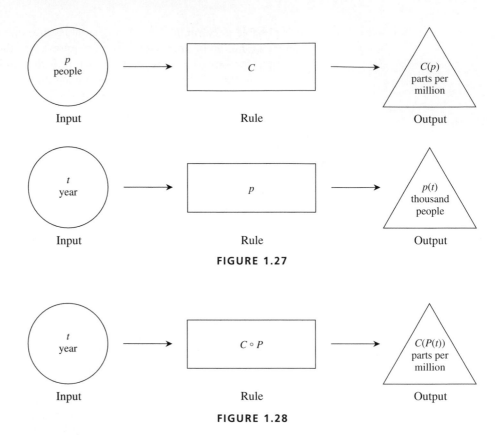

FIGURE 1.27

FIGURE 1.28

Suppose the contamination in the lake can be modeled by $C(p) = \sqrt{p}$ parts per million when the community's population is p people. Suppose also that the population of the community can be modeled by $p(t) = 0.4t^2 + 2.5$ thousand people t years after 1980. Then $P(t) = 1000(0.4t^2 + 2.5)$ people t years after 1980. Thus, the contamination in the lake t years after 1980 can be modeled as

$$(C \circ P)(t) = C(P(t)) = \sqrt{1000(0.4t^2 + 2.5)} \text{ parts per million} \quad \blacksquare$$

We formalize the process of function composition as follows:

Given two functions f and g, we can form their composition if the outputs from one of them, say f, can be used as inputs to the other function, g. In terms of inputs and outputs, we have

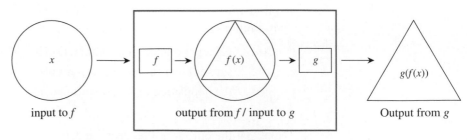

FIGURE 1.29

In this case, we can replace the portion of the diagram within the blue box by forming the composite function $g \circ f$ whose input/output diagram is

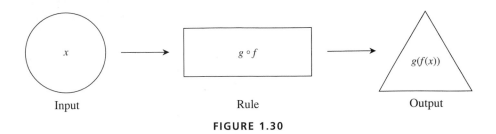

Input   Rule   Output

FIGURE 1.30

Note that the inputs to the composite function $g \circ f$ are the inputs to the "inside" function f, and the outputs are from the "outside" function g. We use composite functions extensively in later chapters.

Piecewise Continuous Functions

One final way in which we will construct functions is putting two or more functions together to create a **piecewise continuous function**. A good example of a piecewise continuous function is the investment example in Section 1.2, where we considered investing $100 at 7% annual interest. We began with a continuous function interpreted discretely at integer values only. We then represented the amount function in another way that better reflected the situation. A graph of the function is shown in Figure 1.18 on page 23.

The formula for the graph is

$$A(t) = \begin{cases} \$100 & \text{when } 0 \le t < 1 \\ \$107 & \text{when } 1 \le t < 2 \\ \$114.49 & \text{when } 2 \le t < 3 \\ \$122.50 & \text{when } 3 \le t < 4 \\ \$131.08 & \text{when } 4 \le t < 5 \\ \$140.26 & \text{when } t = 5 \end{cases}$$

t years after the $100 was invested.

We call this a piecewise continuous function because it is made up of portions of six different continuous functions and is defined for all values of t between 0 and 5.

EXAMPLE 4   *Population of West Virginia*

The population of West Virginia from 1985 through 1993 can be modeled[16] by

$$P(t) = \begin{cases} -23.514t + 3903.667 \text{ thousand people} & \text{when } 85 \le t < 90 \\ 9.1t + 972.6 \text{ thousand people} & \text{when } 90 \le t \le 93 \end{cases}$$

where t is the number of years since 1900.

a. Graph $P(t)$.

b. According to the model, what was the population of West Virginia in 1987? in 1992?

---

16. Based on data from *Statistical Abstract*, 1994.

c. Find the values of $P(85)$ and $P(90)$. Interpret your answers.

Solution:

a. The graph of $P(t)$ is shown in Figure 1.31.

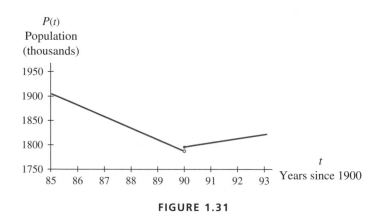

FIGURE 1.31

b. To determine the population in 1987, we substitute 87 for t in the top function.

$$P(87) = -23.514(87) + 3903.667 \approx 1858 \text{ thousand people}$$

To determine the population in 1992, we use the bottom function.

$$P(92) = 9.1(92) + 972.6 \approx 1810 \text{ thousand people}$$

c. $P(85) = -23.514(85) + 3903.667 \approx 1905$ thousand people. In 1985, the population of West Virginia was approximately 1,905,000.

$P(90) = 9.1(90) + 972.6 \approx 1792$ thousand people. In 1990, the population of West Virginia was approximately 1,792,000. ■

Calculus is the mathematics of change that occurs in continuous functions. This is why we have reviewed the basic ideas of functions in the first three sections of this text. We have reviewed the definition of a mathematical function, and we have looked at a few techniques for constructing new functions from basic functions.

Throughout these sections, we have considered mathematical functions as they model the real world. As we embark on our study of calculus, we will consider modeling primarily as a tool for developing the functions we use to analyze the change that occurs in real-world problems. In order to apply calculus to situations that occur in business, economics, agriculture, and the like, we must first be able to describe the situations as continuous or piecewise continuous functions (some having discrete interpretation). The next few sections deal with using different functions to model real-world data. Once we can model data with a function, we can use calculus to analyze that model—especially to determine rates of change and identify maximum, minimum, and inflection points.

1.3   Concept Inventory

■ Inverse functions
■ Adding, subtracting, and multiplying functions
■ Function composition
■ Piecewise continuous functions

1.3   Activities

Swap the inputs and outputs for each of the functions in Activities 1 through 6. Draw an input/output diagram for each new rule. Then write out the new rule in words. Which of the new rules are inverse functions?

1. $R(w)$ = the first-class postal rate (in cents) of a letter weighing w ounces

2. $G(t)$ = your score out of 100 on the first test in this course when you studied t hours the week before the test

3. $B(x)$ = the number of students in this class whose birthday is on the xth day of the year (assuming it is not a leap year)

4. $A(m)$ = the amount (in dollars) you pay for lunch on the mth day of any week

5. $B(t)$ = the amount in an investment account (in dollars) after t years, assuming no withdrawals are made during the t years

6. $D(r)$ = the number of years it takes for an investment to double if the annual percentage rate (APR) is r%

Reverse the inputs and outputs in the function tables in Activities 7 through 10, and determine whether the results are inverse functions.

7. **TABLE 1.29**

Age[17]	Percent with flex schedules at work
16–19	10.6
20–24	12.0
25–34	15.7
35–44	16.5
45–54	15.3
55–64	12.2
over 64	16.4

8. **TABLE 1.30**

Military rank[18]	Basic monthly pay in 1993 (dollars)
Second Lieutenant	1697
First Lieutenant	2311
Captain	2924
Major	3588
Lt. Colonel	4359
Colonel	5371
Brigadier General	6469
Major General	7330
Lt. General	8075
General	9017

9. **TABLE 1.31**

Person's height	Person's weight (pounds)
5′ 3″	139
6′ 1″	196
5′ 4″	115
6′ 0″	203
5′ 10″	165
5′ 3″	127
5′ 8″	154
6′ 0″	189
5′ 6″	143

10. **TABLE 1.32**

Year[19]	Millions of workers with taxable earnings
1985	120
1986	123
1987	126
1988	130
1989	132
1990	133
1991	132
1992	132

17. *Statistical Abstract*, 1994.
18. *Statistical Abstract*, 1994.
19. *Statistical Abstract*, 1994.

TABLE 1.33

Schedule of academic charges	Resident	Nonresident
Full-time academic fee	$1381	$3853
Part-time academic fee (per hour)	112	320
Auditing academic fee (per hour)	56	160
Staff fee (no charge 1–4 semester hours)	112	
Graduate assistant fee	360	360

11. Table 1.33 gives tuition rates at a certain university[20] in the southeast. Note that a full-time student is one taking 12 or more credit hours.

    a. Complete Table 1.34 for a resident undergraduate.

TABLE 1.34

Credit hours	Tuition	Credit hours	Tuition
1		10	
2		11	
3		12	
4		13	
5		14	
6		15	
7		16	
8		17	
9		18	

Let $T(c)$ = tuition (in dollars) for a resident undergraduate taking c credit hours.

    b. Find a formula for $T(c)$.

    c. Graph $T(c)$.

    d. Is $T(c)$ a function?

    e. Give the set of inputs and the set of outputs for $T(c)$.

    f. Draw an input/output diagram for the rule obtained by swapping inputs and outputs of $T(c)$.

    g. Is the rule in part f an inverse function? Why or why not?

12. Table 1.35[21] shows two functions: $I(t)$ is the amount of natural gas imports in trillion cubic feet in year t, and $E(t)$ is the amount of natural gas exports in billions of cubic feet in year t. Show how you could combine the two functions to create a third function giving net trade of natural gas in year t. (Net trade is negative when imports exceed exports).

TABLE 1.35

t	1987	1988	1989	1990	1991
$I(t)$	0.993	1.294	1.382	1.532	1.693
$E(t)$	54	74	107	86	122

13. The percentages of Iowa corn farmers in two communities who had heard about, and who had planted, hybrid seed corn t years after 1924 can be modeled[22] as follows:

Percentage hearing =
$$h(t) = \frac{100}{1 + 128.0427e^{-0.7211264t}} \text{ percent}$$

Percentage planting =
$$p(t) = \frac{100}{1 + 913.7241e^{-0.607482t}} \text{ percent}$$

20. Clemson University 1993–1994 *Undergraduate Announcements.*

21. *Statistical Abstract,* 1994.

22. Based on data from Ryan and Gross, "The Diffusion of Hybrid Seed Corn in Two Iowa Communities," *Rural Sociology,* March 1943.

Write an equation for the percentage of Iowa corn farmers who had heard about but not yet planted hybrid seed corn t years after 1924.

14. Table 1.36[23] represents two functions: $R(t)$ is the average price in dollars of a gallon of regular unleaded gasoline in year t, and $P(t)$ is the purchasing power of the dollar as measured by consumer prices in year t, using 1983 as the base year. Indicate how to combine the two functions to create a third function showing the price of gasoline in constant 1983 dollars.

TABLE 1.36

Year	$R(t)$	$P(t)$
1983	1.24	1.00
1984	1.21	0.96
1985	1.20	0.93
1986	0.93	0.91
1987	0.95	0.88
1988	0.95	0.85
1989	1.02	0.81
1990	1.16	0.77
1991	1.14	0.73
1992	1.13	0.71

15. The number of births[24] to women who are 35 years of age or older can be modeled as

$$n(x) = -0.034x^3 + 1.331x^2 + 9.913x + 164.447$$
thousand births x years after 1980

The ratio of cesarean-section deliveries[25] performed on women in the same age bracket can be modeled as

$$p(x) = -0.183x^2 + 2.891x + 20.215$$
deliveries per 1000 live births x years after 1980

Write an expression for the number of cesarean-section deliveries performed on women 35 years of age or older.

Determine whether the pairs of functions in Activities 16 through 19 can be combined by function composition. If so, give function notation for the new function, and draw and label its input/output diagram.

16. $R(x)$ is the revenue in deutschemarks from the sale of x soccer uniforms.

$D(r)$ is the dollar value of r deutschemarks.

17. $P(c)$ is the profit generated by the sale of c computer chips.

$C(t)$ is the number of computer chips a manufacturer can produce after t hours of production.

18. $C(t)$ is the number of cats in the United States at the end of year t.

$D(c)$ is the number of dogs in the United States at the end of year c.

19. $C(t)$ is the average number of customers in a restaurant on a Saturday night t hours after 4 p.m.

$P(c)$ is the average amount in tips generated by c customers.

In Activities 20 through 23, rewrite each pair of functions as one composite function.

20. $c(x) = 3x^2 - 2x + 5$　　$x(t) = 4 - 6t$

21. $f(t) = 3e^t$　　$t(p) = 4p^2$

22. $h(p) = \dfrac{4}{p}$　　$p(t) = 1 + 3e^{-0.5t}$

23. $g(x) = \sqrt{7x^2 + 5x - 2}$　　$x(w) = 4e^w$

24. The population of North Dakota can be modeled[26] by

$$P(t) = \begin{cases} -7.393t + 1304.857 \text{ thousand people} \\ \quad \text{when } 85 \le t < 91 \\ t + 542 \text{ thousand people} \\ \quad \text{when } 91 \le t \le 93 \end{cases}$$

where t is the number of years since 1900.

a. Find the values of $P(85)$, $P(90)$, and $P(93)$.

b. According to the model, what was the population in 1989?

c. Use the answers to parts a and b to sketch a graph of $P(t)$.

d. Is $P(t)$ a function? Explain.

25. Yearly sales of water skis in the United States (excluding Alaska and Hawaii) can be modeled[27] by

---

23. *Statistical Abstract*, 1994.
24. Based on data from *Statistical Abstract*, 1992.

25. Based on data from *Statistical Abstract*, 1992.
26. Based on data from *Statistical Abstract*, 1994.
27. Based on data from *Statistical Abstract*, 1994.

$$S(x) = \begin{cases} 12.1x - 905.4 \text{ million dollars} \\ \quad \text{when } 85 \leq x \leq 88 \\ -14.8x + 1414.9 \text{ million dollars} \\ \quad \text{when } 89 \leq x \leq 92 \end{cases}$$

where x is the number of years since 1900.

a. Find the values of $S(85)$, $S(88)$, $S(89)$, and $S(92)$.

b. Use the answers to part a to sketch a graph of $S(x)$.

c. Is $S(x)$ a function? Explain.

26. A music club sells CDs for $14.95. When you buy five CDs, the sixth one is free.

    a. Sketch a graph of the cost to buy x CDs for values of x between 1 and 12.

    b. Write a formula for the graph you sketched in part a.

27. A mail-order company charges a percentage of the amount of each order for shipping. For orders of up to $20, the charge is 20% of the order amount. For orders of greater than $20 up to $40, the charge is 18%. For orders of greater than $40 up to $75, the charge is 15%. For orders of greater than $75, the charge is 12%.

    a. Why would it benefit a company to assess the shipping charges described here?

    b. What will shipping charges be on orders of $17.50? $37.95? $75.00? $75.01?

    c. Write a formula for the shipping charge for an order of x dollars.

    d. Sketch a graph of the formula in part c.

    e. Which of the following representations of the shipping charge function do you believe it would be best for the company to put in its catalog? (i) a word description, (ii) a formula as in part c, (iii) a graph, or (iv) some other representation (specify)? Explain the reasons for your choice.

■ 1.4   Linear Functions and Models

Number sequences with a pattern of regular, constant additions are some of the most simple sequences that you will encounter. They are modeled by linear functions because when you plot them, they fall along a line. You may recall discussions about them in your other math classes where they were called *arithmetic progressions*—that is, number sequences that change because a fixed amount is being added to each value to obtain the next value. In Section 2.1, we will consider another familiar number sequence, the *geometric progression*. Geometric progressions are number sequences that change because each value is being multiplied by a fixed quantity to obtain the next value.

Representations of a Linear Model

It is best to start with an example, so let's consider a newspaper delivery team that makes weekly deliveries of newspapers and devotes their Saturday mornings to selling new subscriptions. Suppose that 5 new customers are added each week. If the team starts with 80 customers, how many customers will the team have after 14 weeks? The immediate answer, 150 customers, is obtained by multiplying 5 customers per week by the number of weeks, 14, and then adding the current number of customers, 80.

The data and scatter plot showing the number of customers over the 14-week period give us a more nearly complete picture of the increase in the number of customers. (See Table 1.37 and Figure 1.32.)

TABLE 1.37

Weeks	Number of customers	Weeks	Number of customers
0	80	8	120
1	85	9	125
2	90	10	130
3	95	11	135
4	100	12	140
5	105	13	145
6	110	14	150
7	115		

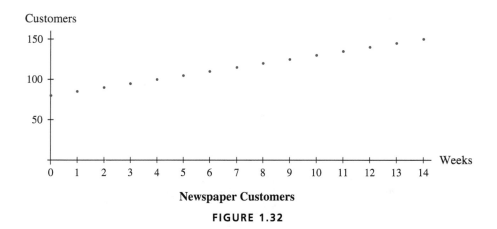

Newspaper Customers

FIGURE 1.32

Let us also consider a continuous model of the situation given by the equation $C = 5w + 80$. Here C represents the number of customers, and w stands for the number of weeks since the team began the subscription drive. Even though our question ("How many customers will we have after 14 weeks?") calls for only a single number as the answer, there may be many reasons to build a model. The model gives us a comprehensive description of the situation and thus is available to help answer other questions about the number of customers.

Note how each of the three representations (data, scatter plot, and model) provides its own particular insight into the nature of the customer base. The number-sequence data in Table 1.37 show the regular pattern of the customer counts and its steady increase. The scatter plot in Figure 1.32 shows the steady climb in the number of subscribers and the constant nature of the growth. Put the edge of a sheet of paper along the scatter plot and observe that the data points lie along a line; thus we use the term **linear** to describe the data. Finally, the algebraic expression given in the mathematical model captures the mathematical relationship between the two variable quantities (weeks and number of customers).

The model can be used in many ways. At the start of the campaign to recruit new customers, team members could take the model as their planned objective and then make plots of each week's result to check their progress. The model could also be used to answer a variety of "what if" questions. For example, if the newspaper had a strike and the team could receive at most 130 papers to deliver, then how many weeks could the strike continue before the team ran into supply problems?

Where should you look for linear models? In general, linear models are appropriate when things are left undisturbed and are allowed to undergo regular, incremental change. In these situations, momentum usually takes hold, and things change by a regular, constant amount; that is, the data are linear. In physics, we have Newton's First Law (the Law of Inertia), which states that "Every body in motion will remain in a state of uniform motion unless acted upon by external forces." In the case of a particle, uniform motion simply refers to motion along a straight line.

Finding a Linear Model

A linear model is determined by two parameters: a starting value and the amount of the incremental change. We shall see that all linear models graph as lines and algebraically appear as

$$f(x) = ax + b$$

where a is the incremental change and b is the starting value.[28] Thus we seek ways to obtain the parameters a and b that connect the output $f(x)$ with the input x. Let us now look at some examples and study them to find systematic ways to determine these parameters.

Suppose you are sole proprietor of a small business that has seen no growth in sales for the last several years. You have noticed, however, that your federal taxes have increased as shown in Table 1.38.

TABLE 1.38

Year	1992	1993	1994	1995	1996	1997
Tax	$2541	$3082	$3623	$4164	$4705	$5246

Upon close examination, you note that taxes increased by the same amount every year.

$$\begin{array}{ccccccccccc} \$2541 & & \$3082 & & \$3623 & & \$4164 & & \$4705 & & \$5246 \\ & \searrow \nearrow & & \searrow \nearrow & & \searrow \nearrow & & \searrow \nearrow & & \searrow \nearrow & \\ & \$541 & & \$541 & & \$541 & & \$541 & & \$541 & \end{array}$$

Taxes increased by $541 per year. We call this the **rate of change** of the tax amount. When the rate of change is constant, the data form a straight-line pattern when plotted (see Figure 1.33).

If the rate of change remains constant, then we can easily predict tax amounts in future years. For example, the tax amount in 1999 for your business is predicted to be

$$\$5246 + 2(\$541) = \$6328$$

From an examination of the data, it is fairly simple to derive an equation for the tax amounts. First, to simplify the problem, we let 1992 be denoted by 0, 1993 by 1, and so on. This process is called **aligning the data**, and we regularly align input data when years are involved (see Table 1.39).

---

28. $y = a + bx$ is an alternate form of the linear model. This form is commonly used by statisticians.

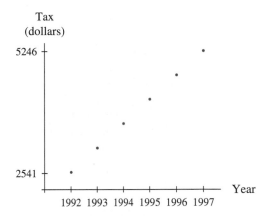

Federal Tax for a Small Business

FIGURE 1.33

TABLE 1.39

Aligned year	0	1	2	3	4	5
Tax	$2541	$3082	$3623	$4164	$4705	$5246

Second, in this example, we further simplify the data by subtracting $2541 (the tax for the first year) from each tax amount. Shifting the data so that the first point is (0,0) allows us to more clearly see any patterns displayed by the data:

TABLE 1.40

Aligned year	0	1	2	3	4	5
Aligned tax	$0	$541	$1082	$1623	$2164	$2705
Rewritten aligned tax	$541(0)	$541(1)	$541(2)	$541(3)	$541(4)	$541(5)

The aligned tax amounts can be written as $541t$, where t is the number of years since 1992.

We now have an equation for the aligned tax values. To get the original tax amounts, we simply reverse the aligning operation. To align, we subtracted $2541, so we will need to add $2541 when we reverse the process.

$$\text{Tax} = 541t + 2541 \text{ dollars}$$

where t represents the number of years since 1992. We call the equation for the tax amount a mathematical model for the data. When the data values increase by a constant amount, as in this situation, we call the model a **linear model**. Finding models that "fit" data is a recurring theme in the next chapter. Such models are important because they often enable us to analyze the results of change. With certain assumptions, they may even allow us to make cautious predictions about the short-term future. For example, to predict the tax amount owed in 2000, we substitute $t = 8$ into the tax model.

$$\text{Tax} = 541(8) + 2541 = \$6869$$

Admittedly, in this instance, an equation is not necessary to make such a prediction, but there are many situations in which it is difficult to proceed without a model.

It is important to understand that when we use mathematical models to make predictions about the short-term future, *we are assuming that future events will follow the same pattern as past events.* This assumption may or may not be true. That does not mean that such predictions are useless, *only that they must be viewed with extreme caution.*

We have already noted that the rate of change of the tax amount is $541 per year. It is common to call the rate of change in a linear model the **slope**, and these terms will be used interchangeably. In the tax example, slope can be thought of as how much taxes increase during a 1-year period. Thinking about slope in this way helps us answer questions such as

■ If you pay taxes twice a year, how much will taxes increase each time?

$$(\$541 \text{ per year})\left(\frac{1}{2}\text{ year}\right) = \$270.50$$

■ How much will taxes increase each time if you pay taxes quarterly?

$$(\$541 \text{ per year})\left(\frac{1}{4}\text{ year}\right) = \$135.25$$

■ How much will taxes increase during the next 3 years?

$$(\$541 \text{ per year})(3 \text{ years}) = \$1623$$

EXAMPLE 1   *Stock Dividends*[29]

Table 1.41 shows dividends paid per share of Houghton Mifflin Company stock in recent years.

TABLE 1.41

Year	1990	1991	1992	1993	1994
Dividend per share (dollars)	0.71	0.75	0.79	0.83	0.87

a. Find the constant rate of change in dividends paid per share.
b. Find an algebraic model for dividends per share.
c. If the rate of change remains constant, when will the dividend first exceed $1.00 per share?

Solution:

a. Dividends increased by $0.04 per year. We could also say that the rate of change is 4 cents per year.

29. Houghton Mifflin Company *Annual Report*, 1994.

b. An algebraic model is

$$\text{Dividend} = 0.04t + 0.71 \text{ dollars}$$

where t is the number of years since 1990.

c. Solving the equation $1.00 = 0.04t + 0.71$ gives $t = 7.25$ years after 1990. The end of 1997 corresponds to $t = 7$, so the dividend per share will exceed $1.00 in 1998. This answer is valid only if the rate of change remains constant through 1998.   ■

In terms of a graphical representation of a linear model, the rate of change, or slope, is a measure of the steepness of a graph, or how rapidly it is rising or falling. This information can help us identify the graph of a particular model. Consider the resale value of a car. The variable t represents the number of years since the car was purchased (see Table 1.42).

TABLE 1.42

t	0	1	2	3	4	5
Car value	$12,000	$10,500	$9000	$7500	$6000	$4500

The rate of change of the value of the car is -$1500 per year. (The negative sign indicates that the value is decreasing.) An equation for the value of the car t years after it was purchased is

$$\text{Value} = 12,000 - 1500t \text{ dollars}$$

This information can also be represented graphically as shown in Figure 1.34.

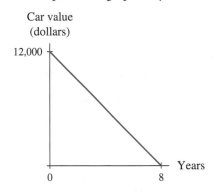

Depreciation of the Value of a Car

FIGURE 1.34

The rate of change of the resale value is negative, so the graph of the value equation decreases. The car had resale value of $12,000 when it was purchased (at time $t = 0$). Therefore, the graph of this equation intersects the vertical axis at 12,000.

Recall that the points where the line crosses the horizontal and vertical axes are called *intercepts*. The vertical axis intercept corresponds to the value of the car when it was purchased. The horizontal axis intercept is the first year in which the car had essentially no value.

Let us further investigate the graphical interpretation of the rate of change, or slope, of a linear model. The directed horizontal distance required to go from one point on the graph to another point on the graph is called the **run**, and the corresponding directed vertical distance is called the **rise** (see Figure 1.35). To travel from the years intercept 8 to the value intercept \$12,000 on the graph requires that you move 12,000 units up ($+\$12,000 =$ rise) and 8 units to the left (-8 years $=$ run). The quotient of the rise divided by the run is $\frac{\$12,000}{-8\ \text{years}} = -\$1500$ per year. Note that this is the same as the rate of change of the value of the car. If you traveled from the value intercept \$12,000 to the years intercept 8, the run would be positive and the rise negative. The quotient of the rise divided by the run would be $\frac{-\$12,000}{8\ \text{years}} = -\$1500$ per year. Note that the quotient $\frac{\text{rise}}{\text{run}}$ is the same either way you compute it. The quotient $\frac{\text{rise}}{\text{run}}$ is equal to the slope of the linear model.

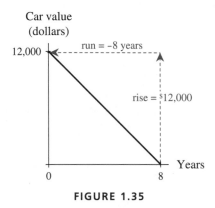

FIGURE 1.35

Although the slope of a particular linear model never changes, the graph of the model may look different when the horizontal or vertical scale is changed. You should therefore always use the same horizontal and vertical views when comparing two different models. The importance of this is shown in Example 2.

EXAMPLE 2   *Energy Production and Consumption*[30]

The two graphs in Figure 1.36 show energy production and energy consumption in the United States from 1975 through 1980.

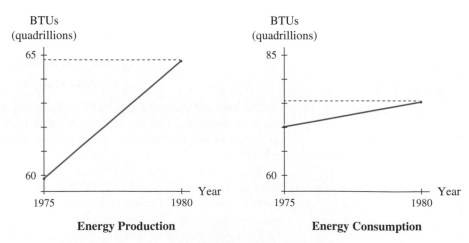

FIGURE 1.36

30. *Statistical Abstract*, 1994.

a. Which of the two graphs appears to be steeper?
b. Calculate the slope of each graph.
c. Which of the two graphs is steeper?
d. Write a linear equation for each graph.

Solution:

a. The energy production graph appears to be steeper.

b.
<table>
<tr><th style="text-align:center">Slope of
energy production</th><th style="text-align:center">Slope of
energy consumption</th></tr>
<tr><td style="text-align:center">$\dfrac{\text{rise}}{\text{run}} \approx \dfrac{64.8 - 59.9 \text{ quadrillion BTU}}{5 \text{ years}}$

$= 0.98$ quadrillion BTU
per year</td><td style="text-align:center">$\dfrac{\text{rise}}{\text{run}} \approx \dfrac{76.0 - 70.6 \text{ quadrillion BTU}}{5 \text{ years}}$

$= 1.08$ quadrillion BTU
per year</td></tr>
</table>

c. The energy consumption graph is the steeper graph because it has a greater positive slope than the energy production graph.

d. If we let x be the number of years since 1975, then the starting value b for energy production is 59.9. The slope a is 0.98, so the equation is

$$\text{Energy production} = 0.98x + 59.9 \text{ quadrillion BTU}$$

where x is the number of years since 1975. Similarly,

$$\text{Energy consumption} = 1.08x + 70.6 \text{ quadrillion BTU}$$

where x is the number of years since 1975.   ■

Examples 1 and 2 illustrate methods of finding the parameters a and b in the linear model $f(x) = ax + b$ for data points that fall on a line. However, real-life data values are seldom perfectly linear. For instance, the tax data we considered earlier are not likely to occur in real-life situations because the revenues of most businesses change from year to year. Consider the following modification to the tax data:

Year	1992	1993	1994	1995	1996	1997
Tax	$2541	$3081	$3615	$4157	$4703	$5242

$540 $534 $542 $546 $539

The changes in output (also called *first differences*) are not constant, but are "nearly constant." How do we get a linear model in this situation? Your calculator or computer will find an equation for the linear model that best fits the data.

Renumber the years so that 1992 is year 0, as shown in Table 1.43.

TABLE 1.43

Aligned year	0	1	2	3	4	5
Tax	$2541	$3081	$3615	$4157	$4703	$5242

Now use your calculator or computer to construct a scatter plot of these data. An examination of the scatter plot reinforces our earlier observation that the data are close to being linear. Next, use the linear regression routine that is built into your calculator or computer to find a linear equation that fits the data. You should find the equation to be

$$\text{Tax} = 540.37143t + 2538.90476 \text{ dollars}$$

where t is the number of years since 1992.

If we are willing to assume that the rate of change of the tax remains constant at the value \$540.37 per year given by this model, then we can use the linear model to predict future tax amounts. For instance, the model predicts the 1998 tax to be $540.37143(6) + 2538.90476 \approx \$5781$ and the 1999 tax to be $540.37143(7) + 2538.90476 \approx \$6322$.

A Word of Caution

When you predict output values for input values that are within the range of data in your scatter plot, you are using a process called **interpolation**. Predicting output values for input values that are beyond the range of the data is called **extrapolation**. Because you do not know how the data will behave outside the range you have plotted, **extrapolation very often results in misleading predictions**.

EXAMPLE 3   *Employee Stock Ownership Plans*

Table 1.44 shows the number of employee stock ownership plans for United States companies.

TABLE 1.44

Year	1981	1983	1985	1987	1989	1991
Number of plans	5680	6456	7402	8514	9385	9888

a. Use a calculator or computer to find a linear model for these data.

b. Use the model to estimate the number of plans in 1984 and 1994. Discuss the accuracy of each estimate.

Solution:

a. A linear model for these data is

$$y = 441.986x + 5235.586 \text{ employee stock ownership plans}$$

where x is the number of years since 1980.

b. By substituting 4 and 14 into the model in part *a*, we estimate that in 1984 there were 7004 plans and that in 1994 there were 11,423 plans. The 1984 value is within the range of the data (interpolation), so we can be more confident in that estimate than in the estimated 1994 value, which is outside the range of the data (extrapolation). ∎

What Is "Best Fit"?

Whenever you use a calculator or computer to fit a linear model to data, you should always use the routine that produces the line of best fit. How do we decide that a particular line best fits the data? After all, there are many lines that can be drawn through the data. Figure 1.37 shows a line fit to some data points.

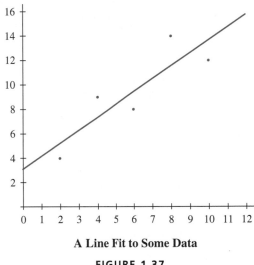

A Line Fit to Some Data

FIGURE 1.37

A visual indication of how well the line fits the data is the extent to which the data points deviate from the line. We determine a numerical measure of this visual observation in the following way. Calculate the amount by which the line misses the data; that is, calculate the vertical distance (the *deviation*) of each data point from the line. These deviations are simply the lengths of the vertical segments shown in Figure 1.38.

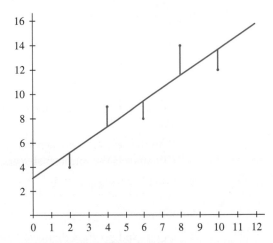

Vertical Deviations of Data About a Line

FIGURE 1.38

Each deviation measures the *error* between the associated data point and the line.

$$\text{Deviation} = \text{error} = y_{\text{data}} - y_{\text{line}}$$

Data points that lie below the line produce negative deviations (errors), and points that lie above the line produce positive deviations (errors). To count a negative error, say -2, on an equal basis with a positive error of 2, we square each error to obtain $(-2)^2 = 4$ and $2^2 = 4$. Then, to obtain a single numerical measure that accounts for all of the squared errors, we simply sum:

$$\begin{aligned} \text{SSE} &= \text{the sum of squared errors} \\ &= (\text{first error})^2 + (\text{second error})^2 + \cdots + (\text{last error})^2 \end{aligned}$$

This number, denoted by *SSE (sum of squared errors)*, is an overall measure of how well the line fits the data. The following example demonstrates how to calculate SSE.

EXAMPLE 4   *Consumer Price Index*[31]

Table 1.45 shows the consumer price index (CPI) from 1980 through 1990. The CPI is 100 times the ratio obtained by comparing the current cost of a specified group of goods and services to the cost of comparable items determined at an earlier date. An index is a ratio of quantities with the same units; thus CPI is a unitless quantity.

TABLE 1.45

Year	CPI (1982–1984 = 100)
1980	82.4
1982	96.5
1984	103.9
1986	109.6
1988	118.3
1990	130.7

a. Explain the meaning of the CPI of 118.3 in 1988. Also explain the CPI of 82.4 in 1980.

b. Find the deviation of each data value from the line $y = 4.5x + 83$, where y is the CPI and x is the number of years since 1980.

c. Calculate SSE for $y = 4.5x + 83$.

Solution:

a. Goods and services that cost $100 from 1982 to 1984 cost $118.30 in 1988 and only $82.40 in 1980.

---

31. *Statistical Abstract*, 1994.

b. **TABLE 1.46**

Year	CPI	y-Value from model	Deviation
1980	82.4	83	−0.6
1982	96.5	92	4.5
1984	103.9	101	2.9
1986	109.6	110	−0.4
1988	118.3	119	−0.7
1990	130.7	128	2.7

c. We obtain SSE by squaring the deviations and summing.

Deviations Squared

0.36

20.25

8.41

0.16

0.49

7.29

SSE = 36.96 ∎

The key idea in the correct use of SSE is this: Smaller values of SSE arise from lines where the errors are small, and larger values of SSE arise from lines where the errors are large. Thus a common strategy for choosing the best-fitting line is to choose the line for which the sum of squared errors (SSE) is as small as possible. Such a line is designated as the line of "best fit," and the procedure for choosing it is called the **method of least squares**. The line obtained by using the method of least squares for the data in Example 4 is $y = 4.47x + 84.571$. The SSE associated with this line is 25.12. This is the smallest SSE possible; therefore, $y = 4.47x + 84.571$ is the line that best fits the CPI data.

While it is possible to use calculus to find a line of best fit according to the method of least squares, we will use the routine in a calculator or computer whenever we wish to fit a line to data. In the next chapter, we will look at other models that can be used to fit data.[32] While the method of least squares is the accepted method of finding a best-fitting line, there are other methods of fitting different kinds of models to data in which SSE does not play a role. Even though it is possible to compute SSE for each of these models, SSE should *not* be used to compare models of different types. Such comparison is not a valid statistical procedure.[33]

Numerical Considerations

When dealing with numerical results, it is important to understand their accuracy and how precise the results need to be. For instance, imagine that you have a bank account with a balance of \$18,532.71 paying interest at a rate of $17\frac{1}{4}\%$ compounded

32. H. Skala, "Will the Real Best Fit Curve Please Stand Up?" Classroom Computer Capsule, *The College Mathematics Journal*, Vol. 27, No. 3, May 1996.
33. *B. Efron*, "Computer-Intensive Methods in Statistical Regression," *SIAM Review*, Vol. 30, No. 3, Sept. 1988.

annually. At the end of the year, the bank calculates your interest to be (18,532.71)(0.1725) = $3196.892475. Would you expect your next bank statement to record the new balance as $21,729.60248? Obviously, when reporting numerical results dealing with monetary amounts, we do not consider partial pennies. The bank reports the balance as $21,729.60. The required precision is only two decimal places.

> You should always round numerical results in a way that *makes sense* in the context of the problem.

Results that represent people or objects should be rounded to the nearest whole number. Results that represent money should be rounded to the nearest cent or, in some cases, to the nearest dollar. Consider, however, a company that reports net sales as shown in Table 1.47.

TABLE 1.47

Year	Net sales (millions of dollars)
1994	256.9
1995	217.4
1996	187.0

A linear model for these data is $y = -34.95x + 69,945.683$ million dollars, where x is the year. If we wished to estimate net sales in 1997, we would substitute $x = 1997$ in the model to obtain $y = \$150.53333333$ million. Should we report the answer as $150,533,333.33 or $150,533,333? Neither! When we are dealing with numerical results, our answer can be only as accurate as the least accurate output data. In this case, the answer would have to be reported as $150.5 million or $150,500,000.

> You should round numerical results to the same accuracy as the given output data.

Although the correct rounding and reporting of numerical results is important, it is even more important to calculate the results correctly. Because you must sometimes round your answers, it is tempting to think that you can round during the calculation process. Don't! Never round a number unless it is the final answer that you are reporting. Rounding during the calculation process may lead to serious errors. Your calculator or computer is capable of working with many digits. Keep them all while you are still working toward a final result.

When you use your calculator or computer to fit a model to data, it finds the parameters in the model to many digits. Although it may be acceptable to your instructor for you to round the coefficients when reporting a model, make sure that you use all of the digits while working with the model. This helps reduce the possibility of round-off error.

For example, suppose that your calculator or computer generates the following model for a set of data showing weekly profit for an airline for a certain route as a function of the ticket price.

$$\text{Weekly profit} = -0.00374285714285x^2 + 2.5528571428572x - 52.71428571429$$
$$\text{thousand dollars}$$

where x is the ticket price in dollars. In the answer key, you would see the model reported to three decimal places:

$$\text{Weekly profit} = -0.004x^2 + 2.553x - 52.714 \text{ thousand dollars}$$

However, if you used the rounded model to calculate weekly profit, your answers would be incorrect because of round-off error. Table 1.48 shows the inconsistencies between the rounded and unrounded models.

TABLE 1.48

Ticket Price	Profit from rounded model	Profit from unrounded model
$200	$298 thousand	$308 thousand
$400	$328 thousand	$370 thousand
$600	$39 thousand	$132 thousand

Never use a rounded model to calculate, and never round intermediate answers during the calculation process.

It would be extremely tedious for everyone if we reported all the digits found by all possible models of graphing calculators or computer software or even listed all digits for one particular type of technology! We therefore adopt the following convention for reporting models and statements of models in the text:

We feel it is important to print the fewest digits possible under the condition that (a) the rounded model visually fits the data, and (b) the rounded model gives values fairly close to answers obtained with the full model. In particular, if there is a difference between the results from the full model and those from the rounded model, that difference should appear only in the last digit for which we can claim accuracy.

Thus the number of decimal places shown in models in the text will vary. However, whenever we calculate with a model, we use all the digits available at that point in the text. If the data are given, the unrounded model will be used in calculations. When a model found earlier is used in a later section and the data are not repeated, all calculations will be done with the rounded model given in that later section. For convenience and consistency, the answer key will report all models with three decimal places in coefficients and six decimal places in exponents.

Also keep in mind that when you are writing models, it is important to write down what the input and output variables represent. Look back to the model we obtained for the small business tax. The model written as Tax = 2538.90476 + 540.37143t is incomplete because no label is given to indicate the units on the tax. Is the tax measured in dollars, cents, or thousand dollars? Tax is *what* is measured and dollars is a label telling *how* it is measured. Always label a model with output units to tell how the output is measured. It is equally important to describe the input variable. If all that had been written was Tax = 2538.90476 + 540.37143t dollars, you might have erroneously predicted your federal taxes in 1996 to be 2538.90476 + 540.37143(96) = $54,415 or even 2538.90476 + 540.37143(1996) = $1,081,120. However, there is a statement with the equation that reads "where t is the number of years since 1992." Because 1999 is the seventh year since 1992, you should correctly predict your federal taxes in 1999 to be 5238.90476 + 540.37143(7) = $6322.

There are three important elements to every model:

1. An equation

2. A label denoting the units on the output

3. A description of what the input variable represents

Finally, keep in mind that a number by itself is likely to be absolutely worthless as an answer to a question. Consider the reaction of a manager to a memo that reads

Sykes:

30

                      Paul

Sykes would probably have no idea what Paul was attempting to communicate. Suppose the next memo from Paul read

Sykes:

Regarding your
question about how
much overtime my
division puts in per
week, we typically
put in 30.

           Paul

Again, Sykes would probably not understand. 30 what? 30 minutes? 30 hours? If Paul had 60 people in his division, it might even mean 30 days. *A number is useless without a label that clearly indicates the units involved!*

In summary, keep in mind the following general guidelines for numerical results.

Guidelines for Numerical Results

1. When *using* a calculator or computer-generated model to obtain answers, do not round the parameters in the model during the calculation process. When you are *reporting* the model, it may be acceptable to your instructor for you to round the parameters to three or four decimal places.

2. When reporting a model, always label units on the output variable and state what the input variable in the model represents. A model is meaningless without a clear description of the variables.

3. Numerical results are worthless if you do not understand what they mean. The first step in understanding is to state the units clearly on all numerical answers. Numerical answers are incomplete without appropriate units.

4. Round final numerical results in a way that *makes sense* in the context of the problem.

5. When you are dealing with numerical results, an answer can be only as accurate as the least accurate piece of output data.

1.4   Concept Inventory

- Algebraic form of a linear model: $f(x) = ax + b$
- Rate of change (slope) of a linear model is constant.
- Calculating slope as $\frac{\text{rise}}{\text{run}}$
- First differences
- Aligning data
- Interpolating and extrapolating
- The method of least squares
- Rounding rules
- Labeling units on answers
- Three elements of a model

1.4   Activities

1. The graph in Figure 1.4.1 shows a corporation's profit in millions of dollars over a period of time.

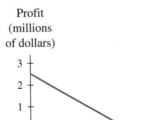

FIGURE 1.4.1

a. Estimate the slope of the graph, and write a sentence explaining the meaning of the slope in this context.

b. What is the rate of change of the corporation's profit during this time period?

c. Identify the horizontal and vertical axis intercepts, and explain their significance to this corporation.

2. The air temperature in a certain location from 8 a.m. to 3 p.m. is shown in the graph in Figure 1.4.2.

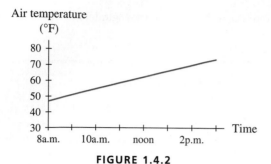

FIGURE 1.4.2

a. Estimate the slope of the graph, and explain its meaning in the context of temperature.

b. How fast is the temperature rising between 8 a.m. and 3 p.m.?

c. What would be the meaning of a horizontal axis intercept for this temperature graph?

3. The total October payroll[34] for all city governments in the United States from 1990 through 1992 can be modeled by

Payroll = 321.5x + 5530.167 million dollars

where x is the number of years after 1990.

a. What is the rate of change of the payroll amount?

b. Sketch an accurate graph of the October payroll.

c. Find the vertical axis intercept, and explain its significance in this context.

4. The number of passengers on Reno Air flights[35] from 1990 through 1992 can be modeled as

$y = 0.365x + 3.068$ million passengers

where x is the number of years since 1990.

a. What is the rate of change of the number of Reno Air passengers from 1990 through 1992?

b. What is the slope of a graph of the model?

5. The revenue for International Game Technology[36] was $114.1 million in 1994 and $99.1 million in 1995. Assume that revenue is decreasing at a constant rate.

a. Find the rate of change of revenue.

b. By how much did revenue decline each quarter of 1995?

c. Assuming that the rate of decrease remains constant, complete Table 1.49.

TABLE 1.49

Year	Revenue (millions of dollars)
1994	
1995	
1996	
1997	

d. Find an equation for revenue in terms of the year.

6. Suppose you bought a Mazda Miata in 1991 for $18,000. In 1993 it was worth $14,200. Assume that the rate at which the car depreciates is constant.

a. Find the rate of change of the value of the car.

b. Complete Table 1.50.

TABLE 1.50

Year	Value
1991	
1993	
1994	
1995	
1996	

c. Find an equation for the value in terms of the year.

d. How much will the value of the car change during a 1-month period? Round your answer to the nearest dollar.

7. Table 1.51 gives the number of gallons of oil in a tank used for heating an apartment complex t days after January 1 when the tank was filled.

TABLE 1.51

t	Oil (gallons)
0	30,000
1	29,600
2	29,200
3	28,800
4	28,400

34. Based on data from *Statistical Abstract*, 1994.
35. Based on data from Reno/Tahoe International Airport.
36. International Game Technology Annual Report, 1995.

a. What is the rate of change of the amount of oil? (Be sure to include units.)

b. How much oil can be expected to be used during any particular week in January?

c. Predict the amount of oil in the tank on January 30. What assumptions are you making when you make predictions about the amount of oil? Are the assumptions valid? Discuss.

d. Find and graph an equation for the amount of oil in the tank.

e. When should the tank be refilled?

8. A jewelry supplier charges a $55 ordering fee and $2.50 per pair of earrings regardless of the order size.

a. If a retailer orders 100 pairs of earrings, what is the cost?

b. What is the average cost per pair of earrings when 100 pairs are ordered?

c. If the retailer marks up the earrings 300% of the average cost, at what price will the retailer sell the earrings? (Sale price = cost + 300% of cost)

d. Find an equation for cost as a function of the number bought.

e. Find an equation for the average cost per pair as a function of the number bought.

f. Find an equation for the sale price based on a 300% markup.

g. If 100 pairs of earrings are purchased, how many must the retailer sell at a 300% markup to recover the cost of the earrings (break-even point)?

9. Table 1.52 shows the enrollment for a university in the southeast from 1965 through 1969.

TABLE 1.52

Year	Students
1965	5024
1966	5540
1967	6057
1968	6525
1969	7028

a. Find a linear model for these data.

b. Estimate the enrollment in 1970. (Round appropriately.)

c. The actual enrollment in 1970 was 8038 students. How far off was your estimate? Do you consider the error to be significant or insignificant? Explain your reasoning.

d. Would it be wise to use this model to predict the enrollment in the year 2000? Explain.

10. You and several of your friends decide to mass-produce "I love calculus and you should too!" T-shirts. Each shirt will cost you $2.50 to produce. Additional expenses include the rental of a downtown building for a flat fee of $675 per month, utilities estimated at $100 each month, and leased equipment costing $150 per month. You will be able to sell the T-shirts at the premium price of $14.50 because they will be in such great demand.

a. Give the equations for monthly revenue and monthly cost as functions of the number of T-shirts sold.

b. How many shirts do you have to sell each month to break even? Explain how you obtained your answer.

11. The price[37] of tickets to a college's home football games is given in Table 1.53.

TABLE 1.53

Year	Price (dollars)
1977	8
1979	9
1981	10
1983	12
1985	13
1987	15
1989	16
1991	18

a. Find a linear model for the price of tickets.

b. Use the model to find the ticket price in 1984, 1986, and 1990. (Round to the nearest dollar.) Were you interpolating or extrapolating to find these prices?

c. Use the model to predict the ticket price in 1994. Is this interpolation or extrapolation?

37. *Statistical Abstract*, 1994.

d. In 1994, tickets cost $20. Was your prediction accurate?

e. Suppose you graduate, get married, and have a child who attends that college. Predict the price you will have to pay to attend a football game with your child during his or her freshman year. Describe in detail how you arrived at your answer.

f. Repeat part *e* assuming that you have a grandchild who attends the same college.

g. What assumptions are you making when you use the linear model to make predictions? Are these assumptions reasonable over long periods of time? Explain.

h. What conclusions can you draw about the validity of extrapolating from a model?

12. You are an employee in the summer at a souvenir shop. The souvenir shop owner wants to purchase 650 printed sweatshirts from a company. The catalog contains the following table and directions to call the company for costs for orders more than 350.

TABLE 1.54

Number purchased	Total cost (dollars)
50	250
100	375
150	500
200	600
250	700
300	825
350	950

The shop owner, who has tried unsuccessfully for a week to contact the company, asks you to estimate the cost for 650 shirts.

a. Find a linear model to fit the data.

b. Use the model to predict the cost for 650 shirts. Note that all costs in the table are integer multiples of $25.

c. Determine the average cost per shirt for 650 shirts.

d. The shop owner is preparing a newspaper advertisement to be published in a week. If the standard markup is 700%, what should the advertised price be?

e. How many of the 650 shirts will need to be sold at the price determined in part *d* in order to pay for the cost of all 650 shirts (break-even point)?

13. The percentage[38] of funding for public elementary and secondary education provided by the federal government during the 1980s is given in Table 1.55.

TABLE 1.55

School year	% Funding
1981–82	7.4
1982–83	7.1
1983–84	6.8
1984–85	6.6
1985–86	6.7
1986–87	6.4
1987–88	6.3
1988–89	6.2
1989–90	6.1

a. Find a linear model for the percentage of funding.

b. What is the rate of change of the model you found?

c. On the basis of the model, estimate the percentage of funding provided by the federal government in the 1993–94 school year.

d. What happened in the early 1990s that might have affected federal funding of education? Would this be more likely to cause your model to overestimate or to underestimate?

e. Judging by the model, determine in what school year federal government funding will first be less than 5% of educational funding. Do you expect this point to be reached sooner or later than predicted? Why?

14. Consumer credit[39] was $742 billion in 1988 and $858 billion in 1993. Assume that consumer credit increases at a constant rate.

a. Find the rate of increase.

b. On the basis of the rate of increase, estimate consumer credit in 1996.

---

38. Based on data from *Statistical Abstract*, 1992 and 1994.
39. *Statistical Abstract*, 1994.

c. Is the assumption that consumer credit increases at a constant rate valid? Explain.

d. When will consumer credit reach $1.5 trillion?

e. Do you expect consumer credit to reach $1.5 trillion sooner or later than your answer in part *d* indicates? Why?

15. A house sells for $73,000 at the end of 1983 and for $97,500 at the end of 1995.

   a. If the market value increased linearly from 1983 through 1995, what was the rate of change of the market value?

   b. If the linear increase continues, what will the market value be in 1998?

   c. In what year might you expect the market value to be $75,000? $100,000?

   d. Find a model for the market value. What does your model estimate for the market value in 1992? Do you believe this estimation? What assumption was made when you created the model? Is this assumption necessarily true?

16. Twenty-two percent of the births[40] in 1985 were to unmarried women. The percent in 1991 was 29.2.

   a. Find the rate of change, assuming that it is constant.

   b. Estimate the percentage of births to unmarried women in 1993.

17. Table 1.56[41] shows 1997 first-class postage for mail up to 11 ounces.

TABLE 1.56

Weight not exceeding	Postage
1 oz	$0.32
2 oz	0.55
3 oz	0.78
4 oz	1.01
5 oz	1.24
6 oz	1.47
7 oz	1.70
8 oz	1.93
9 oz	2.16
10 oz	2.39
11 oz	2.62

a. Draw a scatter plot of these data to determine visually whether a linear model is appropriate.

b. Verify your observations in part *a* by calculating first differences in the postal rates.

c. Find a formula for the postage in terms of weight. Be specific about what the variables represent.

18. The 1993–94 tuition charges for students attending a certain university in the southeast are given in Table 1.57. A full-time student is a student who enrolls in 12 or more credit hours.

TABLE 1.57

Schedule of academic charges	Resident	Nonresident
Full-time academic fee	$1381	$3853
Part-time academic fee (per credit hour)	112	320
Auditing academic fee (per credit hour)	56	160
Staff fee (no charge 1–4 credit hours)	112	
Graduate assistant fee	360	360

Find formulas for the amount of tuition that each of the following students pays when she or he registers for *x* credit hours.

a. Resident student

b. Nonresident student

c. Graduate assistant

19. Explain the difference between using a model for interpolation and using it for extrapolation. Does interpolation always give an accurate picture of what is happening in the real world? Does extrapolation? Why or why not?

20. The population of West Virginia from 1985 through 1993 is shown in Table 1.58.[42]

40. *Statistical Abstract*, 1992 and 1994.
41. Source: Postal Bulletin PB21883A—January 1, 1995, U.S. Government Printing Office.
42. *Statistical Abstract*, 1994.

TABLE 1.58

Year	Population (thousands)
1985	1907
1986	1882
1987	1858
1988	1830
1989	1807
1990	1793
1991	1799
1992	1809
1993	1820

a. Observe a scatter plot of the data. What year is the dividing point that should be used to create a piecewise continuous function?

b. Divide the data in the year you determined in part *a*. Include the dividing point in both data sets. Fit linear models to each set of data, and write the function in correct piecewise continuous function notation. Compare your model with the one given in Example 4 in Section 1.3.

21. The population of North Dakota between 1985 and 1993 is shown in Table 1.59.[43]

TABLE 1.59

Year	Population (thousands)
1985	677
1986	669
1987	661
1988	655
1989	646
1990	639
1991	633
1992	634
1993	635

a. Observe a scatter plot of the data. What year is the dividing point that should be used to create a piecewise continuous function?

b. Divide the data in the year you determined in part *a*. Include the dividing point in both data sets. Fit linear models to each set of data, and write the function in correct piecewise continuous function notation. Compare your model with the one given in Activity 24 in Section 1.3.

22. Membership in the Girl Scouts of the U.S.A. in selected years from 1970 through 1992 is shown in Table 1.60.[44]

TABLE 1.60

Year	Membership (thousands)
1970	3922
1975	3234
1980	2784
1985	2802
1986	2917
1987	2947
1988	3052
1989	3166
1990	3269
1991	3383
1992	3510

a. Look at a scatter plot of the data. Describe the behavior of the data from 1970 through 1980 and from 1985 through 1992.

b. Find linear models for the 1970–1980 data and for the 1985–1992 data. Write the two models as one piecewise model.

c. Use the models to estimate the membership in 1973 and in 1993. Categorize your estimates as interpolation or extrapolation.

23. Chlorofluorocarbons (CFCs) released into the atmosphere are believed to thin the ozone layer. In the 1970s, the United States banned the use of CFCs in aerosols, and the atmospheric release of

43. *Statistical Abstract*, 1994.
44. *Statistical Abstract*, 1994.

CFCs declined. The decline was temporary, however, because of increased CFC use in other countries. In 1987, the Montreal Protocol, calling for a phasing out of all CFC production, was ratified. Table 1.61[45] shows atmospheric release of CFC-12, one of the two most prominent CFCs, from 1974 through 1992.

TABLE 1.61

Year	CFC-12 Atmospheric release (millions of kilograms)
1974	418.6
1976	390.4
1978	341.3
1980	332.5
1982	337.4
1984	359.4
1986	376.5
1988	392.8
1990	310.5
1992	255.3

45. *The True State of the Planet,* ed. Ronald Bailey. (The Free Press, for the Competitive Enterprise Institute, New York: 1995).

a. Look at a scatter plot of the data. Does the behavior of the data seem to support the foregoing description of the atmospheric release of CFCs since the 1970s? Explain.

b. Use three linear models to create a piecewise model for CFC-12 atmospheric release.

c. Sketch a graph of the data and the three-piece model.

d. According to your model, when will there no longer be any atmospheric release of CFC-12?

24. Carefully read the following notice by the American Humane Association, and then answer the questions below. The notice ran in the *Anderson Independent-Mail* the first week of May 1993. For the questions that follow, consider only dog and cat births, not deaths.

a. What is the rate of change of the number of dogs in the United States?

b. What is the rate of change of the number of cats in the United States?

c. Find an equation for the number of dogs t days after the beginning of 1993.

d. Find an equation for the number of cats t days after the beginning of 1993.

e. How many total animals (cats and dogs) do you estimate there were at the end of 1993? Does this agree with the prediction in the article?

Be Kind To Animals Week®
Help Stop U.S. Birthrate of 5,500 Pets Per Hour

Every hour in the United States, more than 2,000 dogs and 3,500 cats are born, compared to 415 humans. The yearly statistics include more than 17 million dogs and 30 million cats. Add these animals to an existing pet population of 54 million dogs and 56 million cats and the total exceeds one billion!

Earlier puberty, multiple births and briefer pregnancies make dogs and cats more prolific than humans. Animal care and control agencies do their best to find loving, responsible owners for as many pets as possible, but there are simply not enough homes for all these animals.

The American Humane Association estimates more than 15 million healthy, friendly dogs and cats will be euthanized this year simply because they are "unwanted".

You can save lives and help solve the pet overpopulation tragedy by neutering or spaying your pet. This will reduce the number of dogs and cats being euthanized by reducing the number of pets being born.

AMERICAN
✛HUMANE
ASSOCIATION

63 Inverness Drive East
Englewood, CO 80112-5117

Provided by the American Humane Association
Be Kind To Animals Week is a registered trademark of the American Humane Association

Chapter 1   Summary

Mathematical Modeling

Mathematical modeling is the process by which we construct a mathematical framework to represent a real-life situation. We analyze this framework with mathematical methods and techniques and then translate and interpret the results back into the context of the real-life situation.

As we apply it in this book, mathematical modeling usually means fitting a line or curve to data. The resulting equation with output label and input description, which we often refer to as the "mathematical model," provides a representation of the underlying relationship between the variable quantities of interest. Typically, we begin with data representing the variables, examine a scatter plot of the data to help determine an appropriate type of model, and then use technology (graphing calculators or computers) to construct the actual model. It is common in many real-life situations for a continuous or piecewise continuous model to have discrete interpretation, so we must be careful when using such models.

As we proceed in subsequent chapters to develop certain concepts and methods of calculus, we apply them to our models with the intent of analyzing the change that is occurring in the relationship between the variables. As you might expect, an important part of the process is to interpret the results of the mathematical investigations in the context of the original situation. This usually calls for mathematical decision making and clear articulation of results.

You should keep in mind that the data from which the models are constructed are sometimes incomplete, often of a size that is smaller than we would prefer, and perhaps of questionable accuracy. Thus the models themselves necessarily reflect these shortcomings. Nevertheless, the models still convey important and valuable information.

Functions and Graphs

The concept of a function is a fundamental mathematical idea that arises in almost every area of mathematics. What is a function? Informally, it is a description of how one thing changes (output) as something else changes (input). More precisely, a function is simply a rule (a formula, prescription, or procedure) that assigns to each input exactly one output. If two differ-ent outputs are assigned to a single input, then the assignment rule is not a function.

Examples are everywhere, and we encounter functions in four ways: through input/output tables, graphs, simple word descriptions, and mathematical equations. With tables, we generally find inputs in the top row (or left-hand column) and the corresponding outputs in the bottom row (or right-hand column). With graphs, it is customary to place inputs along a horizontal axis and outputs along a vertical axis. With equations of the form $y = f(x)$, where $f(x)$ is an algebraic expression in the variable x, we think of x as the input and of y as the output.

Discrete graphs (called scatter plots) show dots (or points) in isolation from one another on a rectangular grid. The horizontal coordinate of any particular dot is the input; the vertical coordinate is the output. Continuous graphs are unbroken curves, and their sets of inputs are assumed to fill up an entire range of values along the horizontal axis. We often begin with a discrete scatter plot and then transform the underlying data into a continuous or piecewise continuous curve— a mathematical model for the data.

To determine whether a graph represents a function, apply the vertical-line test: If no vertical line through an input intersects the graph in two places, then the graph represents a function. If a vertical line through an input intersects the graph in two (or more) places, then the graph does not represent a function. Almost all of the graphs in this book represent functions.

Constructed Functions

Although we have little occasion to work with inverse functions, you should understand what they are. Starting with a given function, say $y = f(x)$ with input x and output y, we reverse the input/output pairs to obtain a new rule, $x = g(y)$. If this new rule is a function (that is, if each input y produces exactly one output x), then we call the new function the inverse of the original function.

The easiest way to combine two functions is to add or subtract them. In order for this to make sense, we must first be certain that both functions have the same set of inputs and that their outputs are given in the same units. Product functions are obtained by multiplying outputs. Again, this gives a meaningful result only when both functions have the same set of inputs and the output units are compatible.

Function composition is a more involved way of combining functions. In order to form the composite $g \circ f$ of two functions f and g (note that f is the inside function and g is the outside function), we require that the outputs of f be used as inputs to g, as in Figure 1.39.

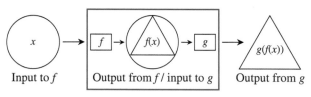

Input to f Output from f / input to g Output from g

FIGURE 1.39

The composite function $g \circ f$ has the diagram shown in Figure 1.40.

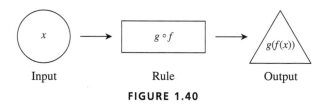

Input   Rule   Output

FIGURE 1.40

Inputs to $g \circ f$ are inputs to the inside function f. Outputs from $g \circ f$ are outputs from the outside function g.

Portions of two or more continuous functions can be combined to form piecewise continuous functions. In piecewise continuous functions, the outputs over different intervals of inputs are determined by different functions. When dealing with piecewise continuous functions, be sure that you use the correct portion of the function when determining the output associated with a given input.

Linear Functions and Models

A linear function models a constant rate of change; its underlying equation is that of a line:

$$y = ax + b$$

In terms of the line, the parameter a is called the slope of the line and is calculated as

$$\text{Slope} = \frac{\text{rise}}{\text{run}}$$

Because the slope of a line is a measure of its rate of increase or decrease, the slope is also known as the rate of change for the linear model. The parameter b appearing in the linear model $y = ax + b$ is simply the output of the model when the input is zero. Together, the parameters a and b completely determine the model.

When inputs to a linear model are evenly spaced, the corresponding outputs have constant first differences. In fact, this feature characterizes linear models. That is, if we consider a process where the first differences of outputs that correspond to evenly spaced inputs are constant, then these input/output pairs for the process lie along a line. Although we rarely encounter real-life situations where the first differences are perfectly constant, it is not uncommon to find nearly constant first differences in processes that are devoid of external influences. This accounts for the widespread occurrence of linear models.

In dealing with numerical results, it is always important to consider the accuracy and meaning of these results. Round the numerical results to the same accuracy as the given data and only in a way that makes sense in the context of the particular problem. Serious errors can influence results when rounding is done during the calculation process or when a rounded model is used to calculate additional results. Every model has three key elements: an equation, output units, and a description of what the input variable represents. Without these three elements, a model is useless.

The Role of Technology

In order to construct mathematical models from data, we must use appropriate tools. Normally, these tools are graphing calculators or microcomputers. To attempt to build the models without these tools—by paper-and-pencil methods only—is impractical.

However, you should clearly understand that our use of technology will simply be a tool in the service of mathematics and that no tool is a substitute for clear, effective thinking. Technology carries only the graphical and numerical computational burden. You will have to perform the mathematical analyses, interpret the results, make the appropriate decisions, and then communicate your conclusions in a clear and understandable manner.

Chapter 1 Review Test

1. The graph in Figure 1.41[46] shows the yearly gain of wetlands in the United States between 1987 and 1994.

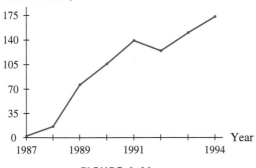

Gain in wetlands
(thousands of acres)

FIGURE 1.41

a. Did the number of acres of wetlands in the United States increase or decrease between 1991 and 1992? Explain.

b. Estimate the slope of the portion of the graph between 1989 and 1991. Interpret your answer.

2. Let $T(m)$ be the temperature in the room in which you are currently located (measured in degrees Fahrenheit by a thermometer in a fixed location) m minutes after you entered the room.

a. Draw and label an input/output diagram for $T(m)$.

b. Is $T(m)$ a function? Why or why not?

c. When the inputs and outputs of $T(m)$ are reversed, is the result an inverse function? Why or why not?

3. The 1994 *Statistical Abstract of the United States* reported that sales of bowling accessories in the United States (excluding Alaska and Hawaii) reached $114 million in 1986 and that in 1993 the amount was $167 million. Assume that sales of bowling accessories increased at a constant rate between 1986 and 1993.

a. Find the rate of increase of sales of bowling accessories between 1986 and 1993.

b. On the basis of the rate of increase found in part *a*, estimate the amount of sales in 1998. Under what conditions is this estimate valid?

4. In *The True State of the Planet*, the statement is made that "the amount of arable and permanent cropland worldwide has been increasing at a slow but relatively steady rate over the past two decades." Selected data between 1970 and 1990 reported in *The True State of the Planet* are shown in Table 1.62.[47]

TABLE 1.62

Year	Cropland (millions of square kilometers)
1970	13.77
1975	13.94
1980	14.17
1985	14.31
1990	14.44

a. Find a linear model for the data.

b. Is the model in part *a* discrete, continuous, or continuous with discrete interpretation?

c. What is the rate of change of your model in part *a*? Write a sentence interpreting the rate of change.

d. Do the data and your model support the statement quoted above? Explain.

e. According to your model, what was the amount of arable and permanent cropland in 1995?

46. *Ibid.*
47. *Ibid.*

Project 1.1

Tuition Fees

Setting

Nearly all students pursuing a college degree are classified as either part-time or full-time. Many colleges and universities charge tuition for part-time students according to the number of credit hours they are taking. They charge full-time students a set tuition regardless of the number of credits they take, except possibly in an overload situation.

Tasks

1. Find the tuition charges at your college or university. Some schools have different tuition rates for different classifications of students (for instance, residents may be charged a different tuition than nonresidents). Pick one classification to model. Write a piecewise linear model for tuition as a function of the number of credit hours a student takes. Consider charges such as matriculation and activity fees that all students must pay as being part of the college tuition.

2. Use your model to generate a table of tuition charges for your school. Compare your table with the published tuition charges. Are there any discrepancies? If so, explain why they might have occurred.

3. Even though you would not pay tuition if you registered for 0 credit hours, your model may have an interpretation at this point. What is that interpretation?

Reporting

Write a letter addressed to the Committee on Tuition. You should explicitly define to whom the model applies and what the variables in the model represent. Include an explanation of how to use the model, a statement about which input values are valid and which are invalid for use in the model, and a discussion of the preceding tasks.

Project 1.2

Finding Data

Setting

Newspapers, journals, and government documents are good sources of data.

Tasks

1. Find (in a newspaper, journal, or government document) a set of data with more than 7 data points that would be reasonably modeled by a linear function.

2. Fit a linear model to the data.

3. What do your model's slope and intercept mean in the context of the data?

Reporting

1. Write a report in which you describe the meaning of the data, discuss the linear model you used to fit the data, and answer the question posed in Task 3. In your report, you should explicitly state the model, define the variables in the model, and state what input values are valid for this model. You should also include a scatter plot and graph of your model. Properly cite the source of your data using correct bibliographic form. Attach a photocopy of the data and the page on which the title of the article or document appears.

2. Prepare a 3–5-minute oral presentation of your report. You should incorporate the use of visual aids to enhance your oral presentation.

Ingredients of Change: Nonlinear Models

Although linear functions and models are among the most frequently occurring ones in nonscience settings, nonlinear models apply in a variety of situations.

For example, sooner or later we are all faced with financial decisions that involve interest-bearing accounts. Some examples are savings accounts, credit cards, automobile loans, and home mortgages. These examples can be described by exponential models—models that exhibit constant percentage change. This chapter begins with a study of exponential models and includes a section on their application to interest-bearing accounts.

However, there are few cases where exponential growth can be sustained indefinitely. Ultimately, various influences act to restrict the growth in some way. Such situations often can be modeled by an S-shaped curve known as a logistic function.

Finally, we consider quadratic and cubic models, along with applications to real-world situations. These are nonlinear models that are represented by the familiar parabolic curves and cubic curves you studied in algebra.

Many situations in life can be modeled by functions whose graphs are not lines. Social changes are often modeled by nonlinear functions. Some questions concerning quantities and change that can be described by functions and analyzed using calculus follow. (1) What is the population of the United States t years after 1990? (2) How quickly was the U.S. population increasing in 1997? (3) What proportion of the population is x years old? (4) Are there more Americans in their fifties or in their twenties? (5) In what decade of life is the highest proportion of Americans?

What other social quantities can be represented by mathematical functions?

2.1   Exponential Functions and Models

In Section 1.4, we studied linear models that arose from data whose output formed an arithmetic progression. We turn now to another familiar number pattern that you may have encountered in previous math classes and examine sequences generated by multiplication by a constant—**geometric progressions**. Although we examine this pattern in a more general setting, you will see that it is closely related to change that is caused by repeated multiplication by a fixed number.

You may not realize it, but the balance of your savings account is determined by repeated multiplication. For example, suppose that you make an initial deposit of $100 into an account that pays 8% per year with interest compounded yearly, and you make no other deposits and no withdrawals. You will have the following annual balances at the anniversary dates of your account.

TABLE 2.1

Year	0	1	2	3	4
Balance	$100.00	$108.00	$116.64	$125.97	$136.05

A little mathematical investigation reveals that each balance is 1.08 times the prior balance. This suggests that the mathematical model that connects the balance B and t, the number of years since the start of the account, is

$$B = 100 \, (1.08)^t \text{ dollars}$$

When we examine a scatter plot of the account balance over the first 20 years that the account is in existence, we see that the points certainly do not fall on a line.

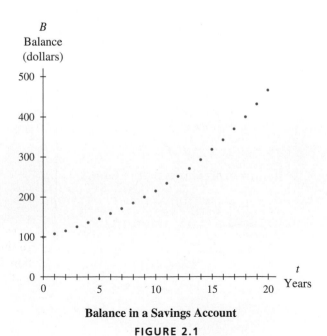

Balance in a Savings Account

FIGURE 2.1

The account balance grows more rapidly in the later years. Why? It is because the multiplier operates on an increasingly large balance as the years advance. The repeated multiplication by 1.08 makes the year variable t appear as an exponent on the 1.08 term in the model; that is, the year value counts the number of times that 1.08 has been used as a multiplier. Because the variable appears in the exponent, we call the resulting growth **exponential growth**.

Note that any change in a quantity that results from repeated multiplication generates an **exponential model**. That is, if we start with an amount a and multiply by a constant factor b each year, then the quantity that we have at the end of t years is given by the exponential model

$$Q(t) = ab^t$$

Applications of Exponential Models

Exponential models have many applications. Our previous example about the savings account balance shows that an exponential model describes how money increases when interest is compounded. Exponential models are also useful in population studies such as those where we are interested in how often a bacterial culture doubles. For instance, in the case of a bacterial culture that starts with 10,000 cells and doubles every hour, we have a population at the end of t hours that is given by the equation

$$P(t) = 10,000(2)^t \text{ cells}$$

Here we start with the initial 10,000 cells and multiply the number of cells by 2 every hour to account for the exploding population.

Similar equations arise in the study of radioactive material where we are interested in *half-lives*. The amount of a radioactive isotope is cut in half during every half-life period k. Thus, to find the amount remaining, we divide the time t by k to determine the number of times that the radioactive level has been halved. If the amount is originally R grams of a radioisotope with a half life of k years, then

$$A(t) = R(0.5)^{\frac{t}{k}} \text{ grams}$$

remain after t years.

As in this example, when the repeated multiplier is less than 1, we see a decreasing amount instead of growth. This situation with a diminishing amount is referred to as **exponential decay**.

Often we do not know what the proper repeated multiplier should be in order to write an exponential model. However, if we are given data, sometimes we can look at percentage differences to determine what the repeated multiplier must be.

Suppose that a small town's population has been dwindling. According to the town's records, the population data from 1988 through 1997 is as shown in Table 2.2.

TABLE 2.2

Year	Population	Year	Population
1988	7290	1993	4805
1989	6707	1994	4420
1990	6170	1995	4067
1991	5677	1996	3741
1992	5223	1997	3442

A scatter plot of the data is shown in Figure 2.2.

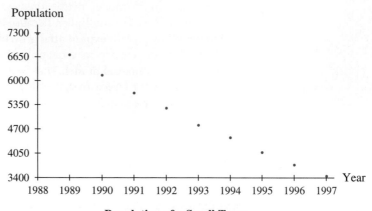

Population of a Small Town

FIGURE 2.2

Let us now develop an algebraic description of the declining population. We can examine the data in greater detail by calculating first differences:

7290   6707   6170   5677   5223   4805   4420   4067   3741   3442

    −583   −537   −493   −454   −418   −385   −353   −326   −299

The first differences show us that the change in population from year to year is not constant. However, if we calculate the percentages that these yearly changes represent, we notice a pattern.

From 1988 to 1989, the population decreased by 583 people. This represents an 8% (approximately) decrease from the 1988 population of 7290 people.

$$\frac{-583 \text{ people}}{7290 \text{ people}} = -0.07997 \approx -8\%$$

From 1989 to 1990, the population decreased by 537 people. This also represents about an 8% decrease from the previous year's (1989) population.

$$\frac{-537 \text{ people}}{6707 \text{ people}} = -0.08007 \approx -8\%$$

In fact, every year the population decreased by approximately 8%.

$$1990 \text{ to } 1991: \quad \frac{-493}{6170} = -0.07990 \approx -8\%$$

$$1991 \text{ to } 1992: \quad \frac{-454}{5677} = -0.07997 \approx -8\%$$

$$1992 \text{ to } 1993: \quad \frac{-418}{5223} = -0.08003 \approx -8\%$$

$$1993 \text{ to } 1994: \quad \frac{-385}{4805} = -0.08012 \approx -8\%$$

$$1994 \text{ to } 1995: \quad \frac{-353}{4420} = -0.07986 \approx -8\%$$

$$\text{1995 to 1996:} \quad \frac{-326}{4067} = -0.08016 \approx -8\%$$

$$\text{1996 to 1997:} \quad \frac{-299}{3741} = -0.07993 \approx -8\%$$

In other words, the population data follow the general pattern

1988:   7290

1989: $7290 - 0.08(7290) = 7290(1 - 0.08) = 7290(0.92)$

1990: $7290(0.92) - 0.08[7290(0.92)] = 7290(0.92)(1 - 0.08) =$
$7290(0.92)(0.92) = 7290(0.92)^2$

1991: $7290(0.92)^2 - 0.08[7290(0.92)^2] = 7290(0.92)^2(1 - 0.08) =$
$7290(0.92)^2(0.92) = 7290(0.92)^3$

Thus the population data can be modeled by the exponential equation

$$P(x) = 7290(0.92)^x \text{ people}$$

where x is the number of years since 1988.

Enter the data points (aligning the years with $x = 0$ in 1988, $x = 1$ in 1989, $x = 2$ in 1990, etc.) and have your calculator or computer fit an exponential model.[1] You should obtain

$$P(x) = 7290.25032(0.91999)^x \text{ people}$$

where x is the number of years since 1988.

Why is this model not identical to the one we developed previously? Remember that as we computed percentage change for each year, we rounded, so the percentage decline was not exactly 8% each year. This rounding contributed to the difference between the models.

If we assume that the population continues to decline by the same percentage (approximately 8%), we can predict that during 1997 (that is, from the beginning of 1997 through the end of 1997), the town's population will decrease by

$$8\% \text{ of } 3442 \text{ people} = (0.08)3442 \text{ people} \approx 275 \text{ people}$$

Thus the town's population in 1998 is approximately

$$3442 - 275 \text{ people} = 3167 \text{ people}$$

This prediction can also be computed using the model for population with input $x = 1998 - 1988 = 10$:

$$P(10) = 7290.25032(0.91999)^{10} \approx 3166 \text{ people}$$

Note the slight difference between the answers. Two factors are involved in this difference. First, when calculating 3167 people, we used rounded figures: 8% instead of 8.001% as in the model. Second, the result 3167 people was calculated from a data point instead of from the population in 1997 generated by the model.

---

1. Whenever you use a calculator or computer to fit an exponential model to data, it is important to align the input values so that they are small in magnitude. If you fail to align the input values appropriately, the numerical computation routine may return an invalid result because of improper scaling, numerical overflow, or round-off errors. For instance, if you use $x = 1988, 1989, 1990$ instead of $x = 0, 1, 2$ in this example, the model returned is likely to be $y = ab^x$, where $a \approx 7.1999 \cdot 10^{75}$. This huge value for a could cause problems computationally. (See the discussion in footnote 2 for an even more serious problem that can occur.)

Because these two factors definitely affect the answer, we adopt the following rule of thumb:

> Once a model has been fitted to data, we will use the model to answer questions rather then using data points or rounded estimates.

Percentage Change in Exponential Models

Linear models exhibit a constant rate of change, but exponential models exhibit a constant *percentage change* (such as the 8% annual growth in an investment and the 8% decline in population). Because calculus is so closely associated with the mathematics of change, exponential models and linear models are important in calculus.

For the exponential model $f(x) = ab^x$ we determine the constant percentage change by calculating $(b - 1) \cdot 100\%$. For example, for the bank account whose balance was modeled by $B = 100(1.08)^t$ dollars t years after the initial investment was made, the constant percentage change on the balance is $(1.08 - 1)100\% = (0.08)100\% = 8\%$. That is, each year the balance of the account increases by 8%. Likewise, for the bacterial growth model $P(t) = 10,000(2)^t$ cells, the constant percentage change is $(2 - 1)100\% = 100\%$; that is, the bacterial population increases by 100% each hour. Finally, for the small town with population given by $P(t) = 7290(0.92)^t$ people t years after 1988, the constant percentage change is $(0.92 - 1)100\% = (-0.08)100\% = -8\%$. Thus the population of the town decreases by 8% each year.

Where do we find constant percentage change—that is, exponential change? We find it in situations where things feed on themselves—when change is determined by the current size. For instance, when a new product is introduced into the economic market place, word-of-mouth advertising often takes place. Each satisfied customer immediately tells another potential customer about the exciting new product. For that reason, exponential models are often used to analyze a product's sales growth.

EXAMPLE 1   *Automotive Tire Sales*

Table 2.3 shows sales data for a tire manufacturer.

TABLE 2.3

Year	Tire sales (thousands of dollars)
1966	23
1970	38.4
1974	64
1978	107
1982	179
1986	299
1990	499
1994	833

a. What is the constant percentage change in tire sales?

b. Use the percentage change to predict tire sales in 1998.

c. Write a model that reflects the constant percentage change in part *a*.

Solution:

a. A scatter plot of the data is shown in Figure 2.3. It is clear that a linear model is not appropriate.

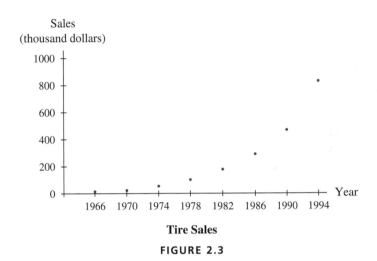

Tire Sales

FIGURE 2.3

Calculating percentage changes, we see that every 4 years, sales have increased by approximately 67%.

$$1966 \text{ to } 1970: \quad \frac{15.4}{23} = 0.66957 \approx 67\%$$

$$1970 \text{ to } 1974: \quad \frac{25.6}{38.4} = 0.66667 \approx 67\%$$

$$1974 \text{ to } 1978: \quad \frac{43}{64} = 0.67188 \approx 67\%$$

$$1978 \text{ to } 1982: \quad \frac{72}{107} = 0.67290 \approx 67\%$$

$$1982 \text{ to } 1986: \quad \frac{120}{179} = 0.67039 \approx 67\%$$

$$1986 \text{ to } 1990: \quad \frac{200}{299} = 0.66890 \approx 67\%$$

$$1990 \text{ to } 1994: \quad \frac{334}{499} = 0.66934 \approx 67\%$$

Thus 67% is the approximate percentage change of the data every 4 years.

b. If the percentage growth remains constant, then sales in 1998 will increase 67% over the 1994 sales.

Amount of increase (rounded to the nearest thousand) = 67% of 1994 sales

= (0.67)($833) ≈ $558 thousand

Prediction for 1998:

1994 sales + amount of increase = 833 + 558 = $1391 thousand

c. As in the town population example, we see that the data follow a general pattern.

1966:   23

1970: $23 + 0.67(23) = 23(1 + 0.67) = 23(1.67)$

1974: $23(1.67) + 0.67[23(1.67)] = 23(1.67)(1 + 0.67) =$
$23(1.67)(1.67) = 23(1.67)^2$

1978: $23(1.67)^2 + 0.67[23(1.67)^2] = 23(1.67)^2(1 + 0.67) =$
$23(1.67)^2(1.67) = 23(1.67)^3$

The sales data can be written as Sales = $23(1.67)^t$ thousand dollars, where $t = 0$ in 1966, $t = 1$ in 1970, $t = 2$ in 1974, and so on.   ∎

Now enter the tire sales data points into your calculator or computer (aligning the years with $t = 0$ in 1966, $t = 4$ in 1970, $t = 8$ in 1974, etc.) and have your calculator or computer fit an exponential model.[2] Did you obtain the following?

Sales = $22.98235(1.13683)^t$ thousand dollars

t years after 1966.

Note that the percentage change in this exponential model is approximately 14%. Were you expecting to see 67%? What happened?

The first model, Sales = $23(1.67)^t$ thousand dollars, had input aligned as $t = 0$ in 1966, $t = 1$ in 1970, $t = 2$ in 1974, etc. Thus the 67% is the percentage growth over 4 years. The second model, Sales = $22.98235(1.13683)^t$ thousand dollars, had input aligned on a yearly basis ($t = 0$ in 1966, $t = 4$ in 1970, $t = 8$ in 1974, etc.). Therefore, the 14% is the *annual* percentage growth.

A graph of the second model and the tire sales data is shown in Figure 2.4.

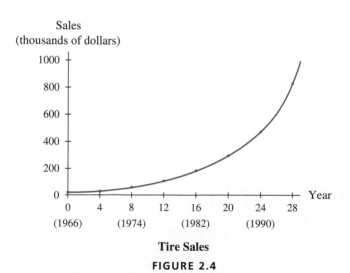

Tire Sales

FIGURE 2.4

2. Remember that whenever you use a calculator or computer to fit an exponential model to data, it is important that the input values be aligned in a manner such as in this example, $t = 0, 4, 8, \ldots$ instead of $t = 1966, 1970, 1974, \ldots$ If you fail to align the input values so that the inputs to the model are small in magnitude, the numerical computation routine may return a ridiculous result because of improper scaling, numerical overflow, or round-off errors. For instance, if you use $t = 1966, 1970, 1974$ in the current situation, the model that is returned is likely to be $y = ab^t$, where $a \approx 0$.

Using the exponential model Sales $= 22.98235(1.13683)^t$ thousand dollars, where t is the number of years since 1966, we calculate the sales for 1998 ($t = 32$) as Sales $\approx$ \$1392 thousand. You may have noticed that this value for sales is slightly higher than the one calculated directly from the data in Example 1. Remember that calculating with the data and a rounded approximation of percentage increase often yields a different answer from that obtained with the model. For this reason, once a model is obtained, we use that model rather than the data to answer questions.

Exponential Growth with Constraints

It is unrealistic to believe that exponential growth can continue forever. In most situations, there are forces that ultimately limit the growth. Here is a situation that may seem familiar to you:

You and a friend are shopping in a music store and find a new compact disk that you are certain will become a hit. You each buy the CD and rush back to campus to begin spreading the news. As word spreads, the total number of CDs sold begins to grow exponentially as in Figure 2.5.

However, this trend cannot continue indefinitely. Eventually, the word will spread to people who have already bought the CD or to people who have no interest in it, and the rate of increase in total sales begins to decline (Figure 2.6). In fact, because there is only a limited number of people who will ever be interested in buying the CD, total sales ultimately must level off. The graph representing the total sales of the CD as a function of time is a combination of rapid exponential growth followed by a slower increase and ultimate leveling off. See Figure 2.7.

S-shaped behavior such as this is common in marketing situations, the spread of disease, the spread of information, the adoption of new technology, and the growth of certain populations. A mathematical model for such an S-shaped curve is called a **logistic model**. Its equation is of the form

$$N(t) = \frac{L}{1 + Ae^{-Bt}}$$

The number L appearing in the numerator is the *leveling-off value* or *horizontal limiting value* for the function.

Suppose that 10 people work in an office. Someone comes to work with a highly contagious virus that causes the person to sneeze once every hour. If an infected person sneezes near an uninfected person, then the uninfected person becomes infected instantly and begins sneezing within the next hour.

Imagine that the spread of the virus is as summarized in Table 2.4 and graphed in Figure 2.8.

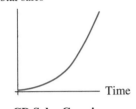

CD Sales Growing Exponentially

FIGURE 2.5

CD Sales Leveling Off

FIGURE 2.6

Total Sales of a CD

FIGURE 2.7

TABLE 2.4

Time	Total number of people infected
0 (before 8 a.m.)	1
1 (8 to 9 a.m.)	2
2 (9 to 10 a.m.)	4
3 (10 to 11 a.m.)	6
4 (11 a.m. to noon)	8
5 (noon to 1 p.m.)	9

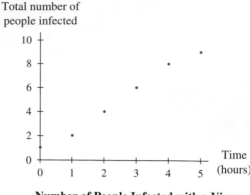

Number of People Infected with a Virus

FIGURE 2.8

Assuming that no one else comes into the office when someone is sneezing and that no one leaves the office and sneezes, there are only 10 people who can get the virus. The maximum potential number of people infected is therefore 10. This means that as time increases, the number infected can approach, but never exceed, 10. Thus $L = 10$ would make sense as a leveling-off, or limiting, value for this situation. A logistic equation with limiting value 10 that fits the above data is

$$\text{Total number infected} = \frac{10}{1 + 9.22111e^{-0.88860t}} \text{ people}$$

t hours after 8 a.m.

However, a logistics model calculated by a calculator or computer using a best-fit technique[3] is

$$\text{Total number infected} = \frac{9.98632}{1 + 9.33812e^{-0.89511t}} \text{ people}$$

t hours after 8 a.m. Note that the limiting value here is $L = 9.98632$, which is approximately, but not exactly, 10 people.

A graph of the model of the total number infected is shown in Figure 2.9. The dotted, horizontal line portrays the horizontal limiting value. Note that the curvature on the left side of this graph is **concave up** ⌣ whereas the curvature on the right side is **concave down** ⌢. The point on the graph at which the concavity changes is called the **inflection point**. The inflection point on the logistic graph modeling the spread of the virus (Figure 2.9) is marked with a black dot. In some situations, inflection points have very important interpretations. We later use calculus to find and help interpret these special points.

---

3. Different calculators and computer software may use slightly different methods for calculating a "best-fit" model. This results in slightly different coefficients and exponents in the models. Recall that our goal in fitting curves to data is to develop models so that we can use calculus to study the general behavior of the data. Slightly different models will not significantly affect our study of the general behavior of the data. From this point on we report a logistic model obtained from a TI-83 calculator.

When you use a calculator or a computer program to find an equation and construct the graph of a logistic model for a set of data, be sure to align the input data. This is important in logistic, as well as exponential, modeling.

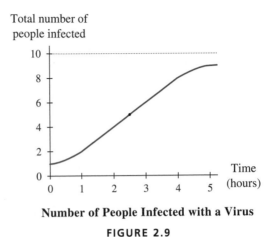

Number of People Infected with a Virus

FIGURE 2.9

EXAMPLE 2   *Swimsuit Sales*

 Table 2.5 shows the total number of swimsuits that a department store sold during the first three quarters of last year:

TABLE 2.5

End of month	Total number of swimsuits
January 31	4
February 28	12
March 31	25
April 30	58
May 31	230
June 30	439
July 31	648
August 31	748
September 30	769

Find a logistic model that fits the data. What is the limiting value of the model?

Solution   A scatter plot of the data (Figure 2.10) suggests that a logistic model is appropriate.

A possible logistic model is

$$S(x) = \frac{786.70445}{1 + 1464.70161e^{-1.26110x}} \text{ swimsuits}$$

by the end of month x, where $x = 1$ in January, $x = 2$ in February, and so on.

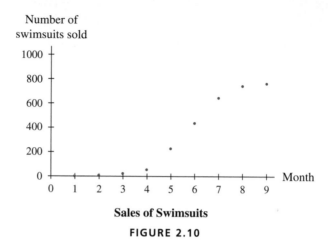

Sales of Swimsuits

FIGURE 2.10

The limiting value of this model is $L = 786.70445$ or, in context, approximately 787 swimsuits. The graph of $S(x)$ on a scatter plot appears to fit fairly well (see Figure 2.11). The dotted line in Figure 2.11 denotes the limiting value.

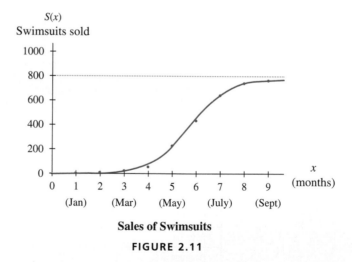

Sales of Swimsuits

FIGURE 2.11

The preceding logistic curves begin near zero and then increase toward a limiting value L. There are also situations where the curve begins near its limiting value and then decreases toward zero.

EXAMPLE 3   *Heights of College Men*

Of a group of 200 college men surveyed, the number who were taller than a given number of inches is recorded in Table 2.6.

TABLE 2.6

Inches	Number of men
65	198
66	195
67	184
68	167
69	139
70	101
71	71
72	43
73	26
74	11
75	4
76	2

Find an appropriate model for the data.

Solution: A scatter plot of the data is shown in Figure 2.12.

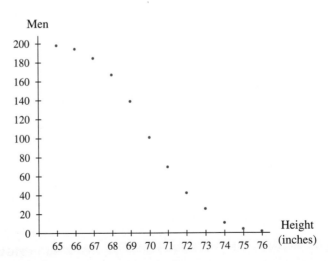

Number of College Men (out of 200) Taller than a Given Height

FIGURE 2.12

A logistic model for the data is

$$M(x) = \frac{205.66709}{1 + 0.02998e^{0.69528x}} \text{ men out of 200 men surveyed}$$

who are taller than $x + 65$ inches.

Recall that a logistic curve is of the form $f(x) = \dfrac{L}{1 + Ae^{-Bx}}$. What happened to the negative sign before the B in Example 3? In that case, $B = -0.69528$. When the formula was written, the negatives canceled.

$$\text{Number of men} = \frac{205.66709}{1 + 0.02998e^{-(-0.69528)x}}$$

$$= \frac{205.66709}{1 + 0.02998e^{0.69528x}}$$

This will always be the case for a logistic curve that decreases from a limiting value.

It is likely that a logistic model will have a limiting value that is lower or higher than the one indicated by the context. This does not mean that the model is invalid, but it does indicate that care should be taken when extrapolating from logistic models.

2.1 Concept Inventory

- Exponential model: $f(x) = ab^x$ (characterized by constant percentage change)
- Exponential growth and decay
- Finding percentage change
- Logistic model: $f(x) = \dfrac{L}{1 + Ae^{-Bx}}$ (exponential in nature with a limiting value)
- Concave up and concave down
- Inflection point
- Importance of aligning data

2.1 Activities

Identify the scatter plots in Activities 1 through 6 as linear, exponential, logistic, or none of these. If you identify the scatter plot as none of these, explain why it cannot be linear, exponential, or logistic.

1.

2.

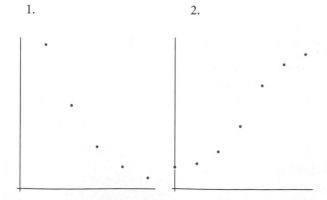

FIGURE 2.1.1   **FIGURE 2.1.2**

3.

4.

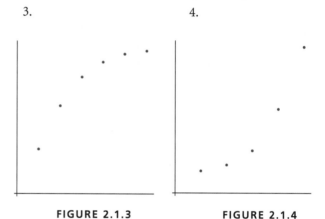

FIGURE 2.1.3   **FIGURE 2.1.4**

5.

6.

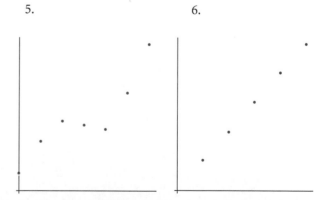

FIGURE 2.1.5   **FIGURE 2.1.6**

For Activities 7 through 10, tell whether the function is an increasing or a decreasing exponential function, and give the constant percentage change.

7. $f(x) = 72(1.05)^x$

8. $K(r) = 33(0.92)^r$

9. $y(x) = 16.2(0.87)^x$

10. $A(t) = 128.57(1.035)^t$

For Activities 11 and 12, find the constant percentage change, and interpret it in context.

11. After h hours, the number of bacteria in a Petri dish during a certain experiment can be modeled as $B(h) = 100(0.61)^h$ thousand bacteria.

12. The membership of a popular club can be modeled by $M(x) = 12(2.5)^x$ members by the end of the xth year after its organization.

For Activities 13 through 16, tell whether the function is an increasing or a decreasing logistic function, and identify the limiting value of the function.

13. $f(x) = \dfrac{100}{1 + 9.8e^{-0.98x}}$

14. $A(t) = \dfrac{1925}{1 + 321.1e^{1.86t}}$

15. $h(g) = \dfrac{39.2}{1 + 0.8e^{0.325g}}$

16. $k(x) = \dfrac{16.5}{1 + 1.86e^{-0.43x}}$

17. Carefully read the newspaper article below (from the *Greenville News* Wednesday, May 12, 1993), and answer the questions that follow.

   a. If the population of the world was 3 billion in 1960, 4 billion in 1975, and 5 billion in 1987, find an exponential model for world population growth.

   b. Use the model to predict the world population in mid-1993 and in 1997. How do your predictions compare with those in the article?

World population mushrooms

WASHINGTON (AP) — World population is growing at the fastest pace ever and virtually all growth is in the Third World, according to a survey released Tuesday by a research group.

The annual survey by the Washington-based Population Reference Bureau said, "We are at a point where, except for the United States, population growth is essentially a Third World phenomenon."

The survey predicted world population will reach 5.5 billion by mid-1993, 40 percent of it in China and India. The bureau said population is growing each year by 90 million, roughly the population of Mexico.

Carl Haub, a demographer who worked on the study, said that world population will grow to 8.5 billion by the year 2025, "only if birth rates continue to come down as expected. If they don't, growth will be even faster."

In an interview, Haub said the world population took 15 years to increase by 1 billion to its 4 billion total in 1975, 12 years to increase to 5 billion in 1987 and is expected to take only 10 years to rise to 6 billion in 1997.

Haub said that if it were not for the relatively high U.S. increase, all growth in the world would be in poor areas.

The survey showed the United States with a growth rate of 0.8 percent a year. This rate "and the world's highest amount of immigration will now produce unexpectedly high growth," it said.

"With a net immigration of about 900,000 per year, the United States effectively absorbs 1 out of every 100 people added to world population each year."

Haub said that most of the immigrants to the United States are from the Third World.

Europe's population has a growth rate of 0.2 percent a year.

Several former communist countries, including Hungary and Bulgaria, already show negative growth rates.

States of the former Soviet Union have been growing at 0.6 percent. But there was a gap between Russia and Ukraine, where population is declining, and the Muslim republics of central Asia, which are growing at more than 2 percent.

The world's fastest growing area is the poorest: sub-Saharan Africa with a growth rate of 3 percent. Population will double in 20 years. Latin America is growing at 1.9 percent.

c. Use the model to predict the world population in 2025. What assumptions are implied when the model is used to make this prediction? What does the article say about these assumptions?

d. Based on the growth rate stated in the article, confirm or refute the assertion that sub-Saharan Africa's population will double in 20 years.

e. Estimate the number of years that it will take Latin America's population to double.

18. When a company ceases to advertise and promote one of its products, sales often decrease exponentially, provided that other market conditions remain constant. At the time that publicity was discontinued for a newly released popular animated film, sales were 520,000 videotapes per month. One month later, videotape sales had fallen to 210,000 tapes per month.

a. What was the monthly percentage decline in sales?

b. Assuming that sales decreased exponentially, give the equation for sales as a function of the number of months since promotion ended.

c. What will sales be 3 months after the promotion ends? 12 months after?

d. Walt Disney has been known to produce videotapes of some of its films for a limited time before completely stopping production. Do you believe that this is a wise business decision?

e. What would be necessary to ensure that the exponential decrease in sales of a product did not occur?

19. In 1992, CDs passed cassettes as the most popular format for pre-recorded music. Table 2.7[4] gives the numbers of CDs (in millions) sold during the early 1990s.

TABLE 2.7

Year	Number of CDs sold (millions)
1991	333
1992	420
1993	500
1994	662

a. Find an exponential model to fit the data.

b. According to the model in part *a*, what was the yearly percentage growth in CD sales from 1991 through 1994?

20. An appliance manufacturer that prides itself on the reliability of its washing machines conducts an extensive survey of 3200 of its customers. The survey asked for the number of years that the washing machine remained in working condition before requiring a major service call (not including routine maintenance). Partial results are given in Table 2.8.

TABLE 2.8

Years before servicing	Number of customers
5	787
10	619
15	487
20	383
25	301
30	237

a. Discuss the appropriateness of linear and exponential models for this data.

b. Find the better model for the data.

c. Use the model to estimate the number of customers surveyed who had washing machines that lasted between 30 and 35 years before service call.

21. The population of Mexico, in millions of people, in selected years from 1921 through 1990 is given in Table 2.9.[5]

TABLE 2.9

Year	Population (millions)
1921	14.335
1930	16.553
1940	19.654
1950	25.791
1960	34.923
1970	48.225
1980	66.847
1990	81.250

4. *USA Today*, USA Snapshots, February 24, 1995. Source: Recording Industry Association of America.

5. SPP and INEGI, Mexican Censuses of Population 1921 through 1990 as reported in Pick and Butler, *The Mexico Handbook*, Westview Press, 1994.

(*Note*: Even though population censuses are normally taken every 10 years in Mexico, the census at the beginning of the 1920s was taken in 1921, not in 1920.)

a. Find an exponential model for the population.

b. Use the model to estimate the population of Mexico in 1993.

c. *The 1993 Information Please Environmental Almanac* reports the population of Mexico to be 88,598,000. How close is this to your estimate?

d. *The 1993 Information Please Environmental Almanac* also reports the growth rate to be between 2.2% and 2.3%. What percentage growth rate does the exponential model represent? What does this tell you about using the exponential model to make predictions?

22. According to *The 1992 Information Please Almanac*, the typical salaries of full professors at public colleges and universities in the United States from 1984 through 1989 were as listed in Table 2.10.

TABLE 2.10

Year	Salary (thousands of dollars)
1984	37.1
1985	39.6
1986	42.3
1987	45.3
1988	47.2
1989	50.1

a. Find both linear and exponential models for the data.

b. Use each model to estimate the typical salaries in 1993, 1996, and 2001.

c. What assumptions are made when extrapolating from each of these models? Are these assumptions necessarily true?

d. Which model do you think is more reasonable and why?

23. The data in Table 2.11 show how postal rates have increased for first-class letters weighing up to 1 ounce.

a. Find an exponential model to fit the data. According to this model, what was the postage in 1995?

TABLE 2.11

Year of rate increase	Rate (cents)
1919	2
1932	3
1958	4
1963	5
1968	6
1971	8
1974	10
1975	13
1978	15
1981	20
1985	22
1988	25
1991	29
1995	32

b. Discuss how well the model fits the data.

c. Disregard the first two data points, and find an exponential model to fit the data from 1958 to 1995. According to this model, what was the postage rate in 1995?

d. Discuss how well this new model fits the data.

e. Use the two models to predict the postage rate in the year 2001. Would you consider either of these predictions to be a good indicator of what may happen? Explain your reasons.

24. The consumer price index[6] for college tuition from 1985 through 1993 is shown in Table 2.12. (1982–1984 = 100)

TABLE 2.12

Year	Index
1985	119.9
1987	141.3
1988	154.6
1989	168.0
1990	182.8
1991	198.0
1992	213.7
1993	228.9

6. *Statistical Abstract*, 1994.

a. Find linear and exponential models to fit the data.

b. What is the yearly percentage increase in the exponential model? What is the yearly increase in the linear model?

c. Graph the two models from 1985 through the year 2000. Which model is more advantageous to the students, and which model is more lucrative for colleges? Explain your reasons.

d. Which model do you believe more accurately reflects what will happen to college tuition in the future? Why?

e. Find the most current consumer price index for college tuition that is available. Compare the two model values for that year with the actual value.

25. Table 2.13[7] below shows emissions of carbon monoxide (in millions of metric tons) in the United States from 1980 through 1992.

TABLE 2.13

Year	Emissions of CO (millions of metric tons)
1980	117.0
1981	111.6
1982	105.4
1983	105.2
1984	102.5
1985	97.9
1986	95.2
1987	90.1
1988	89.9
1989	84.7
1990	83.8
1991	82.3
1992	79.1

a. Find linear and exponential models for the data.

b. What is the yearly percentage decline in CO emissions in the exponential model? What is the yearly decline in CO emissions in the linear model?

c. Graph the two models from 1980 to 2030. Contrast the future behaviors of the two models.

d. What do you believe will happen to CO emissions in the future? Does either model fit your theory?

e. Find the most current CO emissions data available. Compare the two model values for that year with the actual value. Does this additional information change your answer to part d?

26. During the last two decades, computing power has grown enormously. Table 2.14[8] gives the number of transistors (in millions) in Intel processor chips.

TABLE 2.14

Computer	Year	Number of transistors (millions)
The 4004	1971	0.0023
The 386DX	1986	0.275
The 486DX	1989	1.2
The Pentium	1993	3.3
The P6	1995	5.5

a. Find an exponential model to fit the data.

b. According to the model found in part a, what is the annual percentage increase in the number of transistors used in an Intel computer processor chip?

c. Is it reasonable to expect this exponential growth rate to continue into the twenty-first century? Explain your reasoning.

27. Table 2.15[9] gives the number of European, North American, and South American countries that issued postage stamps from 1836 through 1880.

a. Find a logistic model for the data, and discuss how well the model fits.

b. Sketch the graph, and mark where the curve is concave up and where it is concave down. Label the approximate location of the inflection point.

7. *Statistical Abstract*, 1994.
8. *USA Today*, "Intel Unveils P6 Successor to Pentium," February 17, 1995. Source: Intel.
9. "The Curve of Cultural Diffusion," *American Sociological Review*, August 1936, pp. 547–556.

TABLE 2.15

Year	Total number of countries
1840	1
1845	3
1850	9
1855	16
1860	24
1865	30
1870	34
1875	36
1880	37

28. Table 2.16[10] gives the total number of states associated with the national P.T.A. organization from 1895 through 1931.

    a. Find a logistic model to fit the data.

    b. What is the maximum number of states that could have joined the PTA by 1931? How does this number compare to the limiting value given by model in part *a*?

29. The approximate numbers of patents for plow sulkies beginning in 1865 are given in Table 2.17.[11]

    a. What is a *plow sulky*, and to what modern farm implement was it a precursor?

    b. Find a logistic model to fit the data.

    c. Discuss why it is logical that the total number of patents for a new invention would increase according to a logistic equation.

TABLE 2.16

Year	Total number of states
1895	1
1899	3
1903	7
1907	15
1911	23
1915	30
1919	38
1923	43
1927	47
1931	48

TABLE 2.17

Year	Cumulative number of plow sulky patents
1871	200
1877	340
1883	980
1889	1800
1895	2200
1901	2400
1907	2500
1913	2550
1919	2620
1925	2700

10. Hamblin, Jacobsen, and Miller, *A Mathematical Theory of Social Change*, New York: John Wiley & Sons, 1973.
11. *Ibid.*

30. The personnel manager for a contracting company keeps track of the total number of labor hours spent on a construction job each week during the construction. Partial data appear in Table 2.18.

    a. Discuss why a logistic model is appropriate for the data.

    b. Find a logistic model to fit the data.

    c. In which week is the 10,000-labor-hour point reached?

    d. If the company charges $50 per labor hour, what would the total labor cost be for this construction job?

TABLE 2.18

Weeks after start of the construction project	Cumulative labor hours
1	25
4	158
7	1254
10	5633
13	9280
16	10,010
19	10,100

31. In the fall of 1918, an influenza epidemic hit the United States Navy. It spread to the Army, to American civilians, and ultimately to the world. It is estimated that 20 million people had died from the epidemic by 1920. Of these, 550,000 were Americans—over 10 times the number of World War I battle deaths. Table 2.19[12] gives the total numbers of Navy, Army, and civilian deaths that resulted from the epidemic in 1918.

    a. Find logistic equations to fit each set of data. In each case, graph the model on the data.

    b. Compare the limiting values of the models in part *a* with the highest data values in the table. How many more people do the models indicate died from the epidemic after November 30? Do you believe that the limiting values indicated by the models accurately reflect the ultimate number of deaths? Explain.

TABLE 2.19

Week ending	Total deaths Navy	Total deaths Army	Total civilian deaths in 45 major cities
August 31	2		
September 7	13	40	
September 14	56	76	68
September 21	292	174	517
September 28	1172	1146	1970
October 5	1823	3590	6528
October 12	2338	9760	17,914
October 19	2670	15,319	37,853
October 26	2820	17,943	58,659
November 2	2919	19,126	73,477
November 9	2990	20,034	81,919
November 16	3047	20,553	86,957
November 23	3104	20,865	90,449
November 30	3137	21,184	93,641

12. A.W. Crosby, Jr., *Epidemic and Peace 1918*, Westport, Connecticut: Greenwood Press, 1976.

TABLE 2.20

Time	Total number of people not infected
0 (before 8 a.m.)	
1 (8 to 9 a.m.)	
2 (9 to 10 a.m.)	
3 (10 to 11 a.m.)	
4 (11 a.m. to noon)	
5 (noon to 1 p.m.)	

32. Again consider the virus in the office example, except this time look at it from the viewpoint of how many people have *not* yet been infected by time x.

   a. Using the data given in the Table 2.4 on page 79, fill in Table 2.20.

   b. Examine a scatter plot of the "not infected" data, and describe its curvature.

   c. Is the scatter plot increasing to a limiting value or decreasing from a limiting value? How does this compare with the scatter plot of the original "infected" data?

   d. Find a logistic model for the "not infected" data.

   e. Is the limiting value for the "not infected" data the same as the limiting value for the "infected" data? Why?

   f. Is the value of B in the logistic model positive or negative? How is the sign of B related to the curvature of the logistic curve?

33. In 1949, the United States experienced the second worst polio epidemic in its history. (The worst was in 1952.) Table 2.21[13] gives the cumulative number of polio cases diagnosed on a monthly basis.

   a. Observe a scatter plot of the data from January through June. Describe the concavity indicated by the scatter plot. Does this portion of the data appear to be logistic?

   b. Observe a scatter plot of the entire data set given. Does the entire data set appear to be logistic?

   c. Find a logistic curve to fit the data.

   d. Observe the model in part c graphed on the data between January and June. Discuss how well the model fits this portion of the data.

34. The total numbers of visitors to an amusement park that stays open all year are given in Table 2.22.

TABLE 2.21

Month	Total number of polio cases
January	494
February	759
March	1016
April	1215
May	1619
June	2964
July	8489
August	22,377
September	32,618
October	38,153
November	41,462
December	42,375

TABLE 2.22

Month	Cumulative number of visitors by the end of the month (thousands)
January	25
February	54
March	118
April	250
May	500
June	898
July	1440
August	1921
September	2169
October	2339
November	2395
December	2423

13. *Twelfth Annual Report*, The National Foundation for Infantile Paralysis, 1949.

a. Find a logistic model for the data.

b. The park owners have been considering closing the park from October 15 through March 15 each year. How many visitors will they potentially miss by this closure?

35. The rapid advancement in calculator technology over the past 25 years has led to a decrease in the price of a basic, four-function calculator. Suppose that Table 2.23 gives the average price of a four-function calculator from 1970 through 1990.

a. Examine a scatter plot of the data. Is this scatter plot increasing or decreasing? Does there appear to be an inflection point?

b. Find a logistic model to fit the data.

c. In approximately what year does the inflection point appear?

36. Table 2.24 records the volume of sales (in thousands) of a popular movie for the first 18 months after it was released on videocassette.

a. Examine a scatter plot, and describe its curvature.

b. Explain why this type of data could reasonably be modeled by a logistic curve.

TABLE 2.23

Year	Average price (dollars)
1970	70
1972	67
1974	62
1976	56
1978	48
1980	40
1982	30
1984	23
1986	17
1988	13
1990	10

c. Would the value B of a logistic model be positive or negative in this case?

d. Find a logistic model to fit the data.

TABLE 2.24

Months after release	Number of cassettes sold each month (thousands)
1	580
2	565
3	527
4	467
5	321
6	291
7	204
8	131
9	79
10	61
11	31
12	17
13	9
14	4
15	3
16	3
17	2
18	2

■ 2.2   Exponential Models in Finance

Compound Interest

In everyday life, exponential models occur in investment situations. You probably are familiar with the *interest formula*

$$A(t) = P(1 + r)^t$$

where $A(t)$ is the dollar amount accumulated after t years when P dollars are invested at an annual interest rate of $r\%$ compounded annually. (Note that r should be written as a decimal number when used in interest formulas.) If interest is compounded more often than once a year, then the general **compound interest formula** should be used:

Compound Interest Formula

The amount accumulated in an account after t years when P dollars are invested at an annual interest rate of $r\%$ compounded n times a year is

$$A(t) = P\left(1 + \frac{r}{n}\right)^{nt} \text{ dollars}$$

Here $\frac{r}{n}$ is the interest rate that is applied to the balance in the account at the end of each compounding period, and nt is the total number of compounding periods in t years.

Consider \$1000 invested at a 5% annual interest rate compounded monthly. The equation for the accumulated amount $A(t)$ after t years is

$$A(t) = 1000\left(1 + \frac{0.05}{12}\right)^{12t} \text{ dollars}$$

This equation can be rewritten in the exponential form

$$A(t) = 1000(b^t) = 1000[(1.004166667)^{12}]^t \approx 1000(1.05116)^t \text{ dollars}$$

What is the meaning of the value 1.05116 in this formula? Remember, for an exponential model $f(x) = ab^x$, the constant percentage change is $(b - 1) \cdot 100\%$. Thus, in the amount formula $A(t) = 1000(1.05116)^t$ dollars, the constant percentage increase of the amount is $(1.05116 - 1)100\% = (0.05116)100\% = 5.116\%$. That is, the amount invested increases by 5.116% every year. In the investment context, this percentage is called the **effective rate** or **annual percentage yield** (APY). The advertised 5% **annual percentage rate** (APR) is called the **nominal rate**. When comparing compound interest investment opportunities, you should always consider annual percentage yields because the nominal rates do not reflect the compounding periods.

You should be aware that even though the compound interest formula is a continuous function, it has a discrete interpretation because the amount changes only at the actual times of compounding. For instance, we can use the monthly compounding function $A(t) = 1000(1.05116)^t$ dollars after t years to find the amount of the investment at the end of the 3rd month of the 6th year by calculating $A(6.25) \approx$ $1365.95. (This value was calculated using the unrounded value of b.) However, it would be incorrect to use $t = 6.2$ to calculate the amount in the account on the 14th day of the 3rd month of the 6th year, because interest is compounded monthly, not daily.

EXAMPLE 1   *Comparing Annual Percentage Yields*

Which is the better deal: an APR of 6.9% compounded quarterly or an APR of 6.7% compounded monthly?

Solution: For a given principal P, we need to compare the amounts

$$P\left(1 + \frac{0.069}{4}\right)^{4t} \quad \text{and} \quad P\left(1 + \frac{0.067}{12}\right)^{12t}$$

Because $\left(1 + \frac{0.069}{4}\right)^4 \approx 1.07081$ and $\left(1 + \frac{0.067}{12}\right)^{12} \approx 1.06910$, we see that the annual percentage yield for the 6.9% quarterly deal is about 7.08%, compared to only 6.91% APY for the 6.7% monthly deal. You should choose the first of these two options. ■

Continuous Compounding and the Number e

For a given nominal rate, the only factor that influences the annual percentage yield is the number of times the interest is compounded each year. To see the effect of the number of compoundings, consider the following:

Suppose that you invest $1 for 1 year at a nominal rate of 100% compounded n times during the year: Amount $= \$1\left(1 + \frac{1}{n}\right)^n$. As n increases, how does the original $1 grow? Consider Table 2.25, which shows the relationship between n and the accumulated amount.

TABLE 2.25

Compounding occurs	n	Amount
yearly	1	$2
semiannually	2	$2.25
quarterly	4	$2.44
monthly	12	$2.61
weekly	52	$2.69
daily	365	$2.71
every hour	8760	$2.72
every minute	525,600	$2.72
every second	31,536,000	$2.72

Does it surprise you that the amount does not grow infinitely large? In fact, your $1 will not grow to more than $2.72 during the year, regardless of how often interest is compounded. As n gets larger and larger, the unrounded amount is getting closer and closer to the number that we call e. To express the fact that $\left(1 + \frac{1}{n}\right)^n$ is approaching e as n increases without bound, we say that *"the limit, as n approaches infinity, of $\left(1 + \frac{1}{n}\right)^n$ is e."* We write this statement using the following mathematical symbols:

$$\lim_{n\to\infty}\left(1 + \frac{1}{n}\right)^n = e$$

The number e is an *irrational* number, which means that its decimal expansion never repeats and never terminates. A 12-digit decimal approximation to e is 2.71828182846.

Recall that logistic models are characterized by limiting values. We say that "the limit, as t approaches infinity, of $\dfrac{L}{1 + Ae^{-Bt}}$ is L" (provided that B is positive), and we write

$$\lim_{t\to\infty}\frac{L}{1 + Ae^{-Bt}} = L$$

The concept of limits is central to a more rigorous development of calculus. We will say more about limits later.

The limit $\lim_{n\to\infty}\left(1 + \frac{1}{n}\right)^n = e$ represents the unrounded amount to which 1 dollar would grow in a year if the nominal rate were 100% and the number n of compounding periods per year increased without bound; that is, the interest is *compounded continuously*. What if we continuously compounded P dollars for t years at a nominal interest rate r? In other words, what is $\lim_{n\to\infty} P\left(1 + \frac{r}{n}\right)^{nt}$? A little algebra goes a long way. To simplify the algebra, we let $m = \frac{n}{r}$ so that $mr = n$. Then

$$\left(1 + \frac{r}{n}\right)^{nt} = \left(1 + \frac{1}{m}\right)^{mrt}$$

$$= \left[\left(1 + \frac{1}{m}\right)^m\right]^{rt}$$

Also, for a fixed r, because $m = \frac{n}{r}$, $m \to \infty$ as $n \to \infty$. Thus

$$\lim_{n\to\infty} P\left(1 + \frac{r}{n}\right)^{nt} = \lim_{m\to\infty} P\left[\left(1 + \frac{1}{m}\right)^m\right]^{rt}$$

$$= P\left[\lim_{m\to\infty}\left(1 + \frac{1}{m}\right)^m\right]^{rt}$$

$$= Pe^{rt}$$

This result is known as the **continuously compounded interest** formula.

Continuously Compounded Interest Formula

The amount accumulated in an account after t years when P dollars are invested at a nominal interest rate of r% compounded continuously is

$$A(t) = Pe^{rt} \text{ dollars}$$

Note that $A(t)$ is a continuous function with continuous interpretation. Unlike the general compound interest formula (which is a continuous function with discrete interpretation), it is meaningful here to ask questions about the amount in the account at any instant in time.

What is the APY for this type of account? With continuous compounding, the annual percentage yield (APY) is $(e^r - 1) \cdot 100\%$. For example, if \$1000 is invested at a 5% nominal interest rate compounded continuously, then the accumulated amount, $A(t)$, after t years is

$$A(t) = 1000e^{0.05t}$$
$$= 1000(e^{0.05})^t$$
$$\approx 1000(1.05127)^t \text{ dollars}$$

The annual percentage yield is approximately 5.13%.

It is easy to convert from an exponential function of the form ae^{kt} to an exponential function of the form ab^t. We just saw an example of this conversion:

$$1000e^{0.05t} \approx 1000(1.05127)^t$$

Note that the value of a remains the same in either form and that $b = e^k$.

This conversion from ae^{kt} to ab^t gives us a simple way to find the constant percentage change of the function ae^{kt}. (Recall that the constant percentage change of ab^t is $(b - 1) \cdot 100\%$.) We illustrate this concept in Example 2.

EXAMPLE 2   *APY of Continuous Compoundings*

What is the annual percentage yield (APY) on an investment that has an advertised nominal rate of 12.6% compounded continuously?

Solution: For a given principal investment P, we use the continuously compounded interest formula $A(t) = Pe^{rt}$ to determine the amount at any time t (in years). In this case,

$$A(t) = Pe^{0.126t} \text{ dollars after } t \text{ years}$$

Converting from the form ae^{kt} to the form ab^t, we obtain the following equivalent function:

$$A(t) = Pe^{0.126t} = P(e^{0.126})^t \approx P(1.13428)^t \text{ dollars after } t \text{ years}$$

Thus the APY is approximately 13.4%.   ■

It is also possible to convert from an exponential function of the form ab^t to one of the form ae^{kt}. We saw that the value of a remains the same in either form and that $b = e^k$. Thus in order to convert a function, say $f(t) = 120(1.05)^t$, to the form ae^{kt}, we must find the value of k such that $e^k = b$. The correct value for k is $k = \ln 1.05 \approx 0.04879$. The appropriate function of the form e^{kt} is

$$f(t) = 120(1.05)^t \approx 120e^{0.04879t}$$

EXAMPLE 3   *Determining APR*

An investment that has interest compounded continuously has an APY of 9.2%. What is its APR?

Solution: For an investment of P dollars, we set up the function as

$$\text{Amount} = A(t) = P(1.092)^t \text{ dollars after } t \text{ years}$$

Converting this to the form ae^{kt}, we get

$$A(t) = P(1.092)^t = Pe^{(\ln 1.092)t} \approx Pe^{0.08801t}$$

dollars after t years. Thus the APR (when interest is compounded continuously) is approximately 8.8%.   ∎

Present and Future Value

When money is invested in an account that earns interest, we are concerned with its **future value**—that is, its value at some time in the future. The compound interest formulas $A(t) = P\left(1 + \frac{r}{n}\right)^{nt}$ and $A(t) = Pe^{rt}$ are used to determine future values.

EXAMPLE 4   *Future Value*

Suppose that on July 1 of each year, you deposit $1500 into an account that earns interest at the nominal rate of 7.5% compounded monthly. What is the future value of these investments 1 year after the fifth deposit is made?

Solution:

The initial $1500 will grow to $1500\left(1 + \frac{0.075}{12}\right)^{60} \approx \$2179.94$

The second $1500 will grow to $1500\left(1 + \frac{0.075}{12}\right)^{48} \approx \$2022.90$

The third $1500 will grow to $1500\left(1 + \frac{0.075}{12}\right)^{36} \approx \$1877.17$

The fourth $1500 will grow to $1500\left(1 + \frac{0.075}{12}\right)^{24} \approx \$1741.94$

The last $1500 will grow to $1500\left(1 + \frac{0.075}{12}\right)^{12} \approx \$1616.45$

The future value of these investments is the sum

$$\$2179.94 + \$2022.90 + \$1877.17 + \$1741.94 + \$1616.45 = \$9438.40 \quad ∎$$

Often, however, we are interested in how much money should be invested now in order to achieve a desired future value. The amount to be invested now is called the **present value** of the investment. The compound interest formulas are also used to determine present values, as illustrated in Example 5.

EXAMPLE 5   *Present Value*

Find the present value of the $9438.40 in Example 4, assuming that the money is invested at a nominal rate of 7.5% with

a.   monthly compounding for 5 years.

b.   continuous compounding for 5 years.

Solution:

a.   Using an interest rate of 7.5% compounded monthly, we solve the equation

$$9438.40 = P\left(1 + \frac{0.075}{12}\right)^{60}$$

for the present value P. Because this equation is just

$$9438.40 = P(1.453294408)$$

we see that the present value is $P = \frac{9438.40}{1.453294408} \approx \$6494.49$. In other words, you would need to invest \$6494.49 now at 7.5% compounded monthly to have \$9438.40 in 5 years.

b. On the other hand, if we seek the present value of \$9438.40 assuming a rate of 7.5% compounded continuously over 5 years, then we must solve

$$9438.40 = Pe^{0.075(5)} = Pe^{0.375} = P(1.454991415)$$

for P to obtain $P \approx \$6486.91$. Again, this is the amount you would need to invest at 7.5% continuously compounded to have \$9438.40 in 5 years.   ■

2.2   Concept Inventory

- Compound interest formula: $A(t) = P\left(1 + \frac{r}{n}\right)^{nt}$
- Nominal rate, or annual percentage rate (APR)
- Effective rate, or annual percentage yield (APY)
- The number e
- Converting from e^{kt} to b^t
- Converting from b^t to e^{kt}
- Continuously compounded interest formula: $A(t) = Pe^{rt}$
- Present value and future value

2.2   Activities

For Activities 1 through 4, tell whether the exponential function is increasing or decreasing, convert the function from the form ae^{kx} to the form ab^x, and find the constant percentage change of the function.

1. $f(x) = 100e^{0.98x}$

2. $f(x) = 1.92e^{1.02x}$

3. $A(t) = 1000e^{-0.21t}$

4. $P(t) = 294.3e^{-0.97t}$

For Activities 5 through 8, tell whether the exponential function is increasing or decreasing, and convert the function from the form ab^x to ae^{kx}.

5. $f(x) = 39.2(1.29)^x$

6. $f(x) = 62.4(0.93)^x$

7. $h(t) = 1.02(0.62)^t$

8. $g(t) = 0.93(1.05)^t$

9. You invest \$2000 at 4.5% APR compounded quarterly.

a. Give the formula of the form $A = P\left(1 + \frac{r}{n}\right)^{nt}$ to determine the balance in this account after t years. Also write the formula in the form $A = ab^t$.

b. Calculate the amount in the account after 2 years.

c. Find the amount of the investment after 3 years and 9 months.

d. Calculate the amount in the account after 2 years and 2 months.

10. You invest \$10,000 at 8.125% APR compounded monthly.

a. Give the formula of the form $A = P\left(1 + \frac{r}{n}\right)^{nt}$ to determine the balance in this account after t years. Also write the formula in the form $A = ab^t$.

b. Calculate the amount in the account after 1 year.

c. Find the amount of the investment after 5 years.

11. Your credit card statement indicates a finance charge of 1.5% per month on the outstanding balance.

a. What is the nominal rate (APR), assuming that interest is compounded monthly?

b. What is the effective rate of interest (APY)?

12. In order to offset college expenses, at the beginning of your freshman year you obtained a nonsubsidised student loan for \$15,000. Interest on this loan accrues at a rate of 0.739% each month. However, you do not have to make any payments against either the principal or the interest until after you graduate.

a. Write a function giving the total amount you will owe after t years in college.

b. What is the nominal rate?

c. What is the effective rate?

13. Suppose that six years ago you invested $1400 in an account with a fixed interest rate compounded continuously. You do not remember the interest rate, but your end-of-the-year statements for the first five years show the amounts given in Table 2.26.

TABLE 2.26

Year	Dollar amount
1	1489.55
2	1584.82
3	1686.19
4	1794.04
5	1908.80

a. Using the data, estimate the constant percentage change. What is the financial term for this percentage? On the basis of this percentage change, determine the balance at the end of the sixth year.

b. Find an equation of the form $A = Pe^{rt}$ that models the data. What is the financial term for the percent r?

c. How much money will be in the account at the end of the 6th year? Under what circumstances might this amount change?

14. How long would it take an investment to double under each of the following conditions?

a. Interest is 6.3% compounded monthly.

b. Interest is 8% compounded continuously.

c. Interest is 6.85% compounded quarterly.

15. You have $1000 to invest, and you have two options: 4.725% compounded semiannually or 4.675% compounded continuously.

a. Determine the better option by calculating the annual percentage yield for each.

b. Verify your choice of option by comparing the amount the first option would yield with the amount the second option would yield after 2, 5, 10, 15, 25, and 50 years. Does your choice of option depend on the number of years you leave the money invested?

c. By how much would the two options differ after 10 years?

16. Suppose that to help pay for your college education, 15 years ago your parents invested a sum of money that has grown to $25,000 today. How much did they originally invest if

a. the investment earned 7.5% interest compounded monthly?

b. the investment earned 7.5% interest compounded daily?

c. the investment earned 7.5% interest compounded continuously?

17. The formula that is used to calculate the monthly payment of a loan is

$$\frac{r}{12}A = m\left[1 - \left(1 + \frac{r}{12}\right)^{-n}\right]$$

where: A = loan amount
   r = interest rate expressed as a decimal
   n = number of months of the loan
   m = monthly payment

Suppose you are considering buying a $13,500 car. You have $1200 for a down payment and have two options for loans.

   Option A: 10.9% compounded monthly for 3.5 years

   Option B: 10.5% compounded monthly for 4 years

a. Determine the monthly payment for each loan.

b. Determine the total amount paid for each loan.

c. Determine the total amount of interest paid for each loan.

d. What are the advantages and disadvantages of each loan? Which loan would you choose and why?

18. Your parents are hoping to buy a house 5 years from now when you finish college. Because of rising inflation, prices on the housing market are expected to increase by 3% each year over the next 5 years. Estimate how much your parents will have to pay in 5 years for a house that is currently priced at $85,000.

19. A certificate of deposit (C.D.) is bought for $2500 and held for 3 years. What is the future value of the C.D. at the end of the 3 years if it earns interest compounded quarterly at a nominal rate of 6.6%?

20. You were able to save $3000 from your summer job. If you invest it now in an account with 6.7%

annual interest compounded monthly, how much will your account be worth when you graduate?

21. You are saving for a down payment on a car. You worked all summer and want to invest part of your earnings at 6.2% compounded monthly in order to make the down payment when you graduate. How much of your summer earnings should you invest now in order to have $2000 in 2 years?

22. Next summer, you plan to put part of your summer earnings in the bank in order to meet your January tuition payment. How much should you invest in August at 5.5% compounded daily if you need $3600 at the beginning of January?

23. a. How much money would you have to invest now at 10% APR compounded monthly in order to accumulate a nest egg of a quarter of a million dollars over the next 45 years?

    b. Using the answer from part *a* as the principal amount invested at 10% APR compounded monthly, determine the amount to which the principal will grow after 45 years. Why is the answer not $250,000?

24. What is the present value of an investment that will be worth $10,000 in 5 years, assuming that the effective rate is 7.1%?

2.3   Polynomial Functions and Models

Polynomial functions and models have been used extensively throughout the history of mathematics. Their successful use stems from both their presence in certain natural phenomena and their relatively simple application. Prior to the availability of inexpensive computing technology, ease of use made polynomials the most widely applied models because they were often the only ones that could be calculated with pencil-and-paper techniques. However, we no longer have to deal with this restriction, and we call on polynomials only when they are appropriate.

When do polynomials fit? We have already seen that the simplest polynomial model, the linear model, occurs when there is constant, incremental change due to a lack of external influences. The natural step up is to a consistent force for change. Under a consistent force for change, the quadratic model often fits well, and it is encountered in a variety of situations. For instance, gravity is a constant force, and many movements under its influence are, in essence, quadratic. In the business community, sales managers who keep applying pressure sometimes find that their efforts result in numbers that exhibit a quadratic pattern. In fact, situations like these are so common that the curves they generate have a familiar name, *parabolas*. The general parabola equation is one you know well from your algebra classes: $y = ax^2 + bx + c$.

It is more difficult to sense the higher-order polynomial models, so we simply indicate a few of the places where they appear. Cubic models, whose general equation is $y = ax^3 + bx^2 + cx + d$, are widely used in shapes that are required to shift smoothly. They appear in locations such as the sides of ships and in highway designs. In each of these cases, we need to make a smooth transition from a straight line into a curve. The curve that best makes this transition is the cubic. Cubics also occur in some rather unexpected situations, as we shall soon see.

Even though higher-degree polynomials are useful in some situations, we will limit our discussion of polynomial functions and models to the linear, quadratic, and cubic cases.

Quadratic Modeling

A roofing company in Miami keeps track of the number of roofing jobs it completes each month. The data from January through June follow.

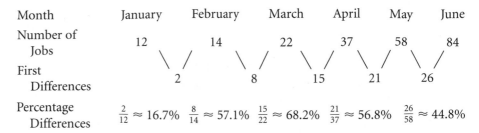

Month	January	February	March	April	May	June
Number of Jobs	12	14	22	37	58	84
First Differences		2	8	15	21	26

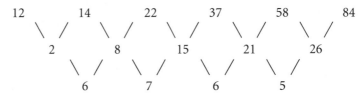

Percentage Differences: $\frac{2}{12} \approx 16.7\%$ $\frac{8}{14} \approx 57.1\%$ $\frac{15}{22} \approx 68.2\%$ $\frac{21}{37} \approx 56.8\%$ $\frac{26}{58} \approx 44.8\%$

The first differences and percentage differences (percentage changes) are not close to being constant, so we can conclude that some model other than a linear model or an exponential model would be appropriate in this case. Note, however, that the differences between the first differences are nearly constant.

We call these **second differences**. When first differences are constant, the data can be modeled by a linear model. When second differences are constant, the data can be modeled by the **quadratic model** $f(x) = ax^2 + bx + c$.

The graph of a quadratic model is a **parabola**. By examining a scatter plot of the roofing company data with January aligned as month 1, we observe a plot (see Figure 2.13) that suggests a portion of the familiar parabolic shape.

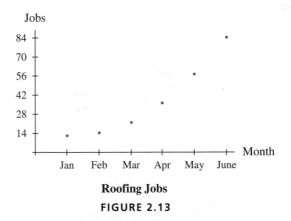

Roofing Jobs

FIGURE 2.13

For this illustration, the rate of change in roofing jobs measures how rapidly the number of jobs is increasing (or possibly decreasing if the trend changes) each month. Unlike the case for a linear model, the rate of change for a quadratic model is *not* constant. We later discuss in detail the rate of change of a quadratic model.

How do we obtain a quadratic model in this situation? After noting that the second differences are nearly, but not quite, constant and observing that the scatter plot suggests that the data are close to quadratic, we use technology to obtain a quadratic model that fits the data, and we draw its graph over the scatter plot.

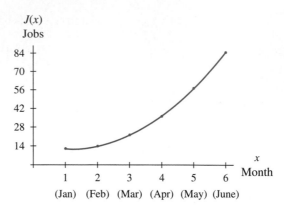

Quadratic Model for Roofing Jobs

FIGURE 2.14

You should find the quadratic model to be

$$J(x) = 3.07143x^2 - 7.01429x + 15.8 \text{ jobs}$$

where $x = 1$ in January, $x = 2$ in February, and so on. Note in Figure 2.14 that this model provides an excellent fit to the data.

If we are willing to assume that this quadratic model continues to model the number of roofing jobs for the next 3 months, how many jobs would we predict the company will have in August? Substituting $x = 8$ into the quadratic equation yields

$$J(8) = 3.07143(8)^2 - 7.01429(8) + 15.8 = 156.2571 \text{ jobs}$$

Note that even though the decimal portion of this answer might represent a roofing job in progress, it makes more sense to round the answer to the nearest whole number and predict 156 jobs in August.

You may notice that the scatter plot of the roofing job data looks as though an exponential model might fit, even though the percentage differences are not perfectly constant. If you fitted an exponential model, it would be

$$J(x) = 6.92293(1.51388)^x \text{ jobs}$$

where $x = 1$ in January, $x = 2$ in February, and so on. A graph of this exponential model on the scatter plot is shown in Figure 2.15.

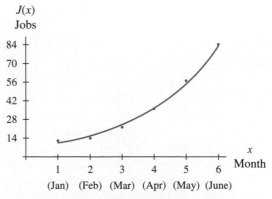

Exponential Model for Roofing Jobs

FIGURE 2.15

The exponential model and the quadratic model both appear to fit well.

Quadratic or Exponential?

It is sometimes the case that two models appear to fit the data equally well. In those cases, you need to consider carefully which model you would prefer to use. (See the hints for model selection in Section 2.4.) Often, however, it is obvious that one model is more appropriate than the other. This is the case in upcoming Example 2.

It is unlikely that you will ever encounter a real data set that conforms perfectly to a quadratic model. Because some real data input values are not evenly spaced, looking at differences between such data is inappropriate. In these cases, considering how well the graph of the function appears to fit the scatter plot may be more helpful in determining which model to use.

EXAMPLE 1   *Unemployment*

The national unemployment rates[14] (in percent) from July 1995 through January 1996 are given in Table 2.27.

TABLE 2.27

Month	Jul	Aug	Sep	Oct	Nov	Dec	Jan
% unemployment	5.70	5.58	5.56	5.50	5.58	5.62	5.80

a. Find a quadratic model to fit the data.
b. Compare the minimum of the model with the minimum of the data.

Solution:

a. First, let us align the data so that July 1995 is 7, August 1995 is 8, and so on. January 1996 will then be 13. See Table 2.28.

TABLE 2.28

Month	7	8	9	10	11	12	13
% unemployment	5.70	5.58	5.56	5.50	5.58	5.62	5.80

A scatter plot of the data suggests a concave-up shape with a minimum in October 1995. See Figure 2.16.

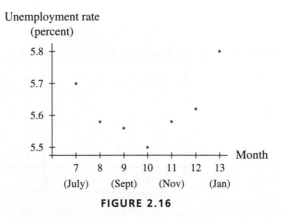

FIGURE 2.16

14. United States Department of Labor.

A quadratic model for the data is

$$U(m) = 0.02476m^2 - 0.48095m + 7.85429 \text{ percent unemployment}$$

where $m = 7$ in July 1995, $m = 8$ in August 1995, and so on.

b. Looking at the quadratic model graphed on the scatter plot (see Figure 2.17), we observe that the minimum of the model is slightly to the left of and higher than the minimum data point. This means that the model slightly overestimates minimum unemployment and that it estimates that minimum unemployment will occur slightly before it does in the table.

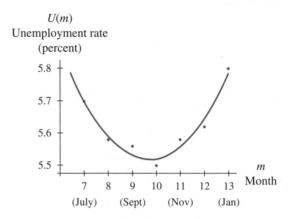

Quadratic Model for Unemployment

FIGURE 2.17   ■

The unemployment data in Example 1 have an obvious minimum, so it is easy to identify as quadratic. However, not all quadratic data sets have an obvious minimum (or, in some cases, a maximum). Sometimes all we see is the left side or right side of a parabola. Sometimes, as in the roofing job data, we must decide whether to use an exponential model or a quadratic model for such data.

EXAMPLE 2   *United States Population*

Table 2.29[15] shows the population of the contiguous states of the United States.

TABLE 2.29

Year	Population (millions)	Year	Population (millions)
1790	3.929	1910	91.972
1810	7.240	1930	122.775
1830	12.866	1950	150.697
1850	23.192	1970	202.229
1870	39.818	1990	247.052
1890	62.948		

15. *Statistical Abstract*, 1993.

Find an appropriate model for the data.

Solution: An examination of the scatter plot shows an increasing, concave-up shape (see Figure 2.18).

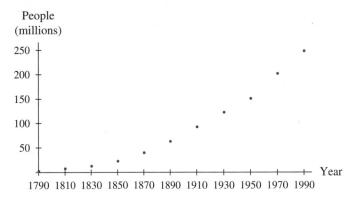

Population of the Contiguous U.S.

FIGURE 2.18

At first, it seems logical to try an exponential model for the data. An exponential model is

$$E(t) = 5.75481(1.02093)^t \text{ million people}$$

where t is the number of years since 1790. However, this model does not seem to fit the data very well, as Figure 2.19 shows.

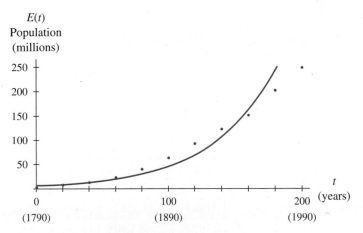

Exponential Model for Population

FIGURE 2.19

Another possibility is a quadratic model. The scatter plot could be the right half of a parabola that opens upward. A quadratic model is

$$Q(t) = 0.00631t^2 - 0.05331t + 4.65226 \text{ million people}$$

where t is the number of years since 1790. When graphed on the scatter plot, this model appears to be a very good fit (see Figure 2.20).

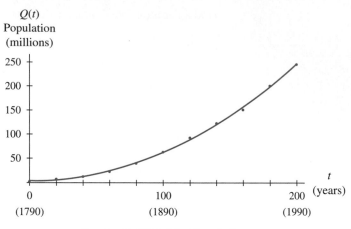

Quadratic Model for Population

FIGURE 2.20

The model

$$Q(t) = 0.00631t^2 - 0.05331t + 4.65226 \text{ million people}$$

where t is the number of years since 1790, appears to be the more appropriate model for the population of the contiguous United States using the given data.   ∎

In the foregoing examples, you have seen data sets that appear to be quadratic. That is, they may be reasonably modeled by a parabola. In each case, the model had a positive coefficient before the squared term. This value is called the **leading coefficient** because it is usually the first one that you write down in the quadratic model $f(x) = ax^2 + bx + c$. Also, in each case, the models appeared to be part of a parabola opening upward. Remember that we call such curvature *concave up*.

Table 2.30 gives the population of Cleveland, Ohio, from 1900 through 1980.

TABLE 2.30

Year	Population
1900	381,768
1910	560,663
1920	796,841
1930	900,429
1940	878,336
1950	914,808
1960	876,050
1970	750,879
1980	573,822

A scatter plot of the data suggests a parabola opening downward (or *concave down*). The scatter plot and the graph of a quadratic model are shown in Figure 2.21.

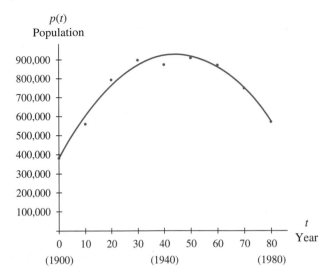

Population of Cleveland, Ohio

FIGURE 2.21

Would you expect the leading coefficient of the model to be positive or negative? The population of Cleveland is given by $p(t) = -279.995t^2 + 24{,}919.057t + 374{,}959.903$ people, where t is the number of years since 1900.

When a quadratic model is concave up, its leading coefficient is positive; when a quadratic model is concave down, its leading coefficient is negative. Curvature will be important in later discussions that involve concepts of calculus.

Cubic Modeling

We saw that when the first differences of a set of evenly spaced data are constant, the data can be modeled perfectly by the linear model $y = ax + b$. Likewise, when the second differences of evenly spaced data are constant, the data can be modeled perfectly by the quadratic model $y = ax^2 + bx + c$. It is possible for the third differences to be constant. In this case, the data can be modeled perfectly by the cubic model $y = ax^3 + bx^2 + cx + d$.

Because in the real world we are extremely unlikely to encounter data that are perfectly cubic, we will not look at third differences. Instead, we will examine a scatter plot of the data to see whether a cubic model may be appropriate.

Figure 2.22 shows the graphs of some cubic equations.

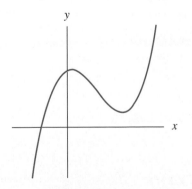

FIGURE 2.22a

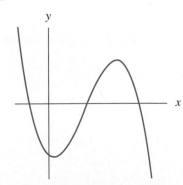

FIGURE 2.22b

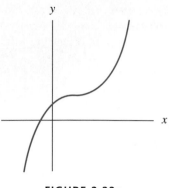

FIGURE 2.22c

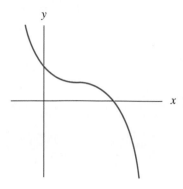

FIGURE 2.22d

In which of the graphs does y continue to increase as x increases? (Graph c) As x increases in Figures 2.22a and 2.22b, which graph shows the greatest change in y? (Graph b) Does y remain constant over an interval of x values in any of the graphs? (No)

The four graphs in Figure 2.22 are typical of all cubic models. That is, every cubic equation has a graph that resembles one of these four.

Figure 2.23 shows scatter plots of data sets that could be modeled by cubic equations.

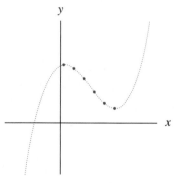

FIGURE 2.23a

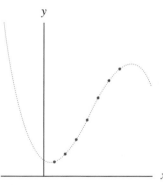

FIGURE 2.23b

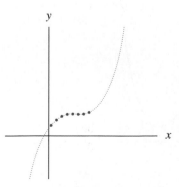

FIGURE 2.23c

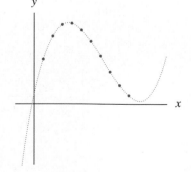

FIGURE 2.23d

You may have already noticed that in every cubic model, the curvature of the graph changes once from concave down to concave up, or vice versa. As we noted with the graph of a logistic curve, the point on the graph at which concavity changes is called the *inflection point*. All cubic models have one inflection point. The approximate location of the inflection point in each of the graphs in Figure 2.24 is marked with a dot. In Chapter 5, we see how calculus can be used to determine the exact location of the inflection point of a cubic model.

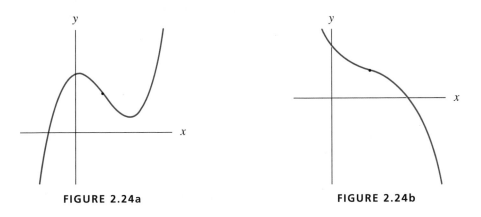

FIGURE 2.24a   FIGURE 2.24b

It is often the case that a portion of a cubic model appears to fit extremely well on a set of data that can be adequately modeled with a quadratic equation. In an effort to keep things as simple as possible, we adopt the following convention:

> If the scatter plot of a set of data fails to exhibit an inflection point, then it is not appropriate to fit a cubic model to the data.

We must be extremely cautious when using cubic models to extrapolate. For the data sets whose scatter plots are shown in Figures 2.23a and 2.23c, the models indicated by the dotted curves appear to follow the trend of the data. However, in Figure 2.23b, it would be possible for additional data to level off (as in a logistic model), whereas the cubic model takes a downward turn. Also, additional data in Figure 2.23d might continue to get closer to the *x*-axis, whereas the cubic model that is fitted to the available data begins to rise.

EXAMPLE 3   *Natural Gas Prices*

The average price in dollars per 1000 cubic feet of natural gas for residential use in the United States from 1980 through 1990 is given in Table 2.31.[16]

---

16. *Statistical Abstract*, 1992.

TABLE 2.31

Year	Price (dollars)	Year	Price (dollars)
1980	3.68	1986	5.83
1981	4.29	1987	5.54
1982	5.17	1988	5.47
1983	6.06	1989	5.64
1984	6.12	1990	5.77
1985	6.12		

Find an appropriate model to fit the data. Would it be wise to use this model to predict future natural gas prices?

Solution: An examination of a scatter plot shows that a cubic model is appropriate. Note that the scatter plot shown in Figure 2.25 appears to be mostly concave down between 1980 and 1986 but then is concave up between 1986 and 1990. That is, there appears to be an inflection point (a change of concavity) near 1986.

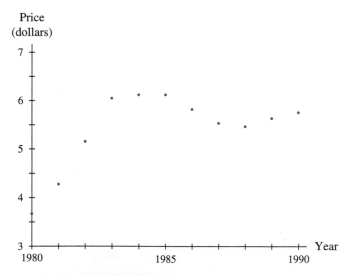

Average Price of 1000 Cubic Feet of Natural Gas

FIGURE 2.25

A cubic model that fits the price of natural gas is

$$\text{Price} = 0.01244x^3 - 74.11132x^2 + 147{,}221.1455x - 97{,}484{,}168.09 \text{ dollars}$$

where x is the year. A graph of the cubic model over the scatter plot is shown in Figure 2.26. The approximate location of the inflection point is marked with a black dot.

Note that the graph is increasing to the right of 1990. Natural gas prices will probably not continue to rise indefinitely as the cubic model does, so it is unwise to use the model to predict future prices of natural gas. Additional data should be obtained to see the pattern past 1990.

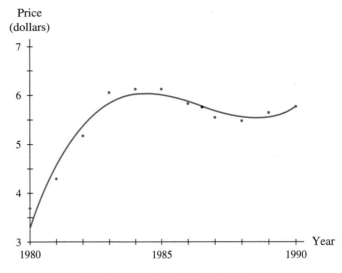

Cubic Model for Natural Gas Price

FIGURE 2.26

We can model natural gas prices with a cubic model that has smaller coefficients by renumbering the years so that x is the number of years since 1980:

$$\text{Price} = 0.01244x^3 - 0.24209x^2 + 1.40699x + 3.44455 \text{ dollars}$$

Although the two models have different inputs, they yield the same results.   ■

2.3   Concept Inventory

■ First and second differences

■ Quadratic model: $f(x) = ax^2 + bx + c$ (constant second differences, no change in concavity)

■ Parabola

■ Cubic model: $f(x) = ax^3 + bx^2 + cx + d$ (one change in concavity)

■ Inflection point

2.3   Activities

Identify the curves in Activities 1–6 as concave up or concave down. In each case, indicate the portion of the horizontal axis over which the part of the curve that is shown is increasing or decreasing.

1.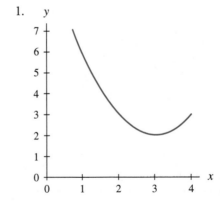

2.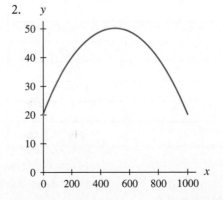

3.
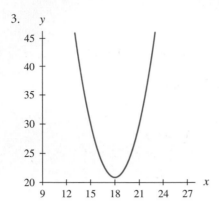

4.
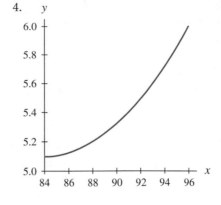

5.
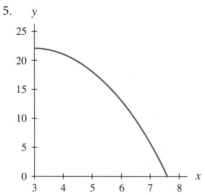

6.
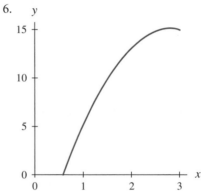

7. A Tomahawk Cruise missile misfires on a training mission in the South Pacific. It goes over the side of the ship and hits the water. Suppose the data showing the height of the missile are collected via telemetry as shown in Table 2.32.

TABLE 2.32

Seconds from launch	Feet above water
0	128
0.5	140
1	144
1.5	140
2	128
2.5	108
3	80
3.5	
4	

a. Without graphing, show that the data in Table 2.32 are quadratic.

b. Without finding an equation, fill in the missing values.

c. Find a quadratic equation for the data.

d. Use the equation to determine when the missile hits the water.

8. A discount store has calculated from past sales data the weekly revenue resulting from the sale of blenders. These amounts are shown in Table 2.33.

TABLE 2.33

Number of blenders sold	Revenue (dollars)
4	114
7	247
10	356
13	418
16	451
19	446
22	412

a. Why does selling 19 blenders result in less revenue than selling 16 blenders?

b. Examine a scatter plot of the data in Table 2.33, and find a quadratic model that fits the data.

c. What is the revenue when 17 blenders are sold? 18 blenders?

d. How could the data in the table and the answers in part c be beneficial to the manager of the store?

9. Suppose that the roofing company discussed in this section operates in Michigan instead of Miami.

a. Using the quadratic model

$$J(x) = 3.07143x^2 - 7.01429x + 15.8 \text{ jobs}$$

where $x = 1$ in January, $x = 2$ in February, and so on, determine the number of jobs in December.

b. Do you feel this answer is reasonable? Why or why not?

c. Would your answer to part b be different if the roofing company operated in your home state?

10. Table 2.34 gives the price in dollars of a round-trip ticket from Denver to Chicago on a certain airline and the corresponding monthly profit, in millions of dollars, for that airline.

TABLE 2.34

Ticket price (dollars)	Profit (millions of dollars)
200	3.08
250	3.52
300	3.76
350	3.82
400	3.70
450	3.38

a. Is a quadratic model appropriate for the data? Explain.

b. Find a quadratic model to fit the data.

c. As ticket price increases, the airline should collect more money. How can it be that when the ticket price reaches a certain amount, profit decreases?

d. At what ticket price will the airline begin to post a negative profit (a net loss)?

11. As listed in *The 1992 Information Please Almanac,* the median age (in years) at first marriage of females in the United States is shown in Table 2.35.

TABLE 2.35

Year	1960	1970	1980	1990
Age	20.3	20.8	22	23.9

a. Refute or defend the following statement: "The data are perfectly quadratic."

b. Without finding an equation, estimate the age that corresponds to the year 2000.

c. Find a quadratic model for the data.

d. Use the model to predict the age in the year 2000. Is it the same as your answer to part b?

e. Find the most current data available for the median age at first marriage of women. Compare the model value with the actual value.

f. When you add the new data to Table 2.35, is a quadratic model still appropriate? Explain.

12. The monthly profit P (in dollars) from the sale of x mobile homes at a dealership is given by Table 2.36.

TABLE 2.36

x	P (dollars)	x	P (dollars)
7	43,700	14	63,000
8	48,000	15	63,750
11	57,750	19	61,750
13	61,750	22	55,000

Examine a scatter plot of the data. Find a quadratic model to fit the data, and graph it on the scatter plot. Does it appear that the model is a good fit?

13. A factory makes 7-millimeter aluminum ball bearings. Company analysts have determined how much it costs to make certain numbers of ball bearings in a single run. These costs are shown in Table 2.37.

TABLE 2.37

Number of ball bearings	Cost (dollars)
500	3.10
1000	4.25
3000	8.95
4500	12.29
7000	18.45

a. Should the data be modeled by using a linear or a quadratic model? What is the model?

b. How much is the overhead for a single run?

c. How much will it cost to make 5000 ball bearings?

d. Ball bearings are made in sets of 100. If the company is planning to make 5000 ball bearings in a run, how much more would it cost them to make one extra set of 100 in that same run? What term is used in economics to express this idea?

e. Ball bearings are sold in cases of 500. Rewrite the cost equation in terms of the number of cases of ball bearings.

14. The factory discussed in Activity 13 sells ball bearings in cases of 500. It charges the prices shown in Table 2.38.

TABLE 2.38

Number of cases	Price (dollars)
1	78.47
10	168.85
20	263.83
100	817.88
200	999.96

a. Find a quadratic model for the data.

b. According to the model, how much revenue is made on 10 cases of ball bearings?

c. Using the cost model found earlier, $C(u) = 1.17627u + 1.87986$ where u is the number of cases of ball bearings and the revenue model you just found, write an expression to represent the profit made on u cases of ball bearings.

d. How much profit does the factory make on the sale of 500, 1000, and 10,000 ball bearings, respectively?

e. How many ball bearings must be sold to make a profit of $800?

f. How many ball bearings must the factory sell at one time to a single customer before it no longer makes a profit from that sale?

15. According to the *HIV/AIDS Surveillance 1992 Year End Edition*, the numbers of AIDS cases diagnosed in the United States from 1988 through 1991 were as shown in Table 2.39.

TABLE 2.39

Year	1988	1989	1990	1991
Number of cases	33,590	39,252	41,008	42,472

a. Find a quadratic model to fit the data, and examine the model graphed on a scatter plot of the data.

b. Do you believe your model is a good fit? Explain.

c. The number of AIDS cases diagnosed in 1992 had not yet been tabulated when the *1992 Year End Report* was published. What does the quadratic model give as the number of cases in 1992? Do you believe that this estimate is reliable?

d. According to your model, will there ever be a year in which over 50,000 AIDS cases will be diagnosed?

e. Do you believe this model gives a good or a poor representation of what will happen? Support your answer with more current data.

16. *The 1992 Information Please Almanac* lists the death rates (number of deaths per thousand people whose age is x) for the United States as follows:

TABLE 2.40

x	Death rate (deaths per thousand people)
40	2.2
45	3.0
50	5.0
55	8.0
60	12.6
65	18.7

a. Find a quadratic model for the data in Table 2.40. Discuss how well the model fits the data.

b. Use the model to complete Table 2.41:

TABLE 2.41

Age	Model prediction	Actual rate
51		5.5
52		6.0
53		6.6
57		9.6
59		11.5
63		16.2
66		20.1
68		23.7
70		28.2
75		42.4
80		65.5

c. What can you conclude about using a model to make predictions?

17. Table 2.42 gives the first-class postage for a 1-ounce letter from 1919 through 1995. (The March 22, 1981, increase is not included in the data.)

TABLE 2.42

Year of rate increase	Postage (cents)
1919	2
1932	3
1958	4
1963	5
1968	6
1971	8
1974	10
1975	13
1978	15
1981	20
1985	22
1988	25
1991	29
1995	32

a. Plot the data, and find a quadratic model for the data. Discuss how well the model fits the data.

b. Delete the first two data points in Table 2.42, and repeat part *a*. Compare the two models, and determine which has the better fit.

c. Use the better model to predict the first-class postage for a 1-ounce letter in the year 2001.

18. From a magazine, newspaper, or some other source, collect data that you feel may be modeled by one of the models we have studied. Find a model, and discuss why you chose the model and how well it fits the data. Do you feel that predictions based on this model are likely to be realistic? Why or why not?

For the graphs in Activities 19 through 24, describe the curvature by indicating the portions of the displayed horizontal axis over which each curve is concave down or concave up. Mark the approximate location of the inflection point on each curve.

19.

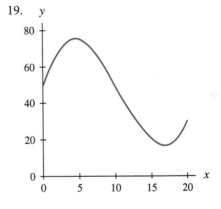

20.

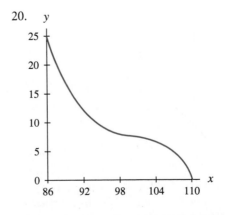

21.

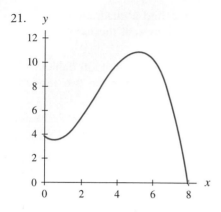

22.

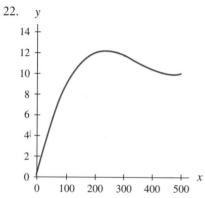

23.

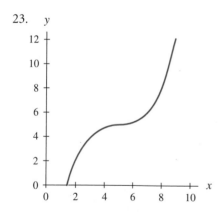

24.

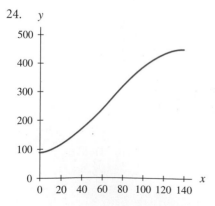

25. The amounts of money (in billions of dollars) spent on pollution control in the United States from 1983 to 1990 are given in Table 2.43.[17]

TABLE 2.43

Year	Amount (billions of dollars)
1983	60.0
1984	66.4
1985	70.9
1986	74.2
1987	76.7
1988	81.1
1989	85.4
1990	90.0

a. Examine a scatter plot of the data in Table 2.43. How does the scatter plot indicate that a cubic model might be appropriate?

b. Find a cubic model to fit the data.

c. What does the model predict that the amount will be in 1999? Do you believe that this prediction is accurate?

26. *The 1992 Information Please Almanac* lists the numbers of live births between the years 1950 and 1985 in the United States to women 45 years of age and older as given in Table 2.44.

TABLE 2.44

Year	Births	Year	Births
1950	5322	1970	3146
1955	5430	1975	1628
1960	5182	1980	1200
1965	4614	1985	1162

a. Examine a scatter plot of the data in Table 2.44. Find a cubic model to fit the data, and graph the model on the scatter plot. Discuss how well the model fits the data.

b. What trend does the model indicate beyond 1985? Do you believe the trend is valid? Explain.

17. *Statistical Abstract*, 1994.

c. Use the model to estimate the numbers of live births in 1986, 1987, and 1988.

d. The actual number of live births in 1986 was 1251. In 1987, 1375 children were born to women 45 years of age and older, and in 1988, the number of such live births was 1427. How accurately did your model predict for these years?

e. Do you believe your model would be an accurate predictor of the number of live births to women 45 years of age and older for the current year? Why or why not?

f. Include the 1986, 1987, and 1988 values in your data set, and examine a scatter plot of the 1950–1988 data. Find a cubic model for these data, and graph the model on the scatter plot.

g. Do you believe the model you found in part *f* would be an accurate predictor of the number of live births to women 45 years of age and older for the current year? Why or why not?

h. Find the number of births to women 45 years of age and older in the most recent year possible. Comment on how close your model estimates this value.

27. Table 2.45 shows the numbers of people who donated to an organization supporting athletics[18] at a certain university in the Southeast from 1975 through 1992.

a. Examine a scatter plot of the data in Table 2.45, find a cubic model to fit the data, and graph it on the scatter plot.

b. What trend does the model indicate beyond 1992? Do you believe that the model mirrors future events?

c. List several factors that might reverse the trend indicated by the model.

28. Table 2.46 shows a manufacturer's total cost (in hundreds of dollars) to produce from 1 to 33 fork lifts per week.

a. Examine a scatter plot of the data in Table 2.46. What characteristics of the scatter plot indicate that a cubic model would be appropriate?

b. Find a cubic model for total manufacturing cost.

c. What does the model predict as the cost to produce 23 fork lifts per week? 35 fork lifts per week?

d. Convert the cubic equation in part *b* for total cost to one for average cost.

e. Graph the average cost function.

f. How could the average cost graph help a production manager make production decisions?

TABLE 2.45

Year	Number of donors	Year	Number of donors
1975	10,706	1984	17,497
1976	8977	1985	18,260
1977	8829	1986	19,032
1978	11,404	1987	20,517
1979	12,730	1988	20,887
1980	13,048	1989	20,455
1981	13,500	1990	18,364
1982	15,587	1991	18,085
1983	16,619	1992	17,235

TABLE 2.46

Weekly production	Total cost (hundreds of dollars)
1	18.5
5	80
9	125
13	160
17	185
21	210
25	225
29	245
33	280

18. Data from the IPTAY association at Clemson University.

29. The purchasing power of the U.S. dollar as measured by producer prices from 1978 through 1991 is given in Table 2.47.[19] (In 1982, one dollar was worth $1.00.)

TABLE 2.47

Year	Purchasing power of $1	Year	Purchasing power of $1
1978	1.43	1985	0.96
1979	1.29	1986	0.97
1980	1.14	1987	0.95
1981	1.04	1988	0.93
1982	1.00	1989	0.88
1983	0.98	1990	0.84
1984	0.96	1991	0.82

a. Examine a scatter plot of the data in Table 2.47, find a cubic model to fit the data, and graph it on the scatter plot. What is the predicted value of the dollar in 1993?

b. What trend does the model indicate beyond 1991? Do you believe that the model will correctly predict future values of the dollar?

c. When was the purchasing power of a dollar at 90% of its 1982 value?

d. How well does the cubic model match the concavity of the data from 1988 through 1991?

e. Divide the data into two subsets, using 1988 as a common data point. Find a piecewise continuous model for the data, and graph it on the scatter plot.

f. Use the piecewise continuous model to estimate the purchasing power of the dollar in 1993.

g. Compare the fit of the model in part *a* with that of the model in part *e*. Which model better reflects the purchasing power of the dollar beyond 1991?

30. Table 2.48[20] gives the numbers of imported passenger cars, in thousands, sold in the United States from 1984 through 1992.

TABLE 2.48

Year	Imported passenger cars sold (thousands)
1984	2439
1985	2838
1986	3245
1987	3196
1988	3004
1989	2699
1990	2403
1991	2038
1992	1938

a. Examine a scatter plot of the data in Table 2.48, and discuss its curvature. Should the data be modeled by a quadratic or a cubic model? What is the model?

b. What does the model predict as the number of imported passenger cars sold in the United States in 1994?

c. Discuss why you feel the model you chose would or would not be a good predictor of the number of imported passenger cars sold in the United States in the year 2000.

d. Find more recent data on imported passenger car sales, and add the values you found to the table. Is the model you chose appropriate for the expanded data set? Explain.

31. The per capita utilization of commercially produced fresh vegetables from 1970 through 1992 is shown in Table 2.49.[21]

TABLE 2.49

Year	Vegetable consumption (pounds per person)
1970	88.1
1975	88.6
1980	92.5
1985	103.0
1988	111.5
1989	115.5
1990	113.3
1991	110.4
1992	109.3

19. *Statistical Abstract*, 1992.
20. *Statistical Abstract*, 1992 and 1994.
21. *Statistical Abstract*, 1994.

a. Observe a scatter plot of the data in Table 2.49, and explain why a piecewise model is needed to describe this data set.

b. Divide the data into two subsets, each sharing a common data point. Fit a model to each subset of data, and combine the models to form a piecewise continuous model.

c. Use your model to estimate the per capita use of fresh vegetables in 1973, 1983, and 1993.

d. List as many reasons as you can for the change in the per capita use of fresh vegetables in the early 1990s.

e. What behavior does your model indicate for per capita use of fresh vegetables beyond 1992?

f. Find more recent data on per capita use of fresh vegetables, and add it to Table 2.49. Is the model you chose appropriate for the expanded data set? Explain.

32. The numbers of nursing students in the United States for selected years between 1975 and 1992 are shown in Table 2.50.[22]

a. Observe a scatter plot of the data in Table 2.50, and explain why a piecewise model is needed to describe the data.

b. Divide the data into two subsets, each sharing a common data point. Fit a model to each subset,

TABLE 2.50

Year	Nursing students (thousands)
1975	250
1980	231
1985	218
1988	185
1989	201
1990	221
1991	238
1992	258

and combine the models to form a piecewise continuous model.

c. Use the model in part *b* to estimate the number of nursing students in 1973, 1983, and 1993.

d. List as many factors as you can that contributed to the change in the trend in the number of nursing students.

e. What current issues in our health care system may affect the number of nursing students in the future?

2.4   Choosing a Model

General Guidelines

The preceding material discussed several types of models that we use in this book to describe two-variable data: linear, quadratic, cubic, exponential, and logistic. Of course, there are others that we have not mentioned and will not use. Although most of our work so far has been concerned with applying a particular type of model to a given data set, it is important to understand that in many real-life situations, it is not always clear which model to apply. Sometimes, none of the five models is appropriate. However, there are some general, common-sense guidelines that we should keep in mind.

1. First of all, it is usually a good idea to *produce a scatter plot* using appropriate scaling for both horizontal and vertical axes. It is important that neither horizontal nor vertical range is too narrow or too wide. If either range is too narrow, then you may not be able to see all the data points. On the other hand, if either

22. *Statistical Abstract*, 1994.

range is too wide, then you may be unable to identify accurately the underlying curvature of the scatter plot.

2. Perhaps the most obvious guideline is to *examine the scatter plot carefully* with an eye toward its general underlying shape relative to the shapes of the models that you know. Try to imagine a smooth curve fitted to the scatter plot:

    a. Does this curve appear to be a straight line? If so, try a linear model.

    b. If the curve does not appear to be a line, then as you look from left to right, does the curve appear to be always concave up or concave down? If so, then a quadratic or exponential model may be appropriate.

    c. What if the curve does not appear to be a line and is not always concave in one direction (up or down)? If there is a single change in concavity, then a cubic or logistic model is suggested because these models have a concavity change (inflection point).

    d. If the data demonstrate a dramatic change in behavior, then it may be appropriate to combine two or more models to form a piecewise continuous model.

3. Suppose that we have narrowed our choices to two models. Which one should we choose? If one model fits the data significantly better than the other one, it should be chosen. If the two models appear to fit the data equally well, then the key is to examine the end behavior of the scatter plot to see whether limiting values are suggested in the context of the situation. Remember, exponential models have one limiting value, logistic models have two limiting values, and cubic and quadratic models have none.

Examining Scatter Plots

Look at the scatter plots in Figure 2.27. None of them appears to be linear, quadratic, or exponential. Can you determine which are cubic and which are logistic?

FIGURE 2.27

a.
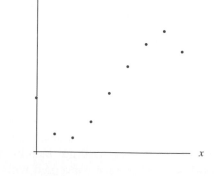

b.
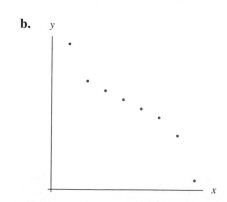

c.
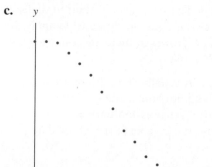

d.
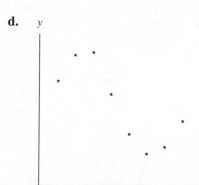

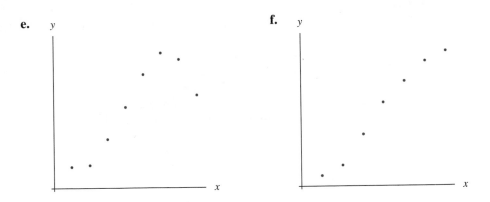

The plots *a* and *b* are cubic (no limiting values). Plots *d* and *e* are also cubic because they do not suggest limiting values. Plots *c* and *f* could be either logistic or cubic. Here you must ask yourself two questions:

1. Are you more concerned with end behavior or with how well the model fits the data?

2. If you are concerned with end behavior, what would the end behavior be?

If you are mainly interested in interpolating, then it would be appropriate to choose the model that better fits the data points. However, if you wish to extrapolate or examine end behavior, you should choose the model that you think would better follow the end behavior pattern provided that the fit is satisfactory. In general, if the context or scatter plot suggests that the data have a limiting value, then use a logistic model. If the data are likely to rise or fall again, then a cubic model would be appropriate.

Consider the scatter plots in Figure 2.28 and see whether you can determine which types of models are appropriate.

FIGURE 2.28

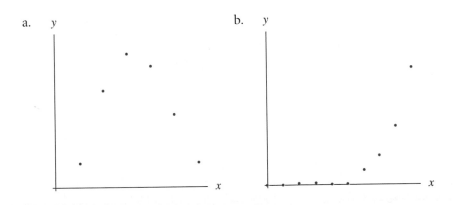

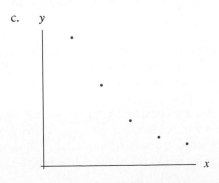

Plot *a* seems to be quadratic, whereas plot *b* appears to be exponential. Plot *c* could be either. Again ask yourself those two questions: (1) Are you more concerned with end behavior or with how well the model fits? (2) If you are concerned with end behavior, what would the end behavior be? Recall that an exponential curve has a limiting value at one end and becomes infinitely large at the other, whereas a quadratic curve becomes infinitely large (either negatively or positively) at each end.

Finally, consider the two scatter plots shown in Figure 2.29.

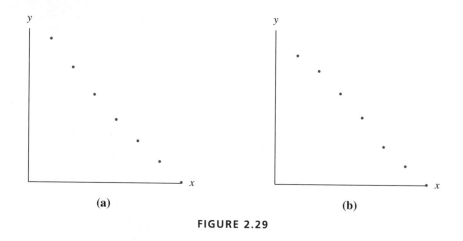

(a)   (b)

FIGURE 2.29

They both appear to be linear, so it makes sense to try a linear model. Examine linear models graphed on the scatter plots, as shown in Figure 2.30.

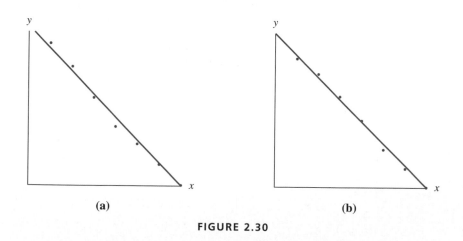

(a)   (b)

FIGURE 2.30

When the lines are graphed over the data, slight curvature can be seen in both cases. You may wish to try one of the other models, but do so with caution. You may sacrifice an easy-to-use linear model without gaining much in the way of increased precision.

When you are considering higher-order polynomial models, our advice is *the simpler the better*. For instance, although scatter plots *a* and *c* in Figure 2.28 could be modeled by either a quadratic or a cubic, you should choose the quadratic. Do not consider a cubic model unless the data exhibit an inflection point.

Four Steps in Choosing a Model

> ### Steps in Choosing a Model
>
> 1. **Look at the curvature of a scatter plot of the data.**
>   - If the points appear to lie in a straight line, try a linear model.
>   - If the scatter plot is curved but has no inflection point, try a quadratic and/or an exponential model.
>   - If the scatter plot appears to have an inflection point, try a cubic and/or a logistic model.
>   - If the scatter plot indicates curvature not described by one of the models we have studied, then consider combining two or more models to form a piecewise continuous model.
>
>   (*Note*: If the input values are equally spaced, it might be helpful to look at first differences versus percentage differences to decide whether a linear or an exponential model would be more appropriate, or to compare second differences with percentage differences to decide whether a quadratic or an exponential model would be better.)
>
> 2. **Look at the "fit" of the possible models.** In Step 1, you should have narrowed the possible models to at most two choices. Compute these models, and graph them on a scatter plot of the data. The one that comes closest to the most points (but does not necessarily go through the most points) is normally the better model to choose.
>
> 3. **Look at the end behavior of the scatter plot.** If Step 2 does not reveal that one model is obviously better than another, consider the end behavior of the data, and choose the appropriate model.
>
> 4. **Consider that there may be two equally good models for a particular set of data.** If that is the case, then you may choose either.

Let us look at some examples where we must apply these steps in choosing a model.

EXAMPLE 1   *Age of Marriage*

The 1992 Information Please Almanac lists the median ages at first marriage of women in the United States as shown in Table 2.51.

TABLE 2.51

Year	1960	1970	1980	1990
Age	20.3	20.8	22	23.9

a. Find the best model for the data.

b. Use this model to predict the median age at first marriage of women in the year 2000.

c. Is this prediction valid?

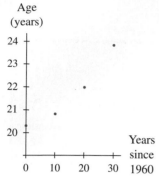

**Median Age of Women
at First Marriage**

FIGURE 2.31

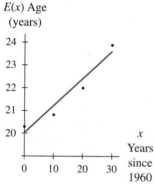

**Exponential Model for
Median Age at Marriage**

FIGURE 2.32

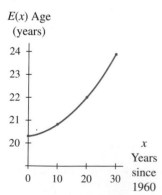

**Quadratic Model for
Median Age at Marriage**

FIGURE 2.33

Solution:

a. **Step 1**: Consider a scatter plot of the data.

The scatter plot shown in Figure 2.31 is concave up, so we should try either an exponential model or a quadratic model.

Step 2: Compare the fit of the models.

An exponential model for the data is

$$E(x) = 20.00024(1.00547)^x \text{ years of age}$$

where x is the number of years after 1960. However, when graphed on the scatter plot, it is obviously a poor fit to the data (see Figure 2.32).

A quadratic model for the data is

$$Q(x) = 0.0035x^2 + 0.015x + 20.3 \text{ years of age}$$

where x is the number of years since 1960. As Figure 2.33 shows, this model fits the data perfectly.

Even though a cubic model always fits any four data points perfectly, it is not appropriate, by the convention we adopted in Section 2.3 (page 109), to use a cubic model because there is no inflection point. Therefore, the best model for these data is the quadratic model.

Steps 3 and 4 are not necessary here, because we found the best model in Step 2. However, note that outside the range of the data, the quadratic model indicates higher median marriage ages for years prior to 1960 and predicts a continued rise in median age in the future. Although this end behavior may not seem correct, the best model for these data is still the quadratic model $Q(x)$. We must keep in mind that this model may not be valid for the years before 1960 or after 1990.

b. According to the model $Q(x)$, the median age of marriage in the year 2000 will be

$$Q(40) = 0.0035(40)^2 + 0.015(40) + 20.3$$
$$= 26.5 \text{ years of age}$$

c. This prediction is probably not valid. Even though the model fits the data perfectly, there is no indication that it is a good predictor of future events and trends. There are many factors other than time that affect the median age of first marriage. ■

Sometimes it is not quite so easy to choose which model is best, as the next example illustrates.

EXAMPLE 2   *Countries Issuing Postage Stamps*

Table 2.52[23] gives the numbers of European, North American, and South American countries that issued postage stamps from 1840 through 1880.

Countries

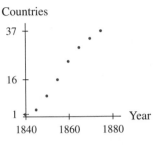

**Countries Issuing
Postage Stamps**

FIGURE 2.34

TABLE 2.52

Years	Number of countries
1840	1
1845	3
1850	9
1855	16
1860	24
1865	30
1870	34
1875	36
1880	37

$L(x)$
Countries

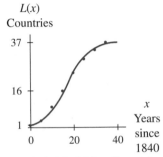

**Logistic Model for Countries
Issuing Postage Stamps**

FIGURE 2.35

$C(x)$
Countries

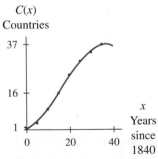

**Cubic Model for Countries
Issuing Postage Stamps**

FIGURE 2.36

Find an appropriate model for the data.

Solution:

a. **Step 1**: Consider the curvature of a scatter plot.

The scatter plot in Figure 2.34 appears to be concave up between 1840 and 1850 or 1855 and then is concave down between 1855 and 1880. Thus there is an apparent inflection point, so either a cubic or a logistic model could be appropriate.

b. **Step 2**: Compare the fit of the two models.

A logistic model is

$$L(x) = \frac{37.19488}{1 + 21.37367e^{-0.18297x}} \text{ countries}$$

where x is the number of years since 1840. It appears to fit the data reasonably well (see Figure 2.35).

A cubic model is

$$C(x) = -0.00120x^3 + 0.06027x^2 + 0.40435x + 0.44444 \text{ countries}$$

where x is the number of years since 1840. The cubic model also appears to fit the data well (see Figure 2.36).

If we graph the two models and the scatter plot at the same time, it appears that the cubic model is a slightly better fit to the data than the logistic model (see Figure 2.37).

23. "The Curve of Cultural Diffusion," *American Sociological Review*, August 1936, pp. 547–556.

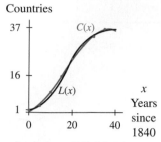

Countries

**Logistic and Cubic Models
for Countries Issuing
Postage Stamps**

FIGURE 2.37

If all you are concerned about is fit, you should choose the cubic model. However, if you would like to predict how many countries may have issued stamps in 1890, you should also consider end behavior.

Step 3: Consider end behavior.

Because the only countries under consideration are those in Europe, North America, and South America, there must be an upper limit on the number of countries issuing stamps. Thus it would be logical to think that a model for these data should approach some limiting value. This end behavior would be better modeled by the logistic function.

In conclusion, if your only concern is the better-fitting model, and you do not wish to use the model to extrapolate, then you should choose

$$C(x) = -0.00120x^3 + 0.06027x^2 + 0.40435x + 0.44444 \text{ countries}$$

where x is the number of years since 1840.

If you are concerned about end behavior more than about a better fit, you should choose

$$L(x) = \frac{37.19488}{1 + 21.37367e^{-0.18297x}} \text{ countries}$$

where x is the number of years since 1840.   ■

As we mentioned in Step 1 of "Steps in Choosing a Model" on page 123, it might be helpful to consider differences when input values are equally spaced. We consider such data in Example 3.

EXAMPLE 3   *Depreciation*

The value of a certain industrial machine depreciates according to the length of time it has been used. Table 2.53 shows the value of this machine after x years of use.

TABLE 2.53

Years of use	0	2	4	6	8	10
Value (dollars)	40,000	36,700	33,670	30,900	28,340	26,000

Find the most appropriate model for the data.

Solution: Consider the curvature of a scatter plot shown in Figure 2.38.

The scatter plot appears to be fairly straight, so we try a linear model. A linear model for this data is

$$V(x) = -1397.85714x + 39,590.95238 \text{ dollars}$$

where x is the number of years that the machine has been in use.

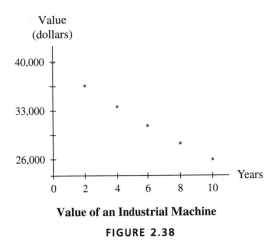

Value of an Industrial Machine

FIGURE 2.38

However, when $V(x)$ is graphed on the scatter plot (see Figure 2.39), it appears that the data may be slightly concave up.

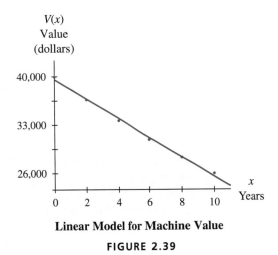

Linear Model for Machine Value

FIGURE 2.39

Because the inputs of the data are equally spaced, we can look at first differences, second differences, and percentage differences to see whether there is sufficient reason to try a curved model.

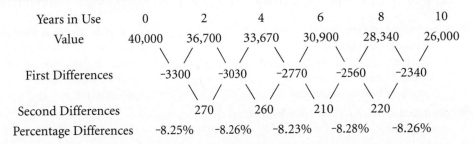

Years in Use	0		2		4		6		8		10
Value	40,000		36,700		33,670		30,900		28,340		26,000
First Differences		-3300		-3030		-2770		-2560		-2340	
Second Differences			270		260		210		220		
Percentage Differences	-8.25%		-8.26%		-8.23%		-8.28%		-8.26%	

Even though none of these are perfectly constant, the percentage differences do seem to be closer to remaining constant than either the first or the second differences. This indicates that an exponential model may be more appropriate than the linear model.

An exponential model is

$$E(x) = 40,002.49083(0.95784)^x \text{ dollars}$$

when the machine has been in use x years. This model appears to be a good fit to the data (see Figure 2.40).

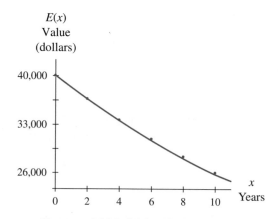

Exponential Model for Machine Value

FIGURE 2.40

Therefore, the exponential model is the most appropriate one for the data.   ■

Be open to the possibility that none of the models we have studied is appropriate for a set of data. Indeed, many sets of data in the real world do not fit neatly into one of our categories. Modeling, even in the simplest sense, is as much an art as a science. Sometimes a creative mind can be as helpful as a mathematical one. In any case, the most beneficial way to learn how to determine which model to use for a data set is by practicing.

2.4   Concept Inventory

- Choosing a model:
  - Consider curvature.
  - Compare the fit of possible models.
  - Consider end behavior.
  - Consider two valid choices.
- Sometimes no model we have studied is appropriate.

2.4   Activities

For each of the scatter plots in Activities 1 through 6, state which model or models are candidates to fit the data. Explain why those models are appropriate, and also explain why the other types of models are not appropriate.

1.
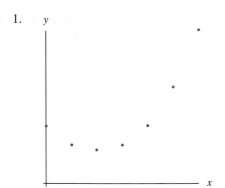

2.
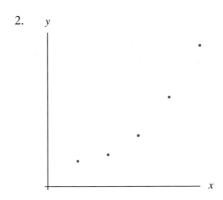

3.
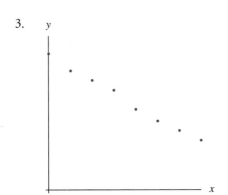

4.
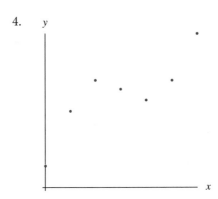

5.
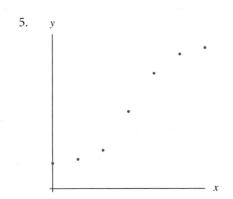

6.
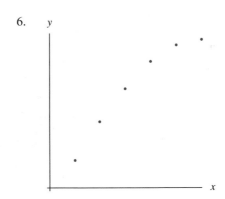

7. Given a data set in which the input values are evenly spaced, discuss how you could determine whether the data are

   a. linear.

   b. quadratic.

   c. exponential.

For each of the data sets in Activities 8 through 12,

examine the data set, and determine the best model. Give the equation, define the variables you use, include units of measure, and explain why you chose that model.

8. The 1991 *Consumer Guide* lists 1991 resale values for Chrysler's Fifth Avenue luxury sedan in excellent condition. The data are shown in Table 2.54.

TABLE 2.54

Model year	Resale value (dollars)
1983	2500
1984	3100
1985	4000
1986	4900
1987	6000
1988	7400
1989	9100

9. *Swimming World* (August 1992) lists the time in seconds that an average athlete takes to swim 100 meters freestyle at age *x*. The data are shown in Table 2.55.

TABLE 2.55

Age (years)	Time (seconds)	Age (years)	Time (seconds)
8	92	22	50
10	84	24	49
12	70	26	51
14	60	28	53
16	58	30	57
18	54	32	60
20	51		

10. The 1992 and 1994 volumes of *Statistical Abstract of the United States* give the number of overseas telephone calls (in millions) in the United States. See Table 2.56.

TABLE 2.56

Year	Number of calls (millions)	Year	Number of calls (millions)
1970	23	1988	706
1980	200	1989	1008
1984	428	1990	1201
1985	412	1991	2279
1986	478	1992	2724
1987	580		

24. *Statistical Abstract*, 1994.

11. The average teacher salaries in the United States, as listed in the 1992 *South Carolina Statistical Abstract*, are shown in Table 2.57.

TABLE 2.57

Year	Average teacher salary
1981	$17,602
1982	19,142
1983	20,715
1984	21,935
1985	23,534
1986	25,206
1987	26,534
1988	28,008
1989	29,547
1990	31,331

12. The consumer price index[24] values for garbage collection from 1987 through 1993 are shown in Table 2.58 (1982–1984 = 100).

TABLE 2.58

Year	CPI
1987	130.3
1988	142.5
1989	155.6
1990	171.2
1991	189.2
1992	207.3
1993	220.5

13. For each of the situations described in Activities 8 through 12, determine a question that could be asked about the data or a prediction that could be made on the basis of the model you found. Determine a reasonable answer to each of your questions.

14. Table 2.59 gives death rates due to lung cancer (deaths per 100,000 males in the United States) from 1930 through 1990.

TABLE 2.59

Year	Death rate (deaths per 100,000 men)
1930	5
1940	11
1950	21
1960	39
1970	59
1980	66
1990	67

a. Discuss briefly why linear, quadratic, cubic, exponential, and logistic models are or are not appropriate for this data set.

b. Of the models listed in part a, choose the one you believe to be the best for these data. Fit the model you chose to the data.

c. Use the model you chose in part b to estimate the death rate in 1995. Do you believe your model is appropriate for this or other extrapolations?

15. Table 2.60 gives the value of manufacturers' shipments of recordings in millions of dollars (based on list price) between 1987 and 1990.

TABLE 2.60

Shipments of LPs		Shipments of CDs	
Year	Value (millions of dollars)	Year	Value (millions of dollars)
1987	793	1987	1593
1988	523.3	1988	2089.9
1989	220.3	1989	2587.8
1990	86.5	1990	3451.6

a. Find models for each set of data.

b. Use the models to estimate values of shipments of LPs and CDs in 1991, 1992, and 1993.

16. Table 2.61 gives the average weight of the brain in human males as a percentage of body weight.

TABLE 2.61

Age (years)	Percentage of body weight accounted for by brain
0	11
2	8
4	7
6	6
8	5
10	4.5
12	4
14	3.5
16	3.25

a. Describe in words what happens to the percentage of body weight accounted for by the brain as a male child grows.

b. Fit a quadratic model and an exponential model to the data.

c. Use each model in part b to determine the age (in months) at which brain weight is 10% of body weight.

d. Use each model in part b to predict the percentage brain weight for an 18-year-old man.

e. Discuss which equation better models the data and why.

17. A college student works for 8 hours without a break assembling mechanical components. Her supervisor records every hour how many total components, N, the student has completed. These figures, which are cumulative, are given in Table 2.62.

TABLE 2.62

Hour	N
1	9
2	15
3	23
4	33
5	44
6	51
7	56
8	58

a. How many components did the student assemble during the fourth hour?

b. If this student started assembling components at 9:00 a.m., how many components were assembled after 1 p.m.?

c. Examine a scatter plot, fit a cubic model, and graph it on the scatter plot. Does the model provide a good fit to the data?

d. According to the cubic model, how many hours will it take the college student to assemble 40 components? (Report your answer in hours and minutes.)

e. What does the model indicate will happen to the number of components after 8 hours? Do you believe the model is a good indicator of future events? Explain.

f. Examine a scatter plot, fit a logistic model, and graph it on the scatter plot. Does the model provide a good fit to the data?

g. What does the model indicate will happen to the number of components after 8 hours? Do you believe the model is a good indicator of future events? Explain.

h. According to the trend exhibited in the logistic model, would it be worthwhile for the student's supervisor to have her work more than 8 hours? Explain.

i. Compare the fit of the logistic model to the fit of a cubic model. Besides fit, what other factors should you consider when deciding which model would be better?

j. Which model would you choose? Why?

18. According to the *HIV/AIDS Surveillance 1992 Year End Edition*, the numbers of AIDS cases diagnosed in the United States from 1988 through 1991 were as shown in Table 2.63.

TABLE 2.63

Year	Number of cases
1988	33,590
1989	39,252
1990	41,008
1991	42,472

25. *Statistical Abstract,* 1992 and 1994.
26. South Carolina Statistical Abstract, 1994.

a. Examine a scatter plot of the data, and discuss its curvature. What type of model would be most appropriate for the data?

b. Discuss why a cubic model might not be appropriate in this case.

c. Find the best model for the data.

19. Table 2.64[25] gives the numbers of imported passenger cars, in thousands, sold in the United States from 1984 through 1991.

TABLE 2.64

Year	Car sales (thousands)
1984	2439
1985	2838
1986	3245
1987	3196
1988	3004
1989	2699
1990	2403
1991	2038

a. Examine a scatter plot of the data, and discuss its curvature. What type of model should you try to fit to these data? What is the model?

b. Use your model from part *a* to estimate the number of imported passenger cars sold in the United States in 1992.

c. The actual number of imported passenger cars sold in 1992 was 1938 thousand cars. How far off was your estimation?

d. Add the data point corresponding to the year 1992 and sales of 1938 thousand to Table 2.64, and repeat parts *a* through *c*.

20. Table 2.65[26] shows the population of South Carolina between 1790 and 1990.

TABLE 2.65

Year	Population (thousands)
1790	249
1900	1340
1950	2117
1980	3122
1990	3487

a. Examine a scatter plot of the data, and discuss its curvature. What types of models might be appropriate for such curvature?

b. Find the two models that are appropriate according to your analysis in part *a*.

c. Discuss the fit of each model from part *b* to the data.

d. Which of the models from part *b* would be more appropriate to model the South Carolina population? Discuss any issues (besides curvature) that lead you to this decision.

21. Chlorofluorocarbons (CFCs) are believed to thin the ozone layer and were banned from use in aerosols in the United States in the 1970s. Table 2.66[27] shows the estimated atmospheric release of CFC-12 (one of the two most prominent CFCs) between 1958 and 1978.

TABLE 2.66

Year	CFC-12 atmospheric release (millions of kilograms)
1958	66.9
1960	89.1
1962	114.5
1964	155.5
1966	195.0
1968	246.5
1970	299.9
1972	349.9
1974	418.6
1976	390.4
1978	341.3

a. Observe a scatter plot of the data. What year do you believe the ban on CFCs in the United States took effect?

b. Divide the data into two subsets that have one data point in common. Fit a model to each set of data, and combine the two to form a piecewise continuous function.

c. Do you believe that the trend beyond 1978 indicated by your model is valid?

d. If your school's library has *The True State of the Planet*, look on page 429 for a more extensive data set. Summarize what happened to CFC-12 release after 1978 and the reasons for the behavior that occurred.

22. The amounts of fish produced by fisheries in the United States for food in selected years between 1970 and 1990 are given in Table 2.67.[28]

TABLE 2.67

Year	Amount (millions of pounds)
1970	2537
1972	2435
1975	2465
1977	2952
1980	3654
1982	3285
1985	3294
1987	3946
1990	7041

a. Discuss the curvature of the data. Why is a cubic model not appropriate? Why is a single quadratic model not appropriate?

b. Divide the data into two subsets that share one data point. Fit models to each set of data, and combine them to form a piecewise continuous model.

c. Discuss factors that might have contributed to the behavior exhibited by the data.

d. Use your model to estimate fish production in 1973, 1983, and 1993.

27. Bailey, Ronald, ed. *The True State of the Planet*, New York: The Free Press for the Competitive Enterprise Institute, 1995.
28. *Statistical Abstract*, 1994.
29. *Statistical Abstract*, 1994.

23. Population data for Iowa in the 1980s and early 1990s are given in Table 2.68.[29]

    a. Examine a scatter plot of the data. Discuss the curvature.

    b. Discuss the appropriateness or inappropriateness of using a cubic model to fit this data set.

    c. Fit a cubic model, and graph it on the scatter plot. Does the graph support or contradict your conclusion in part *b*?

    d. Divide the data into two subsets that share one point. Find a piecewise continuous model for the population of Iowa.

    e. Update the data to the most recent year for which population data are available. Is the model you chose appropriate for the expanded data set?

24. Suppose that you are required to model a set of data that has evenly spaced input values and for which second differences and percentage differences in the outputs are constant. How might you determine which model is best?

TABLE 2.68

Year	Population (thousands)
1980	2914
1985	2830
1986	2792
1987	2767
1988	2768
1989	2771
1990	2777
1991	2790
1992	2803
1993	2814

Chapter 2 Summary

Exponential Functions and Models

Second in importance to linear functions and models are exponential functions and models. Based on the familiar idea of repeated multiplication by a fixed positive multiplier b (the base), the basic exponential function is

$$f(x) = ab^x$$

The parameter a appearing in the function is the output when the input is zero.

Whereas linear functions model situations that have constant rates of change, exponential functions model constant percentage change. Perhaps you can now appreciate the importance of—yet see the difference between—linear and exponential models. Many situations in everyday life undergo change wherein either the rate of change or the percentage change is considered nearly constant.

In terms of the function $f(x) = ab^x$, exponential growth occurs when b is greater than 1, and exponential decline takes place when b is between 0 and 1. In fact, the constant percentage growth or decline is $(b - 1) \cdot 100\%$. In scientific work, it is fairly common to use a standard base, base e, where e is an irrational number whose decimal representation to 12 digits is 2.71828182846. Then the model $f(x) = ab^x$ becomes $f(x) = ae^{kx}$, where $k = \ln b$.

A common situation is to have initial exponential growth followed by a leveling-off approach toward a limiting value L. This is characteristic of logistic growth, modeled by the S-shaped logistic function

$$y = \frac{L}{1 + Ae^{-Bx}}$$

Logistic curves model such everyday occurrences as the accumulated sales of a new fad item and the spread of information. In situations such as these, the parameter B that appears in the exponential term of the denominator is positive, a reflection of initial exponential growth. Logistic curves also occur as inverted, or upside-down, S-shaped curves. In these cases, the parameter B is negative, a reflection of the decline in output toward the horizontal axis as the input values increase.

Logistic models that are S-shaped show curvature that changes from concave up to concave down, whereas logistic models that are inverted S-shapes show curvature that changes from concave down to concave up. In each case, the point where the concavity changes is known as an inflection point. We later develop calculus techniques that enable us to find such points. Inflection points can be of great importance in business and economic decisions.

Exponential Models in Finance

The general compound interest formula is

$$A(t) = P\left(1 + \frac{r}{n}\right)^{nt} \text{ dollars}$$

where $A(t)$ is the amount accumulated in an account after t years when P dollars are invested at an annual interest rate of $r\%$ (expressed as a decimal) compounded n times each year. The advertised annual interest rate $r\%$ is called the annual percentage rate (APR) or the nominal rate. For a given nominal rate r and number n of compounding periods per year, we are also interested in the annual percentage yield (APY), or effective rate, $\left(1 + \frac{r}{n}\right)^n - 1$. If you are an investor, then the annual percentage yield gives a true picture of your actual interest earnings. If you are a borrower who is paying interest, then the effective rate gives a true picture of your actual cost in terms of interest.

If we imagine that the number n of compounding periods per year increases without bound, then we are led to consider the notion of continuous compounding, described by the limit

$$\lim_{n \to \infty} P\left[1 + \frac{r}{n}\right]^{nt} = Pe^{rt}$$

This gives the continuously compounding interest formula $A = Pe^{rt}$. Limits arise often in calculus, and we consider them in more detail later. With continuous compounding, the effective rate is $e^r - 1$.

The term *future value* describes the value of an investment at some future time, and the term *present value* describes how much must be invested now to achieve a desired future value. We return to future value and present value in a later chapter and consider them in a more sophisticated context using calculus.

Polynomial Functions and Models

Polynomial functions and models have a well-established role in calculus. Constructed by the repeated addition of simple terms of the form ax^k, where k is a non-negative integer, they are easily manipulated by paper-and-pencil methods.

A polynomial function of degree n is a function given by an equation of the form

$$y = f(x) = a_n x^n + a_{n-1} x^{n-1} + \cdots + a_1 x + a_0$$

where $a_0, a_1, \cdots, a_n$ are constants, and n is a positive integer.

As special cases, we have

$(n = 1)$ linear functions: $f(x) = ax + b$

$(n = 2)$ quadratic
functions: $f(x) = ax^2 + bx + c$

$(n = 3)$ cubic functions: $f(x) = ax^3 + bx^2 + cx + d$

Even though they may be useful in certain situations, in this text we will not encounter polynomial functions and models beyond cubics.

Quadratic models, whose underlying equation is a quadratic polynomial, have graphs known as parabolas. The parabola $f(x) = ax^2 + bx + c$ opens upward (is concave up) if a is a positive and opens downward (is concave down) if a is negative.

Cubic models, models whose underlying equation is a cubic polynomial, have graphs that resemble one of the four types shown in Figure 2.22 on pages 107–108. Like logistic models, cubic models show a change of concavity at an inflection point. Because of the general "up-down-up" or "down-up-down" behavior of their graphs, we must be especially careful in using cubic models to extrapolate beyond the range of data values on which the models are constructed.

Choosing a Model

Although it is not always clear which (if any) of the models we have discussed apply to a particular real-life situation, it helps to keep in mind a few general, common-sense guidelines. (1) Given a set of discrete data, begin with a scatter plot. The plot will often reveal general characteristics that point the way to an appropriate model. (2) If the scatter plot does not appear to be linear, consider the suggested concavity. One-way concavity (up or down) often indicates a quadratic model, but keep in mind that upward concavity may also suggest an exponential model. If input values are evenly spaced, then second differences that are nearly constant indicate a quadratic model, whereas percentage differences that are nearly constant suggest an exponential model. (3) When a single change in concavity seems apparent, think in terms of cubic or logistic models. But remember, logistic models tend to become flat on each end, whereas cubics do not. Also, logistic models are suggested in cases where the data are cumulative over time with bounds on the extent of the accumulation. Never consider a cubic or a logistic model if you cannot identify an inflection point.

Concluding Perspective: Functions and Models

Although the first two chapters have focused on the construction of elementary mathematical models, it is important that you understand that modeling is *not* our primary objective. In reality, mathematical modeling can be a complicated and highly sophisticated endeavor. We are using basic modeling with elementary curve fitting strictly as a way to obtain functional relationships between variables. Because most information in the real world is collected in discrete form (tables of data), it is important in the study of calculus to obtain functional relationships between the variables in order to study their changing behavior. Our main objective is to apply concepts and methods of calculus to these functions.

Chapter 2   Review Test

1. The number of subscribers to cellular phone service was 91,600 at the end of 1984 and 30 million at the end of 1995[30]. Assume that the number of subscribers grew exponentially between 1984 and 1995.

   a. Find an exponential model for the number of cellular phone subscribers as a function of the number of years since 1984.

   b. What is the percentage change indicated by the model?

   c. According to the model, when will the number of subscribers reach 270 million (the approximate population of the United States in 1997)?

   d. Do you believe an exponential model accurately reflects the future growth of cellular phone subscribers? Explain. If not, what do you think would be a more appropriate model?

2. a. What is the effective rate (APY) for 7.3% compounded monthly?

   b. What is the largest possible effective rate for a nominal rate of 7.3%?

3. The numbers of in-hospital midwife-attended births for selected years between 1975 and 1993 are shown in Table 2.69.[31]

   TABLE 2.69

Year	Births (thousands)
1975	19.7
1981	55.5
1987	98.4
1989	122.9
1990	139.2
1993	196.2

   a. Examine a scatter plot of the data. Discuss the curvature of the data and the possible models that could be used for this data set.

   b. Find the model that you believe is best for this data set.

   c. What issues in our current health care system do you believe will affect the number of in-hospital midwife-attended births in the future? Do you think the model you chose is a good description of the number of in-hospital midwife-attended births in the future?

4. On August 28, 1993, the *Philadelphia Inquirer* reported temperatures from the previous day. Table 2.70 lists these temperatures (in degrees Fahrenheit) from 5 a.m. to 5 p.m.

   TABLE 2.70

Time	°F	Time	°F
5 a.m.	76	noon	90
6 a.m.	75	1 p.m.	91
7 a.m.	75	2 p.m.	93
8 a.m.	77	3 p.m.	94
9 a.m.	79	4 p.m.	95
10 a.m.	83	5 p.m.	93
11 a.m.	87		

   a. Examine a scatter plot of the data. Explain why the curvature indicates that a cubic model is appropriate.

   b. Find a cubic model for the data.

   c. According to the model, what was the temperature at 5:30 p.m.?

   d. According to the model, at approximately what times (on August 27) was the temperature 90 degrees Fahrenheit?

30. As reported in the *Reno Gazette-Journal*, Sept. 15, 1996, page 1E.
31. As reported in the *Reno Gazette-Journal*, August 5, 1996, page 1A.
32. Nancy Cockrell, "Egypt," *International Tourism Reports*, April 1, 1996, page 15.

5. The total hotel capacity in Egypt between 1989 and 1994 is shown in Table 2.71[32].

   a. Examine a scatter plot of the data, and discuss the appropriateness or inappropriateness of each of the following models for this set of data: (i) logistic, (ii) cubic, (iii) quadratic, (iv) exponential, and (v) linear.

   b. Which model would you choose for this data set? Why?

TABLE 2.71

Year	Hotel rooms (thousands)
1989	42.1
1990	47.6
1991	54.0
1992	55.6
1993	58.8
1994	61.5

Project 2.1

Compulsory School Laws

Setting

In 1852, Massachusetts became the first state to enact a compulsory school attendance law. Sixty-six years later, in 1918, Mississippi became the last state to enact a compulsory attendance law. Table 2.72 lists the first 48 states to enact such laws and the year each state enacted its first compulsory school law.

TABLE 2.72

State	Year	State	Year	State	Year
MA	1852	SD	1883	IA	1902
NY	1853	RI	1883	MD	1902
VT	1867	ND	1883	MO	1905
MI	1871	MT	1883	TN	1905
WA	1871	IL	1883	DE	1907
NH	1871	MN	1885	NC	1907
CT	1872	ID	1887	OK	1907
NM	1872	NE	1887	VA	1908
NV	1873	OR	1889	AR	1909
KS	1874	CO	1889	TX	1915
CA	1874	UT	1890	FL	1915
ME	1875	KY	1893	AL	1915
NJ	1875	PA	1895	SC	1915
WY	1876	IN	1897	LA	1916
OH	1877	WV	1897	GA	1916
WI	1879	AZ	1899	MS	1918

Source: Richardson, "Variation in Date of Enactment of Compulsory School Attendance Laws," *Sociology of Education* vol. 53 (July 1980), pp. 153–163.

Tasks

1. Tabulate the cumulative number of states with compulsory school laws for the following 5-year periods:

1852–1856	1887–1891
1857–1861	1892–1896
1862–1866	1897–1901
1867–1871	1902–1906
1872–1876	1907–1911
1877–1881	1912–1916
1882–1886	1917–1921

2. Examine a scatter plot of the data in Task 1. Do you believe a logistic model is appropriate? Explain.

3. Fit a logistic model to the data in Task 1.

4. What do most states in the third column of the original data have in common? Why would these states be the last to enact compulsory education laws?

5. The seventeen states considered to be southern states (below the Mason–Dixon Line) are AL, AR, DE, FL, GA, KY, LA, MD, MO, MS, NC, OK, SC, TN, TX, VA, and WV. Tabulate cumulative totals for the southern states and the northern/western states for these dates:

Northern/Western States Years	Southern States Years
1852–1856	1891–1895
1857–1861	1896–1900
1862–1866	1901–1905
1867–1871	1906–1910
1872–1876	1911–1915
1877–1881	1916–1920
1882–1886	
1887–1891	
1892–1896	
1897–1901	
1902–1906	

6. Examine scatter plots for the two data sets in Task 5. Do you believe that logistic models are appropriate for these data sets? Explain.

7. Find logistic models for each set of data in Task 5. Comment on how well each model fits the data.

8. It appears that the northern and western states were slow to follow the lead established by Massachusetts and New York. What historical event may have been responsible for the time lag?

9. One way to reduce the impact of unusual behavior in a data set (such as that discussed in Task 8) is to group the data in a different way. Tabulate the cumulative northern and western state totals for the following 10-year periods:

   1852–1861
   1862–1871
   1872–1881
   1882–1891
   1892–1901
   1902–1911

10. Fit a logistic curve to the data in Task 9. Comment on how well the model fits the data. Compare models for the data grouped in 10-year intervals and the data grouped in 5-year intervals (Task 6). Does grouping the data differently significantly affect how well the model fits? Explain.

11. Find a model that fits the data for the southern states better than the logistic model. Explain your reasoning.

Reporting

1. Prepare a written report of your work. Include scatter plots, models, and graphs. Include discussions of each of the tasks in this project.

2. (Optional) Prepare a brief (15-minute) presentation on your work.

Project 2.2

Fund-Raising Campaign

Setting

In order to raise funds, the mathematics department in your college or university is planning to sell T-shirts before next year's football game against the school's biggest rival. Your team has volunteered to conduct the fund raiser. Because several other student groups have also volunteered to head this project, your team is to present its proposal for the fund drive, as well as your predictions about its outcome, to a panel of mathematics faculty.

Note: This project is also used as a portion of Project 5.2 on page 319.

Tasks

1. *Getting Started*   Develop a slogan and a design for the T-shirt. Keep in mind that good taste is a concern. Decide on a target market, and determine a strategy to survey (at random) at least 100 students who represent a cross section of the target market to determine the demand for T-shirts (as a function of price) within that market. It is important that your sample survey group properly represent your target market. If, for example, you polled only near campus dining facilities at lunch time, your sample would be biased toward students who eat lunch at such facilities.

   The question you should ask is "How much would you be willing to spend on a T-shirt promoting the big football rivalry? $14, $13, $12, $11, $10, $9, $8, $7, $6, or not interested?" Keep an accurate tally of the number of students who answer in each category. In your report on the results of your poll, you should include information such as your target market; where, when, and how you polled within that market; and why you believe that your polled sample is likely to be a representative cross section of the market.

2. *Modeling the Demand Function*
   a. From the data you have gathered, determine how many students from your sample survey group would buy a T-shirt at $6, $7, $8, and so on.
   b. Devise a marketing strategy, and determine how many students within your target market you can reasonably expect to reach. Assuming that your poll is an accurate indicator of your target population, determine the number of students from your target market who will buy a T-shirt at each of the given prices.
   c. Taking into account the results of your poll and your projected target market, develop a model for demand as a function of price. Keep in mind that your model must make sense for all possible input values.

3. *Modeling Other Functions*   Estimate the cost that you will incur per T-shirt from the partial price listing in Table 2.73. Use the demand function from Task 2 to

create equations for revenue, total cost, and profit as functions of price. (Revenue, total cost, and profit may not be one of the basic models that were discussed in class. They are sums and/or products of the demand function with other functions.)

Reporting

1. Prepare a written report summarizing your survey and modeling. The report should include your slogan and design, your target market, your marketing strategy, the results from your poll (as well as the specifics of how you conducted your poll), a discussion of how and why you chose the model of the demand function, a discussion of the accuracy of your demand model, and graphs and equations for all of your models. Attach your questionnaire and data to the report as an appendix.

2. Prepare a 15-minute oral presentation of your survey, modeling, and marketing strategy to be delivered before a panel of mathematics faculty. You will be expected to have overhead transparencies of all graphs and equations as well as any other information that you consider appropriate as a visual aid. Remember that you are trying to sell the mathematics department on your campaign idea.

TABLE 2.73 The T-Shirt Company: Partial Price Listing

Number of colors	1	2	3	4	5	6
Minimum order						
25	$4.35	$5.10	$5.70	$6.30	$7.00	$7.50
50	$4.25	$5.05	$5.65	$6.25	$6.95	$7.45
100	$4.15	$4.95	$5.55	$6.15	$6.85	$7.35
250	$4.00	$4.80	$5.40	$6.00	$6.65	$7.20
500	$3.75	$4.50	$5.10	$5.70	$6.30	$7.00
750	$3.50	$4.25	$4.90	$5.50	$6.00	$6.50
1000	$3.45	$4.00	$4.65	$5.00	$5.50	$6.00
1500	$3.25	$3.80	$4.45	$4.85	$5.30	$5.80
2000	$3.15	$3.70	$4.30	$4.65	$5.10	$5.60
2500	$3.10	$3.60	$4.20	$4.55	$5.00	$5.45
5000	$3.05	$3.55	$4.15	$4.50	$4.95	$5.40
7500	$3.05	$3.50	$4.10	$4.45	$4.90	$5.35
10,000	$3.00	$3.45	$4.05	$4.40	$4.85	$5.30

Source: Based on data compiled from 1993 prices at Tigertown Graphics, Inc., Clemson, SC.

Describing Change: Rates

Change is everywhere around us and affects our lives on a daily basis. From the explosive growth in the use of personal computers to the decline in the number of family-owned farms, from the shrinking value of a dollar to the growing national debt, change impacts us as individuals, in families, through businesses, and in our government.

Before we can analyze change, we must describe it. How do we describe change? What are the appropriate concepts, and what language do we use? That is the focus of this chapter—to describe change in the language of rates.

We consider several ways to describe change. Starting from the actual change in a quantity over an interval (say, an interval of time), it is a simple step to describe the change as an average rate of change over that interval. The notion of average change, when examined carefully in light of the underlying geometry of graphs, leads to the more subtle and challenging concept of instantaneous change. Indeed, the precise description of instantaneous change in terms of mathematics is one of the principal goals of calculus. Our final descriptions of change will be in terms of percentage change and percentage rates of change. These important ways of describing change appear frequently in business, finance, economics, and government.

There are many situations in biology and physiology that can be modeled by functions. For instance, the weight of a puppy can be modeled as a function of its age. We can use mathematics to answer certain questions about changes in the puppy's weight: (1) How much weight does a puppy gain during its first month of life? (2) How much weight does a puppy gain during the first week after birth? (3) How much weight does a puppy gain during its first day of life? We can also use calculus to answer other questions, such as (4) How quickly is the puppy gaining weight at birth?

■ 3.1   Average Rates of Change

Consider the following statements:

■ The temperature fell ten degrees between 4 p.m. and 6 p.m.

■ The typical Fortune 500 company receives 428 pages a day by FAX compared with 300 pages a day 1 year ago.

■ The airfare for the Los Angeles to New York flight rose $60 in the past 3 months.

Each of these statements says something about how much a quantity changed over time. They can easily be rewritten to reflect how rapidly the quantity changed *on average*:

■ The temperature fell an average of 5° *per hour* between 4 p.m. and 6 p.m.

■ Over the past year, the typical Fortune 500 company saw the number of FAX pages received each day increase by 128 pages. This is an average increase of 10.7 pages per day *each month*.

■ The Los Angeles to New York airfare rose at an average rate of $20 *per month* over the past 3 months.

When we divide the amount that a quantity changes over an interval by the length of the interval, the result is the **average rate of change** of the quantity over that interval.

Average Rate of Change

$$\text{Average rate of change} = \frac{\text{change over an interval}}{\text{length of the interval}}$$

Although average rates of change are useful, they have their limitations. In the case of airfare, it is possible that the price rose $60 in the first month and then remained constant for the following 2 months. However, the average rate of change spreads the $60 increase evenly over the 3-month period of time.

EXAMPLE 1   *Air Temperature*

Consider Table 3.1, which shows temperature values on a typical May day in a certain midwestern city.

TABLE 3.1

Time	Temperature (°F)
7 a.m.	49
8 a.m.	58
9 a.m.	66
10 a.m.	72
11 a.m.	76
12 noon	79
1 p.m.	80
2 p.m.	80
3 p.m.	78
4 p.m.	74
5 p.m.	69
6 p.m.	62

a. What is the average rate of change of temperature between 7 a.m. and 9 a.m.?

b. What is the average rate of change of temperature between 2 p.m. and 6 p.m.?

Solution:

a. To find the average rate of change of temperature between 7 a.m. and 9 a.m.:

   i. Find the amount that the temperature changed between 7 a.m. and 9 a.m.:

$$66\,°F \; - \; 49\,°F = 17\,°F$$

   ii. Divide the change by the length of the time interval.

$$\frac{17\,°F}{2\text{ hours}} = 8.5\,°F \text{ per hour}$$

   Between 7 a.m. and 9 a.m., the temperature rose at an average rate of 8.5° per hour.

b. Similarly, the average rate of change between 2 p.m. and 6 p.m. is

$$\frac{62\,°F - 80\,°F}{4\text{ hours}} = \text{-}4.5\,°F \text{ per hour}$$

   Between 2 p.m. and 6 p.m., the temperature fell at an average rate of 4.5 °F per hour.   ∎

Finding Average Rates of Change Using Secant Lines

You may have noticed that calculating the average rate of change is the same as calculating slope. This observation allows for the easy calculation of average rates of change if you are given a graph. For instance, when plotted, the May daytime temperatures fall in the shape of a parabola (see Figure 3.1).

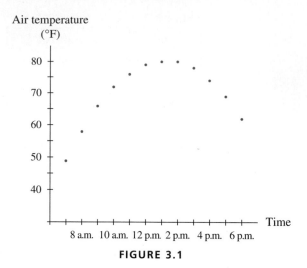

FIGURE 3.1

To find the average rate of change between 9 a.m. and 4 p.m., use a straightedge to carefully draw a line connecting the points at 9 a.m. and 4 p.m. (see Figure 3.2).

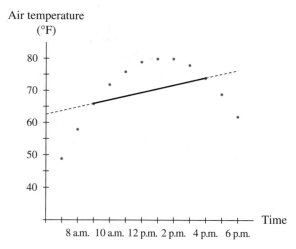

Secant Line Between 9 a.m. and 4 p.m.

FIGURE 3.2

We call a line connecting two points on a scatter plot or a graph a **secant line** (from the Latin *secare*, "to cut") and approximate the slope of the secant line by estimating the rise and the run for a portion of the line (see Figure 3.3). Between 9 a.m. and 4 p.m., the temperature rose at an average rate of 1.1 °F per hour.

$$\frac{\text{Rise}}{\text{Run}} = \frac{\text{change in temperature}}{\text{change in time}} = \frac{8\,°F}{7\ \text{hours}} \approx 1.1\,°F \text{ per hour}$$

It is important to note that this method is imprecise if you are given only a scatter plot or a graph. It gives only an approximation to the slope of the secant line. The method depends on drawing the secant line accurately and then correctly determining the rise and run. Slight variations in sketching are likely to result in slightly different answers. This does not mean that the answers you obtain are incorrect. It simply means that slopes obtained by sketching secant lines are approximations.

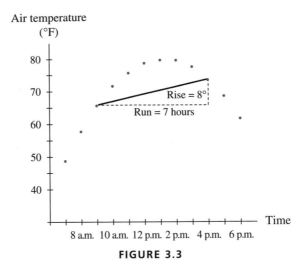

Air temperature
(°F)

Rise = 8°
Run = 7 hours

Time

8 a.m. 10 a.m. 12 p.m. 2 p.m.   4 p.m.   6 p.m.

FIGURE 3.3

Example 1 and the subsequent discussion of air temperature used the term *between* two time values. There are other ways to describe intervals on the input axis, and we take a moment now to discuss one of them.

When we use data that someone else has collected, we often do not know when the data were reported or recorded. It seems logical to assume that yearly (or monthly or hourly and so forth) totals are reported at the *end* of the intervals representing those periods. For instance, the 1992 total covers the period of time from the end of 1991 through the end of 1992 (see Figure 3.4).

Therefore, we adopt the convention that "from *a* through *b*" on the input axis refers to the interval beginning at *a* and ending at *b*. If *a* and *b* are years and the output represents a quantity that can be considered to have been measured at the end of the year, then "from *a* to *b*" means the same thing as "from the end of year *a* through the end of year *b*." The terms "between *a* and *b*" and "from *a* to *b*" have the same meaning as "from *a* through *b*." We use these terminologies in this manner in the remainder of the text.

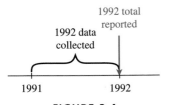

1992 total
reported

1992 data
collected

1991   1992

FIGURE 3.4

EXAMPLE 2   *Interest Income*

The yearly interest income earned by Kelly Services[1] from 1986 through 1994 is shown in Figure 3.5. A smooth curve connecting the points is also shown.

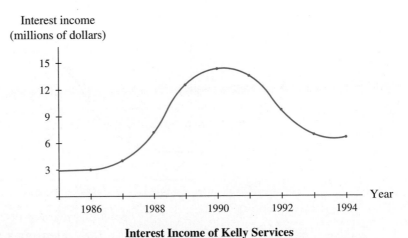

Interest income
(millions of dollars)

15

12

9

6

3

1986   1988   1990   1992   1994

Year

Interest Income of Kelly Services

FIGURE 3.5

1. Kelly Services *1994 Annual Report.*

a. Use a secant line to estimate the average rate of change of interest income from 1987 through 1991.

b. Use a secant line to estimate the average rate of change of interest income between 1992 and 1994, and convert the answer to dollars per quarter.

Solution:

a. First, carefully construct a secant line from the point on the curve in 1987 to the point in 1991, and estimate rise and run (see Figure 3.6).

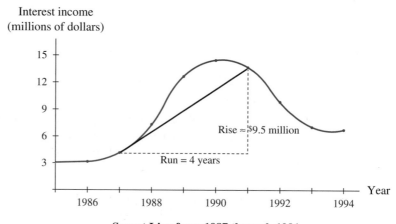

Secant Line from 1987 through 1991

FIGURE 3.6

$$\text{Slope} = \frac{\text{rise}}{\text{run}} \approx \frac{\$9.5 \text{ million}}{4 \text{ years}} \approx \$2.4 \text{ million per year}$$

From 1987 through 1991, Kelly Services' interest income increased at an average rate of 2.4 million dollars per year.

b. To estimate the average rate of change between 1992 and 1994, we again sketch a secant line and estimate rise and run (see Figure 3.7).

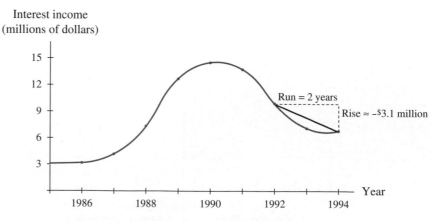

Secant Line Between 1992 and 1994

FIGURE 3.7

$$\text{Slope} = \frac{\text{rise}}{\text{run}} \approx \frac{\text{-\$3.1 million}}{2\text{ years}} = \text{-\$1,550,000 per year}$$

Now, divide by 4 to get the average amount each quarter.

$$\frac{\text{-\$1,550,000}}{4} = \text{-\$387,500 per quarter}$$

Between 1992 and 1994, Kelly Services saw an average decline in interest income of approximately $387,500 per quarter. ∎

Determining Average Rates of Change Using an Equation

It is also possible to determine average rates of change when we are given only an equation. A model for the temperature data on a typical May day in a certain midwestern city is

$$\text{Temperature} = -0.8t^2 + 2t + 79 \,°\text{F}$$

where $t = 0$ at noon.

To calculate the average rate of change between 11:30 a.m. and 6 p.m.:

1. Note that at 11:30 a.m., $t = -0.5$ and that at 6 p.m., $t = 6$.
2. Substitute $t = -0.5$ and $t = 6$ into the equation to obtain the corresponding temperatures.

   At 11:30 a.m.:   Temperature $= -0.8(-0.5)^2 + 2(-0.5) + 79 = 77.8 \,°\text{F}$

   At 6 p.m.:   Temperature $= -0.8(6)^2 + 2(6) + 79 = 62.2 \,°\text{F}$

3. Divide the change in temperature by the change in time, subtracting the earlier temperature from the later temperature.

$$\frac{62.2\,°\text{F} - 77.8\,°\text{F}}{6 - (-0.5)} = \frac{-15.6\,°\text{F}}{6.5\text{ hours}} = -2.4\,°\text{F per hour}$$

Thus, between 11:30 a.m. and 6 p.m., the temperature fell at an average rate of 2.4 °F per hour. Note that the temperature fell 15.6 °F in 6.5 hours; however, when finding an average rate of change, we state the answer (in this case) as the number of degrees per *one* hour.

> In general, average rates of change are always expressed in terms that clearly use the word *average* and specify *output units per input unit.*

Note that in Table 3.1 of temperature data, we were not given a temperature for 11:30 a.m., but we were able to use the continuous model to find an approximation for that temperature. In Example 2, however, the interest income values are end-of-the-year totals, and a continuous model would not be helpful in telling us the interest income at any point during a particular year. Although it is not inappropriate to fit a continuous model to discrete data such as the interest income, we must realize that the input is restricted to integer values that correspond to the ends of certain years. Therefore, the model must be discretely interpreted.

EXAMPLE 3   *Population Density*

The population density of Nevada from 1950 through 1990 can be approximated by the model[2]

$$p(t) = 0.1273(1.05136)^t \text{ people per square mile}$$

where t is the number of years since 1900.

a. Find the average rate of change of population density from 1950 through 1980.

b. Find the average rate of change of population density between 1980 and 1990.

c. Use the fact that the area of Nevada is 110,540 square miles to convert both average rates of change to people per year.

d. On the basis of your answers, what can you conclude about growth in Nevada?

Solution:

a. The average rate of change from 1950 through 1980 is

$$\frac{p(80) - p(50)}{80 - 50} \approx \frac{7.00 - 1.56 \text{ people/mi}^2}{30 \text{ years}} \approx 0.18 \text{ person per square mile per year}$$

b. The average rate of change between 1980 and 1990 is

$$\frac{p(90) - p(80)}{90 - 80} \approx \frac{11.55 - 7.00 \text{ people/mi}^2}{10 \text{ years}} \approx 0.45 \text{ person per square mile per year}$$

c. From 1950 through 1980:

$$\frac{0.18 \text{ person/mi}^2}{\text{year}} \times 110,540 \text{ mi}^2 \approx 20,045 \text{ people per year}$$

From 1980 through 1990:

$$\frac{0.45 \text{ person/mi}^2}{\text{year}} \times 110,540 \text{ mi}^2 \approx 50,287 \text{ people per year}$$

d. In the 10-year period from 1980 through 1990, the population grew by more than twice as much as it had grown in the 30-year period from 1950 through 1980.   ■

3.1   Concept Inventory

■ Average rate of change

■ Secant line

■ Slope of a secant line = average rate of change

3.1   Activities

Rewrite the sentences in Activities 1 through 5 to express how rapidly, on average, the quantity changed over the given interval.

---

2. Based on data from *State Rankings 1992. A Statistical View of the United States.* (Lawrence, Kansas: Morgan Quitno Corporation, 1992).

1. In five trading days, the stock price rose $2.30.

2. The nurse counted 32 heart beats in 15 seconds.

3. The company lost $25,000 during the past 3 months.

4. In 6 weeks, she lost 17 pounds.

5. The unemployment rate has risen 4 percentage points in the past 3 years.

6. In 1993, retail sales of personal watercraft were $618 million. In 1995, sales had risen to $1.1 billion.[3] Find the average rate of change in retail sales of personal watercraft from 1993 through 1995.

7. In 1995, 25,057 hunting licenses were sold in Nevada.[4] That was a decrease from 31,517 licenses sold in 1991. Find the average rate of change in the number of Nevada hunting licenses sold from 1991 through 1995.

8. a. On October 1, 1987, 193.2 million shares were traded on the stock market.[5] On October 30, 1987, 303.4 million shares were traded. Find the average rate of change in the number of shares traded per trading day between October 1 and October 30.

   b. The scatter plot in Figure 3.1.1 shows the number of shares traded each day during October of 1987. On the scatter plot, sketch a line whose slope is the average rate of change between October 1 and October 30.

   c. The behavior of the graph on October 19th and 20th has been referred to as "October Madness." Write a sentence describing how the number of shares traded changed throughout the month. How well does the average rate of change you found in part *a* reflect what occurred throughout the month?

9. The graph in Figure 3.1.2[6] shows the highest elevations above sea level attained by Lake Tahoe (located on the California–Nevada border) from 1982 through 1996.

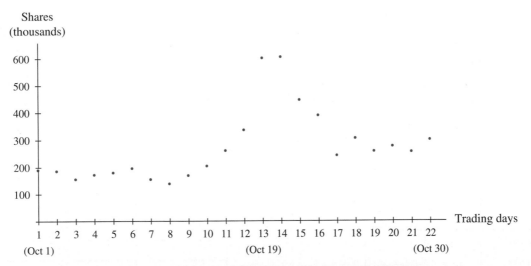

Volume of Dow Jones Industrial Shares Traded in October of 1987

FIGURE 3.1.1

3. Personal Watercraft Industry Association.
4. Nevada Division of Wildlife.
5. *The Dow Jones Averages 1885–1990*, ed. Phyllis S. Pierce. (Homewood, Illinois: Business One Irwin, 1991).
6. Data from Federal Watermaster, United States Department of the Interior.

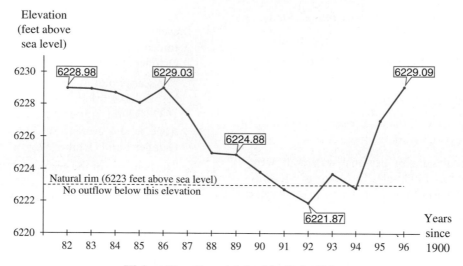

Highest Elevations Attained by Lake Tahoe

FIGURE 3.1.2

a. Sketch a secant line connecting the beginning and ending points of the graph. Find the slope of this line.

b. Write a sentence interpreting the slope in the context of Lake Tahoe levels.

c. Write a sentence summarizing how the level of the lake changed from 1982 through 1996. How well does your answer to part *b* describe the change in the lake level as shown in the graph?

10. Refer once again to the roofing jobs example in Section 2.3. The data are shown in Table 3.2.

TABLE 3.2

Month	Number of jobs
January	12
February	14
March	22
April	35
May	58
June	84

a. Use the data to find the change in the number of roofing jobs from January through June.

b. Use the data to find the average rate of change in the number of roofing jobs from January through June.

c. Fit a model to the data.

d. Use the model to find the change and the average rate of change in the number of roofing jobs from January through June.

e. Are the answers obtained from the data more accurate than those obtained from the model?

11. Imagine that 6 years ago you invested $1400 in an account with a fixed interest rate and with interest compounded continuously. You do not remember the interest rate, but your end-of-the-year statements for the first 5 years yield the data shown in Table 3.3.

TABLE 3.3

End of year	Amount at end of year
1	$1489.55
2	$1584.82
3	$1686.19
4	$1794.04
5	$1908.80

a. Use the data to find the change in the balance from the end of year 1 through the end of year 5.

b. Use the data to find the average rate of change of the balance from the end of year 1 through the end of year 5.

c. Using the data, is it possible to find the average rate of change in the balance from the middle of the fourth year through the end of the fourth year? Explain how it could be done or why it cannot be done.

d. Fit a model to the data, and use the model to find the average rate of change over the last half of the fourth year.

12. Table 3.4 gives the price in dollars of a round trip flight from Denver to Chicago on a certain airline and the corresponding monthly profit (in millions of dollars) for that airline on that route.

TABLE 3.4

Ticket price (dollars)	Profit (millions of dollars)
200	3.08
250	3.52
300	3.76
350	3.82
400	3.70
450	3.38

a. Fit a model to the data.

b. Estimate the average rate of change of profit when the ticket price rises from $200 to $325.

c. Estimate the average rate of change of profit when the ticket price rises from $325 to $450.

13. Refer to the influenza epidemic data in Table 2.19, Activity 31 of Section 2.1.

a. How rapidly (on average) did the number of deaths from influenza increase in the Navy between August 31, 1918, and October 12, 1918?

b. How rapidly did civilian deaths increase from October 5, 1918, through November 2, 1918?

c. Compare the average rates of increase in the Navy and Army between October 12, 1918, and October 19, 1918.

14. A travel agent vigorously promotes cruises to Alaska for several months. The number of cruise tickets sold during the first week and the total (cumulative) sales every 3 weeks thereafter are given in Table 3.5.

TABLE 3.5

Week	Total tickets sold
1	71
4	197
7	524
10	1253
13	2443
16	3660
19	4432
22	4785
25	4923

a. Find the first differences in the numbers of tickets sold, and convert them to average rates of change.

b. When were ticket sales growing most rapidly? How rapidly (on average) were they growing at that time?

c. If the travel agent made a $25 commission on every ticket sold, how rapidly was the agent's commission revenue increasing between weeks 7 and 10?

15. The population of Mexico between 1921 and 1990 is given by the model[7]

Population $= 12.921e^{0.026578t}$ million persons

where t is the number of years since 1921.

a. How much did the population change from 1980 through 1986?

b. How rapidly was the population changing from 1983 through 1985?

---

7. Based on data from SPP and INEGI, Mexican Censuses of Population 1921 through 1990 as reported by Pick and Butler, *The Mexico Handbook*, Westview Press, 1994.

16. The number of AIDS cases diagnosed from 1988 through 1991 can be modeled[8] by

    $$\text{Cases diagnosed} = -1049.50x^2 + 5988.7x + 33{,}770.7 \text{ cases}$$

    where x is the number of years since 1988. Find the change and the average rate of change in cases diagnosed between 1989 and 1991.

17. The graph in Figure 3.1.3 shows the path of the misfired missile from Activity 7 of Section 2.3.

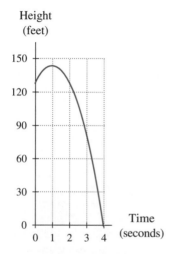

Height of a Missile

FIGURE 3.1.3

   a. Use a secant line to estimate the average rate of change in position between 0 seconds and 2 seconds.

   b. Use a secant line to estimate the average rate of change in position between 2 seconds and 3 seconds.

   c. Convert your answer in part *b* to miles per hour.

18. The graph[9] in Figure 3.1.4 shows private philanthropy giving (in billions of dollars) since 1987.

    Estimate by how much and how rapidly private philanthropy giving grew from 1988 through 1992. What is true about the average rate of change between any two points on a linear graph?

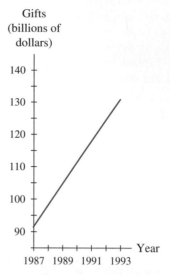

Philanthropy Giving

FIGURE 3.1.4

19. The graph[10] in Figure 3.1.5 models the number of states associated with the national P.T.A. organization from 1895 through 1931.

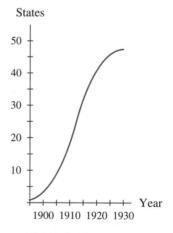

States Joining P.T.A.

FIGURE 3.1.5

   a. Approximately how rapidly was the membership growing from 1905 through 1915?

   b. Approximately how rapidly was the membership growing from 1920 through 1925?

---

8. Based on information in the *HIV/AIDS Surveillance* 1992 Year End Edition.

9. Based on data from *Statistical Abstract*, 1994.

10. Based on data from Hamblin, Jacobson, and Miller, *A Mathematical Theory of Social Change.* (New York: John Wiley & Sons, 1973).

20. Suppose the percentage of high school seniors who have used a graphing calculator in at least one of their math classes can be modeled by

$$\text{Percent} = \frac{100}{1 + 99e^{-0.725t}}$$

where t is the number of years since 1986. Find the average rate of change in graphing calculator use from 1986 through the present year.

3.2   Instantaneous Rates of Change

The Importance of Continuous Models

The air temperature data in Example 1 of Section 3.1 is **discrete** information. Recall that when plotted, discrete information appears as a collection of separate points, as in Figure 3.8.

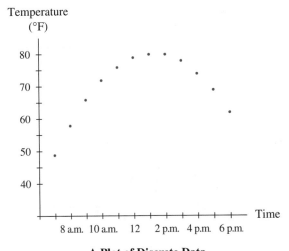

A Plot of Discrete Data

FIGURE 3.8

But when a curve is fitted to the data points (see Figure 3.9), the curve is **continuous;** that is, there are no breaks or gaps in the curve. It can be drawn without lifting the writing instrument from the page.

We often use a continuous curve to model discrete information. We can easily calculate average rates of change from either data or a continuous model. However, one of the advantages of a smooth (no sharp points), continuous model is that it allows us to calculate instantaneous rates of change—something that is impossible with discrete data. With a continuous model, we can apply concepts and methods of calculus to determine such things as maxima and minima, curvature, and inflection points.

The most common example of an instantaneous rate of change is as close as the nearest steering wheel. Suppose that you begin driving north on highway I-81 at the Pennsylvania/New York border at 1:00 p.m. As you drive, you note the time at which you pass each of the indicated mile markers (see Table 3.6).

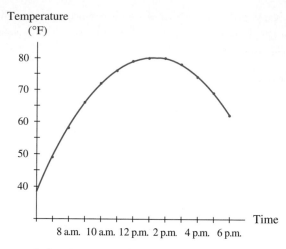

A Continuous Curve Fitted to Discrete Data

FIGURE 3.9

TABLE 3.6

Time	Mile marker
1:00 p.m.	0
1:17 p.m.	19
1:39 p.m.	42
1:54 p.m.	56
2:03 p.m.	66
2:25 p.m.	80
2:45 p.m.	105

These data can be used to determine average rates of change. For example, between mile 0 and mile 19, the average rate of change of distance is 67.1 mph. In this context, the average rate of change is simply the average speed of the car. Average speed between any of the mile markers in the table can be determined in a similar manner. Average speed will not, however, answer the following question:

> If the speed limit is 65 mph and a highway patrol officer with a radar gun clocks your speed at mile post 17, were you exceeding the speed limit by more than 10 mph?

The only way to answer this question is to know your speed at the instant that the radar locked onto your car. This speed is the **instantaneous rate of change**, and your car's speedometer measures that speed in miles per hour.

Just as an average rate of change measures the slope between two points, an instantaneous rate of change measures the slope at a single point. How we measure instantaneous rate of change and why it is useful are important aspects of calculus.

EXAMPLE 1   *NRA Membership*

National Rifle Association membership is rebounding after a slump in the early 1990s. Consider the data[11] in Table 3.7.

TABLE 3.7

Year	1990	1991	1992	1993	1994	1995
NRA membership (millions of members)	2.8	2.6	2.7	3.2	3.5	3.6

For each of the following questions, can the data be used to answer the questions? Or should you fit a continuous model and use it to answer the questions? Or is neither appropriate?

a. What was the NRA membership in 1992?

b. What was the average rate of change in membership from 1991 through 1994?

c. What was the average rate of change in NRA membership during the last half of 1993?

d. How quickly was NRA membership increasing in 1993?

e. When was NRA membership increasing most rapidly?

Solution:

a. The data are better used to report the NRA membership in 1992, because they include the actual 1992 membership. There were 2.7 million members in the NRA in 1992. There is no need to fit a model to answer this question.

b. The data should be used, because membership levels for 1991 and 1994 are included in the table.

c. NRA membership is similar to the cumulative sales discussed in Section 1.2. The growth of membership is not technically continuous, but for practical purposes, we use a continuous model with no restrictions on the output. We are not given the membership in mid-1993, so we should use a model to estimate that membership in order to find the average rate of change. A cubic model fit to the data is

$$m(t) = -0.0398t^3 + 0.3397t^2 - 0.5451t + 2.8087 \text{ million members}$$

t years after 1990. A graph of $m(t)$ is shown in Figure 3.10.

d. This is a question about an instantaneous rate of change and is better answered with the continuous model.

e. We later learn that the NRA membership was increasing most quickly where the slope has a maximum value, *i.e.*, at the inflection point. Again, this is a question about the instantaneous rate of change. The model should be used to answer this question. ∎

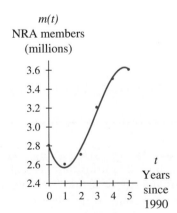

m(t)
NRA members
(millions)

FIGURE 3.10

11. The Associated Press, *Anderson Independent-Mail*, May 20, 1995, page A1.

Tangent Lines and Rates of Change

Refer again to Table 3.1 on page 145 and the fitted curve for the air temperature data in the previous section. The continuous model is shown in Figure 3.11.

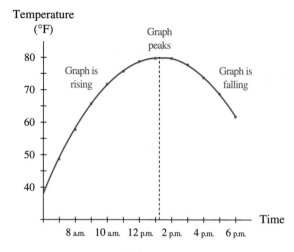

FIGURE 3.11

The graph reaches its peak value at approximately 1:15 p.m. Reading from left to right, the graph is rising until it reaches its peak at 1:15 p.m. and then is falling after 1:15 p.m. The slope of the graph is positive at each point on the left side of the peak and is negative at each point on the right side. The slope of the graph is zero at the top of the parabola. Note that the graph levels off as you move from 7 a.m. to 1:15 p.m. It is not as steep at 1 p.m. as it is at 7 a.m., so the slope at 1 p.m. is smaller than the slope at 7 a.m. In fact, at each point on the graph there is a different slope, and we need to be able to measure that slope in order to find the instantaneous rate of change at each point.

Instantaneous Rate of Change

The **instantaneous rate of change** at a point on a curve is the slope of the curve at that point.

In precalculus mathematics, the concept of slope is intrinsically linked with lines. In terms of lines, slope is a measure of the tilt of a line. Now we wish to measure how tilted a graph is at a point. It may come as no surprise to you to learn that we still must rely on lines to measure the slopes at points on a graph.

Associated with each point on a smooth graph is a line called a **tangent line** (from the Latin word *tangere*, "to touch"). A tangent line at a point on a graph touches that point and is tilted exactly the way that the graph is tilted at the **point of tangency**. The slope of the tangent line at a point is a measure of the slope of the graph at that point.

> The slope of a graph at a point is the slope of the tangent line at that point.

In Figure 3.12, tangent lines are drawn at 7 a.m., 12 noon, and 4 p.m. Can you see that the tangent lines are tilted to match the tilt of the graph at each point? The tangent lines at points *A* and *B* are tilted up, so the slope at these points is positive. The slope at point *C* is negative, because the tangent line at point *C* is tilted down. Even though the tangent line at point *C* has the least slope of these three tangents, it is *steeper* than the tangent line at point *B* because the magnitude (absolute value) of its slope is larger than that of the tangent to the curve at point *B*. That is, the temperature is falling faster at 4 p.m. than it is rising at noon.

Examine Figure 3.12 carefully. The slope of tangent line *A* is 10 °F per hour. (A method for calculating this slope will be discussed later.) Therefore, the slope of the graph at 7 a.m. is also 10 °F per hour. This is the same as saying that the instantaneous rate of change of the temperature at 7 a.m. is 10 °F per hour. In other words, at 7 a.m., the temperature is rising 10 °F per hour.

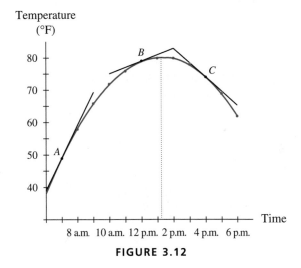

FIGURE 3.12

Similarly, the following statements can be made.

- The slope of tangent line *B* is 2 °F per hour.
- The slope of the graph at 12 noon is 2 °F per hour.
- The instantaneous rate of change of the temperature at 12 noon is 2 °F per hour.
- At 12 noon, the temperature is rising 2 °F per hour.

And,

- The slope of tangent line *C* is -4.4 °F per hour.
- The slope of the graph at 4 p.m. is -4.4 °F per hour.
- The instantaneous rate of change of the temperature at 4 p.m. is -4.4 °F per hour.
- At 4 p.m., the temperature is falling 4.4 °F per hour.

We summarize the results of this discussion in the following way:

> Given a function $f(x)$ and a point P on the graph of $f(x)$, the instantaneous rate of change at point P is the slope of the graph at P and is the slope of the tangent line to the graph at P (provided the slope exists).

EXAMPLE 2   *NRA Membership*

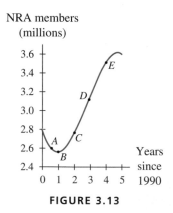

NRA members (millions)

Years since 1990

FIGURE 3.13

Using a cubic model (shown in Figure 3.13) to fit the National Rifle Association membership data given in Example 1, we can obtain the information that follows.

■ The slope of the tangent line to the graph at A is -0.24 million members per year.

■ The slope of the graph is zero at point B.

■ The instantaneous rate of change of the NRA membership at point C is 340,000 members per year.

■ NRA membership is increasing the fastest at point D. The rate of fastest increase is 0.42 million members per year.

■ The slope of the tangent line to the graph at point E is 260,000 members per year.

Using this information, answer the following questions.

a. At which of the indicated points is the slope of the graph (i) the greatest? (ii) the least?

b. At which of the indicated points is the steepness of the graph (i) the greatest? (ii) the least?

c. Arrange the indicated points in order from the point at which the slope of the graph has the smallest magnitude to the point at which the slope of the graph has the largest magnitude.

Solution:

a. The numerical values of the slopes, in million members per year, at the indicated points are

   A: -0.24 B: 0 C: 0.34 D: 0.42 E: 0.26

   The greatest value occurs at point D (the inflection point) and the least where the only negative slope occurs, at point A.

b. The steepness of the graph is a measure of how much the graph is tilted at a particular point. The direction of tilt is considered in the slope but not when describing steepness. Thus the steepness at each of the indicated points is

   A: 0.24 B: 0 C: 0.34 D: 0.42 E: 0.26

   The graph is steepest at point D. The degree of steepness is least at point B.

c. The magnitude of the slope is the absolute value of the slope, *i.e.*, the steepness. The slope of the graph has the smallest magnitude at point B. Then, the magnitude of the slope increases from 0 to 0.24 at point A, to 0.26 at point E, to 0.34 at point C, and to the largest value, 0.42, at point D. ■

3.2   Concept Inventory

- Discrete and continuous
- Instantaneous rate of change
- Tangent line
- Slope of tangent line = instantaneous rate of change
- Slope and steepness of a tangent line

3.2   Activities

1. In your own words, describe the difference between

   a. discrete and continuous.

   b. average rate of change and instantaneous rate of change.

   c. secant lines and tangent lines.

2. What are some advantages of using a continuous model instead of discrete data? What are some disadvantages?

3. Using Table 3.6, the time/mileage table given in this section, verify that the average speed of the car from mile marker 0 to mile marker 19 is 67.1 mph.

4. Using Table 3.6, determine the average speed (in mph) from:

   a. milepost 66 to milepost 80.

   b. milepost 80 to milepost 105.

   c. What might account for the difference in speed?

5. How are average rates of change and instantaneous rates of change measured graphically?

6. At each labeled point on the graph in Figure 3.2.1, determine whether the slope is positive, negative, or zero.

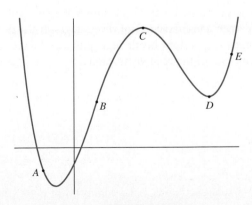

FIGURE 3.2.1

7. a. At each labeled point on the graph in Figure 3.2.2, determine whether the instantaneous rate of change is positive, negative, or zero.

   b. Is the graph steeper at point A or at point B?

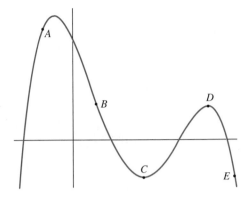

FIGURE 3.2.2

8. Discuss the slopes of the following graphs.

   a. **FIGURE 3.2.3a**

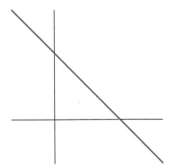

   b. **FIGURE 3.2.3b**

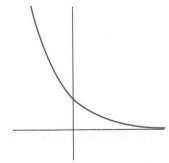

   c. **FIGURE 3.2.3c**

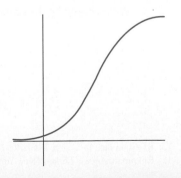

9. a. In graph c of Activity 8, estimate where the slope is greatest. Mark that point on the graph.

   b. In the graph in Figure 3.2.4, estimate where the output is falling most rapidly. Mark that point on the graph.

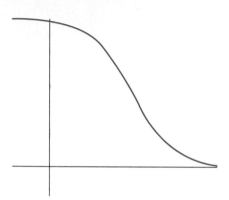

FIGURE 3.2.4

10. The graph in Figure 3.2.5 shows the average yearly amount (in dollars) spent on groceries by a family of four in Baltimore, Maryland. The slope of the curve at point A (year = 1991) is 95.

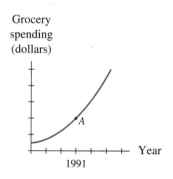

**Average Yearly Amount
Spent on Groceries**

FIGURE 3.2.5

a. What should be the units on the slope at point A?

b. How rapidly was the dollar amount growing in 1991?

c. What is the slope of the tangent line at point A?

d. What is the instantaneous rate of change of the amount at point A?

11. The growth of a pea seedling as a function of time can be modeled by two quadratics as shown in Figure 3.2.6.[12] The slopes at the labeled points are (in ascending order) -4.2, 1.3, and 5.9.

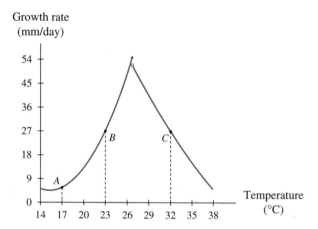

Growth Rate of a Pea Seedling

FIGURE 3.2.6

a. Match the slopes with the points A, B, and C.

b. What are the units on the slopes for each of these points?

c. How quickly is the growth rate changing with respect to temperature at 23 °C?

d. What is the slope of the tangent line at 32 °C?

e. What is the instantaneous rate of change of the growth rate of pea seedlings at 17 °C?

12. The graph in Figure 3.2.7 shows the survival rate (percentage surviving) of three stages in the development of a flour beetle (egg, pupa, and larva) as a function of the relative humidity.[13]

12. Based on data in George L. Clarke, *Elements of Ecology* (New York: Wiley, 1954).
13. Chapman, *Animal Ecology* (New York: McGraw-Hill, 1931).

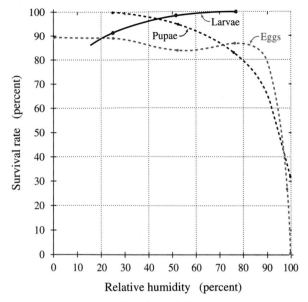

FIGURE 3.2.7

Fill in each of the following blanks with the appropriate stage (eggs, pupae, or larvae).

a. At 60% relative humidity, the instantaneous rate of change of the survival rate of _____ is approximately zero.

b. An increase in relative humidity improves the survival rate of _____ and reduces the survival rate of _____.

c. At 97% relative humidity, the survival rate of _____ is declining faster than that of _____.

d. Any tangent lines drawn to the survival curve for _____ will have negative slope.

e. Any tangent lines drawn to the survival curve for _____ will have positive slope.

f. At 30% relative humidity, the survival rates for _____ and _____ are changing at approximately the same rate.

g. At 65% relative humidity, the survival curves for _____ and _____ have approximately the same slope.

■ 3.3   Tangent Lines

Lines Tangent to Circles

Because the slope of the tangent line at a particular point gives the instantaneous rate of change of the output at that point, it is important that you obtain experience with tangent lines before proceeding further. We begin with the simplest of all tangent lines to draw: lines tangent to a circle.

> **Line Tangent to a Circle**
>
> We construct the line tangent to a point P on a circle as follows (refer to Figure 3.14):
>
> 1. Draw a circle with center O, and draw point P on the circle.
>
> 2. Connect O and P.
>
> 3. Using a T-square or see-through straight edge, draw a line segment through P perpendicular to the line segment $\overline{OP}$.

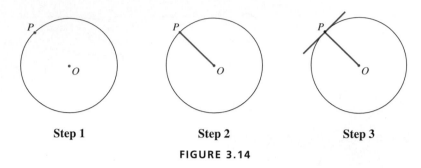

| Step 1 | Step 2 | Step 3 |

FIGURE 3.14

Note that the slope of the tangent line at P measures the slope of the circle at P.

EXAMPLE 1 *Calculating Slope on a Circle*

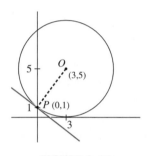

FIGURE 3.15

Consider the circle that passes through the point $P(0, 1)$ and has center $O(3, 5)$. See Figure 3.15. Find the slope of the circle at P.

Solution: To find the slope of the circle at P, it is necessary to find the slope of the tangent line at P. The slope of the line segment $\overline{OP}$ is $\frac{5-1}{3-0} = \frac{4}{3}$. The tangent line is perpendicular to $\overline{OP}$, so its slope is the negative reciprocal of $\frac{4}{3}$. Thus the slope of the circle at P is $\frac{-3}{4}$.   ∎

Although it is possible to find the exact slope at a point on a circle by using the process in Example 1, it is also important to be able to estimate the slope.

EXAMPLE 2 *Estimating Slope on a Circle*

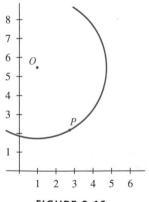

FIGURE 3.16

Estimate the slope of the circle in Figure 3.16 at point P.

Solution: Carefully construct the tangent line. Estimate the rise and the run for a portion of the line. The base of the triangle in Figure 3.17 is drawn from approximately 1 to approximately $6\frac{1}{4}$. This gives a run of $5\frac{1}{4}$. Similarly, the height of the triangle is drawn from about $1\frac{1}{4}$ to 4, giving a rise of $2\frac{3}{4}$. The slope of the circle at P (which is equal to the slope of the tangent at P) $= \frac{\text{rise}}{\text{run}} \approx \frac{2\frac{3}{4}}{5\frac{1}{4}} \approx 0.52$.

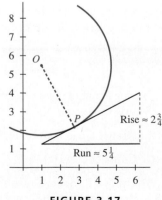

FIGURE 3.17   ∎

Lines Tangent to a Curve

Lines tangent to circles are easy to construct, but what about lines tangent to other curves? To understand other tangent line constructions, it is helpful to understand the relationship between secant lines and tangent lines. We begin with an example on a circle.

EXAMPLE 3   *Secant and Tangent Lines on Circles*

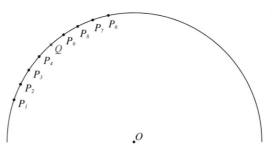

FIGURE 3.18

a.   On the semicircle shown in Figure 3.18, draw the secant lines through P_1 and Q, P_2 and Q, P_3 and Q, and P_4 and Q. Draw the secant lines through P_6 and Q, P_7 and Q, P_8 and Q, and P_9 and Q.

b.   Make use of the center O to construct the tangent line at Q.

c.   Which secant lines most closely approximate the tangent line?

d.   Where could you place a point P_5 so that the secant through P_5 and Q would be even closer in inclination to the tangent line?

Solution:

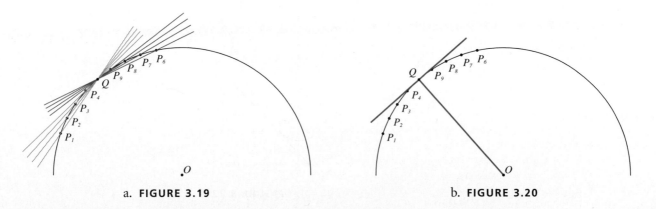

a.   **FIGURE 3.19**   b.   **FIGURE 3.20**

c. The line through P_4 and Q and the line through P_9 and Q most closely approximate the tangent line.

d. P_5 should be placed closer to Q than either P_4 or P_9. ■

If you draw a secant line through Q and a point P on the curve near Q, then the closer P is to Q, the more closely the secant line approximates the tangent line at Q. You can think of the tangent line at Q as the limiting form of the secant lines between P and Q as P gets closer and closer to Q.

Line Tangent to a Curve

The tangent line at a point Q on a continuous graph is the limiting position of the secant lines between point Q and a point P as P approaches Q along the graph (if the limiting position exists).

This relationship between secant lines and tangent lines holds true for any curve, not just for circles. Consider the graph of $y = x^2 + 2$ shown in Figure 3.21 and the secant lines constructed using points to the right and the left of Q.

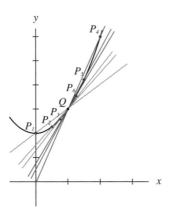

FIGURE 3.21

The tangent line at Q will have a slope a little greater than the slope of the secant line through P_3 and Q and a little less than the slope of the secant line through P_6 and Q. The correctly drawn tangent line at Q is shown in Figure 3.22.

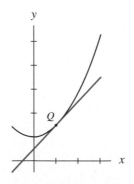

FIGURE 3.22

Although thinking of a tangent line as a limiting value of secant lines is vital to your understanding of calculus, it is important for you to have an intuitive feel for tangent lines and to be able to sketch them without first drawing secant lines.

Consider the curve in Figure 3.23 and the lines through points A, B, and C.

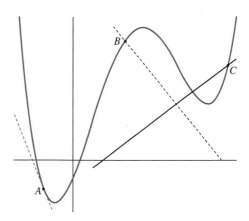

FIGURE 3.23

Which of the three lines shown in Figure 3.23 is tangent to the curve? The line through B is the easiest to spot as an example of a line that is not tangent.

> Recall that tangent lines must be tilted in the same way that the graph is tilted at the point of tangency.

The graph has a positive slope at B, but the line drawn through B has a negative slope. The line through B also violates a general rule about tangent lines:

> Lines tangent to a smooth curve do not "cut through" the graph of the curve at the point of tangency and lie completely on *one side* of the graph near the point of tangency except in the following two cases:
>
> 1. At an inflection point on a smooth curve
>
> 2. At any point on a line

These two exceptions are dealt with after Example 4. For cases in which the exceptions do not apply, we can determine on which side of the curve the tangent line should lie by noting the concavity. If the curve is concave up at the point of tangency, then the tangent line will lie below the curve near the point of tangency. If the curve is concave down at the point of tangency, then the tangent line will lie above the curve near the point of tangency. See Figure 3.24.

Refer again to Figure 3.23. At point C, the curve is concave up. The tangent line should lie below the curve. However, to the left of C, the tangent line lies above the curve, so the line through C is not a tangent line.

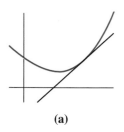

(a)

If the curve is ***concave up*** at the point of tangency, then the tangent line will lie ***below*** the curve near the point of tangency.

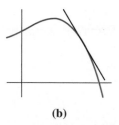

(b)

If the curve is ***concave down*** at the point of tangency, then the tangent line will lie ***above*** the curve near the point of tangency.

FIGURE 3.24

The line at A is the only tangent line of the three. It touches the curve at A; it is tilted in the same way that the graph is tilted at A; and it lies below the curve which is concave up at A.

The correct tangent lines at A, B, and C are shown in Figure 3.25.

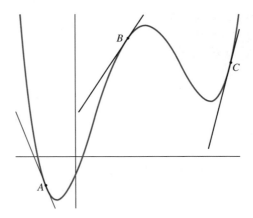

FIGURE 3.25

EXAMPLE 4   *Weight Loss*

A woman joins a national weight-loss program and begins to chart her weight on a weekly basis. Figure 3.26 shows a continuous model of her weight from when she began the program through 7 weeks into the program.

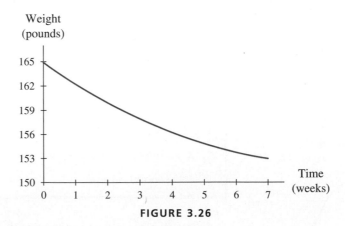

FIGURE 3.26

   a. Carefully sketch a line tangent to the curve at 5 weeks.

   b. Estimate the slope of the tangent line at 5 weeks.

   c. How quickly was the woman's weight declining 5 weeks after the beginning of the program?

   d. What is the slope of the curve at 5 weeks?

Solution:

a.

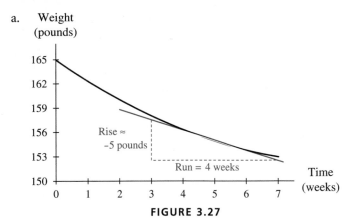

FIGURE 3.27

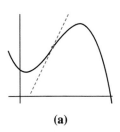

(a)

Although this line lies above the concave-down portion and below the concave-up part of the graph, it is not a tangent line because it is not tilted in the same way that the graph is tilted at the point.

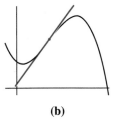

(b)

This tangent line is correctly drawn at an inflection point.

FIGURE 3.28

b.   Slope $= \frac{\text{rise}}{\text{run}} \approx \frac{-5 \text{ pounds}}{4 \text{ weeks}} = -1.25$ pounds per week

c.   The woman's weight was declining by approximately 1.25 pounds per week after 5 weeks in the program.

d.   The slope of the curve at 5 weeks is approximately -1.25 pounds per week.   ∎

As we mentioned in the general rule for tangent lines, there are exceptions to the principle that tangent lines do not cut through the graph and lie on only one side of the graph. At a point of inflection, the graph is concave up on one side and concave down on the other. As you might expect, the tangent line lies above the concave-down portion of the graph and below the concave-up part. To do this, the tangent line must cut through the graph. It does so at the point of inflection. When drawing tangent lines at inflection points, be careful to make sure that the tangent is tilted to match the tilt of the graph at the point of tangency. See Figure 3.28.

Let us also examine the case where the graph itself is a line. Consider the graph of a line with point *P* on that line. Recall that the tangent line at *P* touches the graph at *P* and is tilted in the same direction as the tilt of the graph. The only way to draw such a tangent is to draw the line itself. Because every line has the same slope at every point, every point has the same tangent line, and the slope of the tangent line is exactly the slope of the original line. This fact leads us to a very important concept—that of *local linearity*.

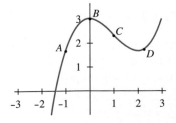

FIGURE 3.29

Local Linearity

With current technology, we can easily obtain close-up views of a portion of a graph. It may surprise you to discover that for any smooth, continuous graph, we will eventually see a line as we look closer and closer. For example, consider the graph in Figure 3.29. Close-ups of the graph 1/100 unit away from each labeled point in both horizontal and vertical directions are shown in Figure 3.30.

FIGURE 3.30

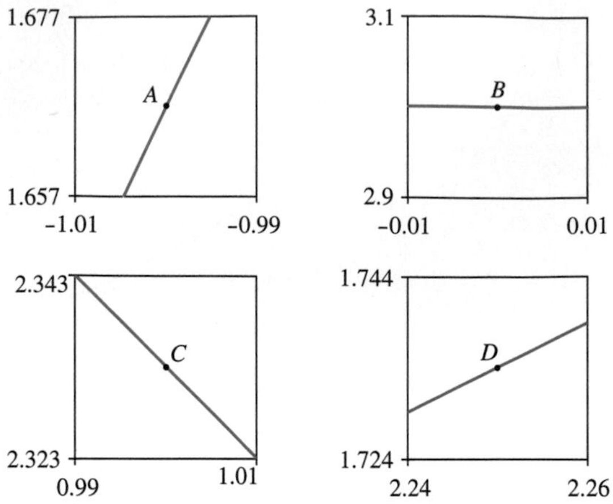

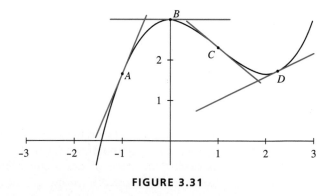

In addition to being close-ups of the curve, the lines in Figure 3.30 are the tangent lines at points *A*, *B*, *C*, and *D*. We call this phenomenon the principle of *local linearity*.

Local Linearity

If we look closely enough near any point on a smooth curve, the curve will look like a line; in fact, it will look like the tangent line at that point.

The tangent lines at points *A*, *B*, *C*, and *D* are shown in Figure 3.31. Do you see that these are the same as the lines in Figure 3.30?

FIGURE 3.31

Local linearity is one additional way to help us visualize and draw tangent lines.

Where Does the Instantaneous Rate of Change Exist?

Our discussion of tangent lines would not be complete without mention of piecewise functions. Consider the data shown in Table 3.8 and the scatter plot and the model (see Figure 3.32) of the amount of fish (in millions of pounds) produced for human food by fisheries in the United States.[14]

---

14. *Statistical Abstract*, 1994.

TABLE 3.8

Year	Amount of fish (millions of pounds)
1970	2537
1972	2435
1975	2465
1977	2952
1980	3654
1982	3285
1985	3294
1987	3946
1990	7041

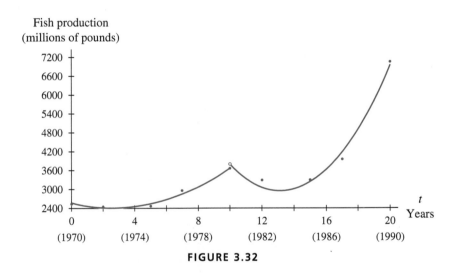

FIGURE 3.32

$$f(t) = \begin{cases} 22.204t^2 - 108.431t + 2538.603 \text{ million pounds} & \text{when } 0 \le t \le 10 \\ 85.622t^2 - 2253.951t + 17{,}772.387 \text{ million pounds} & \text{when } 10 < t \le 20 \end{cases}$$

where t is the number of years after 1970.

Consider the tangent line at the point where we chose to divide the data (1980). If we used the idea of a limiting position of secant lines, then we would have to conclude that we cannot draw a line tangent to the model at $t = 10$, because secant lines drawn using points to the left of $t = 10$ are not approaching the same position as those drawn using points to the right of $t = 10$.

If we use the principle of local linearity and zoom in close to the point at $t = 10$, then we see something similar to Figure 3.33. Recall that if we zoom close enough to a point on a smooth curve, we see a line that is, indeed, the tangent line. In this case, we do not have a smooth curve, and we see two lines. Again we conclude that there is no tangent line at $t = 10$. Does this mean that there is no instantaneous rate of change in the production of fish in 1980? No, it simply means that we cannot use our piecewise continuous model to calculate the rate of change in 1980.

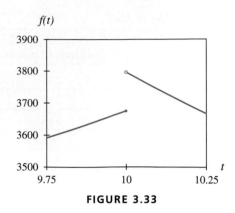

FIGURE 3.33

It is possible for a piecewise function to have a tangent line at a point where the pieces join. This occurs only if a limiting position of secant lines exists; in other words, zooming in close to the point reveals a single line. This is a rare occurrence, and the rarity of it is one of the limitations of piecewise models. It is usually not possible to construct a tangent line at the point(s) where the function is divided into pieces.

As we saw in the model for fish production, it is possible for a continuous or piecewise continuous function to have points at which we cannot sketch a tangent line. Even so, it is possible for such functions to have rates of change at all other points.

To help clarify the relationship between continuous and noncontinuous functions and rates of change, consider the graphs shown in Figure 3.34.

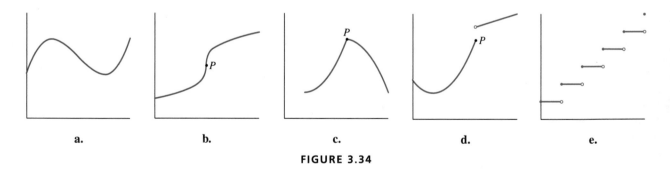

FIGURE 3.34

The graph in Figure 3.34a is continuous everywhere and has a rate of change at every point. You may think that the same thing is true for the graph in Figure 3.34b; however, because a tangent line drawn at P is vertical, the run is zero. Therefore, the slope of the tangent line does not exist at that point. The graph in Figure 3.34c is also continuous; however, we cannot draw a tangent line at P, because secant lines drawn with points on the right and left do not approach the same slope.

The graph in Figure 3.34d has a break at P and, therefore, is not continuous at P. This graph is similar to the fish production graph, and the slope does not exist at the break in the function. The slope does exist at all other points on the graph. The graph in Figure 3.34e is discontinuous at many points. There is no tangent line that can be drawn at these points, but as in Figure 3.34d, the function has an instantaneous rate of change at all other points.

The graphs in Figure 3.34d and e illustrate a general rule relating continuity and rates of change.

> If a function is not continuous at a point, then the instantaneous rate of change does not exist at that point.

If you keep in mind the relationships among instantaneous rates of change, slopes of tangent lines, slopes of secant lines, and local linearity, then you should have little difficulty determining the times when the instantaneous rate of change does not exist.

Approximating Instantaneous Rates of Change

Although a continuous or piecewise continuous model is necessary to calculate an instantaneous rate of change, it is possible to *estimate* a rate of change from a set of data without taking the time to construct a model. Quick approximations are sometimes as helpful as a more involved calculation.

Consider the data in Table 3.9,[15] which show the percentage of hotel/motel occupancy for a tourist town.

TABLE 3.9

Month	Percent occupied
Aug. 1995	82
Sept. 1995	82
Oct. 1995	78
Nov. 1995	69
Dec. 1995	56
Jan. 1996	63
Feb. 1996	71
Mar. 1996	76
Apr. 1996	72
May 1996	74
June 1996	79
July 1996	78

To estimate how quickly the occupancy rate was changing in January of 1996, simply find the slope of the line between the points on either side of January.

$$\frac{71\% - 56\%}{2 \text{ months}} = \frac{15 \text{ percentage points}}{2 \text{ months}} = 7.5 \text{ percentage points per month}$$

In January of 1996, the occupancy rate was growing approximately 7.5 percentage points per month.

We call this estimate a **symmetric difference quotient**. We use the term *symmetric* because we choose the closest point on either side of our point of interest and the same distance away. This technique should not be used if the data are not

---

15. For the Reno-Sparks, Nevada, area as reported in the *Reno Gazette-Journal*, August 27, 1996, page 1E.

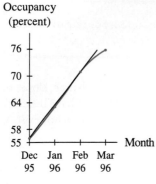

Occupancy
(percent)

FIGURE 3.35

equally spaced around the point at which we desire to estimate the rate of change. We use the term *difference quotient* because we find the difference in the inputs and outputs of the two points and then divide to find the slope that is used to estimate the rate of change.

To understand why a symmetric difference quotient is often a very accurate estimate, consider the data points between December of 1995 and March of 1996 as shown in Figure 3.35. A cubic model has been drawn through the data points, and the line connecting the points in December and February is also shown. Imagine a line tangent to the cubic graph drawn at January. The slope of the line that we found is just slightly smaller than the slope of the tangent line.

Symmetric difference quotients are useful if you want a fast estimate or if you cannot construct a model for a set of data.

3.3   Concept Inventory

- The tangent line is the limiting position of secant lines.
- Tangent lines lie beneath a concave-up graph.
- Tangent lines lie above a concave-down graph.
- Local linearity
- Situations in which the instantaneous rate of change does not exist
- Symmetric difference quotient

3.3   Activities

1. Why are tangent lines important?

2. Explain in your own words how to tell visually whether a line is a tangent line to a smooth graph.

3. Which of the lines drawn on the graph in Figure 3.3.1 are *not* tangent lines?

4. On Figure 3.3.2, draw secants through P_1 and Q, P_2 and Q, and P_3 and Q. Repeat for the points P_4 and Q, P_5 and Q, and P_6 and Q. Then draw the tangent line at Q.

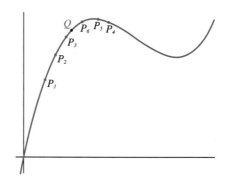

FIGURE 3.3.2

5. Draw secants through P_1 and Q, P_2 and Q, and P_3 and Q on the graph in Figure 3.3.3. Repeat for the points P_4 and Q, P_5 and Q, and P_6 and Q. Then draw the tangent line at Q.

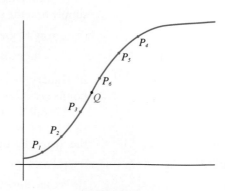

FIGURE 3.3.3

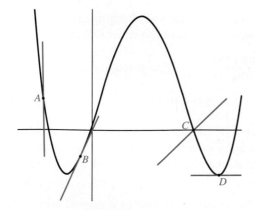

FIGURE 3.3.1

6. Explain in your own words the relationship between secant lines and tangent lines.

7. a. Is the graph shown in Figure 3.3.4 concave up, concave down, or neither (an inflection point) at A, B, C, and D?

   b. Should the tangent lines lie above or below the curve at each of the indicated points?

   c. Carefully draw tangent lines at the labeled points in Figure 3.3.4.

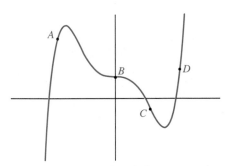

FIGURE 3.3.4

   d. At which of the labeled points is the slope of the tangent line positive? At which of the labeled points is the slope of the tangent line negative? Do any of the labeled points appear to be inflection points?

8. a. Is the graph shown in Figure 3.3.5 concave up, concave down, or neither at A, B, C, and D?

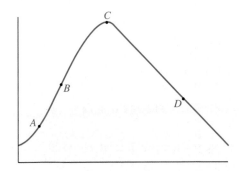

FIGURE 3.3.5

   b. Should the tangent lines lie above or below the curve at each of the indicated points?

   c. Carefully draw tangent lines at the labeled points.

   d. At which of the labeled points is the slope of the curve positive? At which of the labeled points is the slope of the curve negative? Do any of the labeled points appear to be inflection points?

Use carefully drawn tangent lines to estimate the slopes at the labeled points in Activities 9 through 12.

9.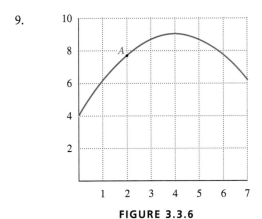

FIGURE 3.3.6

10.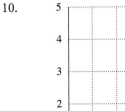

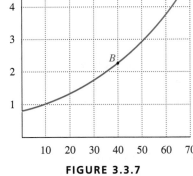

FIGURE 3.3.7

11.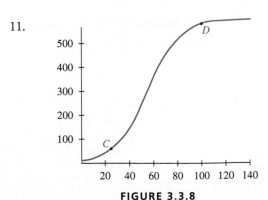

FIGURE 3.3.8

12.

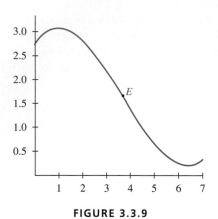

FIGURE 3.3.9

13. Figure 3.3.10 shows per capita consumption of whole milk and lowfat milk in the United States from 1985 through 1992.[16]

Per capita consumption
(gallons per person)

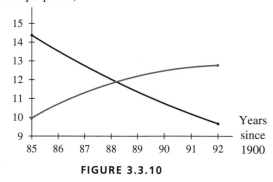

FIGURE 3.3.10

a. Which graph do you think represents whole milk consumption and which represents lowfat milk consumption?

b. To approximate the rates of increase and decrease in milk consumption in 1987 for the two types of milk, sketch tangent lines to both curves, and estimate the slopes.

14. The effects of temperature on the percentage of grasshoppers' eggs from West Australia that hatch is shown in the graph in Figure 3.3.11.[17]

a. What is the optimum hatching temperature?

b. What is the slope of the tangent line at the optimum temperature?

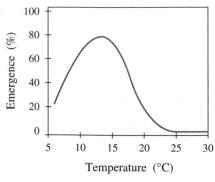

FIGURE 3.3.11

c. Sketch tangent lines at 10°C, 17°C, and 22°C, and estimate the slopes at these points.

d. Where does the inflection point appear to be on this graph?

15. Predictions for the U.S. resident population from 1997 through 2050, as reported by *Statistical Abstract* for 1994, can be approximated by the model

$$p(t) = 2370.14580t + 39{,}789.95719 \text{ thousand people}$$

where t = the number of years since 1900. A graph of $p(t)$ is shown in Figure 3.3.12.

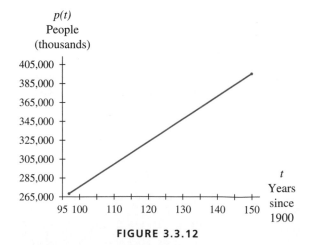

FIGURE 3.3.12

a. Sketch a tangent line at $t = 120$, and find its slope.

b. What is true about any line tangent to this model?

16. Based on data from *Statistical Abstract*, 1994.
17. George L. Clarke, *Elements of Ecology* (New York: Wiley, 1954).

c. What is the slope of any line tangent to this graph?

d. What is the slope at every point on the graph of this model?

e. According to the model, what is the instantaneous rate of change of the predicted population in any year from 1997 through 2050?

16. The graph in Figure 3.3.13 shows employment in Slovakia from 1948 through 1980.[18]

a. To estimate how rapidly employment in agriculture and forestry was declining in 1958, sketch a tangent line at the appropriate point, and estimate its slope.

b. To estimate the instantaneous rate of change in industry employment in 1962, sketch a tangent line at the appropriate point, and estimate its slope.

c. Why is it not possible to sketch a tangent line to the industry graph at 1974?

17. The number of nursing students (in thousands) in the United States from 1975 through 1992 can be modeled[19] by the graph in Figure 3.3.14.

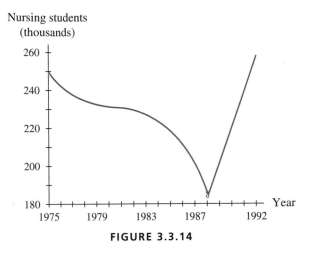

FIGURE 3.3.14

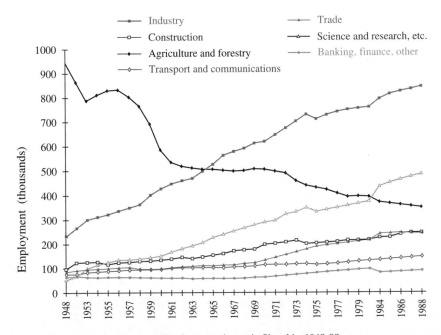

Employment change in Slovakia, 1948-88
(source: elaborated from *Statisticka rŏcenta* SUSR, various dates)

FIGURE 3.3.13

18. A. Smith, "From Convergence to Fragmentation," *Environment and Planning*, vol. 28, 1996.
19. Based on data from *Statistical Abstract*, 1994.

Draw tangent lines, if possible, to estimate how quickly the number of nursing students was changing in the indicated years. If it is impossible to do so, explain why.

a. 1980

b. 1988

c. 1990

18. The annual amounts paid to stockholders of Houghton Mifflin Company between 1990 and 1994 are shown in Table 3.10.[20] Figure 3.3.15 shows a cubic model for the data.

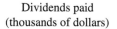

TABLE 3.10

Year	Dividends paid (thousands)
1990	$10,121
1991	$10,746
1992	$11,037
1993	$11,475
1994	$12,026

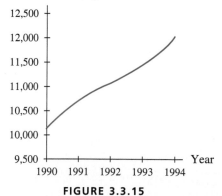

FIGURE 3.3.15

a. Use a symmetric difference quotient to estimate the instantaneous rate of change of dividends paid in 1991.

b. Sketch a tangent line to the model in 1991, and use it to estimate the instantaneous rate of change in 1991.

19. The values of United States exports from 1983 through 1993 are given in Table 3.11.[21] Figure 3.3.16 shows a model for the data.

TABLE 3.11

Year	Exports (billions of dollars)
1983	200.5
1985	213.1
1987	254.1
1989	363.8
1991	421.7
1993	464.8

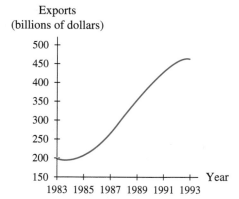

FIGURE 3.3.16

a. Use a symmetric difference quotient to estimate how quickly the value of U.S. exports was growing in 1991.

b. Sketch a line tangent to the model in 1991, and use it to estimate the instantaneous rate of change in 1991.

20. *Houghton Mifflin Company Annual Report*, 1994.

21. *Statistical Abstract*, 1994.

3.4   Derivatives

Derivative Terminology and Notation

By now, you should be comfortable with the concepts of average rates of change and instantaneous rates of change. Let's summarize the differences between these two rates of change.

Average Rates of Change	Instantaneous Rates of Change
■ measure how rapidly (on average) a quantity changes over an interval	■ measure how rapidly a quantity is changing at a point
■ can be obtained by calculating the slope of the secant line between two points	■ can be obtained by calculating the slope of the tangent line at a single point
■ require discrete data points, a continuous curve, or a piecewise continuous curve	■ require a continuous or piecewise continuous curve to calculate

Because instantaneous rates of change are so important in calculus, we commonly refer to them simply as **rates of change**. The calculus term for instantaneous rate of change is **derivative**. It is important to understand that the following phrases are equivalent.

> All of the following phrases have the *same* meaning.
>
> ■ instantaneous rate of change
> ■ rate of change
> ■ slope of the curve
> ■ slope of the tangent line
> ■ derivative

Even though we consider all these phrases synonymous, we must keep in mind that the last three phrases have specific mathematical definitions and so may not exist at a point on a model. However, the rate of change does have an interpretation at that point in context. In such cases, we will have to estimate the rate of change by using a symmetric difference quotient, as we saw in Section 3.3, or some other estimation technique.

There are also several symbolic notations that are commonly used to represent the rate of change of a continuous function $G(t)$. In this book, we use three different but equivalent symbolic notations:

$\dfrac{dG}{dt}$	This is read, *"dee G–dee t,"* *"the rate of change of G with respect to t,"* or *"the derivative of G with respect to t."*
	(or)
$G'(t)$	This is read, *"G prime of t,"* or *"the rate of change of G with respect to t,"* or *"the derivative of G with respect to t."*
	(or)
$\dfrac{d}{dt}[G(t)]$	This is read, *"dee-dee-t of G of t,"* *"the rate of change of G with respect to t,"* or *"the derivative of G with respect to t."*

Suppose that $G(t)$ is your grade out of 100 points on the next calculus test when you study t hours during the week before the test. The graph of $G(t)$ may look like that shown in Figure 3.36.

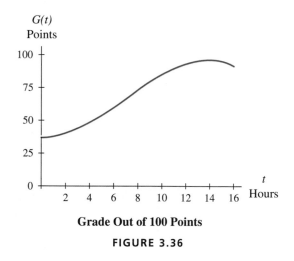

Grade Out of 100 Points

FIGURE 3.36

Note how the grade changes as your studying time increases. The grade slowly improves during the first 2 hours. The longer you study, the more rapidly the grade improves until you have studied approximately 7 hours. After 7 hours, the grade improves at a slower rate. Your grade peaks after 14 hours of studying and then actually declines. What might explain the decline?

Let us compare the rates at which your grade is increasing at $t = 1$ hour and $t = 4$ hours of study. Tangent lines at $t = 1$ and $t = 4$ are shown in Figure 3.37. After 1 hour of studying, your grade is increasing at a rate of approximately 1.7 points per hour. (We will show you how to calculate, not estimate, this rate in a later section.) This value is the slope of the curve when $t = 1$ hour. After 4 hours of studying, your grade is increasing at a rate of approximately 5.2 points per hour. Can you see that the graph is steeper when $t = 4$ than when $t = 1$? The grade is improving more rapidly after 4 hours than it is after 1 hour. In other words, a small amount of additional study is more beneficial if you have already studied 4 hours than it is when you have studied only 1 hour.

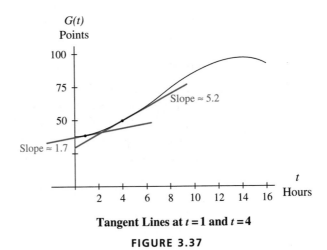

Tangent Lines at $t = 1$ and $t = 4$

FIGURE 3.37

These rates can be summarized with the following notation:

$$\frac{dG}{dt} = 1.7 \text{ points per hour when } t = 1 \text{ hour, or}$$

$$G'(1) = 1.7 \text{ points per hour, and}$$

$$\frac{dG}{dt} = 5.2 \text{ points per hour when } t = 4 \text{ hours, or}$$

$$G'(4) = 5.2 \text{ points per hour.}$$

EXAMPLE 1   *Interpreting Derivatives*

Interpret the following two mathematical statements in the context of the above illustration on studying.

a. $\frac{dG}{dt} = 6.4$ points per hour when $t = 7$ hours

b. $G'(12) = 3.0$ points per hour

Solution:

a. The first statement says that when you have studied 7 hours, your grade is improving by 6.4 points per hour. As we later learn, this is the point of greatest slope—that is, the time when a small amount of additional study will benefit you the most.

b. The second statement says that after 12 hours of study, your grade is improving by 3.0 points per hour. Does this mean that at 12 hours of study, your grade is less than at 7 hours of study? No! It simply means that a small amount of additional study time beyond 12 hours does not result in as many extra points on your test as the same amount of time produces after you have studied only 7 hours. ∎

EXAMPLE 2   *More Interpretation*

Interpret and explain these two mathematical statements in terms of studying time and grades.

a. The derivative of G with respect to t is 0 points per hour when $t = 14$ hours.

b. The slope of the tangent line when $t = 15$ hours is approximately -2 points per hour.

Solution:

a. The first statement says that after you have studied 14 hours, your grade will no longer be improving. A glance back at Figure 3.37 shows that you have reached your best possible score; more study will not improve your grade.

b. The second statement tells you that after 15 hours of study, your grade is actually declining. Additional study will only hurt your grade.   ■

Make sure that you understand that these statements tell you nothing about what your grade is—they tell you only how quickly it is changing.

EXAMPLE 3   *Drug Concentration*

$C(h)$ is the average concentration (in nanograms per milliliter, ng/mL) of a drug in the blood stream h hours after the administration of a dose of 360 mg. On the basis of the following information, sketch a graph of $C(h)$.

$$C(0) = 125 \text{ ng/mL} \quad C'(0) = 0 \text{ ng/mL per hour}$$
$$C(4) = 215 \text{ ng/mL} \quad C'(4) = 37 \text{ ng/mL per hour}$$

The concentration of the drug is increasing most rapidly after 4 hours. The maximum concentration, 380 ng/mL, occurs after 10 hours. Between $h = 10$ and $h = 24$, the concentration declines at a constant rate of 16 ng/mL per hour. The concentration after 24 hours is 31 ng/mL higher than it was when the dose was administered.

Solution: The information about $C(h)$ at various values of h simply locates points on the graph of $C(h)$. Plot the points $(0, 125)$, $(4, 215)$, $(10, 380)$, and $(24, 156)$.

Because $C'(0) = 0$, the curve has a horizontal tangent at $(0, 125)$. The point of most rapid increase, $(4, 215)$, is an inflection point. The graph is concave up to the left of that point and concave down to the right. The maximum concentration occurs after 10 hours, so the highest point on the graph of $C(h)$ is $(10, 380)$. Concentration declining at a constant rate between $h = 10$ and $h = 24$ means that that portion of C is a line with slope $= -16$.

One possible graph is shown in Figure 3.38. Compare each statement about $C(h)$ to the graph.

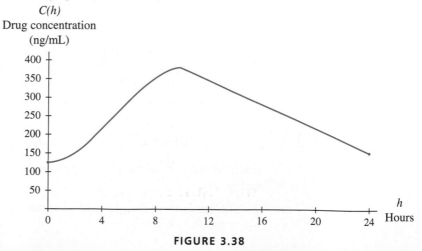

FIGURE 3.38

Note in Figure 3.38 that the tangent line to $C(h)$ at $h = 10$ does not exist. We therefore cannot assign a value to $C'(10)$ because $C'(h)$ does not exist at $h = 10$. However, the maximum concentration occurs at 10 hours, and on the basis of that, we can estimate that the rate of change at that time is zero (even though there is not a horizontal tangent line at $h = 10$ on the graph in Figure 3.38).   ■

Approximating with Derivatives

Remember that derivatives are simply slopes of tangent lines. Return to Figure 3.36, and consider the portion of the grade function graph between 8 and 15 hours studied and the tangent line at $t = 11$ hours (see Figure 3.39).

If we magnify the boxed-in portion of this graph (between $t = 11$ and $t = 12$ hours), as seen in Figure 3.40, then we obtain the view of the grade function graph and tangent line at $t = 11$ shown in Figure 3.41.

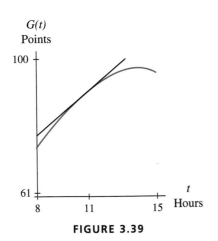

FIGURE 3.39

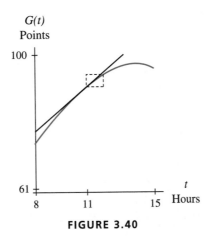

FIGURE 3.40

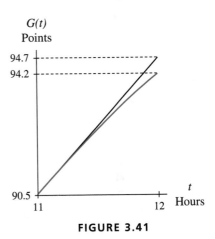

FIGURE 3.41

It so happens that after 11 hours, the derivative (slope) is 4.2 points per hour, and the grade is 90.5 points. What is the grade after 12 hours? It is tempting to reason that if the grade after 11 hours is 90.5 and is increasing by 4.2 points per hour, then after 1 more hour of study, the grade would be $90.5 + 4.2 = 94.7$ points. However, this is not correct, because as Figure 3.41 shows, the grade after 12 hours is 94.2 points. It is the tangent line, not the grade graph, that reaches 94.7 points at 12 hours.

It is common practice to use tangent lines to estimate function outputs, but this must be done carefully. For example, it is certainly proper to say that, on the basis of a score of 90.5 points and a slope of 4.2 points per hour at 11 hours, the grade will be *approximately* $90.5 + 4.2 = 94.7$ points at 12 hours. You may also say that at 11.5 hours, the grade will be *approximately* $90.5 + 4.2\left(\frac{1}{2}\right) = 92.6$ points. But you must also be certain, in making such statements, to make it clear that the values are only *approximations* and are not exact values. As you can see from the graph in Figure 3.41, such approximations should be made only at points that are relatively close to the point of tangency. We shall have more to say about approximating with derivatives in a later chapter.

EXAMPLE 4   *Unemployment Rates*

A graph modeling[22] the unemployment rates in Somerset County, New Jersey, from 1983 through 1992 is shown in Figure 3.42.

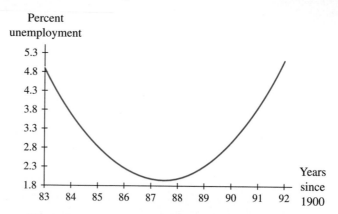

Unemployment Rates in Somerset County, New Jersey

FIGURE 3.42

a. Sketch the tangent line at $x = 85$, and estimate the slope.

b. Estimate the unemployment rate in 1985.

c. Using only your answers to parts *a* and *b*, estimate the unemployment rate in 1986.

Solution:

a.

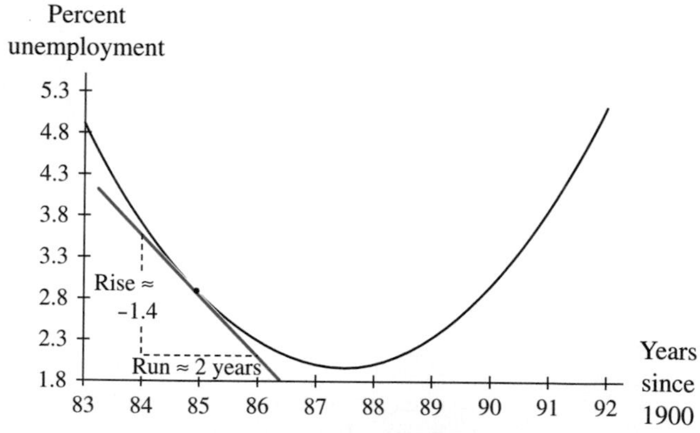

Tangent Line at 1985

FIGURE 3.43

$$\text{Slope} = \frac{\text{rise}}{\text{run}} \approx \frac{-1.4 \text{ percentage points}}{2 \text{ years}} = -0.7\% \text{ points per year}$$

b. From the graph, it appears that unemployment was approximately 2.8% in 1985.

22. Based on data from the New Jersey Department of Labor.

c. The slope at $x = 85$ is approximately -0.7% points per year, so between 1985 and 1986, the unemployment rate dropped approximately 0.7% points.

$$2.8\% - 0.7\% = 2.1\% \text{ unemployment in } 1986$$

In 1986, the unemployment rate was approximately 2.1%.   ∎

You should be aware that in part *c* of Example 4, we were using the tangent line at $x = 85$ to estimate the unemployment rate in 1986. Figure 3.44 shows a close-up of the graph and tangent line near the point of tangency. Because the tangent line lies below the graph, the approximation that the tangent line yields underestimates the actual value. The unemployment rate, according to the graph, was close to 2.3%.

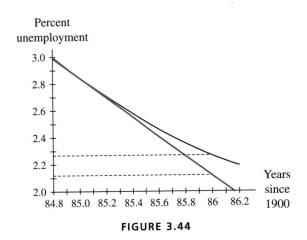

FIGURE 3.44

Does *Instantaneous* Refer to Time?

We saw that the instantaneous rate of change at a point P on the graph of a continuous function is the slope of the tangent line to the graph at P. If the function inputs are measured in units of time, then it is certainly natural to use the word *instantaneous* when describing rates of change, because each point P on the graph of the function corresponds to a particular instant in time. In fact, the use of the word *instantaneous* in this context arose precisely from the historical need to understand how rapidly distance traveled changes as a function of time. Today, we are accustomed to referring to the rate of change of distance traveled as a function of time as *speed*.

You should be aware, however, that the use of the word *instantaneous* in connection with rates of change does not necessarily mean that time units are involved. For example, suppose that a graph depicts profit (in dollars) resulting from the sale of a certain number of used cars. In this case, the slope of the tangent line at any particular point (the instantaneous rate of change) expresses how rapidly profit is changing per car. The units of change are dollars per car; no time units are involved.

You should also remember that instantaneous rates of change, like average rates of change, are always expressed in output units per input unit. Without proper units, a number that purports to describe a rate of change is meaningless.

EXAMPLE 5   *Temperature in the Polar Night Region*

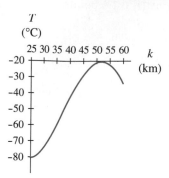

FIGURE 3.45

The graph in Figure 3.45[23] shows the temperature, T, of the polar night region (in °C) as a function of the number of kilometers above sea level, k.

a. Sketch the tangent line at 45 km, and estimate its slope. Label the answer with units.

b. What is the derivative notation for your answer to part *a*? Write a sentence interpreting the meaning of the slope in context.

Solution:

a.

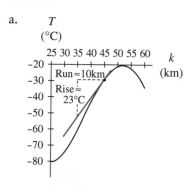

Tangent Line at 45km

FIGURE 3.46

$$\text{Slope} = \frac{\text{rise}}{\text{run}} \approx \frac{23°C}{10 \text{ km}} = 23°C \text{ per kilometer}$$

b. Correct derivative notations include $\frac{dT}{dk}$, $\frac{d}{dk}[T(k)]$, and $T'(k)$. At 45 km above sea level, the temperature of the atmosphere is increasing 2.3 °C per kilometer. In other words, the temperature changes by approximately 2.3 °C between 45 and 46 km above sea level.   ∎

3.4   Concept Inventory

■ Derivative = rate of change = slope of tangent line

■ Derivative of f with respect to $x = \frac{df}{dx} = f'(x) = \frac{d}{dx}[f(x)]$

■ Interpreting derivatives

■ Approximating with derivatives

3.4   Activities

1. Suppose that $P(t)$ equals the number of miles from an airport that a plane has flown after t hours.

a. What are the units on $\frac{d}{dt}[P(t)]$?

b. What common word do we use for $\frac{d}{dt}[P(t)]$?

2. Let $B(t)$ be the balance in a mutual fund t years after the initial investment. Assume that no deposits or withdrawals are made during the investment period.

a. What are the units on $\frac{dB}{dt}$?

b. What is the financial interpretation of $\frac{dB}{dt}$?

3. Let $W(t)$ be the number of words per minute (wpm) that a student in a typing class can type after t weeks in the course.

a. Is it possible for $W(t)$ to be negative? Explain.

b. What are units of $W'(t)$?

23. "Atmospheric Exchange Processes and the Ozone Problem," in *The Ozone Layer*, ed. Asit K. Biswas, Institute for Environmental Studies, Toronto. Published for the United Nations Environment Program by Pergamon Press, Oxford, 1979.

c. Is it possible for $W'(t)$ to be negative? Explain.

4. Imagine that $C(f)$ equals the number of bushels of corn produced on a tract of farm land when f pounds of fertilizer are used.

   a. What are the units on $C'(f)$?

   b. Is it possible for $C'(f)$ to be negative? Explain.

   c. Is it possible for $C(f)$ to be negative? Explain.

5. Suppose that $F(p)$ is the weekly profit (in thousands of dollars) that an airline makes on its Boston to Washington D.C. flights when the ticket price is p dollars. Interpret the following:

   a. $F(65) = 15$

   b. $F'(65) = 1.5$

   c. $\frac{dF}{dp} = -2$ when $p = 90$

6. Let $T(p)$ be the number of tickets from Boston to Washington D.C. that a certain airline sells in 1 week when the price of each ticket is p dollars. Interpret the following:

   a. $T(65) = 1750$

   b. $T'(65) = -20$

   c. $\frac{dT}{dp} = -2$ when $p = 95$

7. On the basis of the following information, sketch a possible graph of $t(x)$.

   ■ $t(3) = 7$

   ■ $t(4.4) = t(8) = 0$

   ■ $\frac{dt}{dx} = 0$ at $x = 6.2$

   ■ The graph of $t(x)$ has no concavity changes.

8. Using the information that follows, sketch a possible graph of $m(t)$.

   ■ $m(0) = 3$

   ■ $\frac{d}{dt}[m(t)] = 0.34$

9. Suppose that $W(t)$ equals your weight t weeks after you begin a diet. Interpret the following:

   a. $W(0) = 167$

   b. $W(12) = 148$

   c. $\frac{dW}{dt} = -2$ when $t = 1$

   d. $\frac{dW}{dt} = -1$ when $t = 9$

   e. $W'(12) = 0$

   f. $W'(15) = 0.25$

g. On the basis of the information in parts a through f, sketch a possible graph of $W(t)$.

10. Suppose that $G(v)$ equals the fuel efficiency in miles per gallon (mpg) of a car going v miles per hour. Give the practical meaning of the following statements.

    a. $G(55) = 32.5$

    b. $\frac{dG}{dv} = -0.25$ when $v = 55$

    c. $G'(45) = 0.15$

    d. $\frac{d}{dv}[G(51)] = 0$

11. $P(b)$ is the percentage of all births to single mothers in the United States in year b from 1940 through 1993. Using the following information,[24] sketch a graph of $P(b)$.

    ■ $P(1940) \approx 4\%$

    ■ $P'(b)$ is never zero.

    ■ $P(b) \approx 12\%$ when $b = 1970$

    ■ $P(1993)$ is about 17 percentage points more than $P(1970)$.

    ■ The average rate of change of $P(b)$ between 1970 and 1980 is 0.6 percentage points per year.

    ■ The line tangent to the graph of $P(b)$ lies below the graph at all points between 1940 and 1993.

12. $E(x)$ is the public day school enrollment (in millions of students) in the United States for secondary school pupils x years since 1940.

    a. Use the following information[25] to sketch a graph of $E(x)$.

       ■ $E(10) = 5.7$ million students

       ■ $E(30) = 13$ million students

       ■ $E'(7) = 0$ million students per year

       ■ The slope of the secant line joining $E(0)$ and $E(30)$ is 0.214.

       ■ Prior to 1990, the graph peaks in June of 1983, when approximately 16.7 million students were enrolled in secondary schools.

       ■ Between 1940 and 1990, enrollment was increasing the most rapidly at $x = 25$.

    b. What does your graph in part a indicate will happen to the number of secondary students enrolled in public day schools past 1990? Do

---

24. Based on data from L. Usdansky, "Single Motherhood: Stereotypes vs. Statistics," *The New York Times*, February 11, 1996, Section 4, page E4.

25. Based on data appearing in *Datapedia of the United States*, Bernan Press, 1994.

you feel this prediction is valid? Explain.

13. Let $P(x)$ be the profit in dollars that a fraternity makes selling x T-shirts.

   a. Is it possible for $P(x)$ to be negative? Explain.

   b. Is it possible for $P'(x)$ to be negative? Explain.

   c. If $P'(200) = -1.5$, is the fraternity losing money? Explain.

14. Let $M(t)$ be the number of members in a political organization t years after its founding. What are the units on $\frac{d}{dt}[M(t)]$?

15. Let $D(r)$ be the time in years that it takes for an investment to double if interest is continuously compounded at an annual rate of $r\%$. (Here r is expressed as a percentage, not a decimal.)

   a. What are the units on $\frac{dD}{dr}$?

   b. Why does it make sense that $\frac{dD}{dr}$ is always negative?

   c. Give the practical interpretation of the following:

   i. $D(9) = 7.7$

   ii. $\frac{dD}{dr} = -2.77$ when $r = 5$

   iii. $\frac{dD}{dr} = -0.48$ when $r = 12$

16. Let $U(t)$ be the number of people unemployed in a country t months after the election of a new president.

   a. Draw and label an input/output diagram for $U(t)$.

   b. Is $U(t)$ a function? Why or why not?

   c. Interpret the following facts about $U(t)$ in statements describing the unemployment situation.

   i. $U(0) = 3,000,000$

   ii. $U(12) = 2,800,000$

   iii. $U'(24) = 0$

   iv. $\frac{dU}{dt} = 100,000$ when $t = 36$

   v. $\frac{dU}{dt} = -200,000$ when $t = 48$

   d. On the basis of the information in part c, sketch a possible graph of the number of people unemployed during the first 48 months of the president's term. Label numbers and units on the axes.

17. The graph[26] in Figure 3.4.1 depicts the average salaries (in thousands of dollars) of full professors in the United States.

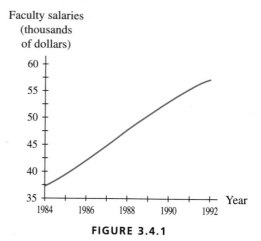

Faculty salaries (thousands of dollars)

FIGURE 3.4.1

   a. Sketch a secant line connecting the points in 1986 and 1992. What information does the slope of this line give?

   b. Sketch a line tangent to the curve in 1986. What information does the slope of this line give?

   c. Estimate the derivative in 1986. Interpret your answer.

   d. Estimate the rate at which the salary of full professors was rising in 1992.

18. The scatter plot and graph in Figure 3.4.2 depict the number of customers that a fast-food restaurant serves each hour on a typical weekday.

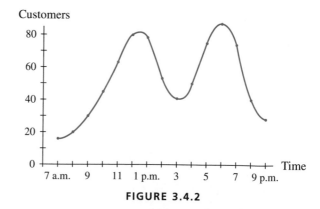

Customers

FIGURE 3.4.2

   a. Estimate the average rate of change between 7 a.m. and 11 a.m. Interpret your answer.

   b. Estimate the instantaneous rate of change at 4:30 p.m. Interpret your answer.

   c. List the factors that might affect the accuracy of your answers to parts a and b.

26. Based on data from the *1992 Information Please Almanac* (Boston: Houghton Mifflin).

19. Refer once more to the function $G(t)$, your grade out of 100 points on the next calculus test when you study t hours during the week before the test. The graph of $G(t)$ is shown in Figure 3.4.3.

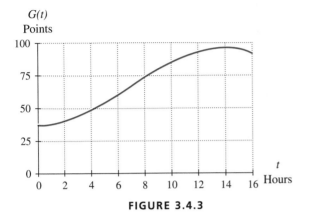

FIGURE 3.4.3

a. Carefully draw tangent lines at 4 hours and 11 hours. Estimate the slope of each tangent line.

b. Compare your answers with the slopes given on pages 181 and 183. How accurate are your estimates?

c. Estimate the average rate of change between 4 hours and 10 hours. Interpret your answer.

20.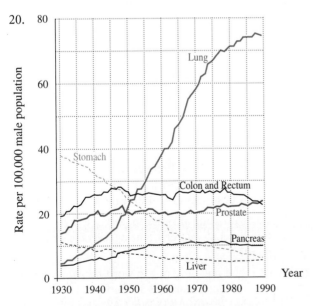

Cancer Death Rates by Site, Males, United States, 1930-89

Rates are adjusted to the 1970 US census population

Source: American Cancer Society, available on reproduction sheet (5005.93)

FIGURE 3.4.4

a. Figure 3.4.4 is a graph of rates of death from cancer among U.S. males.
   i. Estimate how rapidly the number of deaths due to lung cancer was increasing in 1970 and in 1980.
   ii. Estimate the rate of change of deaths due to liver cancer in 1980.
   iii. Estimate the slope of the stomach cancer curve in 1960.

b. Describe in detail the behavior of the lung cancer curve from 1930 to 1990. Explain why the lung cancer curve differs so radically from the other curves shown.

c. List as many factors as you can that might affect a cancer death rate curve.

21.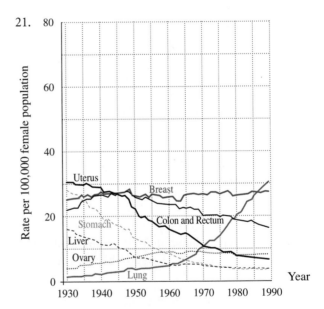

Cancer Death Rates by Site, Females, United States, 1930-89

Rates are adjusted to the 1970 US census population

Source: American Cancer Society, available on reproduction sheet (5005.93)

FIGURE 3.4.5

a. Figure 3.4.5 is a graph of rates of death from cancer among U.S. females. Estimate how rapidly the number of deaths due to lung cancer was increasing in 1970 and 1980.

b. Compare the lung cancer death rate curves for males and females. What do you think will happen to the two curves during the next 30 years?

22.

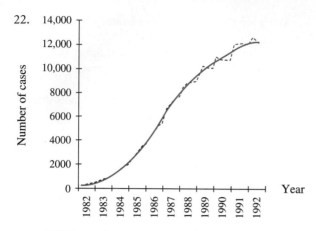

**AIDS cases by quarter-year of diagnosis, adjusted
for reporting delays, January 1982 through
September 1992, United States**

Source: HIV/AIDS Surveillance Report (February 1993)

FIGURE 3.4.6

The function $C(t)$ equals the number of AIDS cases
diagnosed each quarter in year t as represented by
the smoothed curve shown in Figure 3.4.6.

a. Estimate and interpret the derivative $\frac{dC}{dt}$ at the
end of the first quarter of 1984.

b. Estimate and interpret the derivative $C'(1987)$.

c. Estimate and interpret $C(1987)$.

d. Estimate the derivative of $C(t)$ at the end of the
first quarter of 1992.

e. Give an approximation for the number of AIDS
cases diagnosed at the end of the first quarter of
1993.

23. The United States loses wetlands every year as they
are drained and used for farmland. Table 3.12[27]
shows the number of acres of wetlands lost each
year between 1987 and 1994.

Use a symmetric difference quotient to estimate
the rate of change of wetlands losses in 1992. Interpret your answer.

TABLE 3.12

Year	Wetlands losses (thousands of acres)
1987	135
1988	131
1989	126
1990	121
1991	116
1992	112
1993	107
1994	102

24. Table 3.13[28] shows batting averages for a major
league baseball player from 1985 through 1995.

TABLE 3.13

Year	Batting average
1985	0.260
1986	0.271
1987	0.282
1988	0.295
1989	0.310
1990	0.317
1991	0.320
1992	0.323
1993	0.325
1994	0.324

a. Use symmetric difference quotients to estimate
the rates of change of this player's batting average in 1986 and 1992. Interpret your answers.

b. What do your answers to part *a* tell you about
this baseball player's batting average improvement?

25. What is the meaning of the word *derivative* in the
world's financial markets? To find out, see the article entitled "Derivatives? What Are Derivatives?" in
Newsweek, March 13, 1995, page 50.

---

27. *The True State of the Planet*, ed. Ronald Bailey (New York: The Free Press for the Competitive Enterprise Institute, 1995).

28. From a Fleer baseball card of Kevin Mitchell.

■ **3.5**   Percentage Change and Percentage Rates of Change

Imagine that you are a stockholder in a company and that at a quarterly meeting, the CEO reports a drop in profit of $1 million over the last quarter. How concerned should you be?

If the company has annual profits in the billions of dollars, a drop of $1 million is hardly something to be very concerned about. However, if the company is a small business with an annual profit of $1.2 million, a quarterly drop in profit of $1 million is devastating. In order for the drop in profit to be meaningful, it must be accompanied by more information about profit.

A common way to describe change in a meaningful context is to express the change as a percentage of a total. Consider Table 3.14, which shows quarterly earnings for a business.

TABLE 3.14

Quarter ending	Earnings (millions)
March 1994	$27.3
June 1994	$28.9
Sept. 1994	$24.6
Dec. 1994	$32.1
March 1995	$29.4
June 1995	$27.7

Note that from the third quarter of 1994 to the fourth quarter of the same year, earnings rose by $7.5 million. To calculate the corresponding percentage change, simply divide the increase by the earnings in the third quarter and multiply by 100.

$$\frac{\$7.5 \text{ million}}{\$24.6 \text{ million}} \times 100\% \approx 0.305 \times 100\% = 30.5\%$$

From the third to the fourth quarter, earnings rose 30.5%.

Similarly, from the first quarter of 1995 to the second quarter of 1995, earnings fell by $1.7 million. This represents a percentage change of

$$\frac{-\$1.7 \text{ million}}{\$29.4 \text{ million}} \times 100\% \approx -0.058 \times 100\% = -5.8\%$$

During this period, earnings fell 5.8%.

In the activities at the end of this section, you will be given an opportunity to calculate percentage change using tables, graphs, and equations. Regardless of how a function is represented, **percentage change** can be found by dividing the change in a function over an interval by the function value at the beginning of the interval.

Percentage Change

$$\text{Percentage change} = \frac{\text{change in a function over an interval}}{\text{value of the function at the beginning of the interval}} \times 100\%$$

Similarly, **percentage rate of change** can be found by dividing the rate of change at a point by the function value at the same point.

Percentage Rate of Change

$$\text{Percentage rate of change} = \frac{\text{rate of change at a point}}{\text{value of the function at that point}} \times 100\%$$

EXAMPLE 1   *Sales*

The graph in Figure 3.47 shows sales (in thousands of dollars) for a small business from 1988 through 1996.

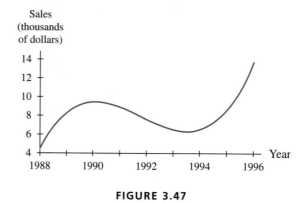

FIGURE 3.47

a. Estimate the rate of change of sales in 1992.

b. Estimate the percentage rate of change of sales in 1992.

Solution:

a. A tangent line is drawn at 1992 as shown in Figure 3.48.

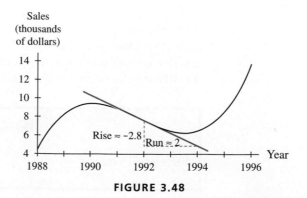

FIGURE 3.48

The slope is estimated to be

$$\frac{-\$2.8 \text{ thousand}}{2 \text{ years}} = -\$1.4 \text{ thousand per year} = -\$1400 \text{ per year}$$

In 1992, sales were falling at a rate of approximately $1400 per year.

b. We can express this as a percentage rate of change if we divide by the sales in 1992. It appears from the graph that the sales in 1992 were approximately $7.5 thousand dollars, or $7500. Therefore, the percentage rate of change in 1992 was approximately

$$\frac{-\$1400 \text{ per year}}{\$7500} \times 100\% \approx -0.187 \times 100\% \text{ per year} = -18.7\% \text{ per year}$$

In 1992, sales were falling approximately 18.7% per year. Expressing the rate of change of sales as a percentage of sales gives a much clearer picture of the impact of the decline in sales. The business was experiencing a reduction in sales of nearly 20% per year in 1992.   ■

EXAMPLE 2   *Toys R Us Growth*

Table 3.15[29] shows the number of Toys R Us stores from 1984 through 1994.

TABLE 3.15

Year	1984	1986	1988	1990	1992	1994
Stores	213	338	522	712	918	1115

a. Determine the average rate of change in the number of stores from 1984 through 1994.

b. Determine the percentage change in the number of stores from 1984 through 1994.

c. Find a model for the data.

d. Use the model to estimate the average rate of change and percentage change in the number of stores from 1989 through 1993.

Solution:

a. Average rate of change $= \dfrac{1115 - 213 \text{ stores}}{1994 - 1984}$

$$= \frac{902 \text{ stores}}{10 \text{ years}} = 90.2 \text{ stores per year}$$

b. Percentage change $= \dfrac{1115 - 213 \text{ stores}}{213 \text{ stores}} \times 100\% \approx 423\%$

From 1984 through 1994, the number of stores increased by 423%. Expressing the change as a percentage of the number of stores in 1984 gives a much clearer picture of the relative magnitude of the increase in Toys R Us stores. We see that the number of stores in 1994 was more than five times what it was in 1984.

29. Toys R Us *1995 Annual Report.*

c. The data appear to be essentially linear. A linear model for the data is

$$N(x) = 92x - 7551.67 \text{ stores}$$

where x is the number of years since 1900.

d. Average rate of change $= \dfrac{N(93) - N(89) \text{ stores}}{93 - 89 \text{ years}}$

$$= \dfrac{368 \text{ stores}}{4 \text{ years}} = 92 \text{ stores per year}$$

(Note that because the model is linear, any average rate of change will be the slope of the model—in this case, 92 stores per year.)

$$\text{Percentage change} = \dfrac{N(93) - N(89)}{N(89)} \times 100\%$$

$$= \dfrac{368 \text{ stores}}{636.33 \text{ stores}} \times 100\% \approx 57.8\%$$

From 1989 through 1993, the number of stores increased by approximately 57.8%. ■

Describing change and rates of change in terms of percentages is a useful tool that is widely applied in business and industry.

3.5 Concept Inventory

■ Percentage change

■ Percentage rate of change

3.5 Activities

1. Kelly Services saw an increase in gross profit from $184.2 million in 1984 to $463.0 million in 1994.[30] Calculate the change, average rate of change, and percentage change in Kelly's profit from 1984 through 1994.

2. Bankruptcy filings in New Hampshire fell from 3848 in 1991 to 3172 in 1995.[31] Calculate the change, average rate of change, and percentage change in bankruptcy filings from 1991 through 1995.

3. During the fourth quarter of 1994, LCI International Telecommunications generated $135 million in revenue, and revenue was increasing by approximately $12.1 million per quarter.[32]

   a. What is the rate of change of revenue for the fourth quarter of 1994?

   b. What is the percentage rate of change of revenue for the fourth quarter of 1994?

4. Upon beginning a program of weight training, a student weighs 140 pounds, and his weight is increasing by 3 pounds per month.

   a. What is the rate of change of the student's weight with respect to time?

   b. What is the percentage rate of change of the student's weight?

5. Table 3.16[33] gives imports of natural gas (in billions of cubic feet) for selected years from 1970 through 1991.

---

30. Kelly Services *Annual Report*, 1994.
31. *San Francisco Examiner*, February 20, 1996, A-14.
32. LCI *Annual Report*, 1994.
33. *Statistical Abstract*, 1994.

TABLE 3.16

Year	Natural gas imports (billions of cubic feet)
1970	821
1973	1033
1975	953
1980	985
1983	920
1984	843
1985	950
1986	750
1987	993
1988	1294
1989	1382
1990	1532
1991	1693

a. Use the data to calculate the change in natural gas imports during the following time intervals: (i) 1970 through 1986, (ii) 1986 through 1991, and (iii) 1989 through 1991.

b. Express the changes in part *a* as percentage changes.

c. Use the data to calculate the average rates of change for the same time intervals. Discuss the limitations of the average rate of change as a description of the change in natural gas imports between 1970 and 1986.

d. What would be needed in order to calculate the instantaneous rate of change of imports of natural gas in 1990?

e. Use a symmetric difference quotient to estimate the rate of change of imports of natural gas in 1990.

6. The purchasing power of the dollar as measured[34] by producer prices from 1978 through 1991 is given in Table 3.17. (In 1982, one dollar was worth $1.00.)

a. Use the data to calculate the change in the purchasing power of the dollar during the following time intervals: (i) 1978 through 1991, (ii) 1984 through 1985, (iii) 1985 through 1986, and (iv) 1987 through 1991.

TABLE 3.17

Year	Purchasing power of $1
1978	1.43
1979	1.29
1980	1.14
1981	1.04
1982	1.00
1983	0.98
1984	0.96
1985	0.96
1986	0.97
1987	0.95
1988	0.93
1989	0.88
1990	0.84
1991	0.82

b. Express the changes in part *a* as percentage changes. Write a sentence interpreting each answer.

c. Calculate the average rate of change and the percentage change in purchasing power of the dollar from 1980 through 1990.

d. What is the practical meaning to consumers if the percentage change in the purchasing power of the dollar is negative? positive? zero?

7. Table 3.18[35] gives the number of imported passenger cars (in thousands) sold in the United States between 1984 and 1992.

TABLE 3.18

Year	Number of imported passenger cars sold (thousands)
1984	2439
1985	2838
1986	3245
1987	3196
1988	3004
1989	2699
1990	2403
1991	2058
1992	1938

34. *Statistical Abstract*, 1992.
35. *Statistical Abstract*, 1992 and 1994.

a. Calculate the percentage change in imported-car sales from 1984 through 1992.

b. Calculate the average rate of change in imported-car sales from 1984 through 1992.

c. Find an appropriate model for the data. Determine the percentage change and average rate of change in the imported-cars sales predicted by the model from 1984 through 1992. Compare your answer to the values obtained in parts *a* and *b*.

d. How well do the statistics in parts *a*, *b*, and *c* describe what happened to imported-car sales in the United States from 1984 through 1992?

8. The amount of money (in billions of dollars) spent on pollution control from 1983 through 1991 in the United States is given in Table 3.19.[36]

TABLE 3.19

Year	Amount (billions of dollars)
1983	60.002
1984	66.445
1985	70.941
1986	74.178
1987	76.672
1988	81.081
1989	85.407
1990	89.996
1991	91.456

a. Find the change in the amount spent on pollution control from 1983 through 1991.

b. Find the percentage change from 1983 through 1991.

c. Find the average rate of change from 1983 through 1991.

d. What would be needed to calculate the instantaneous rate of change of spending on pollution control in 1986?

e. Use a symmetric difference quotient to estimate the rate of change of spending on pollution control in 1986.

9. The number of general aviation aircraft accidents from 1975 through 1992 can be modeled[37] by

$$\text{Number of accidents} = -123.7746x + 4057.6633$$

where x is the number of years since 1975.

a. Calculate the change and percentage change in accidents from 1976 through 1992.

b. Interpret the change and percentage change from part *a*.

c. Calculate the average rate of change from 1976 through 1992.

10. U.S. factory sales of electronics (in billions of dollars) from 1986 through 1990 can be modeled[38] by

$$\text{Sales} = 220 - 44.43e^{-0.3912t} \text{ billion dollars}$$

where t is the number of years since 1986.

a. Calculate the change and percentage change in factory sales of electronics from 1986 through 1990.

b. Calculate the average rate of change from 1986 through 1990.

11. The number of personal computers (in millions of units) sold to colleges and universities from 1981 through 1988 can be modeled[39] by

$$\text{Number of PCs} = 0.00595238x^2 + 0.04095238x + 0.025 \text{ million PCs}$$

where x is the number of years since 1981.

a. Find the change and percentage change in PCs sold to colleges and universities from 1984 through 1988.

b. Find the average rate of change from 1984 through 1988.

12. The percentage of married couples with unmarried children living at home from 1960 through 1990 is shown in the graph[40] in Figure 3.5.1.

36. *Statistical Abstract*, 1994.
37. Based on data from *Statistical Abstract*, 1994.
38. Based on data from *Statistical Abstract*, 1992.
39. Based on data from *Statistical Abstract*, 1992.
40. Based on data from *Statistical Abstract*, 1994.

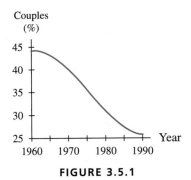

FIGURE 3.5.1

a. Estimate the change and percentage change from 1970 through 1990.

b. Sketch a secant line between the points at 1970 and 1990. What information does the slope of this line give?

c. Sketch a tangent line to the graph at 1970. What information does the slope of this line give?

d. Estimate the average rate of change from 1970 through 1990.

e. Estimate the rate of change and percentage rate of change in 1970.

13. The death rates (deaths per 1000 people of age x) for people of ages 40 through 65 in the United States in 1991 are shown in the graph[41] in Figure 3.5.2.

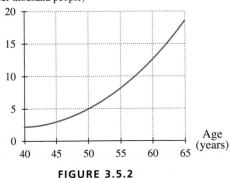

FIGURE 3.5.2

a. Sketch a secant line between the points at $x = 50$ and $x = 60$. What information does the slope of this line give?

b. Sketch a tangent line to the graph at $x = 55$. What information does the slope of this line give?

c. Estimate the rate of change and percentage rate of change in death rate for a 55-year-old person. Interpret your answers in the context of death rates.

14. The graph[42] in Figure 3.5.3 shows the number of Visa credit cards (in millions) from 1985 through 1992.

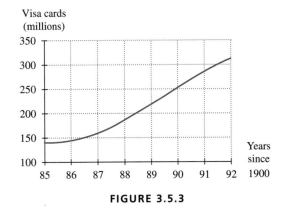

FIGURE 3.5.3

a. Estimate the change, percentage change, and average rate of change in the number of Visa cards from 1985 through 1992.

b. Estimate the rate of change and percentage rate of change in the number of Visa cards in 1987. Interpret your answer.

c. From 1987 through 1988, did the number of credit cards change by more or less than your answer in part *b*? Explain your reasoning.

d. Estimate the rate of change and percentage rate of change in the number of credit cards in 1991. Interpret your answer.

e. From 1991 through 1992, did the number of credit cards change by more or less than your answer in part *d*? Explain your reasoning.

f. In approximately what year was the number of cards increasing most rapidly? What is the mathematical name for this point on the graph?

15. The graph[43] in Figure 3.5.4 shows the number of births (in thousands) in the United States to

41. Based on data from *Statistical Abstract*, 1994.
42. Source: Card Companies, *Credit Card News*.
43. Based on data from *Statistical Abstract*, 1992.

women under 15 years of age from 1970 through 1991.

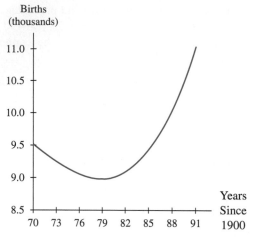

FIGURE 3.5.4

a. Estimate the change and percentage change in the number of such births from 1980 through 1990.

b. Estimate the rate of change and percentage rate of change in 1985.

16. Examine a news magazine, a financial magazine, or the business section of a newspaper, and find at least three statements that contain information written as a percentage change or percentage rate of change. Be sure to give the source for each statement.

Chapter 3   Summary

This chapter is devoted to describing change: the underlying concepts, the language, and proper interpretations.

Average Rates of Change

The change in a quantity over an interval is a difference of output values. Apart from describing the actual change in a quantity that occurs over an interval, the simplest description of change is the average rate of change over an interval, defined as

$$\text{Average rate of change} = \frac{\text{change over an interval}}{\text{length of the interval}}$$

In the reporting of average rates of change, the word *average* must be used, and it must be accompanied by a numerical value expressed in terms of output units per input unit.

The numerical description of an average rate of change has an associated geometric interpretation—namely, the slope of the secant line joining two points on the graph that correspond to the two endpoints of the interval. Such slopes can be determined from discrete scatter plots, continuous models, or piecewise continuous models.

Instantaneous Rates of Change

Whereas average rates of change indicate how rapidly a quantity changes (on average) over an interval, instantaneous rates of change indicate how rapidly a quantity is changing at a point. Familiar examples include speed (how rapidly distance traveled is changing as a function of time at an instant) and marginal profit from economics (how rapidly profit is changing as a function of the number of items sold at a particular sales level).

The instantaneous rate of change at a point on a graph is simply the slope of the tangent line to the graph at that point. It describes how quickly the output is increasing or decreasing at that point. Given this geometric perspective, it is clear that the use of the word *instantaneous* does not necessarily mean that time units are involved (*i.e.*, we may not be speaking of change at an instant in time.)

Tangent Lines

Average rates of change are slopes of secant lines. However, instantaneous rates of change are slopes of tangent lines.

What are tangent lines? The tangent line to a graph at a point P is the limiting position of nearby secant lines—that is, secant lines through P and nearby points on the graph. With the exceptions of the tangent line to a continuous graph at an inflection point and the tangent line to each point on a line, tangent lines to a graph do not cut through the graph at the point of tangency. Rather, they include the point of tangency and lie completely on one side of the graph near the point of tangency. A tangent line reflects the tilt, or slope, of the graph at the point of tangency.

Tangent lines can be used to determine the rate of change of most points on continuous graphs, the notable exceptions being at sharp points and at points where the tangent is vertical. Tangent lines can also be used to measure rates of change for piecewise continuous functions, though not usually at the division points.

Derivatives

The word *derivative* has a totally different meaning in mathematics from its meaning in the world's financial markets. In mathematics, *derivative* is the calculus term for (instantaneous) rate of change. Accordingly, all of the following terms are synonymous: derivative, instantaneous rate of change, rate of change, slope of the curve, and slope of the tangent line to the curve.

Three common ways of symbolically referring to the derivative of a function, $G(x)$, are $\frac{dG}{dx}$, $\frac{d}{dx}[G(x)]$, and $G'(x)$. You should remember to include the proper units on derivatives: output units per input unit. Without the correct units, a derivative is a meaningless number.

A continuous or piecewise continuous model is needed to calculate values of derivatives. However, if we are given a data set, then we can quickly estimate a rate of change by finding the slope between the nearest points an equal distance away from, and on either side of, our point of interest. This slope, which is called a symmetric difference quotient, usually gives a good approximation to the rate of change.

Percentage Change and Percentage Rates of Change

Because some expressions of change do not always adequately describe the magnitude of the change, it is common, especially in business, economics, finance, and government, to describe change in terms of percent.

Percentage change is simply the change over an interval divided by the value at the beginning of the interval, where the result is expressed as a percent. It is readily understood by people who have never studied calculus. On the other hand, percentage rate of change is a concept of calculus, defined as the rate of change at a point divided by the value of the function at that point, where the result is expressed as a percent per input unit.

Chapter 3   Review Test

1. Answer the following questions about the graph shown in Figure 3.49.

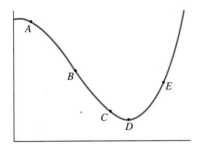

FIGURE 3.49

a. List the labeled points at which the slope appears to be
   (i) negative, (ii) positive, and (iii) zero.

b. If *B* is the inflection point, what is the relationship between the steepness at *B* and the steepness at the points *A*, *C*, and *D* on the graph?

c. For each of the labeled points, will a tangent line at the point lie above or below the graph?

d. Sketch tangent lines at points *A*, *B*, and *E*.

e. Suppose the graph represents the speed of a roller coaster (in feet per second) as a function of the number of seconds after the roller coaster reached the bottom of the first hill.
   i. What are the units on the slopes of tangent lines? What common word is used to describe the quantity measured by the slope in this context?
   ii. When, according to the graph, was the roller coaster slowing down?
   iii. When, according to the graph, was the roller coaster speeding up?
   iv. When was the roller coaster's speed the slowest?
   v. When was the roller coaster slowing down most rapidly?

2. Enrollment in a university's Spanish courses for 4 years is shown in Table 3.20.[44]

TABLE 3.20

School year	Number of students
1991–92	1444
1992–93	1501
1993–94	1511
1994–95	1722

a. Find the average rate of change in enrollment between the 91–92 school year and the 94–95 school year. Interpret your answer.

b. Find the percentage change in enrollment between the last 2 years given in the table. Interpret your answer.

c. Do you think the answers to parts *a* and *b* indicate an increased interest in Spanish at this university? Explain.

3. The graph in Figure 3.50 shows a model[45] of the average weekly earnings for all private-sector workers between 1986 and 1994 in constant 1982 dollars.

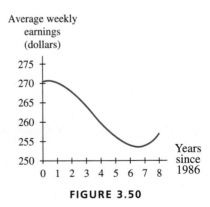

FIGURE 3.50

---

44. University of Nevada, Reno, Department of Foreign Languages and Literature.
45. Based on data from the Bureau of Labor Statistics.

a. Sketch a secant line between 1986 and 1989. Describe the information that the slope of this line provides.

b. Sketch a tangent line in 1993. Describe the information that the slope of this line provides.

c. Estimate the rate of change in average weekly earnings in 1993. Interpret your answer.

d. Estimate the average rate of change between 1986 and 1989. Interpret your answer.

4. $T(x)$ is the number of seconds that it takes an average athlete to swim 100 meters free style at age x years.

a. Write sentences interpreting (i) $T(22) = 49$ and (ii) $T'(22) = -0.5$.

b. What does a negative derivative indicate about a swimmer's time?

5. The amount of money used for research projects at a university from 1987 through 1995 can be modeled[46] by

$$R(t) = 6.062t + 21.753 \text{ million dollars}$$
$$t \text{ years after 1987}$$

a. What is the rate of change of $R(t)$?

b. Find the change, average rate of change, and percentage change in research funds indicated by the model between 1990 and 1995.

c. Assess the validity of the following statements.

   i. The rate of change of a linear model is always constant.

   ii. The percentage rate of change of a linear model is always constant.

   iii. The average rate of change of a linear model is always constant.

---

46. Based on data from University of Nevada, Reno.

Project 3.1

Fee-Refund Schedules

Setting

Some students at many colleges and universities enroll in courses and then later withdraw from them. Such students may have part-time status upon withdrawing. Part-time students have begun questioning the fee-refund policy, and a public debate is taking place. The Higher Education Commission has scheduled hearings on the issue. The Board of Trustees has hired your firm as consultants to help them prepare their presentation. Recently, the student senate passed a resolution condemning the current fee-refund schedule. Then, the associate vice president issued a statement claiming that further erosion of the university's ability to retain student fees would reduce course offerings.

Tasks

1. *Preparing Alternative Plans* Examine the current fee-refund schedule for your college or university. Present a graph and formula for the current refund schedule. Critique the refund schedule.

   Create alternative fee-refund schedules that include at least two quadratic plans (one concave up and one concave down), an exponential plan, a logistic plan, a no-refund plan, and a complete-refund plan. (Hint: Linear models have constant first differences. What is true about quadratic and exponential models?) For each plan, present the refund schedule in a table, in a graph, and with an equation. Critique each plan from the students' viewpoint and from that of the administration.

   Select the nonlinear plan that you believe to be the best choice from both the students' and the administration's perspective. Outline the reasons for your choice.

2. *Discussing Rates of Change* Estimate the rate of change of your model for withdrawals after 1 week, 3 weeks, and 5 weeks. Interpret the rates of change in this context. How might the rate of change influence the administration's view of the model you chose? Would the administration consider a different model more advantageous? If so, why? Why did you not propose it as your model of choice?

Reporting

1. Prepare a written report of your results for the Board of Trustees. Include scatter plots, models, and graphs. Include in an appendix the reasoning that you used to develop each of your models.

2. Prepare a press release for the college or university to use when it announces the adoption of your plan. The press release should be succinct and should answer the questions Who, What, When, Where, and Why. Include the press release in your report to the Board.

3. (Optional) Prepare a brief (15-minute) presentation on your work. You will be presenting it to members of the Board of Trustees of your college or university.

Project 3.2

Doubling Time

Setting

Doubling time is defined as the time it takes for an investment to double. Doubling time is calculated by using the compound interest formula $A = P\left(1 + \frac{r}{n}\right)^{nt}$ or the continuously compounded interest formula $A = Pe^{rt}$. An approximation to doubling time can be found by dividing 72 by $100r$. This approximating technique is known as the **Rule of 72**.

Dr. C. G. Bilkins, a nationally known financial guru, has been criticized for giving false information about doubling time and the Rule of 72 in seminars. Your team has been hired to provide mathematically correct information for Dr. Bilkins to use in future seminar presentations.

Tasks

1. *Evaluating Doubling-Time Rules* Construct a table of doubling times for interest rates from 2% to 20% (in increments of 2%) when interest is compounded annually, semiannually, quarterly, monthly, and daily. Construct a table of doubling-time approximations for interest rates of 2% through 20% when using the Rule of 72. Devise similar rules for 71, 70, and 69. Then construct tables for these rules. Examine the tables and determine the best approximating rule for interest compounded semiannually, quarterly, monthly, and daily. Justify your choices. For each interest compounding listed above, compare percent errors when using the Rule of 72 and when using the rule you choose. Percent error is $100 \times (estimate - true\ value)/(true\ value)$. Comment on when the rules overestimate, when they underestimate, and which is preferable.

2. *Evaluating the Sensitivity of Doubling Time* Dr. Bilkins is interested in knowing how sensitive doubling time is to changes in interest rates. Estimate rates of change of doubling times at 2%, 8%, 14%, and 20% when interest is compounded quarterly. Interpret your answers in a way that would be meaningful to Dr. Bilkins, who knows nothing about calculus.

Reporting

1. Prepare a written report for Dr. Bilkins in which you discuss your results in Tasks 1 and 2. Be sure to discuss whether Dr. Bilkins should continue to present the Rule of 72 or present other rules that depend on the number of times interest is compounded.

2. Prepare a document for Dr. Bilkins's speechwriter to insert into the seminar presentation (which Dr. Bilkins reads from a Teleprompter). It should include (1) a brief summary of how to estimate doubling time using an approximation rule and (2) a statement about the error involved in using the approximation. Keep in mind that Dr. Bilkins wishes to avoid further allegations of misinforming seminar participants. Also include a brief statement summarizing the sensitivity of doubling time to fluctuations in interest rates. Include the document in your written report.

3. (Optional) Prepare a brief (15-minute) presentation of your study. You will be presenting it to Dr. Bilkins and the speechwriter. Your presentation should be only a summary, but you need to be prepared to answer any technical questions that may arise.

Determining Change: Derivatives

We have described change in terms of rates: average rates, instantaneous rates, and percentage rates. Of these three, instantaneous rates are the most important in our study of calculus.

In this chapter, we turn our attention to determining instantaneous rates of change. Because instantaneous rates of change are slopes of tangent lines, we first consider the numerical calculation of these slopes. After some initial calculations of slopes of tangent lines using slopes of approximating nearby secant lines, we draw slope graphs. Slope graphs, the graphs of certain derived functions known as derivatives, tell us a great deal about the changing behavior of the underlying functions.

We are accustomed to using equations and formulas to produce graphs, so slope graphs should be no exception. We therefore engage in numerical investigations that lead us to discover slope formulas for a variety of simple functions. We also obtain an analytic description—a formula of sorts—for the derivative of an arbitrary function.

Finally, we consider some rules for derivatives: the Sum Rule, the Chain Rule, and the Product Rule. These rules provide the foundation needed to work with more complicated functions that we often encounter in the course of real-life investigations of change.

Using calculus (specifically, derivatives), we can answer questions about instantaneous rates of change (such as velocity). During a storm, it is important for meteorologists to have answers to the following questions: (1) How much rain has fallen? (2) How quickly is rain falling? (3) How quickly is the rate of rainfall increasing? (4) What is the wind speed? (5) How quickly is wind speed changing?

At least three of these questions involve rates of change. Are there other rates of change that might be of interest to a meteorologist?

4.1   Numerically Finding Slopes

Finding Slopes by the Numerical Method

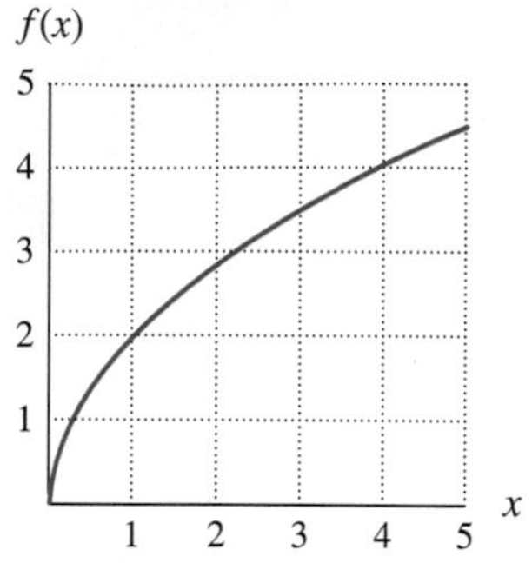

$f(x)$

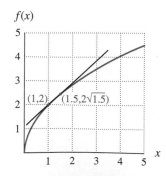

A Graph of $f(x) = 2\sqrt{x}$

FIGURE 4.1

By now you should have a firm graphical understanding of rates of change. However, sketching tangent lines is an imprecise method of determining rates of change. Although approximations are often sufficient, there are times when we need to find an exact answer.

Consider the relatively simple problem of finding the slope of $y = 2\sqrt{x}$ at $x = 1$. Part of the graph of $f(x) = 2\sqrt{x}$ is shown in Figure 4.1. Take a few moments to sketch carefully a tangent line on the graph at $x = 1$ and estimate its slope. You should find that the tangent line at $x = 1$ has slope approximately 1.

Another (more precise) method of estimating the slope of $y = 2\sqrt{x}$ at $x = 1$ uses a technique introduced in Section 3.3 of Chapter 3. Recall that the tangent line at a point is the limiting position of secant lines through the point of tangency and other increasingly close points. In other words, *the slope of the tangent line is the limiting value of the slopes of nearby secant lines.* (See Figure 4.2.)

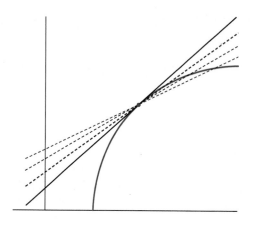

**The Tangent Line is the Limiting Position
of Nearby Secant Lines.**

FIGURE 4.2

$f(x)$

**A Secant Line Through $x = 1$
and $x = 1.5$**

FIGURE 4.3

To illustrate, we begin by finding the slope of the secant line on $y = 2\sqrt{x}$ through $x = 1$ and $x = 1.5$. (Note that $x = 1.5$ is an arbitrarily chosen value that is close to $x = 1$.) A graph of the secant line is shown in Figure 4.3. Its slope is calculated as follows:

Point at $x = 1$: $\left(1, 2\sqrt{1}\right) = (1, 2)$

Point at $x = 1.5$: $\left(1.5, 2\sqrt{1.5}\right) \approx (1.5, 2.449489743)$

$$\text{Slope} \approx \frac{2.449489743 - 2}{1.5 - 1} = 0.8989794856$$

This value is an approximation to the slope of the tangent line at $x = 1$. To obtain a better approximation, we must choose a point closer to $x = 1$ than $x = 1.5$, say $x = 1.1$. (This is also an arbitrary choice.) See Figure 4.4.

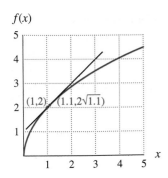

$f(x)$

A Secant Line Between
$x = 1$ and $x = 1.1$

FIGURE 4.4

Point at $x = 1$: $\left(1, 2\sqrt{1}\right) = (1, 2)$
Point at $x = 1.1$: $\left(1.1, 2\sqrt{1.1}\right) \approx (1.1, 2.097617696)$

$$\text{Slope} \approx \frac{2.097617696 - 2}{1.1 - 1} = 0.9761769634$$

This is a better approximation to the slope of the tangent line at $x = 1$ than the one from the previous calculation. To get an even better approximation, we need only choose a closer point, such as $x = 1.01$.

Point at $x = 1$: $\left(1, 2\sqrt{1}\right) = (1, 2)$
Point at $x = 1.01$: $\left(1.01, 2\sqrt{1.01}\right) \approx (1.01, 2.009975124)$

$$\text{Slope} \approx \frac{2.009975124 - 2}{1.01 - 1} = 0.9975124224$$

We also use $x = 1.001$.

Point at $x = 1$: $\left(1, 2\sqrt{1}\right) = (1, 2)$
Point at $x = 1.001$: $\left(1.001, 2\sqrt{1.001}\right) \approx (1.001, 2.00099975)$

$$\text{Slope} \approx \frac{2.00099975 - 2}{1.001 - 1} = 0.9997501248$$

As we choose points increasingly close to $x = 1$, what do you observe about the slopes of the secant lines?

$x = 1.5$	Slope ≈ 0.8989794856
$x = 1.1$	Slope ≈ 0.9761769634
$x = 1.01$	Slope ≈ 0.9975124224
$x = 1.001$	Slope ≈ 0.9997501248

The slopes of the secant lines appear to be approaching 1. This is the slope of the line tangent to the function $y = 2\sqrt{x}$ at $x = 1$. In this case, the two methods for estimating the slope of the tangent line yield similar results. However, Example 1 shows that calculating the slopes of nearby secant lines generally yields a much more precise result than sketching the tangent line and estimating its slope.

EXAMPLE 1   *The 1949 Polio Epidemic*

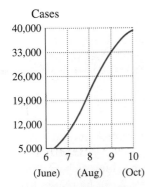

Cases

Number of Polio
Cases in 1949

FIGURE 4.5

The number of polio cases in 1949 can be modeled[1] by the equation

$$\text{Cases} = \frac{42{,}183.911}{1 + 21{,}484.253e^{-1.248911t}}$$

where $t = 1$ on January 31, 1949, $t = 2$ on February 28, 1949, etc. How rapidly was the number of polio cases growing at the end of August 1949?

Solution: First, note that the question asks for the slope of the tangent line at $t = 8$. One method of determining the slope of the tangent line is to sketch the tangent line and estimate its slope. On the graph in Figure 4.5, draw the tangent line at $t = 8$, and estimate the slope.

1. Data from *Twelfth Annual Report*, The National Foundation for Infantile Paralysis, 1949.

If you accurately sketched the tangent line and were careful in reading two points off that line, you should have found that the slope of the tangent is somewhere between 12,000 and 14,000 cases per month. To obtain a more precise estimate of the slope of the tangent line, we calculate slopes of nearby secant lines. Instead of choosing close points to the right of $t = 8$, we choose increasingly close points to both the right and the left of the point where $t = 8$: (8, 21,262.9281). You should verify each of the following computations.

Points to the left	Points to the right
Point at $t = 7.9$: (7.9, 19,946.95986)	Point at $t = 8.1$: (8.1, 22,577.56655)
Slope $= \dfrac{19{,}946.95986 - 21{,}262.9281}{7.9 - 8}$	Slope $= \dfrac{22{,}577.56655 - 21{,}262.9281}{8.1 - 8}$
Slope $\approx 13{,}159.68$	Slope $\approx 13{,}146.38$
Point at $t = 7.99$: (7.99, 21,131.22196)	Point at $t = 8.01$: (8.01, 21,394.62098)
Slope $= \dfrac{21{,}131.22196 - 21{,}262.9281}{7.99 - 8}$	Slope $= \dfrac{21{,}394.62098 - 21{,}262.9281}{8.01 - 8}$
Slope $\approx 13{,}170.62$	Slope $\approx 13{,}169.28$
Point at $t = 7.999$: (7.999, 21,249.75795)	Point at $t = 8.001$: (8.001, 21,276.09819)
Slope $= \dfrac{21{,}249.75795 - 21{,}262.9281}{7.999 - 8}$	Slope $= \dfrac{21{,}276.09819 - 21{,}262.9281}{8.001 - 8}$
Slope $\approx 13{,}170.19$	Slope $\approx 13{,}170.05$
Point at $t = 7.9999$: (7.9999, 21,261.61112)	Point at $t = 8.0001$: (8.0001, 21,264.24514)
Slope $= \dfrac{21{,}261.61112 - 21{,}262.9281}{7.9999 - 8}$	Slope $= \dfrac{21{,}264.24514 - 21{,}262.9281}{8.0001 - 8}$
Slope $\approx 13{,}170.13$	Slope $\approx 13{,}170.12$

Whether points to the left or right of $t = 8$ are chosen, it seems clear that the slopes are approaching approximately 13,170. (The slope is 13,170.12218 correct to five decimal places, but recall that we give answers that are meaningful in context.)

Note that this numerical method of calculating slopes of nearby secant lines in order to estimate the slope of the tangent line at $t = 8$ gives a much more precise answer than graphically estimating the slope of the tangent line. (Be certain that you keep enough decimal places in your calculations to be able to see the trend.) We conclude that at the end of August 1949, the number of polio cases was growing at the rate of 13,170 cases per month. ∎

Finding Slopes by Using the Algebraic Method

We also use slopes of nearby secant lines to find slopes of tangent lines in a slightly more sophisticated way. Consider finding the slope of $f(x) = x^2 + 3x$ at $x = 2$. We begin with the point (2, 10). Instead of choosing close x-values such as 1.9, 2.1, 2.01, and so on, we simply call the close value $x = 2 + h$. (Note that if $h = 0.1$, then $x = 2.1$; if $h = 0.01$, $x = 2.01$; if $h = -0.01$, $x = 1.99$, and so on.) The output value that corresponds to $x = 2 + h$ is

$$f(2 + h) = (2 + h)^2 + 3(2 + h) = 4 + 4h + h^2 + 6 + 3h = 10 + 7h + h^2$$

Next we find the slope of the secant line between the point of tangency $(2, 10)$ and the close point $(2 + h, 10 + 7h + h^2)$.

$$\text{Slope of the secant line} = \frac{(10 + 7h + h^2) - 10}{(2 + h) - 2} = \frac{7h + h^2}{h} = 7 + h$$

We now have a formula for the slope of the secant line through the points at $x = 2$ and $x = 2 + h$. We can apply the formula to obtain slopes at points increasingly close to $x = 2$.

Close point to left	h	Slope $= 7 + h$	Close point to right	h	Slope $= 7 + h$
1.9	−0.1	6.9	2.1	0.1	7.1
1.99	−0.01	6.99	2.01	0.01	7.01
1.999	−0.001	6.999	2.001	0.001	7.001
1.9999	−0.0001	6.9999	2.0001	0.0001	7.0001

As we choose points closer and closer to $x = 2$, what happens to h? Clearly, h becomes increasingly close to zero. What happens to the slopes of the secant lines? The slopes become increasingly close to 7. We summarize this with the following statement:

At $x = 2$, the slope of the tangent line is $\lim_{h \to 0} (7 + h) = 7$.

This method of finding a formula for the slope of a secant line in terms of h and then determining the limiting value of the formula as h approaches zero is important because it always yields the *exact* slope of the tangent line. Unfortunately, the method is difficult to use for exponential and logistic functions.

EXAMPLE 2   *Foreign-Born U.S. Residents*

The percentage of people residing in the United States who were born abroad can be modeled[2] by

$$f(x) = 0.00006x^3 - 0.00456x^2 - 0.11021x + 14.75657 \text{ percent}$$

where x is the number of years since 1910. (*Note:* The model is based on data from 1910 through 1990.)

a. Find $\frac{df}{dx}$ at $x = 80$ by writing an expression for the slope of a secant line in terms of h and then evaluating the limit as h approaches 0.

b. Interpret $\frac{df}{dx}$ at $x = 80$ in the context given.

Solution:

a. First, we find the output of $f(x)$ when $x = 80$.

$$f(80) = 7.47577$$

---

2. Based on data from *World Almanac and Book of Facts*, ed. Robert Farrighetti, (Funk and Wagnalls Corporation, 1995).

Second, we write an expression for $f(x)$ when $x = 80 + h$.

$$f(80 + h) = 0.00006(80 + h)^3 - 0.00456(80 + h)^2$$
$$- 0.11021(80 + h) + 14.75657$$

We simplify this expression as much as possible, using the facts that

$$(80 + h)^2 = 6400 + 160h + h^2$$

and

$$(80 + h)^3 = (80 + h)^2 (80 + h) = 512,000 + 19,200h + 240h^2 + h^3$$

Thus

$$f(80 + h) = 7.47577 + 0.31219h + 0.00984h^2 + 0.00006h^3$$

Third, the slope of the secant line through the two close points $(80, f(80))$ and $(80 + h, f(80 + h))$ is

$$\text{Slope of secant} = \frac{f(80 + h) - f(80)}{(80 + h) - 80}$$

$$= \frac{(7.47577 + 0.31219h + 0.00984h^2 + 0.00006h^3) - 7.47577}{(80 + h) - 80}$$

Again, simplify this as much as possible.

$$\text{Slope of secant line (continued)} = \frac{0.31219h + 0.00984h^2 + 0.00006h^3}{h}$$

$$= \frac{h(0.31219 + 0.00984h + 0.00006h^2)}{h}$$

$$= 0.31219 + 0.00984h + 0.00006h^2$$

Finally, we find the slope of the tangent line (the derivative) at $x = 80$ by evaluating the limit of the slope of the secant line as h approaches 0.

$$\frac{df}{dx} = \text{slope of tangent line} = \lim_{h \to 0} (0.31219 + 0.00984h + 0.00006h^2)$$

$$= 0.31219$$

b. In 1990, the percentage of the United States population who were foreign-born was increasing 0.31 percentage points per year. ∎

Finding Slopes of Piecewise Functions

Remember that when we model data piecewise, we usually end up with one or more points on our function where the graph is not smooth. Often our function is not even continuous. The derivative of the function does not exist at a point where the function is not continuous or at a point where the function is continuous but is not smooth. In order for the derivative to exist at a point where the function is continuous, the slopes that we find by approximating with secants through points to the left and through points to the right must be equal. For example, the function represented in Figure 4.6 is defined by

$$f(x) = \begin{cases} x^2 - 8x + 20 & \text{when } 0 \le x \le 5 \\ 2x - 5 & \text{when } 5 < x \end{cases}$$

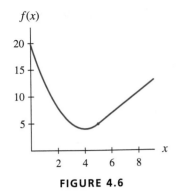

FIGURE 4.6

To determine whether the rate of change of $f(x)$ at $x = 5$ exists, we must first determine whether the function is continuous at $x = 5$. We do this by evaluating the two pieces of the function at $x = 5$.

$$(5)^2 - 8(5) + 20 = 5$$
$$2(5) - 5 = 5$$

We see that both pieces of the function join at $(5, 5)$, so the function is continuous.

Next we must determine if the function is smooth at $x = 5$. To do so, we must check the limiting value of the slopes of secant lines for points to the left of $x = 5$ and the limiting value for points to the right of $x = 5$ to see whether they are equal. The equation to the left of $x = 5$ is quadratic, and we use this in our approximations from the left.

Close point to left	Slope
(4.9, 4.81)	1.9
(4.99, 4.9801)	1.99
(4.999, 4.998001)	1.999

The limit from the left appears to be 2.

Now let us approximate the slope from the right of $x = 5$ by using the equation $f(x) = 2x - 5$.

Close point to right	Slope
(5.1, 5.2)	2
(5.01, 5.02)	2
(5.001, 5.002)	2

The limit from the right appears to be 2, which is the same as the limit from the left, so the derivative is 2 at $x = 5$. In this case, even though the function is defined as piecewise, it is continuous and smooth where the split occurs. The slope of a tangent line drawn at $x = 5$ is 2. In Example 3, we see a function that is not continuous where the two pieces of the function are split.

EXAMPLE 3   *Unemployment in Michigan*

On July 18, 1996, the *Lansing State Journal* reported the Michigan unemployment figures given in Table 4.1.

TABLE 4.1

	1995						1996				
Month	July	Aug	Sept	Oct	Nov	Dec	Jan	Feb	Mar	Apr	May
Number unemployed (thousands of people)	263	250	247	244	246	243	238	233	222	222	225

a. Align the data so that $x = 0$ in July 1995, and find a piecewise model for the data. Split the data in February 1996.

b. Sketch a graph of the piecewise model.

c. Do the two pieces of the model intersect in February 1996?

d. Does the derivative of the piecewise model exist at $x = 7$?

e. Estimate the rate of change of unemployment in February 1996.

Solution:

a. Michigan unemployment can be modeled by

$$U(x) = \begin{cases} -0.323x^3 + 3.632x^2 - 13.889x + 262.227 & \text{when } 0 \leq x < 7 \\ \text{thousand people} \\ 3.5x^2 - 61.9x + 494.4 \text{ thousand people} & \text{when } 7 \leq x \leq 10 \end{cases}$$

where x represents the number of months since July 1995.

b.

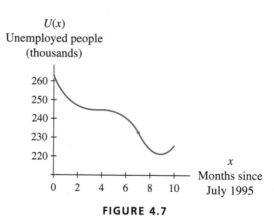

FIGURE 4.7

c. If we look at a magnified view of the graph with input values from $x = 6.8$ to $x = 7.4$ and output values from 228 to 235, we see how the two pieces of the model behave near $x = 7$.

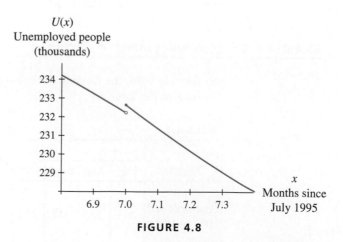

FIGURE 4.8

From Figure 4.8, we see that the two pieces of the model do not intersect at $x = 7$. The function is not continuous at $x = 7$.

We could also determine that the function is not continuous at $x = 7$ by evaluating the two pieces of the unrounded function at $x = 7$. The cubic portion of $U(x)$ between $x = 0$ and $x = 7$ evaluated at $x = 7$ is 232.106, while the quadratic portion of $U(x)$ evaluated at $x = 7$ is 232.6. The two outputs must be the same for the function to be continuous.

d. Because the two pieces of this model do not intersect at $x = 7$, the model is not continuous at $x = 7$. Hence the derivative of the model does not exist at that point.

e. Even though the derivative does not exist at $x = 7$ because our piecewise model is not continuous at that point, we can still estimate the rate of change at $x = 7$ by finding a different model that allows for the calculation of the slope at $x = 7$ or by using a symmetric difference quotient (introduced in Section 3.3). We use the slope of the line through the data points (6, 238) and (8, 222) to estimate the rate of change at $x = 7$.

$$\text{Slope} = \frac{238 - 222 \text{ thousand people}}{6 - 8 \text{ months}} = \text{-8 thousand people per month}$$

This tells us that unemployment was decreasing by approximately 8 thousand people per month in February 1996. ∎

4.1   Concept Inventory

- Slope of tangent line = limiting position of slopes of secant lines
- Numerical method of estimating the slope of a tangent line
- Algebraic method of finding the slope of a tangent line
- Derivative does not exist at a point if the function is not continuous at that point.
- Derivative does not exist at a point if the limiting value of slopes of nearby secant lines from the left does not equal the limiting value of slopes of secant lines from the right.

4.1   Activities

1. The graph[3] in Figure 4.1.1 depicts the number of AIDS cases diagnosed between 1988 and 1991. The equation is Cases $= -1049.5x^2 + 5988.7x + 33{,}770.7$ where x is the number of years since 1988.

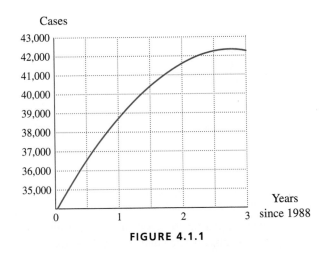

Cases

FIGURE 4.1.1

a. Sketch the tangent line at $x = 2$, and estimate its slope.

b. Choose at least three increasingly close points, and numerically estimate the slope at $x = 2$.

c. Interpret your answer to part b as a rate of change.

3. Based on data from *HIV/AIDS Surveillance 1992 Year End Edition*.

2. Refer to the model in Activity 1.

   a. Find the number of AIDS cases when $x = 2$.

   b. Write an expression for the number of cases when $x = 2 + h$.

   c. Write a simplified formula for the slope of the secant line connecting the points at $x = 2$ and $x = 2 + h$.

   d. What is the limiting value of the slope formula in part c as h approaches 0?

   e. Interpret your answer to part d.

   f. How do your answers to parts a and b of Activity 1 and part d of Activity 2 compare? Which method is the most accurate and why?

3. The balance in a savings account is shown in the graph in Figure 4.1.2 and is given by the equation Balance $= 1500(1.0407)^t$ dollars, where $t =$ the number of years since the principal was invested.

Account balance
(dollars)

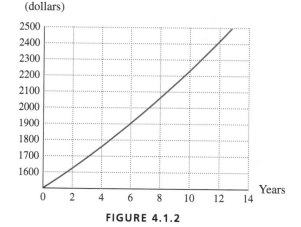

FIGURE 4.1.2

   a. Using only the graph, estimate how rapidly the balance is growing after 10 years.

   b. Use the equation to investigate numerically the rate of change of the balance when $t = 10$.

   c. Which of the two methods (part a or part b) is more accurate? Support your answer by listing the different estimations that had to be made during each method.

4. The time it takes an average athlete to swim 100 meters freestyle at age x years can be modeled[4] by the equation $T(x) = 0.181x^2 - 8.463x + 147.376$ seconds.

   a. Use one of the methods discussed in this section to find the rate of change of the time for a 13-year-old swimmer to swim 100 meters freestyle.

   b. Is a 13-year-old swimmer's time improving or getting worse?

5. Refer to the model in Activity 4.

   a. Find the swim time when $x = 13$.

   b. Write a formula for the swim time when $x = 13 + h$.

   c. Write a simplified formula for the slope of the secant line connecting the points at $x = 13$ and $x = 13 + h$.

   d. What is the limiting value of the slope formula in part c as h approaches 0?

   e. How do your answers to part a of Activity 4 and part d of Activity 5 compare? Which method is the more accurate one and why?

6. U.S. factory sales (in billions of dollars) of electronics from 1986 through 1990 can be modeled[5] by the equation

   $$S(t) = 220 - 44.43e^{-0.3912t} \text{ billion dollars}$$

   where $t =$ the number of years since 1986.

   a. Find the derivative of $S(t)$ when $t = 4$ by using one of the methods discussed in this section.

   b. Interpret your answer to part a.

   c. Which method did you use to find $S'(4)$? Why?

7. The percentage of households with VCRs can be modeled[6] by the equation

   $$P(t) = \frac{72.5}{1 + 75.473e^{-0.6486t}} \text{ percent}$$

   where t is the number of years since 1980. A graph of the model is shown in Figure 4.1.3.

4. Based on data from *Swim World*, August 1992.
5. *Statistical Abstract*, 1992.
6. Based on data from *Statistical Abstract*, 1994.

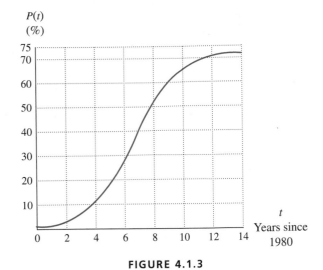

FIGURE 4.1.3

a. Use the graph to estimate $\frac{dP}{dt}$ when $t = 7$.

b. Use the equation to investigate $P'(7)$ numerically.

c. Interpret your answer to part *b*.

d. Discuss the advantages and disadvantages of using the two methods (in parts *a* and *b*) for finding derivatives.

8. Table 4.2[7] gives the number of cotton gins in operation in the early 1900s.

TABLE 4.2

Year	Cotton gins	Year	Cotton gins
1900	29,214	1910	26,234
1901	29,254	1911	26,349
1902	30,948	1912	26,279
1903	30,218	1913	24,749
1904	30,337	1914	24,547
1905	29,038	1915	23,162
1906	28,709	1916	21,624
1907	27,592	1917	20,351
1908	27,598	1918	19,259
1909	26,669	1919	18,815

a. Use a symmetric difference quotient to estimate the rate of change in 1915.

b. Fit a quadratic model to the data.

c. Use the model to estimate numerically the derivative in 1915.

d. Compare your answers to parts *a* and *c*, and discuss whether part *a* or part *c* gives a more accurate representation of the change that occurred in 1915.

9. The total disposable personal income (in billions of 1987 dollars) in the United States[8] is given for certain years in Table 4.3.

TABLE 4.3

Year	TDPI (billions of 1987 dollars)
1975	2355.4
1976	2440.9
1977	2512.6
1978	2638.4
1979	2710.1
1980	2733.6
1981	2795.8
1982	2820.4
1983	2893.6
1984	3080.1
1985	3162.1
1986	3261.9
1987	3289.6
1988	3404.3
1989	3464.9
1990	3516.5
1991	3509.0
1992	3585.1

a. Use a symmetric difference quotient to estimate the rate of change of total disposable personal income in 1986.

---

7. *Cotton Ginnings in the United States Crop of 1970*, U.S. Department of Commerce, Economics and Statistics Administration, A30–90.

8. Case, Karl and Ray Fair, *Principles of Microeconomics*, 3rd ed. (Englewood Cliffs, N.J.: Prentice-Hall, 1994).

b. Fit a linear model to this data set. What is the slope of the model? Compare this slope to your answer in part *a*, and discuss the information each contributes to your understanding of the change in total disposable personal income.

10. Table 4.4[9] gives the number of personal computers used in kindergarten through 12th grade classrooms from 1981 through 1988.

TABLE 4.4

Year	Number of PCs (millions)
1981	0.10
1982	0.21
1983	0.40
1984	0.59
1985	0.87
1986	1.17
1987	1.48
1988	1.76

a. Use a symmetric difference quotient to estimate how rapidly the number of computers in classrooms was growing in 1986.

b. Find a quadratic model to fit the data.

c. Use the model to estimate numerically how rapidly the number of computers in classrooms was growing in 1986.

d. Compare your answers to parts *a* and *c*. Which answer is more accurate? Support your choice.

11. The consumer price index[10] (CPI) for college tuition between 1985 and 1993 is shown in Table 4.5 (1982–1984 = 100).

a. Use a symmetric difference quotient to approximate the rate of change of the consumer price index for college tuition in 1990.

b. Fit a quadratic model to the data.

c. Use the model and at least three close points to estimate numerically the rate of change of the consumer price index for college tuition in 1990.

TABLE 4.5

Year	CPI
1985	119.9
1987	141.3
1988	154.6
1989	168.0
1990	182.8
1991	198.0
1992	213.7
1993	228.9

d. Compare and contrast the ease and accuracy of the two methods you used in parts *a* and *c*.

12. Using the equation you found in part *b* of Activity 10 with coefficients rounded to three decimal places:

a. Find the number of computers in classrooms in 1986.

b. Write a formula in terms of *h* for the number of computers in classrooms a little after 1986.

c. Write a simplified formula for the slope of the secant line connecting the points at 1986 and a little after 1986.

d. What is the limiting value for the slope formula as *h* approaches 0?

e. Interpret your answer to part *d*.

f. How do your answers to part *c* of Activity 10 and part *d* of Activity 12 compare? Which method is the more accurate one? To support your answer, list the places at which estimations were made during the execution of each method.

13. Using the equation found in part *c* of Activity 11 with coefficients rounded to three decimal places:

a. Find the consumer price index for college tuition in 1990.

b. Write a formula in terms of *h* for the consumer price index of college tuition a little after 1990.

c. Write a simplified formula for the slope of the secant line connecting the points at 1990 and a little after 1990.

d. What is the limiting value for the slope formula as *h* approaches 0?

9. *Statistical Abstract*, 1992.
10. *Statistical Abstract*, 1994.

e. Interpret your answer to part *d.*

f. Which method of finding a rate of change (part *a* of Activity 11, part *c* of Activity 11, or Activity 13) should you use in each of the following situations?

  i. You are concerned most with accuracy.

  ii. You want a quick, rough estimate.

  iii. You want a fairly good estimate without taking much time.

14. Discuss the advantages and disadvantages of finding rates of change graphically, numerically, and algebraically. Include in your discussion a brief description of when each method might be appropriate to use.

15. Cattle prices (for choice 450-pound steer calves) from October 1994 through May 1995 can be modeled[11] by

$$p(m) = \begin{cases} -0.0025m^2 + 0.0305m + \\ \quad 0.8405 \text{ dollars per pound} \quad \text{when } 0 \le m < 3 \\ -0.028m + 0.996 \quad \text{when} \\ \quad \text{dollars per pound} \quad 3 \le m \le 7 \end{cases}$$

where *m* is the number of months since October 1994.

a. What is the limiting value of slopes of secant lines of $p(m)$ from the left of $m = 3$?

b. What is the limiting value of slopes of secant lines of $p(m)$ from the right of $m = 3$?

c. What do your answers to parts *a* and *b* tell you about the derivative of $p(m)$ at 3?

d. Estimate the rate of change of cattle prices in January of 1995.

16. The capacity of jails in a southwestern state has been increasing since 1990. The average daily population of one of the jails can be modeled[12] by

$$j(t) = \begin{cases} 8.101t^3 - 55.53t^2 + \\ \quad 128.8t + 626.8 \text{ inmates} \quad \text{when } 0 \le t \le 5 \\ 18.8t + 800.6 \text{ inmates} \quad \text{when } 5 < t \end{cases}$$

where *t* is the number of years since 1990.

a. What is the limiting value of slopes of secant lines of $j(t)$ from the left of $t = 5$?

b. What is the limiting value of slopes of secant lines of $j(t)$ from the right of $t = 5$?

c. What do your answers to parts *a* and *b* tell you about the derivative at $t = 5$?

d. Estimate the rate of change of jail population in 1995.

17. Again consider the cattle price function in Activity 15.

a. Use the algebraic method to find the limit of the slope of secants to $p(m)$ from the left of $m = 3$.

b. Use the algebraic method to find the limit of the slope of secants from the right of $m = 3$.

18. Again consider the inmate population function in Activity 16.

a. Use the algebraic method to calculate the limit of slopes of secant lines from the left of $t = 5$.

b. Use the algebraic method to calculate the limit of slopes of secant lines from the right of $t = 5$.

19. Let $R(x) = 1.02^x$ deutschemarks be the revenue from the sale of *x* mountain bikes. On March 12, 1996, *R* deutschemarks were worth $D(R) = \frac{R}{1.4807}$ American dollars. Assume that this conversion applies today.

a. Write a function for revenue in dollars from the sale of *x* mountain bikes.

b. What is the revenue in deutschemarks and in dollars from the sale of 400 mountain bikes?

c. How quickly is revenue (in dollars) changing when 400 mountain bikes are sold?

20. Refer to the functions $R(x)$ and $D(R)$ in Activity 19.

a. Write a function giving average revenue per mountain bike for the sale of *x* mountain bikes in deutschemarks.

b. Write a function for average revenue in dollars.

c. How quickly is average revenue (in dollars) changing when 400 mountain bikes are sold?

---

11. Based on data from the National Cattleman's Association.
12. Based on data from Washoe County Jail, Reno, Nevada.

4.2   Drawing Slope Graphs

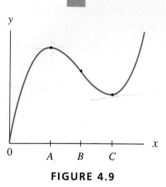

FIGURE 4.9

Every smooth, continuous curve with no vertical tangent lines has a slope associated with each point on the curve. When these slopes are plotted, they also form a smooth, continuous curve. We call the resulting curve a **slope graph**, **rate-of-change graph**, or **derivative graph**.

What do we know about the slopes of the graph shown in Figure 4.9?

1. At *A* the graph is at a maximum, and at *C* the graph is at a minimum, so the slope is zero at these points.

2. Between 0 and *A*, the graph is increasing, so the slopes will be positive.

3. Between *A* and *C*, the graph is decreasing, so the slopes will be negative.

4. At *B*, the graph has an inflection point. This is the point at which the graph is decreasing most rapidly—that is, the point at which the slope is most negative.

5. To the right of *C*, the graph is again increasing, so the slopes will be positive.

We record this information as indicated in Figure 4.10.

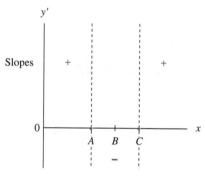

FIGURE 4.10

On the basis of this information about the slopes, we sketch the shape of the slope graph, which appears in Figure 4.11.

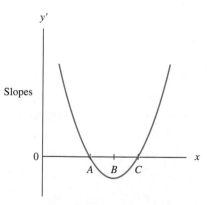

FIGURE 4.11

We do not know the specifics of the slope graph—how far below the horizontal axis it dips, where it crosses the vertical axis, how steeply it rises to the right of *C*, and so on. However, we do know its basic shape.

EXAMPLE 1   *Slope Graph of a Logistic Curve*

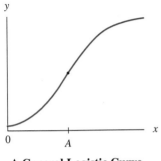

A General Logistic Curve

FIGURE 4.12

The graph in Figure 4.12 is a general logistic curve. Sketch its slope graph.

Solution: First we note that the logistic curve in Figure 4.12 is always increasing. Thus its slope graph is always positive (above the horizontal axis). We also note that even though there is no relative maximum or relative minimum, the logistic curve does level off at both ends. Thus its slope graph will be near zero at both ends.

Finally, we note that the logistic curve has its steepest slope at *A* because this is the location of the inflection point. Therefore, the slope graph is greatest (has a maximum) at this point. (See Figure 4.13.) We sketch the slope graph as shown in Figure 4.14.

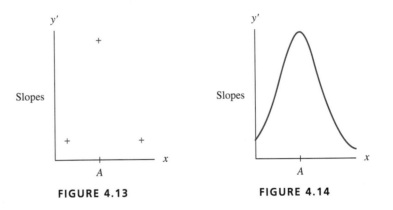

FIGURE 4.13   **FIGURE 4.14**   ■

In the preceding discussion and example, the graphs we used had no labeled tick marks on the horizontal or vertical axes. In such cases, it is not possible to estimate the value of the slope of the graph at any given point. Instead, we sketch the general shape of the slope graph by observing the critical points and general behavior of the graph, such as

1. Points at which maxima and minima occur
2. The regions over which the graph is increasing or decreasing
3. Points of inflection
4. Places at which the graph appears to be horizontal or leveling off

We can also consider the relative magnitudes of the slopes. This technique is demonstrated in Example 2.

EXAMPLE 2   *Using Relative Magnitudes to Sketch Slope Graphs*

The height of a plant often follows the general trend shown in Figure 4.15. Draw a graph depicting the growth rate of the plant.

Solution: The slopes at *A*, *B*, and *C* are all positive. Is the slope at *A* smaller or larger than that at *B*? It is larger, so the slope graph at *A* should be higher than it is at *B*. The graph at *C* is not as steep as it is at either *A* or *B*, so the slope graph should be lower at *C* than at *B*. (See Figure 4.16.)

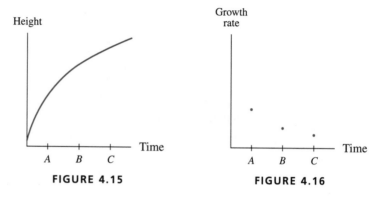

FIGURE 4.15   FIGURE 4.16

Where on the graph is the slope steepest? The answer is at the left endpoint. Where is the graph least steep? You can see that it is at the right endpoint. Add these observations to your plot. (See Figure 4.17.)

Now sketch the slope graph according to your plot, and be sure to include the appropriate labels on the horizontal and vertical axes. (See Figure 4.18.)

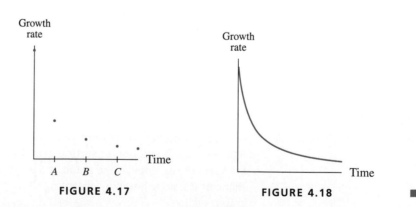

FIGURE 4.17   FIGURE 4.18   ■

A Detailed Look at the Slope Graph

When a graph has labeled tick marks on both the horizontal and the vertical axes or an equation for the graph is known, it is possible to estimate the values of slopes at certain points on the graph. However, it would be tedious to calculate the slope graphically or numerically for every point on the graph. In fact, because there are infinitely many points on a continuous curve, it is not possible. Instead, we calculate the slope at a few special points, such as inflection points.

Consider Figure 4.9, which we saw at the beginning of this section. It is shown again in Figure 4.19, but this time the graph has labeled tick marks on both the horizontal and the vertical axes.

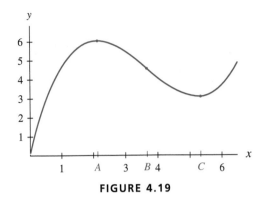

FIGURE 4.19

We know that the slope graph crosses the horizontal axis at A and C and that a minimum occurs on the slope graph below the horizontal axis at B. Before, we did not know how far below the axis to draw this minimum. Now that there are tick marks on the axes, we can graphically estimate the slope at the inflection point and use that estimate to help us sketch the slope graph.

By drawing the tangent line at B and estimating its slope, we find that the minimum of the slope graph occurs approximately 1.4 units below the horizontal axis. (See Figure 4.20.)

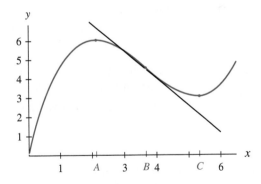

**Slope of Tangent Line at B
is approximately –1.4**

FIGURE 4.20

If we estimate the slopes at two additional points, say at $x = 1$ and $x = 6$, then we can produce a fairly accurate sketch of the slope graph.

Table 4.6 is a list of estimated slope values.

TABLE 4.6

x	1	A	B	C	6
Slope	2.7	0	–1.4	0	1.8

Plotting these points and sketching the slope graph give us the graph in Figure 4.21.

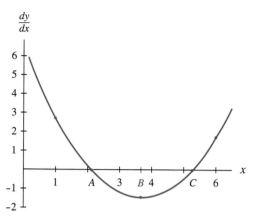

FIGURE 4.21

If we do not have a continuous curve but have a scatter plot of data, then we can sketch a rate-of-change graph by first sketching a smooth curve that fits the scatter plot. Then we can draw the slope graph of that smooth curve.

EXAMPLE 3 *Population of Cleveland, Ohio*

TABLE 4.7

Year	Population
1810	57
1820	606
1830	1076
1840	6071
1850	17,034
1860	43,417
1870	92,829
1880	160,146
1890	261,353
1900	381,768
1910	560,663
1920	796,841
1930	900,429
1940	878,336
1950	914,808
1960	876,050
1970	750,879
1980	573,822
1990	505,616

In 1797, the Lorenzo Carter family built a cabin on Lake Erie where today the city of Cleveland, Ohio, is located. Table 4.7 gives population data[13] for Cleveland from 1810 through 1990.

a. Sketch a smooth curve representing population. Your curve should have no more inflection points than the number suggested by the scatter plot.

b. Sketch a graph representing the rate of change of population.

Solution:

a. Draw a scatter plot of the population data, and sketch the smooth curve. (See Figure 4.22.)

b. The population graph is fairly level in the early 1800s, so the slope graph will begin near zero. The smooth sketch increases during the 1800s and early 1900s until it peaks in the 1940s. Thus the slope graph will be positive until the mid-1940s, at which time it will cross the horizontal axis and become negative. Population decreased from the mid-1940s onward.

There appear to be two inflection points. The point of most rapid growth appears around 1910, and the point of most rapid decline appears near 1975. These are the years in which the slope graph will be at its maximum and at its minimum, respectively. By drawing tangent lines at 1910 and at 1975 and estimating their slopes, we find that population was increasing by approximately 20,000 people per year in 1910 and was decreasing by about 13,000 people per year in 1975.

13. U.S. Department of Commerce, Bureau of the Census.

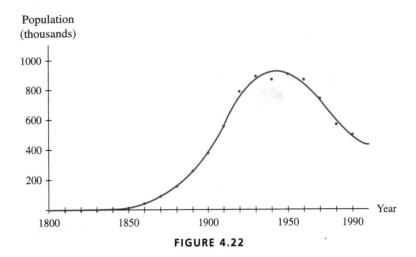

FIGURE 4.22

Now we use all the information from this analysis to sketch the slope graph shown in Figure 4.23.

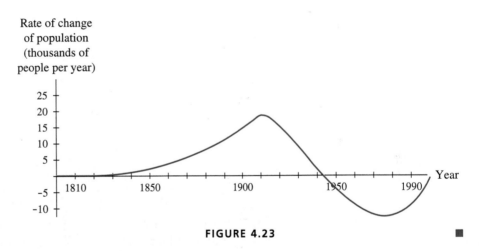

FIGURE 4.23 ■

Points of Undefined Slope

It is possible for a function to have a point at which the slope does not exist. Remember that the slope of a tangent line is the limit of slopes of approximating secant lines and that we should be able to use secant lines through points either to the left or to the right of the point at which we are estimating the slope to find this limit.[14] If the limits from the left and from the right are not the same, then we say that the derivative does not exist at that point. We depict the nonexistence of the derivative at such points on the slope graph by drawing an open dot on each piece of the slope graph. This is illustrated in Figure 4.24a.

It is possible for there to be a point on a graph at which the derivative does not exist although the slope from the left and the slope from the right are the same. This occurs when the function is not continuous at that point (see Figure 4.24b). If a function is not continuous at a point, then its slope is undefined at that point.

14. It is worth noting that in the case in which a function is defined only to the right or only to the left of a point, then the derivative at that point can be found using secant lines through points only to the right or left.

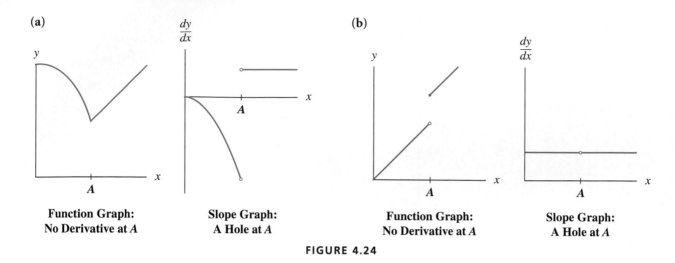

(a)

Function Graph:
No Derivative at *A*

Slope Graph:
A Hole at *A*

(b)

Function Graph:
No Derivative at *A*

Slope Graph:
A Hole at *A*

FIGURE 4.24

Also, points at which the tangent line is vertical (that is, the slope calculation results in a zero in the denominator) are considered to have an undefined slope. One such function is shown in Figure 4.25.

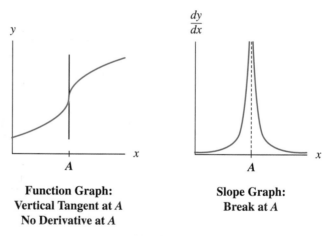

Function Graph:
Vertical Tangent at *A*
No Derivative at *A*

Slope Graph:
Break at *A*

FIGURE 4.25

Most of the time when there is a break in the slope graph, it is because the original function was piecewise as shown in Figure 4.24. You should be careful when drawing slope graphs of piecewise continuous functions. Figure 4.25 showed a smooth, continuous function with a point at which the derivative was undefined. We do not often encounter such phenomena in real-life applications, but they could happen.

4.2   Concept Inventory

- Slope graph = rate-of-change graph = derivative graph
- Increasing function ⇒ positive slopes
- Decreasing function ⇒ negative slopes
- Maximum or minimum of function ⇒ zero slope
- Inflection point ⇒ maximum or minimum point on slope graph or point of undefined slope
- Points of undefined slope

4.2   Activities

In Activities 1 through 10, list as many facts as you can about the slopes of the graphs. Then, on the basis of those facts, sketch the slope graph of each function.

1.
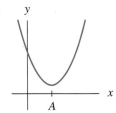

2.
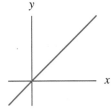

3.
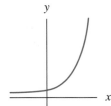

4.
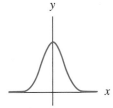

5.
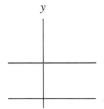

6.
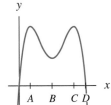

7.
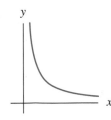

8.
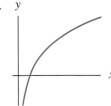

9.
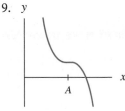

10.
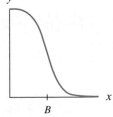

11. The graph in Figure 4.2.11 gives the membership in a campus organization during its first year.

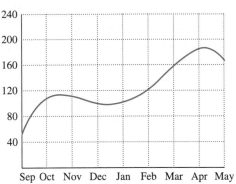

Members

FIGURE 4.2.11

a. Estimate the average rate of change in the membership during the academic year.

b. Estimate the instantaneous rates of change in September, November, February, and April.

c. On the basis of your answers to part *b*, sketch a rate-of-change graph. Label the units on the axes.

d. The membership of the organization was growing most rapidly in September. Not including that month, when was the membership growing most rapidly? What is this point on the membership graph called?

e. Why was the result of the calculation in part *a* of no use in part *c*?

12. The scatter plot in Figure 4.2.12 depicts the number of calls placed each hour since 2 a.m. to a sheriff's department.[15]

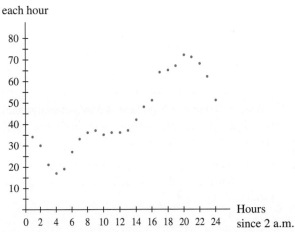

Calls each hour

FIGURE 4.2.12

15. Greenville, South Carolina Sheriff's Department.

a. Sketch a smooth curve through the scatter plot with no more inflection points than the number suggested by the scatter plot.

b. At what time(s) is the number of calls a minimum? a maximum?

c. Are there any other times when the graph appears to have a zero slope? If so, when?

d. Estimate the slope of your smooth curve at any inflection points.

e. Use the information in parts *a* through *d* to sketch a graph depicting the rate of change of calls placed each hour. Label the units on both axes of the rate-of-change graph.

13. The capacity of jails in a southwestern state has been increasing since 1990. The average daily population of one jail can be modeled[16] by

$$j(t) = \begin{cases} 8.101t^3 - 55.53t^2 + 128.8t + 626.8 \text{ inmates} & \text{when } 0 \le t \le 5 \\ 18.8t + 800.6 \text{ inmates} & \text{when } 5 < t \end{cases}$$

where t is the number of years since 1990. Figure 4.2.13 shows a graph of this model.

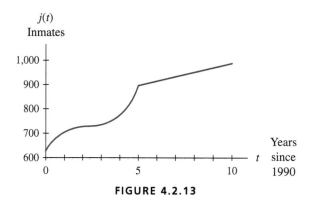

FIGURE 4.2.13

Sketch the slope graph of $j(t)$. (*Hint:* Numerically estimate $j'(t)$ at $t = 0$ and $t = 4.5$ in order to sketch the slope graph accurately.) Label both the horizontal and the vertical axes.

14. Cattle prices (for choice 450-pound steer calves) from October 1994 through May 1995 can be modeled[17] by

$$p(m) = \begin{cases} -0.0025m^2 + 0.0305m + 0.8405 \text{ dollars per pound} & \text{when } 0 \le m < 3 \\ -0.028m + 0.996 \text{ dollars per pound} & \text{when } 3 \le m \le 7 \end{cases}$$

where m is the number of months since October 1994. A graph of $p(m)$ is given in Figure 4.2.14.

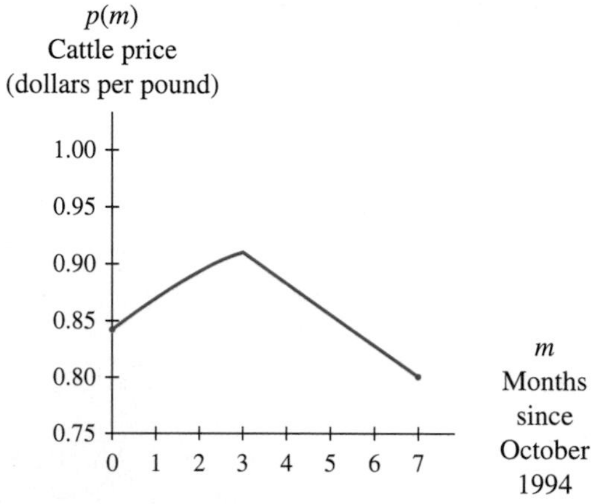

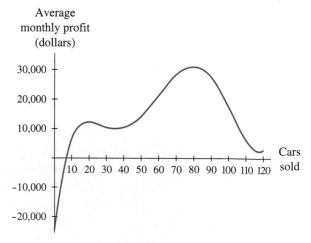

FIGURE 4.2.14

Sketch a slope graph of $p(m)$. (*Hint:* Numerically estimate $p'(m)$ at $m = 0$ and $m = 2.5$ in order to sketch the slope graph accurately.) Label the horizontal and vertical axes.

15. A graph depicting the average monthly profit for Slim's Used Car Sales for the previous year is shown in Figure 4.2.15.

FIGURE 4.2.15

16. Based on data from Washoe County Jail, Reno, Nevada.

17. Based on data from the National Cattleman's Association.

a. Estimate the average rate of change in Slim's average monthly profit if the number of cars he sells increases from 40 to 70 cars.

b. Estimate the instantaneous rates of change at 20, 40, 60, 80, and 100 cars.

c. On the basis of your answers to part *b*, sketch a rate-of-change graph. Label the units on the axes.

d. For what number of cars sold between 20 and 100 is average monthly profit increasing most rapidly? For what number of cars sold is average monthly profit decreasing most rapidly? What is the mathematical term for these points?

e. Why was the result of the calculation in part *a* of no use in part *c*?

16. Using the graph in Figure 3.4.4 on page 189, carefully estimate the rate of change in deaths due to lung cancer in 1940, 1960, and 1980. Use this information to sketch an accurate rate-of-change graph for deaths due to lung cancer. Label the units on both axes of the derivative graph.

In Activities 17 through 20, indicate the input values for which the graph has no derivative. Explain why the derivative does not exist at those points. Sketch a derivative graph for each of the functions.

18.

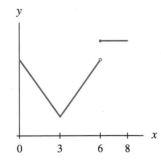

19.

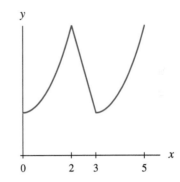

17.

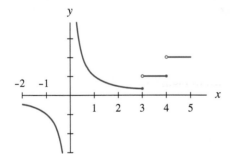

20.

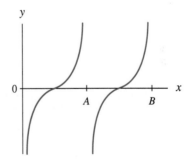

■ **4.3   Slope Formulas**

In our discussion of rates of change, we have rarely referred to equations or formulas. This was deliberate. It is easy to get caught up in algebraic manipulations and lose sight of the fundamental relationships among secant lines, tangent lines, slopes, and rates of change. In the previous section, we discussed the rate-of-change graph, or slope graph, for smooth, continuous functions as well as for piecewise continuous functions. It may have occurred to you that when the original function is given by an equation, then its slope graph should also have an equation. We explore that idea in this section.

We begin with lines. What is the equation for the slope graph of $y = ax + b$? This line has constant slope represented by a in the equation $y = ax + b$, so the equation of its slope graph is simply $y' = a$. In Figure 4.26, the slope graphs are shown as blue lines.

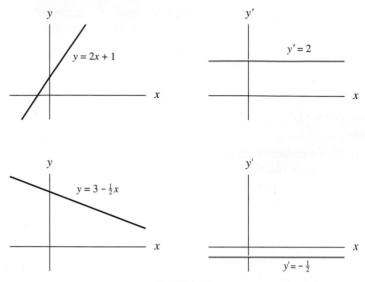

FIGURE 4.26

What is the rate-of-change equation for a horizontal line? Because the slope of a horizontal line is zero, the slope equation is $y' = 0$. In Figure 4.27, the slope graphs are shown in blue.

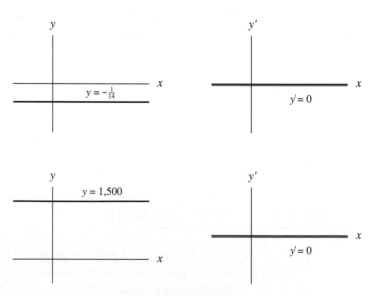

FIGURE 4.27

Using Derivative Notation for Slope Formulas

Remember that the word *derivative* means the same as *slope* and *rate of change*. We thus restate our conclusions about slope formulas for lines using derivative notation:

Derivative of a Linear Function

If $y = ax + b$, then $\dfrac{dy}{dx} = a$.

If $y = b$, then $\dfrac{dy}{dx} = 0$.

EXAMPLE 1   *Cricket's Chirping*

The frequency of a cricket's chirp is affected by air temperature and can be modeled by

$$C(t) = 0.212t - 0.309 \text{ chirps per second}$$

when the temperature is $t°$F. Write a formula for the rate of change of a cricket's chirping speed with respect to a change in temperature.

Solution: The frequency of a cricket's chirp is changing by

$$C'(t) = 0.212 \text{ chirps per second per degree Fahrenheit}$$

no matter what the temperature t is.   ∎

Now consider the quadratic function $y = x^2$. We could use one of the methods discussed in Section 4.1 to obtain the slopes at various points on this function. However, because of the sometimes tedious nature of those methods, we choose to use technology to construct a table of values for the slopes of $y = x^2$ at the indicated values of x (rounding y' to integers). These slope values are shown in Table 4.8.

TABLE 4.8

x	-3	-2	-1	0	1	2	3
$y = x^2$	9	4	1	0	1	4	9
$y' = \dfrac{dy}{dx}$	-6	-4	-2	0	2	4	6

Do the values in the table suggest a formula for y' in terms of x? The slope is always twice the x-value, which suggests that $y' = \frac{dy}{dx} = 2x$. This is, in fact, the correct formula. The slope formula for $y = x^2$ is indeed $y' = 2x$. (See Figure 4.28.)

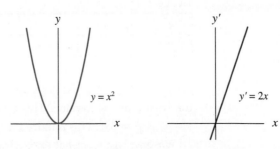

Graph   **Slope Graph**

FIGURE 4.28

We repeat this procedure for the function $y = e^x$ (rounding to three decimal places). The results are shown in Table 4.9.

TABLE 4.9

x	-3	-2	-1	0	1	2	3
$y = e^x$	0.050	0.135	0.368	1.000	2.718	7.389	20.086
$y' = \dfrac{dy}{dx}$	0.050	0.135	0.368	1.000	2.718	7.389	20.086

This function is unique in that the values for the rates of change are precisely the same as the values for the function. *This function is its own derivative!* In other words, if $y = e^x$, then $\frac{dy}{dx} = y' = e^x$. The slope graph of $y = e^x$ coincides with the graph of the original function.

Derivative of e^x

If $y = e^x$, then $\dfrac{dy}{dx} = e^x$

It is slightly more difficult to discover a rate-of-change equation for the function $y = x^3$. Again, we construct a table of values (rounded to integers). (See Table 4.10.)

TABLE 4.10

x	-3	-2	-1	0	1	2	3
$y = x^3$	-27	-8	-1	0	1	8	27
$y' = \dfrac{dy}{dx}$	27	12	3	0	3	12	27

First, note that each of the y'-values is a multiple of 3. In Table 4.11, 3 has been factored from each slope value.

TABLE 4.11

x	-3	-2	-1	0	1	2	3
y'	3(9)	3(4)	3(1)	3(0)	3(1)	3(4)	3(9)

Next, note that after 3 is factored from each slope value, the remaining value is just the square of the x-value. This investigation suggests that the slope equation for $y = x^3$ is $\frac{dy}{dx} = 3x^2$. This is indeed the correct slope formula. (See Figure 4.29.)

We have just seen that the derivative of $y = x^2$ is $y' = 2x$ and that the derivative of $y = x^3$ is $y' = 3x^2$. In each of these slope formulas, the power on x in the original

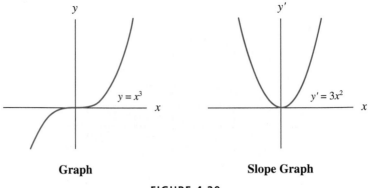

Graph   **Slope Graph**

FIGURE 4.29

function is reduced by 1 and the *x*-term is multiplied by the original power on *x*. These are special cases of the following general formula:

Simple Power Rule for Derivatives

If $y = x^n$, then $\dfrac{dy}{dx} = nx^{n-1}$

where n is any nonzero real number.

We observed that the derivative of $y = e^x$ is $\frac{dy}{dx} = e^x$. Does this rule apply to exponential functions that have bases different from *e*? In other words, if $y = b^x$, is $\frac{dy}{dx} = b^x$? We begin by exploring the derivative of $y = 2^x$ with a table of values (rounded to five decimal places for convenience):

TABLE 4.12

x	$y = 2^x$	$y' = \dfrac{dy}{dx}$
-3	0.125	0.08664
-2	0.25	0.17329
-1	0.5	0.34657
0	1	0.69315
1	2	1.38629
2	4	2.77259
3	8	5.54518

It is obvious from Table 4.12 that the derivative of $y = 2^x$ is *not* $y' = 2^x$. However, that the derivative values are multiples of the 2^x values. (See Table 4.13.)

TABLE 4.13

$y = 2^x$	y'
0.125	$(0.693147)(0.125)$
0.25	$(0.693147)(0.25)$
0.5	$(0.693147)(0.5)$
1	$(0.693147)(1)$
2	$(0.693147)(2)$
4	$(0.693147)(4)$
8	$(0.693147)(8)$

We find the factor 0.693147 by dividing the y'-values by the corresponding y-values. It may seem that the multiplier 0.693147 is arbitrary because it is not a familiar number, but that is not the case. You should verify that $0.693147 \approx \ln 2$. We state the formula for the derivative of $y = 2^x$ as $y' = (\ln 2)\, 2^x$.

Now we construct a table of values for the derivative of $y = 3^x$ (again rounding to five decimal places).

TABLE 4.14

x	$y = 3^x$	$y' = \dfrac{dy}{dx}$
-3	0.03704	0.04069
-2	0.11111	0.12207
-1	0.33333	0.36620
0	1	1.09861
1	3	3.29584
2	9	9.88751
3	27	29.66253

Again, the derivative is not immediately obvious from the table values. However, the derivative values are multiples of the values of 3^x. The multiplier, correct to five decimal places, is 1.09861, which is approximately $\ln 3$. We state the formula for the derivative of $y = 3^x$ as $y' = (\ln 3)\, 3^x$.

The two derivative formulas $\frac{d}{dx}(2^x) = (\ln 2)2^x$ and $\frac{d}{dx}(3^x) = (\ln 3)3^x$ are special cases of the general derivative formula for exponential functions. The derivative of $y = b^x$ is $y' = (\ln b)b^x$ if $b > 0$. In fact, the rule $\frac{d}{dx}(e^x) = e^x$ is also a special case of this formula.

> ### Derivative of b^x
>
> If $y = b^x$ with the real number $b > 0$, then $\dfrac{dy}{dx} = (\ln b)b^x$.

We develop slope formulas for several other specific functions in the Activities section.

EXAMPLE 2   *Finding Derivatives*

Find derivative formulas for

a. $y = x^5$ b. $y = 17x + 3$ c. $y = 1.2^x$ d. $y = -12$

Solution:

a. Apply the Power Rule to $y = x^5$. The derivative is $y' = 5x^4$.
b. The function $y = 17x + 3$ is a linear function, so its derivative is $y' = 17$.
c. Apply the derivative rule for $y = b^x$ to $y = 1.2^x$. We find that the derivative is $y' = (\ln 1.2)1.2^x$.
d. The derivative of a horizontal line at $y = -12$ is $y' = 0$.   ∎

A General Formula for Derivatives

Recall our method for finding slopes of tangent lines in Section 4.1: We chose close points and determined the limiting value of the slopes of nearby secant lines both numerically and analytically. With a simple modification, we can apply this technique to find formulas of derivatives.

To illustrate, consider $y = x^2$. Because we desire a general equation for any x-value, we use (x, x^2) as the point of tangency. This is the same idea as the algebraic method in Section 4.1, but there we worked with a numerical value for x. Next we choose a close point. We use $x + h$ as the x-value of the close point and find the y-value by substituting $x + h$ into the function: $y = (x + h)^2 = x^2 + 2xh + h^2$. Thus the original point is (x, x^2), and a close point is $(x + h, x^2 + 2xh + h^2)$. Now we find the slope of the secant line between these two points.

$$\text{Slope of the secant line} = \frac{(x^2 + 2xh + h^2) - x^2}{(x + h) - x} = \frac{2xh + h^2}{h} = 2x + h$$

Finally, we determine the limiting value of the secant line slope as h approaches 0.

$$\lim_{h \to 0}(2x + h) = \lim_{h \to 0} 2x + \lim_{h \to 0} h = 2x$$

Therefore, the slope formula for $y = x^2$ is $\frac{dy}{dx} = 2x$.

This method can be generalized to obtain a formula for the derivative of an arbitrary function $f(x)$.

Four-Step Method to Find $f'(x)$

Given a function $f(x)$, the equation for the derivative can be found as follows:

1. Begin with a typical point $(x, f(x))$.

2. Choose a close point $(x + h, f(x + h))$.

3. Write a formula for the slope of the secant line between the two points.

$$\text{Slope} = \frac{f(x + h) - f(x)}{(x + h) - x} = \frac{f(x + h) - f(x)}{h}$$

It is important at this step to simplify the slope formula algebraically until there is no h in the denominator.

4. Evaluate the limit of the slope as h approaches 0.

$$\lim_{h \to 0} \frac{f(x + h) - f(x)}{h}$$

This limiting value is the derivative formula at each input where the limit exists.

Thus we have the following derivative formula (slope formula, rate-of-change formula) for an arbitrary function $f(x)$:

Derivative Formula

If $y = f(x)$, then the derivative $\frac{dy}{dx}$ is given by the formula

$$\frac{dy}{dx} = \lim_{h \to 0} \frac{f(x + h) - f(x)}{h}$$

provided the limit exists.

EXAMPLE 3   *Women in the Civilian Labor Force*

The number of women in the United States civilian labor force[18] from 1930 through 1990 can be modeled as

$$W(t) = 0.011t^2 + 0.101t + 10.817 \text{ million women}$$

where t is the number of years since 1930. Use the limit definition of the derivative to develop a formula for the rate of change of the number of women in the labor force.

Solution:

Step 1. A typical point is $(t, 0.011t^2 + 0.101t + 10.817)$.

Step 2. A close point is $(t + h, 0.011(t + h)^2 + 0.101(t + h) + 10.817)$.

18. *Information Please Almanac, Atlas, and Yearbook, 1996* (Boston: Houghton Mifflin, 1996).

We rewrite the output of close point before we proceed:

$$0.011(t + h)^2 + 0.101(t + h) + 10.817$$
$$= 0.011(t^2 + 2th + h^2) + 0.101(t + h) + 10.817$$
$$= 0.011t^2 + 0.022th + 0.011h^2 + 0.101t + 0.101h + 10.817$$

Step 3. Write the slope of the secant line between the two points from Steps 1 and 2 (simplifying until there is no h in the denominator).

$$\text{Slope} = \frac{[(0.011t^2 + 0.022th + 0.011h^2 + 0.101t + 0.101h + 10.817) -}{(0.011t^2 + 0.101t + 10.817)] \div [(t + h) - t]}$$

$$= \frac{0.022th + 0.011h^2 + 0.101h}{h}$$

$$= \frac{h(0.022t + 0.011h + 0.101)}{h}$$

$$= 0.022t + 0.011h + 0.101$$

Step 4. Evaluate the limit of the slope as h approaches 0.

$$W'(t) = \lim_{h \to 0}(0.022t + 0.011h + 0.101)$$

$$= 0.022t + 0.101$$

Thus the number of women in the civilian labor force was increasing by

$$W'(t) = 0.022t + 0.101 \text{ million women per year}$$

t years after 1930. ∎

4.3   Concept Inventory

- Derivative formulas
   If $y = b$, then $y' = 0$. (Horizontal line)
   If $y = ax + b$, then $y' = a$. (Line)
   If $y = e^x$, then $y' = e^x$.
   If $y = b^x$, then $y' = (\ln b)b^x$. (Exponential function)
   If $y = x^n$, then $y' = nx^{n-1}$. (Power Rule)
- Four-step method to finding the derivative formula of a function $f(x)$

4.3   Activities

For each of the functions whose graphs are given in Activities 1–6, first sketch the derivative graph and then give the slope equation.

1. 2.

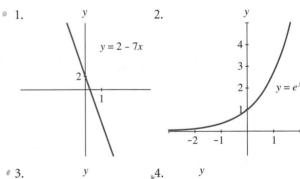

3. 4.

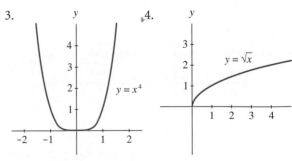

5.
6.

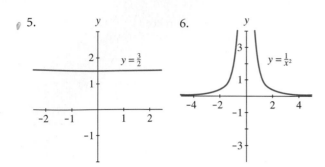

7. For the first couple of hours after yeast dough has been kneaded, it approximately doubles in volume every 42 minutes. If we prepare 1 quart of yeast dough and let it rise in a warm room, then its growth can be modeled by the function

$$V = e^h \text{ quarts}$$

where h is the number of hours the dough has been allowed to rise.

a. How many minutes will it take the dough to attain a volume of 2.5 quarts?

b. Write a formula for the rate of growth of the yeast dough.

c. How quickly is the dough expanding after 24 minutes, after 42 minutes, and after 55 minutes? Report your answers in quarts per minute.

8. Table 4.15 shows the metabolic rate[19] of a typical 18- to 30-year-old male according to his weight.

TABLE 4.15

Weight (pounds)	Metabolic rate (kilocalories per day)
88	1291
110	1444
125	1551
140	1658
155	1750
170	1857
185	1964
200	2071

a. Find a formula for a typical man's metabolic rate.

b. Write the derivative of the formula you found in part a.

c. What does the derivative in part b tell you about a man's metabolic rate if that man weighs 110 pounds? if he weighs 185 pounds?

9. The population of Hawaii[20] can be modeled by

$$P(t) = 15.48t + 485.4 \text{ thousand people}$$

t years after 1950.

a. Write the derivative formula for $P(t)$.

b. How many people lived in Hawaii in 1970?

c. How quickly was Hawaii's population changing in 1990?

10. The value of a $1000 investment in an account with 4.3% interest compounded continuously can be modeled as

$$A = e^{0.043t} \text{ thousand dollars after } t \text{ years}$$

a. Write the rate-of-change formula for the value of the investment. (*Hint:* Let $b = e^{0.043}$, and use the rule for $y = b^t$.)

b. How much is the investment worth after 5 years?

c. How quickly is the investment growing after 5 years?

d. What is the percentage rate of growth after 5 years?

11. We have seen that the derivative of $y = b^x$ is $\frac{dy}{dx} = (\ln b)b^x$ as long as $b > 0$. We have also seen that the derivative of $y = e^x$ is $\frac{dy}{dx} = e^x$.

a. Show that the derivative formula for $y = e^x$ is a special case of the derivative formula for $y = b^x$ by applying the formula for $y = b^x$ to $y = e^x$ and then reconciling the result with the known derivative formula for $y = e^x$.

b. Use the derivative formula for $y = b^x$ to find a formula for the derivative of an exponential function of the form $y = e^{kx}$, where k is some known constant.

19. Smolin, L., and M. Grosvenor, *Nutrition: Science and Applications* (Saunders College Publishing, 1994).

20. George T. Kurian, *Datapedia of the United States, 1790–2000* (Latham, Maryland: Bernan Press, 1994).

12. a. When the inputs and outputs to the function $y = e^x$ are reversed, the resulting rule is also a function—the inverse function of $y = e^x$. Sketch $y = e^x$ and its inverse function on the same axes.

   b. The name of the inverse function for $y = e^x$ is the **natural logarithm function**, denoted by $y = \ln x$. Use your calculator or computer to graph $y = e^x$ and $y = \ln x$ on the same axes. Compare these graphs to your sketches in part *a*.

   c. Construct a table of values for the slopes of $y = \ln x$.

TABLE 4.16

x	$\dfrac{1}{4}$	$\dfrac{1}{3}$	$\dfrac{1}{2}$	1	2	3	4
$y' = \dfrac{dy}{dx}$							

   d. State a formula for the derivative of $y = \ln x$.

13. To determine what effect a constant multiplier has on rate-of-change formulas, use your calculator or computer to complete the following tables.

   a. The function is $y = 7x^2$.

TABLE 4.17

x	-3	-2	-1	0	1	2	3
$y' = \dfrac{dy}{dx}$							

   b. The function is $y = 4e^x$.

TABLE 4.18

x	-2	-1	0	1	2
$y' = \dfrac{dy}{dx}$					

   c. The function is $y = -\frac{1}{2}x^3$.

TABLE 4.19

x	-3	-2	-1	0	1	2	3
$y' = \dfrac{dy}{dx}$							

   d. Compare each of these tables (Tables 4.17–4.19) with the tables for $y = x^2, y = e^x$ and $y = x^3$ that were given in this section (Tables 4.8–4.10). Determine formulas for the derivatives of $y = kx^n$ and $y = ke^x$, where k is an arbitrary constant.

14. On the basis of your answers to Activity 12 and your conclusions in part *d* of Activity 13, what are the formulas for the derivatives of $y = kb^x$ and $y = k \ln x$, where k is an arbitrary constant?

15. Rewrite $y = \frac{k}{x}$ as $y = kx^{-1}$ and find a formula for $\frac{dy}{dx}$. Simplify the formula.

In Activities 16 through 22, use the four-step method outlined near the end of this section to show that each statement is true.

16. The derivative of $y = 3x - 2$ is $\frac{dy}{dx} = 3$.

17. The derivative of $y = 15x + 32$ is $\frac{dy}{dx} = 15$.

18. The derivative of $y = ax + b$ is $\frac{dy}{dx} = a$.

19. The derivative of $y = 3x^2$ is $\frac{dy}{dx} = 6x$.

20. The derivative of $y = ax^2 + bx + c$ is $\frac{dy}{dx} = 2ax + b$.

21. The derivative of $y = -3x^2 - 5x$ is $\frac{dy}{dx} = -6x - 5$.

22. The derivative of $y = -2x^3$ is $\frac{dy}{dx} = -6x^2$.
   (*Hint:* $(x + h)^3 = x^3 + 3x^2h + 3xh^2 + h^3$)

23. An object is dropped off a building. Ignoring air resistance, we know from physics that its height above the ground t seconds after being dropped is given by

$$\text{Height} = -16t^2 + 100 \text{ feet}$$

   a. Use at least three increasingly close points to estimate numerically how rapidly the object is falling after 1 second.

   b. Use the four-step method to verify that your answer to part *a* is correct.

24. Clinton County, Michigan, is mostly flat farmland partitioned by straight roads (often gravel) that run either north/south or east/west. A tractor driven north on Lowell Road from the Schafer's farm is

$$d(m) = 0.28m + 0.6 \text{ miles}$$

   north of Howe Road m minutes after leaving the farm's drive.

   a. How far is the Schafer's drive from Howe Road?

b. Use the four-step method to show that the tractor is moving at a constant speed.

c. How quickly (in miles per hour) is the tractor moving?

25. The number of licensed drivers[21] between the ages of 16 and 24 in 1995 is given in Table 4.20.

TABLE 4.20

Age (years)	Number of drivers (millions)
16	1.4
17	2.1
18	2.5
19	2.7
20	2.8
21	3.0
22	3.4
23	3.6
24	3.7

a. Find a quadratic model for the number of licensed drivers as a function of age. Round the coefficients of the model to three decimal places.

b. Use the limit definition of the derivative to develop a formula for the derivative of the rounded model.

26. The data in Table 4.21 give the estimated number of 16- to 24-year-old female licensed drivers[22] in 1994 according to their age.

TABLE 4.21

Age (years)	Number of drivers (millions)
16	0.7
17	1.0
18	1.2
19	1.3
20	1.4
21	1.5
22	1.7
23	1.7
24	1.8

a. Find a quadratic model to fit these data. Round the coefficients in the model to four decimal places.

b. Use the limit definition of the derivative to develop the derivative formula for the rounded model.

c. Use the derivative formula in part *b* to find the slope of the model in part *a* when the input is 24 years of age.

21. *The World Almanac and Book of Facts, 1996*, 1996.
22. *The World Almanac and Book of Facts, 1995*, 1995.

4.4   The Sum Rule

Before we consider the main topic of this section, here is a list of slope formulas from Section 4.3 and its Activities.

Simple Derivative Rules

Rule Name	Function	Derivative
Constant Rule	$y = b$	$\dfrac{dy}{dx} = 0$
Linear Function Rule	$y = ax + b$	$\dfrac{dy}{dx} = a$
Power Rule	$y = x^n$	$\dfrac{dy}{dx} = nx^{n-1}$
e^x Rule	$y = e^x$	$\dfrac{dy}{dx} = e^x$
Exponential Rule	$y = b^x,\ b > 0$	$\dfrac{dy}{dx} = (\ln b)b^x$
Natural Log Rule	$y = \ln x,\ x > 0$	$\dfrac{dy}{dx} = \dfrac{1}{x}$

The Constant Multiplier Rule

Also, in the Activities, we investigated the effect of multiplying these simple functions by a constant k and then taking their derivatives.

1. $y = kx^n$ $\dfrac{dy}{dx} = knx^{n-1}$
2. $y = ke^x$ $\dfrac{dy}{dx} = ke^x$
3. $y = kb^x$ $\dfrac{dy}{dx} = k(\ln b)b^x$
4. $y = k \ln x$ $\dfrac{dy}{dx} = \dfrac{k}{x}$

The four formulas in this list are special cases of the **Constant Multiplier Rule**.

Constant Multiplier Rule

If $y = kf(x)$, then $\dfrac{dy}{dx} = kf'(x)$.

The Constant Multiplier Rule can be illustrated graphically. Figure 4.30 shows a graph of $y = x^2$ and the tangent line at $x = 1$. Note that the slope of the tangent line at $x = 1$ is 2.

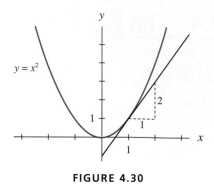

FIGURE 4.30

Figure 4.31 shows a graph of $y = 3x^2$ and the tangent line at $x = 1$. Compare the slope of $y = x^2$ at $x = 1$ with the slope of $y = 3x^2$ at $x = 1$.

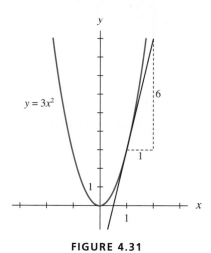

FIGURE 4.31

The slope of $y = 3x^2$ at $x = 1$ is three times the slope of $y = x^2$ at $x = 1$. That is precisely what the Constant Multiplier Rule states.

EXAMPLE 1   *Credit Card Payment*

If credit card purchases are not paid off by the due date on the credit card statement, finance charges are applied to the remaining balance. In July 1996, one major credit card company had a daily finance charge of 0.05425% on unpaid balances. Assume that the unpaid balance is $1 and that no new purchases are made.

a. Find an exponential function for the credit balance d days after the due date.

b. How much is owed after 30 days?

c. Write the derivative formula for the function from part *a*.

d. How quickly is the balance changing after 30 days?

e. Repeat parts *a* through *d*, assuming that the unpaid balance is $2000.

Solution:

a. Recall that the constant b in an exponential function $f(x) = ab^x$ is $(1 + \text{percent-age growth})$ and that the constant a is the value of $f(0)$. Thus we use the function

$$f(d) = 1(1.0005425^d) = 1.0005425^d \text{ dollars}$$

to represent the balance due d days after the due date.

b. Thirty days after the due date, the balance is $f(30) \approx \$1.02$.

c. According to the derivative rules we saw in this chapter, the derivative of $y = b^x$ is $y' = (\ln b)b^x$. Thus the derivative formula for our function $f(d)$ is

$$f'(d) = (\ln 1.0005425)1.0005425^d \text{ dollars per day}$$

after d days.

d. We use the derivative formula $f'(d)$ evaluated at $d = 30$ to find the rate of change of the balance. After 30 days, the balance is increasing 0.0006 dollar per day. That may not seem like much, but what happens when the unpaid balance is much larger than $1?

e. If the unpaid balance is $2000, the balance-due function is

$$f(d) = 2000(1.0005425^d) \text{ dollars}$$

after d days. Specifically, the amount due after 30 days is $2032.81.

According to the Constant Multiplier and Exponential Rules, the derivative of the balance function, $f(d)$, is

$$f'(d) = 2000(\ln 1.0005425)1.0005425^d \text{ dollars per day.}$$

After 30 days, the credit balance is increasing $1.10 per day.   ∎

The Sum Rule

We now use a similar graphical argument to determine the slope formula for the sum of two functions.

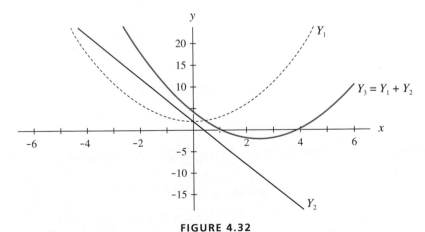

FIGURE 4.32

Figure 4.32 shows the graphs of $Y_1 = x^2 + 2$ and $Y_2 = -5x + 2$. It also shows the graph of the sum function $Y_3 = Y_1 + Y_2 = x^2 - 5x + 4$. We want to compare the slopes of Y_1, Y_2, and Y_3. The lines tangent to the graphs at $x = 1.5$ are drawn in Figure 4.33a, and the tangent lines at $x = 4$ are drawn in Figure 4.33b.

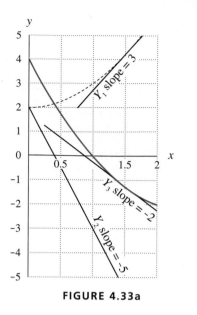

FIGURE 4.33a   **FIGURE 4.33b**

The slope of Y_1 at $x = 1.5$ is 3. The slope of Y_2 at $x = 1.5$ is -5. (The slope of Y_2 at any point is -5 because Y_2 is a linear function.) As you might have suspected, the slope of Y_3 at $x = 1.5$ is $3 + (-5) = -2$. Similarly, at $x = 4$, the slope of Y_1 is 8, the slope of Y_2 is -5, and the slope of Y_3 is 3. It is not a coincidence that the slope of Y_3 is the sum of the slopes of Y_1 and Y_2. This property for functions that are sums of other functions is called the **Sum Rule.**

The Sum Rule

If $h(x) = f(x) + g(x)$, then $\dfrac{dh}{dx} = \dfrac{df}{dx} + \dfrac{dg}{dx}$.

The Sum Rule is especially useful for finding derivatives of quadratic and cubic models, because these models are sums of simple power functions.

EXAMPLE 2   *Finding Derivatives*

Find the derivatives of

a. $y = 5x^5$ b. $f(x) = 4x^2 + 17x + 3$
c. $g(t) = 39(1.2^t)$ d. $A(b) = 7e^b - 12$

Solution:

a. Apply the Constant Multiplier and Power Rules:

$$y' = 5(5x^4) = 25x^4$$

b. Apply the Sum, Constant Multiplier, Power, and Constant Rules:

$$\frac{df}{dx} = 4(2x) + 17 + 0 = 8x + 17$$

c. Apply the Constant Multiplier and Exponential Rules:

$$\frac{dg}{dt} = 39[(\ln 1.2)1.2^t] = 39(\ln 1.2)1.2^t$$

d. Apply the Sum, e^x, Constant Multiplier, and Constant Rules:

$$\frac{dA}{db} = 7(e^b) - 0 = 7e^b$$ ∎

EXAMPLE 3   *Average Fuel Consumption*

Table 4.22[23] gives average yearly fuel consumption per car in the United States from 1980 through 1990.

TABLE 4.22

Year	Fuel consumption (gallons per car)	Year	Fuel consumption (gallons per car)
1980	591	1986	526
1981	576	1987	514
1982	566	1988	509
1983	533	1989	509
1984	536	1990	505
1985	525		

Find a model for the data, and use it to approximate how rapidly fuel consumption was declining in 1990.

Solution: On the basis of a scatter plot of the data, we fit a quadratic model.

Fuel consumption $= g(t) = 0.775t^2 - 140.460t + 6868.818$ gallons per car

where $t = 80$ in 1980, 81 in 1981, and so on. Applying the Sum, Power, and Constant Rules, we find that the derivative of $g(t)$ is

$$\frac{dg}{dt} = 2(0.775)t - 140.460 + 0$$

so

$$\frac{dg}{dt} = 1.55t - 140.460 \text{ gallons per car per year}$$

t years after 1900.

Evaluating the derivative at $t = 90$ gives -0.96 gallons per car per year as the rate of change of fuel consumption in 1990. Thus we estimate that in 1990, the average fuel consumption per car was decreasing at a rate of 0.96 gallons per car per year.   ∎

---

23. *Statistical Abstract*, 1992.

4.4 Concept Inventory

- Simple derivative rules
- Constant Multiplier Rule
- Sum Rule

4.4 Activities

Give the derivative formulas for the functions in Activities 1 through 10.

1. $y = 7$

2. $f(x) = -3x + 4$

3. $g(x) = 3e^x$

4. $y = 4 \ln x$

5. $h(x) = 7x^2 - 12x + 13$

6. $y = 1 + 4.9876e^x$

7. $g(x) = 17(4.962)^x$

8. $f(x) = 0.0127x^3 + 9.4861x^2 - 0.2649x + 128.9800$

9. $f(x) = 100,000\left(1 + \frac{0.05}{12}\right)^{12x}$ (*Hint:* Rewrite $f(x)$ as ab^x.)

10. $y = \dfrac{-9}{x^2}$

11. The graph in Figure 4.4.1 shows the temperature values (°F) on a typical May day in a certain midwestern city.

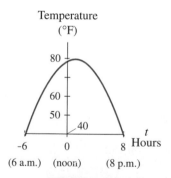

Temperature (°F)

FIGURE 4.4.1

The equation of the graph is

$$\text{Temperature} = -0.8t^2 + 2t + 79 \text{ °F,}$$

where $t = 0$ at noon. Use the derivative formula to verify each of the following statements.

a. The graph is not as steep at 1:30 p.m. as it is at 7 a.m.

b. The slope of the tangent line at 7 a.m. is 10 °F per hour.

c. The instantaneous rate of change of the temperature at 12 noon is 2 °F per hour.

d. At 4 p.m. the temperature is falling by 4.4 °F per hour.

12. A graph representing a test grade (out of 100 points) as a function of hours studied is given in Figure 3.36 on page 180. The equation of the graph is

$$G(t) = -0.044t^3 + 0.918t^2 + 38.001 \text{ points after}$$
$$t \text{ hours of study.}$$

Use the equation to verify each of the following statements.

a. $\frac{dG}{dt} = 1.704$ points per hour when $t = 1$ hour.

b. $G'(4) = 5.232$ points per hour.

c. The slope of the tangent line when $t = 15$ hours is approximately -2 points per hour.

13. The number of personal computers (in millions of units) sold to colleges and universities from 1981 through 1988 can be modeled[24] by the equation

$$P = 0.00595x^2 + 0.04095x + 0.025 \text{ millions PCs}$$

where x equals the number of years since 1981.

a. Find the number of PCs sold in 1990. Do you believe this extrapolation to be reasonable? Explain.

b. Find the rate of change of the number of PCs in 1988.

c. Find the percentage rate of change in 1988.

14. The number of general aviation aircraft accidents from 1985 through 1990 can be modeled[25] by the equation $A(x) = 2713.8095 - 120.8571x$ accidents, where x is the number of years since 1985.

a. How rapidly did the number of accidents rise or fall from 1985 through 1990?

b. How rapidly was the number of accidents rising or falling in 1987?

c. What was the percentage rate of change in the number of accidents in 1985? in 1987? in 1990?

15. The number of live births to U.S. women 45 years and older between 1950 and 1985 can be modeled[26] by the equation

24. Based on data from *Statistical Abstract*, 1992.
25. Based on data from *Statistical Abstract*, 1992.
26. Based on data from *The 1992 Information Please Almanac*.

$$B(x) = 0.431x^3 - 88.903x^2 + 5868.697x$$
$$- 119,801.648 \text{ births}$$

where $x = 50$ in 1950, 60 in 1960, and so on.

a. Was the number of live births rising or falling in 1955? in 1980?

b. How rapidly was the number of live births rising or falling in 1955 and in 1980?

16. Dairy company managers have found that it costs them approximately $c(u) = 3250 + 75(\ln u)$ dollars to produce u units of dairy products each week. They also know that it costs them approximately $s(u) = 50u + 1500$ dollars to ship u units. Assume that the company ships its products once each week.

a. Write the formula for the total weekly cost of producing and shipping u units.

b. Write the formula for the rate of change of the total weekly cost of producing and shipping u units.

c. How much does it cost the company to produce and ship 5000 units in 1 week?

d. What is the rate of change of total production and shipping costs at 5000 units?

17. The managers of Windolux, Inc. have modeled some cost data and found that if they produce x storm windows each hour, the cost (in dollars) to produce one window is given by the function

$$C(x) = 0.0146x^2 - 0.7823x + 46.9125 + \frac{49.6032}{x}$$

Windolux sells its storm windows for $175 each. (You may assume that every window made will be sold.)

a. Write the formula for the profit made from the sale of one storm window when Windolux is producing x windows each hour.

b. Write the formula for the rate of change of profit.

c. What is the profit made from the sale of a window when Windolux is producing 80 windows each hour?

d. How rapidly is profit from the sale of a window changing when 80 windows are produced each hour?

18. A publishing company estimates that when a new book by a best-selling American author first hits the market, its sales can be predicted by the equation $n(x) = 68,952.921\sqrt{x}$, where $n(x)$ represents the total number of books sold in the United States by the end of the xth week. The number of books sold abroad by the end of the xth week can be modeled by $a(x) = -0.039x^2 + 125.783x$ books.

a. Write the formula for the total number of books sold in the United States and abroad by the end of the xth week.

b. Write the formula for the rate of change of the total number of books sold.

c. How many books will be sold by the end of the first year (that is, after 52 weeks)?

d. How rapidly are books selling at the end of the first year?

19. The joint population of the United States and Canada can be modeled[27] as

$$n(x) = 0.0426x^2 + 2.049x + 251.796 \text{ million people}$$

for the period from 1980 through 1994 where x is 0 in 1980. The population of Mexico for the same period can be modeled[28] as

$$m(x) = 68.738(1.0213)^x \text{ million people}$$

x years after 1980.

a. Find the formula for the combined population of the United States, Canada, and Mexico from 1980 through 1994.

b. Find the formula for the rate of change of the combined population of the United States, Canada, and Mexico.

c. According to the models, what was the combined population of the United States, Canada, and Mexico in 1994?

d. How rapidly was the combined population of the United States, Canada, and Mexico changing in 1994?

20. When working with sums of models, why is it important to make sure that the input values of the models correspond? (That is, why must you align all models in the sum the same way?)

---

27. Based on data from *Statistical Abstract*, 1994.
28. Based on data from SPP and INEGI, Mexican Censuses of Population 1921 through 1990 as reported in Pick and Butler, *The Mexico Handbook*, Westview Press, 1994.

21. The purchasing power of the dollar as measured[29] by producer prices from 1978 through 1991 is given in Table 4.23. (In 1982, one dollar was worth $1.00.)

TABLE 4.23

Year	Purchasing power of $1
1978	1.43
1979	1.29
1980	1.14
1981	1.04
1982	1.00
1983	0.98
1984	0.96
1985	0.96
1986	0.97
1987	0.95
1988	0.93
1989	0.88
1990	0.84
1991	0.82

a. Observe a scatter plot to determine an appropriate model. Find a model to fit the data.

b. Using only the values given in the table, estimate the rate of change in 1979.

c. Use the model to find the rate of change in 1979, and compare it with your answer to part b.

d. Using the model, find the percentage rate of change in 1990.

22. An artisan makes hand-crafted painted benches to sell at a craft mall on weekends. Her weekly revenue and costs (not including labor) are given in Table 4.24.

a. Find models for revenue and cost.

b. Construct a model for profit from the revenue and cost models in part a.

TABLE 4.24

Number of benches sold each week	Weekly revenue (dollars)	Weekly cost (dollars)
1	300	57
3	875	85
5	1375	107
7	1750	121
9	1975	143
11	1950	185
13	1700	213

c. Write the derivative formula for profit.

d. Find and interpret the rate of change of profit when she sells 6 benches.

e. Repeat part c for 9 benches and 10 benches.

f. What does the information in part e tell you about the number of benches she should produce each week?

23. Table 4.25[30] below gives the number of imported passenger cars (in thousands) sold in the United States from 1984 through 1992.

TABLE 4.25

Year	Number of imported passenger cars sold (thousands)
1984	2439
1985	2838
1986	3245
1987	3196
1988	3004
1989	2699
1990	2403
1991	2038
1992	1938

29. *Statistical Abstract*, 1992.
30. *Statistical Abstract*, 1992 and 1994.

a. Find a model to fit the data.

b. Write the formula for the derivative of the model in part *a*.

c. Use your model to find how rapidly the number of imported-car sales was changing in 1992.

d. Use the model to predict the number of imported cars sold in 1993.

e. What was the percentage rate of change in the number of imported cars sold in 1984?

24. Production costs (in dollars per hour) for a company to produce between 10 and 90 units per hour are given in Table 4.26.

TABLE 4.26

Units	Cost (dollars per hour)
10	150
20	200
30	250
40	400
50	750
60	1400
70	2400
80	3850
90	5850

a. Graph a scatter plot of the data, and discuss its curvature.

b. Consider the cost for producing 0 units to be $0. Include (0, 0) in the data, and find the first differences of the cost data. Convert each of these differences to average rates of change. Use the average rates of change to argue that this data set has an inflection point.

c. Find a cubic model for production costs. Include the point (0, 0).

d. Convert the model in part *c* to one for the average cost per unit produced.

e. Find the slope formula for average cost.

f. How rapidly is the average cost changing when 15 units are being produced? 35 units? 85 units?

25. The amount spent on pollution abatement and control from 1983 through 1989 (in millions of dollars) is given in Table 4.27.[31]

TABLE 4.27

Year	Amount (millions of dollars)
1983	61,779
1984	68,929
1985	74,636
1986	78,717
1987	81,486
1988	86,063
1989	91,348

a. Fit a model to the data.

b. Write the rate-of-change formula for this model.

c. Use your model to find the rate of change and percentage rate of change of spending in 1986.

d. Use your model to estimate the amount spent in 1990.

26. An individual has $45,000 to invest: $32,000 will be put into a low-risk mutual fund averaging 6.2% interest compounded quarterly, and the remainder will be invested in a bond fund averaging 9.7% interest compounded continuously.

a. Find an equation for the total amount in the two investments.

b. Give the rate-of-change equation for the combined amount.

c. How rapidly is the combined amount of the investments growing after 6 months? after 15 months?

27. In 1985, the Bureau of the Census released data on the median family income for the past 40 years.[32] (Median income means that half of American families make more than this value and half make less).

---

31. *Statistical Abstract*, 1992.

32. U.S. Bureau of the Census, Current Population Reports, p. 60. No. 151 and 1994 *Statistical Abstract of the United States.*

TABLE 4.28

Year	Median family income (dollars)
1947	14,095
1957	18,326
1967	24,680
1977	27,440
1980	26,500
1984	26,433
1992	36,812

a. Fit a model to the data.

b. Find a formula for the rate of change of median family income.

c. Find the rates of change and percentage rates of change in 1972, 1980, 1984, and 1992.

d. Do you think the above rates of change and percentage rates of change affected the re-election campaigns of Presidents Nixon (1972), Carter (1980), Reagan (1984), and Bush (1992)?

28. Table 4.29[33] shows the total emission, in millions of metric tons of nitrogen oxides, NO_x, in the United States from 1940 through 1990.

TABLE 4.29

Year	NO_x emissions (millions of metric tons)
1940	6.9
1950	9.4
1960	13.0
1970	18.5
1980	20.9
1990	19.6

a. Discuss the behavior of the data. Would you expect other pollutants (such as carbon monoxide) to exhibit similar behavior over the same time interval?

b. Fit a model to the data.

c. On the basis of the model, how rapidly were emissions changing in 1950, 1980, and 1990?

d. Use your model to estimate the level of emissions in 1995. Do you believe your estimation is accurate? Explain.

29. The consumer price index[34] for all items from 1985 through 1991 is given in Table 4.30. [From 1982 to 1984, the consumer price index (CPI) was 100.]

TABLE 4.30

Year	CPI	Year	CPI
1985	107.6	1989	124.0
1986	109.6	1990	130.7
1987	113.6	1991	136.2
1988	118.3		

a. Fit quadratic and cubic models to the data. Explain which model you believe is the better one to describe the consumer price index.

b. Compare the rates of change of the CPI in 1992 given by the two models.

30. The following data give the value of a $2000 investment after 2 years in an account whose annual percentage yield is $r\%$ (in decimals).

APY	Value (dollars)	APY	Value (dollars)
0.02	2080.80	0.08	2332.80
0.03	2121.80	0.09	2376.20
0.04	2163.20	0.10	2420.00
0.05	2205.00	0.11	2464.20
0.06	2247.20	0.12	2508.80
0.07	2289.80		

a. Find a model for the value of a $2000 investment after two years as a function of the interest rate (in decimals).

33. *Statistical Abstract*, 1992.
34. *Statistical Abstract of the U.S.*, 1994.

b. Find the rate of change of the value of the investment when the annual percentage yield is 4%.

c. Repeat part *a*; however, instead of entering the interest rate *r*% in decimals, enter the rate in whole numbers so that $r = 2$ at 2%, $r = 3$ at 3%, etc. Compare this model with the model found in part *a*.

d. Use the model from part *c* to find the rate of change of the value of the investment when the annual percentage yield is 4%. Compare this answer to that of part *b*.

31. The value of a $1000 investment after 10 years in an account whose interest rate is *r*% (where *r* is input in decimals) compounded continuously is

$$A(r) = 1000e^{10r} \text{ dollars}$$

a. Write the rate of change function for the value of the investment.

b. Determine the rate of change of the value of the investment at 7% interest. Discuss why the rate of change appears to be 100 times as large as expected.

c. If the rate *r*% is input as a percentage instead of in decimals, the function for the value of a $1000 investment after 10 years in an account whose interest rate is *r*% compounded continuously is

$$A(r) = 1000e^{0.1r} \text{ dollars}$$

where $r = 1.00$ when the interest rate is 1%, $r = 1.25$ when the interest rate is 1.25%, etc. Write the rate of change function for the value of the investment. Compare this rate of change function to that in part *a*.

d. Determine the rate of change of the value of the investment at 7% interest. Compare this answer to that of part *b*.

4.5   The Chain Rule

The First Form of the Chain Rule

It is well known that high levels of carbon dioxide (CO_2) in the atmosphere are linked to increasing populations in highly industrialized societies. This is because large urban environments consume enormous amounts of energy and CO_2 is a natural by-product of the (often incomplete) consumption of that energy.

Imagine that in a certain large city, the level of CO_2 in the air is linked to the size of the population by the equation $C(p) = \sqrt{p}$, where the units of $C(p)$ are parts per million (ppm) and *p* is the population. Also suppose that the population is growing quadratically and can be modeled by the equation $p(t) = 400t^2 + 2500$ people, where *t* is the number of years since 1980. Note that *C* is a function of *p*, and *p* is a function of *t*. Thus indirectly, *C* is also a function of *t*. Suppose we want to know the rate of change of the CO_2 concentration *with respect to time* in 1993—that is, how rapidly the CO_2 concentration is rising or falling in 1993. The notation for this rate of change is $\frac{dC}{dt}$, and the units are ppm per year.

If we take the derivative of $C(p)$, we get $\frac{dC}{dp} = \frac{1}{2\sqrt{p}}$ ppm per person. But this is not the rate of change that we want because $\frac{dC}{dp}$ is the rate of change *with respect to population*, not time. The question now becomes "How do we transform ppm per person to ppm per year?"

If we knew the rate of change of population with respect to time (people per year), then we could multiply as indicated to get the desired units:

$$\left(\frac{\text{ppm}}{\text{person}}\right)\left(\frac{\text{people}}{\text{year}}\right) = \frac{\text{ppm}}{\text{year}}$$

Population is given as a function of time, so its derivative is the rate of change that we need: $\frac{dp}{dt} = 800t$ people per year. Thus

$$\left(\frac{dC}{dp}\right)\left(\frac{dp}{dt}\right) = \frac{dC}{dt} \quad \text{or}$$

$$\left(\frac{1}{2\sqrt{p}}\frac{\text{ppm}}{\text{person}}\right)\left(800t\frac{\text{people}}{\text{year}}\right) = \frac{dC}{dt}\frac{\text{ppm}}{\text{year}}$$

Because $\frac{dC}{dt}$ is a rate of change *with respect to time*, it is standard procedure to write the derivative formula in terms of t. Recall that $p(t) = 400t^2 + 2500$ people, where t is the number of years since 1980. Substituting $400t^2 + 2500$ for $p(t)$ in the equation for $\frac{dC}{dt}$, we have

$$\frac{dC}{dt} = \frac{1}{2\sqrt{400t^2 + 2500}}(800t) \text{ ppm/year}$$

Now we substitute $t = 13$ (for 1993) to obtain our desired result:

$$\frac{dC}{dt} = \frac{1}{2\sqrt{400(13)^2 + 2500}}[800(13)] \approx 19.64 \text{ ppm/year}$$

In 1993, the CO_2 concentration was increasing by approximately 19.64 ppm per year.

The method used to find $\frac{dC}{dt}$ in the situation above is called the **Chain Rule** because it links together the derivatives of two functions to obtain the derivative of their composite function.

The Chain Rule (Form 1)

If C is a function of p and p is a function of t, then

$$\frac{dC}{dt} = \left(\frac{dC}{dp}\right)\left(\frac{dp}{dt}\right)$$

EXAMPLE 1   *Violin Production*

Let $A(v)$ denote the average cost to produce a student violin when v violins are produced, and let $v(t)$ represent the number (in thousands) of student violins produced t years after 1990. Suppose that 10 thousand student violins are produced in 1998 and that the average cost to produce a violin at that time is \$42.10. Also, suppose that in 1998 the production of violins is increasing by 100 violins per year and the average cost of production is decreasing by 1.5 cents per violin.

a. Describe the meaning and give the value of each of the following in 1998:
      i. $v(t)$ ii. $v'(t)$ iii. $A(v)$ iv. $A'(v)$

b. Calculate the rate of change with respect to time of the average cost for student violins in 1998.

Solution:

a.   i. There are $v(8) = 10$ thousand violins produced in 1998.
      ii. The rate of change of violin production in 1998 is $v'(8) = 0.1$ thousand violins per year. That is, $\frac{dv}{dt} = 100$ violins per year.

iii. The average cost to produce a violin is $A(10) = \$42.10$ when 10 thousand violins are produced.

iv. When 10 thousand violins are produced, the average cost is changing at a rate of $A'(10) = -\$0.015$ per violin. That is, $\frac{dA}{dv} = -\$0.015$ per violin.

b. The rate of change with respect to time of the average cost to produce a student violin in 1998 is

$$\frac{dA}{dt} = \frac{dA}{dv} \cdot \frac{dv}{dt}$$

$$= (-\$0.015 \text{ per violin})(100 \text{ violins per year})$$

$$= -\$1.5 \text{ per year}$$

In 1998, the average cost to produce a violin is declining by $1.50 per year. ∎

The Second Form of the Chain Rule

Recall the discussion at the beginning of this section concerning CO_2 pollution and population. We were given two functions, $C(p)$ and $p(t)$, and then asked to find the derivative $\frac{dC}{dt}$. You may wonder why we did not substitute the expression for population into the CO_2 equation before finding the derivative:

$$C(p) = \sqrt{p} \text{ with } p(t) = 400t^2 + 2500$$

so

$$C(p(t)) = \sqrt{400t^2 + 2500}$$

This process, called function composition (see Section 1.3), allows us to express C directly as a function of t. If we now take the derivative, we get $\frac{dC}{dt}$, which is exactly what we needed all along! The reason we did not do this before is that we did not know a formula for finding the derivative of a composite function. However, we can now use the Chain Rule to obtain a formula. First, we review some terminology from Section 1.3.

Because $p(t)$ was substituted into $C(p)$ to create the composite function $C(p(t))$, we call $p(t)$ the inside function and $C(p)$ the outside function. Next, recall the Chain Rule process:

$$\frac{dC}{dt} = \left(\frac{dC}{dp}\right)\left(\frac{dp}{dt}\right)$$

$$= \left(\frac{1}{2\sqrt{p}}\right)(800t)$$

$$= \left(\frac{1}{2\sqrt{400t^2 + 2500}}\right)800t$$

The first term, $\frac{1}{2\sqrt{400t^2 + 2500}}$, is simply the derivative of $\sqrt{p}$ with p replaced by $400t^2 + 2500$. This is the derivative of the outside function with the inside function substituted for p. The second term, $800t$, is the derivative of $p(t)$, the inside function. This leads us to a second form of the Chain Rule. If a function is expressed as a result of function composition (that is, it is a combination of an inside function and an outside function), then its slope formula can be found as follows:

$$\text{Slope formula of composite function} = \begin{pmatrix} \text{derivative of the} \\ \text{outside function} \\ \text{with the inside} \\ \text{function untouched} \end{pmatrix} \begin{pmatrix} \text{derivative} \\ \text{of the inside} \\ \text{function} \end{pmatrix}$$

Mathematically, we state this form of the Chain Rule as follows:

The Chain Rule (Form 2)

If a function $f(x)$ can be expressed as the composition of two functions $h(g)$ and $g(x)$—that is, if

$$f(x) = (h \circ g)(x) = h(g(x))$$

then its slope formula is

$$\frac{df}{dx} = f'(x) = h'(g(x)) \cdot g'(x)$$

In Example 2, we consider three somewhat different forms of composed functions, identify the inside function and the outside function for each, and use the Chain Rule to find formulas for the derivatives.

EXAMPLE 2   *Using the Chain Rule*

Write the derivatives (with respect to x) for the following three functions.

a. $y = e^{x^2}$ b. $y = (x^3 + 2x^2 + 4)^{\frac{1}{2}}$ c. $y = \dfrac{3}{4 - 2x^2}$

Solution:

a. We can think of $y = e^{x^2}$ as composed of an outside function $y = e^p$ and an inside function $p = x^2$. The derivative of the outside function is e^p. (This exponential function is its own derivative.) Form 2 of the Chain Rule instructs us to leave the inside function untouched (that is, in its original form) so instead of e^p appearing in the derivative, the first expression in the slope formula is e^{x^2}. The second expression in the slope formula is the derivative of the inside function, $2x$. The final answer is the product of these two functions.

$$\frac{dy}{dx} = \left(e^{x^2}\right)(2x) = 2x \, e^{x^2}$$

b. The inside function of $y = (x^3 + 2x^2 + 4)^{\frac{1}{2}}$ is $p = x^3 + 2x^2 + 4$, and the outside function is $y = p^{\frac{1}{2}}$. The derivative of the outside function is $\frac{1}{2}p^{-\frac{1}{2}}$; with p untouched, this becomes $\frac{1}{2}(x^3 + 2x^2 + 4)^{-\frac{1}{2}}$. The derivative of the inside function is $3x^2 + 4x$. Thus the Chain Rule gives

$$\frac{dy}{dx} = \frac{1}{2}(x^3 + 2x^2 + 4)^{-\frac{1}{2}}(3x^2 + 4x)$$

c. The function $y = \frac{3}{4 - 2x^2}$ can be thought of as the composition of the outside function $y = \frac{3}{p}$ and the inside function $p = 4 - 2x^2$. The derivative of the outside function is $\frac{-3}{p^2}$ or $\frac{-3}{(4 - 2x^2)^2}$. The derivative of the inside function is $-4x$. The derivative of the composite function is then

$$\frac{dy}{dx} = \left(\frac{-3}{(4 - 2x^2)^2}\right)(-4x) = \frac{12x}{(4 - 2x^2)^2}$$ ∎

One common use of the Chain Rule is to find the derivative of a logistic function.

EXAMPLE 3   VCRs

The percentage of households with VCRs can be modeled[35] by

$$P(t) = \frac{72.5}{1 + 75.473e^{-0.486t}} \text{ percent}$$

where t is the number of years since 1980. Find the rate-of-change formula for $P(t)$.

Solution: The function $P(t)$ can be rewritten as

$$P(t) = 72.5(1 + 75.473e^{-0.486t})^{-1}$$

In this form, it is easy to see that $72.5u^{-1}$ is the outside function and the inside function is $u = 1 + 75.473e^{-0.486t}$. Further, we can split u into an outside and inside function with $1 + 75.473e^v$ as the outside function and $v = -0.486t$ as the inside function.

Now, the derivative of $P(t)$ is

$$P'(t) = (\text{derivative of } 72.5u^{-1})(\text{derivative of } u)$$
$$= (\text{derivative of } 72.5u^{-1})[(\text{derivative of } 1 + 75.473e^v)(\text{derivative of } v)]$$
$$= (-72.5u^{-2})[(75.473e^v)(-0.486)]$$
$$= \frac{-72.5(75.473e^v)(-0.486)}{u^2}$$
$$\approx \frac{2659.291e^v}{u^2}$$

Next, substitute $u = 1 + 75.473e^{-0.486t}$ and $v = -0.486t$ back into the expression to obtain the derivative in terms of t.

$$\frac{dP}{dt} \approx \frac{2659.291e^{-0.486t}}{(1 + 75.473e^{-0.486t})^2} \text{ percentage points per year}$$ ∎

---

35. Based on data from *Statistical Abstract*, 1994.

4.5   Concept Inventory

■ Function composition

■ Inside and outside functions

■ Chain Rule:

$$\frac{dC}{dt} = \frac{dC}{dp} \cdot \frac{dp}{dt} \qquad \text{(Form 1)}$$

$$C'(t) = C'(p(t)) \cdot p'(t) \quad \text{(Form 2)}$$

4.5   Activities

1. The functions $f(x)$ and $x(t)$ can be composed to form $f(x(t))$. If $x(2) = 6, f(6) = 140, x'(2) = 1.3$, and $f'(6) = -27$, give the values of

   a. $f(x(2))$

   b. $\dfrac{df}{dx}$ when $x = 6$

   c. $\dfrac{dx}{dt}$ when $t = 2$

   d. $\dfrac{df}{dt}$ when $t = 2$

2. The functions $g(v)$ and $v(x)$ can be composed to form $g(v(x))$. If $v(88) = 17, v'(88) = 1.6, g(17) = 0.04$, and $g'(17) = 0.005$, give the values of

   a. $g(v(88))$

   b. $\dfrac{dv}{dx}$ when $x = 88$

   c. $\dfrac{dg}{dv}$ when $x = 88$

   d. $\dfrac{dg}{dx}$ when $x = 88$

3. An investor has been buying gold at a constant rate of 0.2 troy ounce per day. The investor currently owns 400 troy ounces of gold. If gold is currently worth $395.70 per troy ounce, how quickly is the value of the investor's gold increasing per day?

4. A gas station owner is unaware that one of the underground gasoline tanks is leaking. The leaking tank currently contains 600 gallons of gas and is losing 3.5 gallons per day. If the value of the gasoline is $1.51 per gallon, how much potential revenue is the station losing per day?

5. Let $R(x)$ be the revenue in deutschemarks from the sale of x units of a commodity, and let $D(r)$ be the dollar value of r deutschemarks. On March 12, 1996, 10,000 deutschemarks were worth $6750, and the rate of change of the dollar value was $0.675 per deutschemark. On the same day, sales were 476 units, producing a revenue of 10,000 deutschemarks, and revenue was increasing by 2.6 deutschemarks per unit. Identify the following values on March 12, 1996, and write a sentence interpreting each value.

   a. $R(476)$

   b. $D(10,000)$

   c. $\dfrac{dR}{dx}$

   d. $\dfrac{dD}{dr}$

   e. $\dfrac{dD}{dx}$

6. Suppose that $V(t)$ is the volume of mail (in thousands of pieces) processed at a post office on the tth day of the current year and that $E(v)$ is the number of employee-hours needed to process v thousand pieces of mail. On January 1 of this year, 150 thousand pieces of mail were processed, and that number was decreasing by 200 items per day. The rate of change of the number of employee-hours is a constant 12 hours per thousand pieces of mail. Identify the following quantities on January 1 of this year, and write a sentence interpreting each value.

   a. $V(1)$

   b. $\dfrac{dV}{dt}$

   c. $\dfrac{dE}{dv}$

   d. $\dfrac{dE}{dt}$

7. The population of a city in the northeast is given by $p(t) = \dfrac{130}{1 + 12e^{-0.02t}}$ thousand people, where t is the number of years since 1985. The number of garbage trucks needed by the city can be modeled by $g(p) = 2p - 0.001p^3$, where p is the population in thousands. Find the value of the following in 1995:

   a. $p(t)$

   b. $g(p)$

   c. $\dfrac{dp}{dt}$

   d. $\dfrac{dg}{dp}$

   e. $\dfrac{dg}{dt}$

   f. Interpret the answers to parts a–e.

8. Let $r(x) = 1.019^x$ deutschemarks be the revenue from the sale of x mountain bikes. On March 12, 1996, r deutschemarks were worth $D(r) = \frac{r}{1.4807}$ American dollars. On the same day, sales were 476 mountain bikes. Identify the following quantities on March 12, 1996:

   a. $r(x)$

   b. $D(r)$

   c. $\dfrac{dr}{dx}$

   d. $\dfrac{dD}{dr}$

   e. $\dfrac{dD}{dx}$

   f. Interpret the answers to parts a–e.

Rewrite each pair of functions in Activities 9 through 12 as a single composite function, and then find the derivative of the composite function.

9. $c(x) = 3x^2 - 2 \qquad x(t) = 4 - 6t$

10. $f(t) = 3e^t \qquad t(p) = 4p^2$

11. $h(p) = \dfrac{4}{p}$ $p(t) = 1 + 3e^{-0.5t}$

12. $g(x) = \sqrt{7 + 5x}$ $x(w) = 4e^w$

For each of the composite functions in Activities 13 through 27, identify an outside function and an inside function, and find the derivative of the composite function.

13. $f(x) = (3.2x + 5.7)^5$

14. $f(x) = (5x^2 + 3x + 7)^{-1}$

15. $f(x) = \sqrt{x^2 - 3x}$ 16. $f(x) = \sqrt[3]{x^2 + 5x}$

17. $f(x) = \ln(35x)$ 18. $f(x) = (\ln 6x)^2$

19. $f(x) = \ln(16x^2 + 37x)$ 20. $f(x) = e^{3.785x}$

21. $f(x) = 72.378e^{0.695x}$ 22. $f(x) = e^{4x^2}$

23. $f(x) = 1 + 58.32e^{0.0856x}$ 24. $f(x) = \dfrac{8}{(x - 1)^3}$

25. $f(x) = \dfrac{350}{4x + 7}$

26. $f(x) = \dfrac{112}{1 + 18.370e^{0.695x}} + 7.39$

27. $f(x) = \dfrac{3706.5}{1 + 8.976e^{-1.243x}} + 89,070$

28. U.S. factory sales of electronics in billions of dollars from 1986 through 1990 can be modeled[36] by the equation

$$s(t) = 220 - 44.43e^{-0.3912t} \text{ billion dollars}$$

where t is the number of years since 1986.

   a. Write the rate-of-change formula for sales.

   b. How rapidly were sales of electronics growing in 1986? in 1990?

   c. What was the percentage rate of change of sales in 1989?

29. Imagine that you invest $1500 in a savings account at 4% annual interest compounded continuously.

   a. Write an equation for the balance in the account after t years.

   b. Write an equation for the rate of change of the balance.

   c. What is the rate of change of the balance at the end of 1 year? 2 years?

   d. Do the rates of change in part c tell you how much interest your account will earn over the next year? Explain.

30. The tuition at a private 4-year college from 1980 through 1990 is given in Table 4.31.

TABLE 4.31

Year	Tuition (dollars)
1980	4057
1981	4434
1982	4847
1983	5298
1984	5790
1985	6329
1986	6918
1987	7561
1988	8264
1989	9033
1990	9873

   a. Fit an exponential model of the form $f(x) = ab^x$ to the data.

   b. Convert the model you found in part a to an exponential model of the form $f(x) = ae^{kx}$.

   c. Find rate-of-change formulas for both models.

   d. Use both models to find the rate of change in 1990. How do your answers compare?

31. The national debt is a much bigger concern today than it was some years ago. If you take the national debt and divide it by the number of people living in the United States, you obtain the "per capita national debt." The per capita national debt over the past several years can be modeled[37] as

$$P(x) = 5609.82e^{0.1133x} \text{ dollars}$$

where x is the number of years since 1980.

   a. Find the rate-of-change formula for the model.

   b. Complete Table 4.32.

TABLE 4.32

Year	1980	1985	1990
Per capita national debt			
Rate of change			
Percentage rate of change			

36. Based on data from *Statistical Abstract*, 1992.
37. Based on data from *Statistical Abstract*, 1992.

c. Per capita national debt can be interpreted as the amount each person in the United States would have to contribute to the United States government in order to pay off the national debt. In what year do you expect to graduate from college? According to the model, how much will your share of the national debt be by the time you graduate?

d. Note that the percentage rate of change is constant. Why (mathematically) is this not surprising?

32. The percentage of households with TVs who subscribe to cable from 1970 through 1990 can be modeled[38] by the logistic equation

$$P(t) = \frac{100}{1 + 14.96e^{-0.1527t}} \text{ percent}$$

where t is the number of years since 1970.

a. Write the rate-of-change formula for the percentage of households with TVs who subscribe to cable.

b. How rapidly was the percentage growing in 1989?

c. According to the model, what will happen to the percentage of cable subscribers in the long run? Do you believe that the model is a correct predictor of the long-term behavior? Explain.

33. Dispatchers at a sheriff's office[39] record the total number of calls received since 5 a.m. in 3-hour intervals. Total calls for a typical day are given in Table 4.33.

a. Is a cubic or a logistic model more appropriate for this data set? Explain.

b. Fit the more appropriate model to the data.

c. Find the rate-of-change formula for the model.

d. Evaluate the rate of change at noon, 10 p.m., midnight, and 4 a.m. Interpret the rates of change.

e. Discuss how rates of change can help a sheriff's office schedule dispatchers for work each day.

TABLE 4.33

Time	Total calls since 5 a.m.
8 a.m.	81
11 a.m.	167
2 p.m.	301
5 p.m.	495
8 p.m.	738
11 p.m.	1020
2 a.m.	1180
5 a.m.	1225

34. A model[40] for the number of states associated with the national P.T.A. organization is

$$m(x) = \frac{49}{1 + 36.0660e^{-0.206743x}} \text{ states}$$

x years after 1895.

a. Write the derivative of $m(x)$.

b. How many states had national P.T.A. membership in 1902?

c. How rapidly was the number of states joining the P.T.A. growing in 1890? in 1915? in 1927?

35. Civilian deaths due to the influenza epidemic in 1918 can be modeled[41] as

$$c(t) = \frac{93,700}{1 + 5095.9634e^{-1.097175t}} \text{ deaths}$$

t weeks after August 31, 1918.

a. How rapidly was the number of deaths growing on September 28, 1918?

b. What percentage increase does the answer to part *a* represent?

c. Repeat parts *a* and *b* for October 26, 1918.

d. Why is the percentage change for parts *b* and *c* decreasing even though the rate of change is increasing?

38. Based on data from *Statistical Abstract*, 1992.
39. Greenville, South Carolina, Sheriff's Office.
40. Based on data from Hamblin, Jacobsen, and Miller, *A Mathematical Theory of Social Change*. (New York: John Wiley & Sons, 1973).
41. Based on data from A.W. Crosby, Jr., *Epidemic and Peace 1918*. (Westport, Connecticut: Greenwood Press, 1976).

36. A manufacturing company has found that it can stock no more than 1 week's worth of perishable raw material for its manufacturing process. When purchasing this material, however, the company receives a discount based on the size of the order. Company managers have modeled the cost data and found that it costs the company approximately $C(t) = 196.25 + 44.45 \ln t$ dollars to produce t units per week. Each quarter, improvements are made to the automated machinery to help enhance production. The company has kept a record of the average units per week that were produced in each quarter since January 1990. These data are given in Table 4.34.

TABLE 4.34

Quarter	Units per week
Jan–Mar 1990	2000
Apr–June 1990	2070
July–Sept 1990	2160
Oct–Dec 1990	2260
Jan–Mar 1991	2380
Apr–June 1991	2510
July–Sept 1991	2660
Oct–Dec 1991	2820
Jan–Mar 1992	3000
Apr–June 1992	3200
July–Sept 1992	3410
Oct–Dec 1992	3620
Jan–Mar 1993	3880
Apr–June 1993	4130
July–Sept 1993	4410
Oct–Dec 1993	4690

a. Find an appropriate model for production per week x quarters after January 1990.

b. Use the company cost model along with your production model to write an expression modeling cost per unit as a function of the number of quarters since January 1990.

c. Use your model to predict the company's cost per week for each quarter of 1994.

d. Carefully study the cost graph from January 1990 to January 1995. According to this graph, will cost ever decrease?

e. Find an expression for the rate of change of the cost function. Look at the graph of the rate-of-change function. According to this graph, will cost ever decrease?

37. A dairy company's records reveal that it costs the company about $C(u) = 3250.23 + 74.95 \ln u$ dollars per week to produce u units each week. Consumer demand has been increasing, so the company has been increasing production to keep up with demand. Table 4.35 indicates the production of the company, in units per week, from 1980 through 1993.

TABLE 4.35

Year	Production (units per week)
1980	5915
1981	5750
1982	5940
1983	6485
1984	7385
1985	8635
1986	10,245
1987	12,210
1988	14,530
1989	17,200
1990	20,230
1991	23,610
1992	27,345
1993	31,440

a. Describe the curvature of the scatter plot of the data in Table 4.35. What types of models could be used to fit these data?

b. Find the most appropriate model.

c. Use the company's cost model along with your production model to write an expression modeling cost per week as a function of the number of years since 1980.

d. Write the rate-of-change function of the cost function you found in part c.

e. Use your model to estimate the company's cost per week in 1992, 1993, 1994, and 1995. Also, estimate the rates of change for those same years.

f. Carefully study the cost graph from 1980 to 2000. According to this graph, will cost ever decrease? Why or why not?

g. Look at the slope graph of the cost function from 1980 to 2000. According to this graph, will cost ever decrease? Why or why not?

38. The marketing division of a large firm has found that it can model the effectiveness of an advertising campaign by $S(u) = 0.75\sqrt{u} + 1.8$, where $S(u)$ represents sales in millions of dollars when the firm invests u thousands of dollars in advertising.

The firm plans to invest $u(x) = -2.3x^2 + 53.2x + 249.8$ thousand dollars each year x years from now.

a. Write the formula for predicted sales x years from now.

b. Write the formula for the rate of change of predicted sales x years from now.

c. What will be the rate of change of sales in 2001?

39. When you are composing functions, why is it important to make sure that the output of the inside function agrees with the input of the outside function?

4.6   The Product Rule

It is fairly common to construct a new function by multiplying two functions. For example, revenue is demand multiplied by price. If both demand and price are given by functions, then revenue is given by the product of the two functions. How to find rates of change for product functions is the topic of this section.

Suppose that the enrollment in a university is given by a function $E(t)$ and the percentage (expressed as a decimal) of students who are from out of state is given by a function $P(t)$. In both functions, t is the year corresponding to the beginning of the school year (that is, for the 1996–97 school year, t is 1996) because school enrollment figures are stated for the beginning of the fall term. Note that the product function $N(t) = E(t) \cdot P(t)$ gives the number of out-of-state students in year t. For example, if in the current year, enrollment began at 17,000 students with 30% of those from out of state, then the number of out-of-state students is calculated as $(17,000)(0.30) = 5100$.

Suppose that, in addition to enrollment being 17,000 with 30% from out of state, the enrollment is decreasing at a rate of 1420 students per year $\left(\frac{dE}{dt} = -1420\right)$, and the percentage of out-of-state students is increasing at a rate of 1.5 percentage points per year $\left(\frac{dP}{dt} = 0.015\right)$. How rapidly is the number of out-of-state students changing?

Because $N(t)$, the number of out-of-state students, is the product of $E(t)$ and $P(t)$, there are two rates to consider. First, of the 1420 students per year by which enrollment is declining, 30% of the students are from out of state. The product $(-1420$ students per year$)(0.30) = -426$ gives the number by which the out-of-state students are decreasing. Thus, as a consequence of the decline in enrollment, the number of out-of-state students is declining by 426 students per year.

On the other hand, of the 17,000 students enrolled, the percentage of out-of-state students is growing by 1.5% per year. Thus the product $(17,000$ students$) \cdot (0.015$ per year$) = 255$ gives the number by which the out-of-state students are increasing. The increasing percentage of out-of-state students results in an increase of 255 out-of-state students per year.

To find the overall rate of change in the number of out-of-state students, we simply add the rate of change due to decline in enrollment and the rate of change due to increase in percentage.

$$-426 \text{ students} + 255 \text{ students} = -171 \text{ students}$$
$$\text{per year} \qquad \text{per year} \qquad \text{per year}$$

We interpret this result as follows: *As a result of declining enrollment and increasing percentage of students from out of state, the number of out-of-state students is declining by 171 per year.*

The steps to obtain this rate of change can be summarized in the equation

$$\begin{aligned} \text{Rate of change of} \\ \text{out-of-state students} \end{aligned} \quad \begin{aligned} &= (-1420 \text{ students per year})(0.30) \\ &+ (17{,}000 \text{ students})(0.015 \text{ per year}) \end{aligned}$$

$$= -171 \text{ out-of-state students per year}$$

Expressed in terms of the functions $N(t)$, $E(t)$, and $P(t)$, this equation can be written as

$$\frac{dN}{dt} = \left(\frac{dE}{dt}\right)P(t) + E(t)\left(\frac{dP}{dt}\right)$$

This result is known as the **Product Rule** and can be stated as follows:

If a function is the product of two functions, that is,

$$\text{Product function} = \begin{pmatrix} \text{first} \\ \text{function} \end{pmatrix}\begin{pmatrix} \text{second} \\ \text{function} \end{pmatrix}$$

then

$$\begin{aligned} \text{Derivative} \\ \text{of product} \\ \text{function} \end{aligned} = \begin{pmatrix} \text{derivative} \\ \text{of first} \\ \text{function} \end{pmatrix}\begin{pmatrix} \text{second} \\ \text{function} \end{pmatrix} + \begin{pmatrix} \text{first} \\ \text{function} \end{pmatrix}\begin{pmatrix} \text{derivative} \\ \text{of second} \\ \text{function} \end{pmatrix}$$

The Product Rule

If $f(x) = g(x) \cdot h(x)$, then

$$\frac{df}{dx} = \left(\frac{dg}{dx}\right)h(x) + g(x)\left(\frac{dh}{dx}\right)$$

EXAMPLE 1   *Egg Production*

The industrialization of chicken (and egg) farming brought improvements to the production rate of eggs. Consider a chicken farm that has 1000 laying hens, each of which lays an average of 42 eggs each month. By selling or buying hens, the farmer can decrease or increase production. Also, by selective breeding and genetic research, it is possible that over a period of time the farmer can increase the average number of eggs that each hen lays.

a. How many eggs does the farm produce in a month?

b. Suppose the farmer increases the number of hens by 12 hens per month and increases the average number of eggs laid by each hen by 1 egg per month. By how much will the farmer's production be increasing?

Solution:

a. The farmer's current monthly production is

$$(1000 \text{ hens})(42 \text{ eggs per hen}) = 42{,}000 \text{ eggs}$$

b. The farmer will be increasing production by

$$(1000 \text{ hens})(1 \text{ egg per hen per month}) +$$
$$(42 \text{ eggs per hen})(12 \text{ hens per month}) = 1504 \text{ eggs per month} \qquad \blacksquare$$

EXAMPLE 2   *Revenue from the Sale of Compact Discs*

A music store has determined from a customer survey that when the price of each CD is $x, the number of CDs sold monthly can be modeled by the function

$$N(x) = 6250 \, (0.92985)^x \text{ CDs}$$

Find and interpret the rates of change of revenue when CDs are priced at $10, $12, $13.75, and $15.

Solution: Revenue is quantity sold times price. In this case, the monthly revenue $R(x)$ is given by

$$R(x) = N(x) \cdot x = 6250(0.92985)^x \cdot x \text{ dollars}$$

where x dollars is the selling price. Using the Product Rule, we find that the rate of change equation is

$$\frac{dR}{dx} = \frac{d}{dx}(N(x)) \cdot x + N(x) \cdot \frac{d}{dx}(x)$$
$$= 6250(\ln 0.92985)(0.92985)^x \cdot x + 6250(0.92985)^x \cdot 1$$
$$\text{dollars of revenue per dollar of price}$$

where x dollars is the selling price. Evaluating $\frac{dR}{dx}$ at the indicated values of x yields Table 4.36.

TABLE 4.36

Price	Rate of change of revenue (to nearest dollar)
$10.00	823
$12.00	332
$13.75	0
$15.00	−191

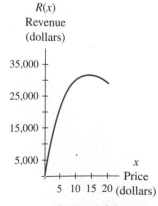

$R(x)$
Revenue
(dollars)

35,000

25,000

15,000

5,000

x

Price

5 10 15 20 (dollars)

Figure 4.34

At $10, revenue is increasing by $823 per $1 of CD price. In other words, increasing the price results in an increase in revenue. Similarly, at $12, revenue is increasing by $332 per $1 of CD price. At $13.75, revenue is neither increasing nor decreasing. This is the price at which revenue has reached its peak. Finally, at $15, revenue is declining by $191 per $1 of CD price.

The graph of the revenue function is shown in Figure 4.34. Review the statements above about how the revenue is changing as they are related to the graph. ∎

Often product functions are formed by multiplying a quantity function by a function that indicates the proportion of that quantity for which a certain statement is true. This is illustrated in Example 3.

EXAMPLE 3 *European Tourists*

The number of foreigners who traveled to the United States between 1984 and 1991 can be modeled by $f(t) = \dfrac{18,000}{1 + 36.02e^{-0.854t}} + 25,000$, where t is the number of years since 1984. Suppose that during the same time period, the proportion (percentage expressed as a decimal) of foreign travelers to the United States who were from Western Europe is given by

$$p(t) = (1.429 \cdot 10^{-5})t^2 - (2.234 \cdot 10^{-3})t + 0.08955$$

where t is the number of years since 1984.

a. Find a formula for the number of Western European tourists to the United States.

b. Find the derivative of the formula in part *a*.

c. Find the number of Western European tourists to the United States in 1989, and determine how rapidly that number was changing in that year.

Solution:

a. The number $N(t)$ of Western European tourists to the United States is given by the product function $N(t) = f(t) \cdot p(t)$.

$$N(t) = \left(\dfrac{18,000}{1 + 36.02e^{-0.854t}} + 25,000\right)[(1.429 \cdot 10^{-5})t^2$$
$$- (2.234 \cdot 10^{-3})t + 0.08955]$$

b. To use the Product Rule, we need the derivatives $f'(t)$ and $p'(t)$.

$$f'(t) = \dfrac{-18,000(36.02e^{-0.854t})(-0.854)}{(1 + 36.02e^{-0.854t})^2} = \dfrac{553,699.44e^{-0.854t}}{(1 + 36.02e^{-0.854t})^2} \text{ tourists per year, and}$$

$$p'(t) = (2.858 \cdot 10^{-5})t - 2.234 \cdot 10^{-3}$$

(Note that we did not label $p'(t)$ with units. If $p(t)$ had been expressed as a percentage, then the units would be percentage points per year. Expressed as a decimal, $p(t)$ is actually a proportion which is a unitless number. While it is possible to label $p'(t)$ as hundredths of a percentage point per year, we choose to state the derivative of $p(t)$ without a label.)

Thus, by the Product Rule,

$$N'(t) = \dfrac{553,699.44e^{-0.854t}}{(1 + 36.02e^{-0.854t})^2}[(1.429 \cdot 10^{-5})t^2 - (2.234 \cdot 10^{-3})t + 0.08955]$$
$$+ \left(\dfrac{18,000}{1 + 36.02e^{-0.854t}} + 25,000\right)[(2.858 \cdot 10^{-5})t - 2.234 \cdot 10^{-3}]$$

Western European tourists per year t years after 1984.

c. The number of Western European tourists in 1989 is $N(5)$.

$$N(5) = (36,971.0792)(0.07874) \approx 2911 \text{ tourists}$$

The rate of change of the number of Western European tourists in 1989 is $N'(5)$.

$$N'(5) \approx (3424.193)(0.07874) + (36{,}971.0792)(-0.00209)$$
$$\approx 269.62 - 77.27$$
$$\approx 192 \text{ Western European tourists per year}$$

In 1989, there were approximately 2911 tourists from Western Europe, and that number was growing by about 192 tourists per year. ■

4.6   Concept Inventory

Product Rule

If $f(x) = g(x) \cdot h(x)$, then $\dfrac{df}{dx} = \dfrac{dg}{dx} \cdot h(x) + g(x) \cdot \dfrac{dh}{dx}$

4.6   Activities

1. Find $h'(2)$ if $h(x) = f(x) \cdot g(x)$, $f(2) = 6$, $f'(2) = -1.5$, $g(2) = 4$, and $g'(2) = 3$.

2. Find $r'(100)$ if $r(t) = p(t) \cdot q(t)$, $p(100) = 4.65$, $p'(100) = 0.5$, $q(100) = 160$, and $q'(100) = 12$.

3. Let $h(t)$ be the number of households in a city, and let $c(t)$ be the proportion (expressed as a decimal) of households in that city that own a computer. In both functions, t is the number of years since 1995.

   a. Write sentences interpreting the following mathematical statements.

   i. $h(2) = 75{,}000$ ii. $h'(2) = -1200$
   iii. $c(2) = 0.52$ iv. $c'(2) = 0.05$

   b. If $N(t) = h(t) \cdot c(t)$, what are the input and output of $N(t)$?

   c. Find the values of $N(2)$ and $N'(2)$. Interpret your answers.

4. Let $D(x)$ be the demand (in units) for a new product when the price is x dollars.

   a. Write sentences interpreting the following mathematical statements.

   i. $D(6.25) = 1000$ ii. $D'(6.25) = -50$

   b. Give a formula for the revenue $R(x)$ generated from the sale of the product when the price is x dollars.

   c. Find $R'(x)$ when $x = 6.25$. Interpret your answer.

5. The value of one share of a company's stock is given by $S(x) = 15 + \frac{2.6}{x + 1}$ dollars x weeks after it is first offered. An investor buys some of the stock each week and owns $N(x) = 100 + 0.25x^2$ shares after x weeks. The value of the investor's stock after x weeks is given by $V(x) = S(x) \cdot N(x)$.

   a. Find and interpret the following:

   i. $S(10)$ and $S'(10)$
   ii. $N(10)$ and $N'(10)$
   iii. $V(10)$ and $V'(10)$

   b. Give a formula for $V'(x)$.

6. The number of students in an elementary school t years after 1996 is given by $S(t) = 100 \ln (t + 5)$ students. The yearly cost to educate one student can be modeled by $C(t) = 1500(1.05)^t$ dollars per student.

   a. What are the input and output of the function $F(t) = S(t) \cdot C(t)$?

   b. Find and interpret the following:

   i. $S(3)$ ii. $S'(3)$
   iii. $C(3)$ iv. $C'(3)$
   v. $F(3)$ vi. $F'(3)$

   c. Find a formula for $F'(t)$.

7. A wheat farmer is converting to corn because he believes that corn is a more lucrative crop. It is not feasible for him to convert all his acreage to corn at once. In the current year, he is farming 500 acres of corn and is increasing that number by 50 acres per year. As he becomes more experienced in growing corn, his output increases. He currently harvests 130 bushels of corn per acre, but the yield is increasing by 5 bushels per acre per year. When both the increasing acreage and the increasing yield are considered, how rapidly is the total number of bushels of corn increasing per year?

8. A point guard for an NBA team averages 15 free-throw opportunities per game. He currently hits 72% of his free throws. As he improves, the number of free-throw opportunities decreases by 1 free throw per game, while his percentage of hits increases by 0.5 percentage points per game. When his decreasing free throws and increasing percentage are taken into account, what is the rate of

change in the number of free-throw points that this point guard makes per game?

9. Two candidates are running for mayor in a small town. The campaign committee for candidate A has been taking weekly telephone polls to assess the progress of the campaign. Currently there are 17,000 registered voters, 48% of whom are planning to vote. Of those planning to vote, 57% will vote for candidate A. Candidate B has begun some serious mud slinging, which has resulted in increasing public interest in the election and decreasing support for candidate A. Polls show that the percentage of people who plan to vote is increasing by 7 percentage points per week, while the percentage who will vote for candidate A is declining by 3 percentage points per week.

   a. If the election were held today, how many people would vote?

   b. How many of those would vote for candidate A?

   c. How rapidly is the number of votes that candidate A will receive changing?

Find derivative formulas for the functions in Activities 10 through 20.

10. $f(x) = (x + 5)e^x$

11. $f(x) = (3x^2 + 15x + 7)(32x^3 + 49)$

12. $f(x) = 2.5(0.9)^x(\ln x)$

13. $f(x) = (12.8893x^2 + 3.7885x$
    $+ 1.2548)[29.685(1.7584)^x]$

14. $f(x) = (5x + 29)^5(15x + 8)$

15. $f(x) = (5.7x^2 + 3.5x + 2.9)^3(3.8x^2 + 5.2x + 7)^{-2}$

16. $f(x) = \dfrac{2.97x^3 + 3.05}{2.71x + 15.29}$

17. $f(x) = \dfrac{12.624(14.831)^x}{x^2}$

18. $f(x) = (8x^2 + 13)\dfrac{39}{1 + 15.29e^{-0.0954x}}$

19. $f(x) = (79.32x)\left(\dfrac{1984.32}{1 + 7.68e^{-0.859347x}} + 1568\right)$

20. $f(x) = \ln(15.7x^3)(e^{15.7x^3})$

21. During the first 8 months of last year, a grocery store raised the price of a certain brand of tissue paper from $1.19 to $1.54. Consequently, sales declined. The price and number sold each month are shown in Table 4.37.

TABLE 4.37

Month	Price	Number sold
Jan	$1.19	279
Feb	$1.25	277
Mar	$1.29	272
Apr	$1.34	266
May	$1.38	257
June	$1.45	247
July	$1.48	236
Aug	$1.54	221

   a. Find models for price and number sold as functions of the month.

   b. From the models in part a, construct an equation for revenue.

   c. Use the equation to find the revenue in August and the projected revenue in September.

   d. Would you expect the rate of change of revenue to be positive or negative in August? Why?

   e. Give the rate-of-change formula for revenue.

   f. How rapidly was revenue changing in February, August, and September?

22. A music store has determined that the number of CDs sold monthly is approximately

$$\text{Number} = 6250(0.9286)^x \text{ CDs}$$

where x is the price in dollars.

   a. Give an equation for revenue as a function of price.

   b. If each CD costs the store $7.50, find an equation for profit as a function of price.

   c. Find formulas for the rates of change of revenue and profit.

   d. Complete Table 4.38.

TABLE 4.38

Price	Rate of change of revenue	Rate of change of profit
$13		
$14		
$20		
$21		
$22		

e. What does the chart tell the store manager about the price corresponding to the highest revenue?

f. What is the price corresponding to the highest profit?

23. The population of the United States in millions as a function of the year is given by Table 4.39.[42]

TABLE 4.39

Year	Population (millions)
1960	179.3
1970	203.3
1980	226.5
1985	238.0
1990	248.7
1991	252.1
1992	255.1
1993	257.9

a. Determine the best model for the data.

b. The percentage of people in the United States who live in the midwest can be modeled[43] by the logistic curve $m(t) = \dfrac{6.2}{1 + 0.0678e^{0.144t}} + 23$, where t is the number of years since 1960. Write an expression for the number of people who live in the midwest t years after 1960.

c. Find an expression for the rate of change of the population of the midwest.

d. How rapidly was the population of the midwest changing in 1980, 1985, and 1990?

24. Production costs for a company to produce between 10 and 90 units per hour are given in Table 4.40.

a. Find an exponential model for production costs.

b. Find the slope formula for production costs.

c. Convert the model in part a to one for the average cost per unit produced.

d. Find the slope formula for average cost.

TABLE 4.40

Units	Cost (dollars)
10	150
20	200
30	250
40	400
50	750
60	1400
70	2400
80	3850
90	5850

e. How rapidly is the average cost changing when 15 units are being produced? 35 units? 85 units?

f. Examine the slope graph for average cost. Is there a range of production levels for which average cost is decreasing?

g. Determine the point at which average cost begins to increase. (That is, find the point at which the rate of change of average cost changes from negative to positive.) Explain how you found this point.

25. Table 4.41[44] gives the number (in millions) of men 65 or older in the United States and the percentage of men age 65 or older living below the poverty level.

TABLE 4.41

Year	Men 65 years or older (millions)	Percentage below poverty level
1970	8.3	20.2
1980	9.9	11.1
1985	11.0	8.7
1990	12.3	7.8
1993	12.8	8.9

a. Determine the best model for each set of data.

---

42. *Statistical Abstract*, 1992 and 1994.
43. Based on data from *Statistical Abstract*, 1992.
44. *Statistical Abstract*, 1994.

b. Write an expression for the number of men who are 65 or older and are living below the poverty level.

c. How rapidly was the number of male senior citizens living below the poverty level changing in 1990 and in 1993?

26. The number of households (in millions) with TVs in the United States is given by Table 4.42.[45]

TABLE 4.42

Year	Households (millions)	Year	Households (millions)
1970	59	1987	87
1975	69	1988	89
1980	76	1989	90
1984	84	1990	92
1985	85	1991	93
1986	86		

a. Find a logistic model for the data.

b. Table 4.43[46] gives, for 1978 through 1991, the percentages of households with TVs that also have VCRs.

TABLE 4.43

Year	Percentage	Year	Percentage
1978	0.3	1985	20.8
1979	0.5	1986	36.0
1980	1.1	1987	48.7
1981	1.8	1988	58.0
1982	3.1	1989	64.6
1983	5.5	1990	71.9
1984	10.6	1991	71.9

Align the data so that the input values correspond with those in the model for part a (that is, if 1980 is $x = 10$ in part a, then you want 1980 to be $x = 10$ here also.) Find a logistic model for the data.

c. Find a model for the number of households with VCRs.

d. Find the derivative equation for the number of households with VCRs.

e. How rapidly was the number of households with VCRs growing in 1980? in 1985? in 1990?

27. The number of births[47] (in thousands) to women who are 35 years or older is given by $n(x) = -0.03406x^3 + 1.33145x^2 + 9.91340x + 164.44689$, and the number of cesarean-section deliveries[48] (out of every 100 births) performed on women in the same age bracket is given by $p(x) = -0.18333x^2 + 2.89121x + 20.21545$, where x is the number of years since 1980.

a. Write an expression for the number of cesarean-section deliveries performed on women who are 35 years or older.

b. How rapidly was the number of cesarean sections increasing in 1980? in 1985? in 1989?

c. Suppose that the average income for an obstetrician performing a cesarean section x years after 1980 can be modeled as $c(x) = 1500 (1.07496)^x$ dollars. What was the total income that obstetricians made from performing cesarean sections on women who were 35 years or older in 1987?

d. Write an expression for the rate of change of the total income made by obstetricians in the scenario above.

e. Why do you suppose that the percentage of cesarean-section deliveries has increased substantially over the past 20 years?

28. The amount of money (in billions of dollars) spent on pollution control from 1983 through 1991 in the United States is given in Table 4.44.[49]

TABLE 4.44

Year	Amount (billions of dollars)
1983	60.002
1984	66.445
1985	70.941
1986	74.178
1987	76.672
1988	81.081
1989	85.407
1990	89.996
1991	91.456

45. *Statistical Abstract*, 1992.
46. *Statistical Abstract*, 1992.
47. Based on data from *Statistical Abstract*, 1992.
48. Based on data from *Statistical Abstract*, 1992.
49. *Statistical Abstract*, 1994.

The purchasing power of the dollar, as measured by producer prices from 1978 through 1991, is given in Table 4.45.[50] (In 1982, one dollar was worth $1.00.)

TABLE 4.45

Year	Purchasing power of $1
1978	1.43
1979	1.29
1980	1.14
1981	1.04
1982	1.00
1983	0.98
1984	0.96
1985	0.96
1986	0.97
1987	0.95
1988	0.93
1989	0.88
1990	0.84
1991	0.82

a. Fit models to both sets of data. (Remember to align such that both models have the same input values.)

b. Use these models to determine a new model for the amount, measured in constant 1982 dollars, spent on pollution abatement and control.

c. Use your new model to find the rate of change and percentage rate of change in 1986 and 1990.

d. Why might it be of interest to consider an expenditure problem in constant dollars?

29. Table 4.46[51] shows the number of students enrolled in the ninth through twelfth grades and the number of dropouts from those same grades in South

Carolina for each school year from 1980–1981 through 1989–1990.

TABLE 4.46

School year	Enrollment	Dropouts
1980–81	194,072	11,651
1981–82	190,372	10,599
1982–83	185,248	9314
1983–84	182,661	9659
1984–85	181,949	8605
1985–86	182,787	8048
1986–87	185,131	7466
1987–88	183,930	7740
1988–89	178,094	7466
1989–90	172,372	5768

a. Find a model for enrollment and a cubic model for the number of dropouts.

b. Use the two models that you found in part a to construct an equation for the percentage of high school students who dropped out each year.

c. Find the rate-of-change formula of the percentage of high school students who dropped out each year.

d. Look at the rates of change for each school year from 1980–1981 through 1989–1990. In which school year was the rate of change smallest? When was it greatest?

e. Are the rates of change positive or negative? What does this say about high school attrition in South Carolina?

30. A house painter has found that the number of jobs that he has per year is decreasing in inverse proportion to the number of years he has been in business. That is, the number of jobs he has each year

50. *Statistical Abstract*, 1992.
51. Compiled from *Rankings of the Counties and School Districts of South Carolina*.

can be modeled by $j(x) = \frac{104.25}{x}$, where x is the number of years since 1990. He has also kept a ledger of how much, on average, he was paid for each job. His income per job is presented in Table 4.47.

TABLE 4.47

Year	Income per job (dollars)
1990	430
1991	559
1992	727
1993	945
1994	1228
1995	1597
1996	2075

a. Find an exponential model for his income per job.

b. Write the formula for the painter's total income per year.

c. Write the formula for the rate of change of the painter's income each year.

d. What was the painter's total income in 1996?

e. How rapidly was the painter's income changing in 1996?

31. When you are working with products of models, why is it important to make sure that the input values of the two models correspond? (That is, why must you align both models in the same way?)

32. We have discussed three ways to find rates of change: graphically, numerically, and algebraically. Discuss the advantages and disadvantages of each method. Explain when it would be appropriate to use each method.

Chapter 4 Summary

Numerically Finding Slopes

Instantaneous rates of change are slopes of tangent lines. It is important to remember that when we are given a graph, sketching a tangent line and estimating its slope will give only an approximation to the desired rate of change. If we have no equation to associate with the graph, then such an approximation is the best that we can do, and we should therefore be as precise as possible in our work. However, when we have an equation $y = f(x)$ to associate with the curve, we can improve our graphical approximations with numerical calculations.

Because the tangent line at a point is the limiting position of secant lines through the point of tangency and other increasingly close points, the slope of the tangent line is the limiting value of the slopes of nearby secant lines. If the point of tangency is (x_0, y_0) on the graph, then we can use the equation $y = f(x)$ to calculate numerically the slopes of secant lines through the point (x_0, y_0) and nearby points $(x_0 + h, f(x_0 + h))$ where, for example, $h = \pm 0.1$, ± 0.01, ± 0.001, ± 0.0001, and so on. However, it is important to understand that these nearby points are arbitrarily chosen; other nearby points may reveal the slope of the tangent line more quickly.

Drawing Slope Graphs

The smooth, continuous graphs that we use to model real-life data have slopes (derivatives) at every point on the graph except at points that have vertical tangent lines. When these slopes (derivatives) are plotted, they usually form a smooth, continuous graph—the slope graph (rate-of-change graph, or derivative graph) of the original graph. We can also obtain slope graphs for piecewise continuous models, although the slope graph is usually not defined at the points where the model is divided. Slope graphs tell us a great deal about the change that is occurring on the original graph.

For example, what does the slope graph shown in Figure 4.35 tell us about the change occurring on the original graph?

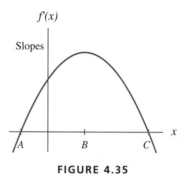

FIGURE 4.35

- At points A and C, the slopes are 0, so the original graph has horizontal tangents at these points and is therefore undergoing no change.

- To the left of point A and to the right of point C, the slopes are negative, so the original graph is decreasing to the left of A and to the right of C.

- Between A and C, the slopes are positive, so the original graph is increasing between A and C.

- The slopes increase until they reach their maximum value at B and then decrease from B to C. This tells us that the original graph undergoes its maximum rate of change at point B.

Figure 4.36 shows a possibility for the original graph.

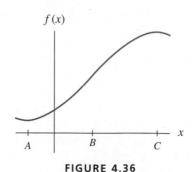

FIGURE 4.36

Slope Formulas

Until now, we have deliberately avoided using algebraic formulas, except in the various models. This was an intentional decision. We did not want to obscure the fundamental underlying relationships among secant lines, tangent lines, and rates of change. However, when a graph or a model is described by a formula, there is also a formula that describes its slope.

We discovered simple slope formulas through the numerical investigations of Section 4.3. Others were developed in the activities that follow Section 4.3. Here is a list of slope (derivative) formulas that you should know. Remember, if $y = f(x)$, then the symbolic representation for the derivative is $\frac{dy}{dx}$ or $f'(x)$.

Function	Derivative
$y = b$	$\frac{dy}{dx} = 0$
$y = ax + b$	$\frac{dy}{dx} = a$
$y = x^n$	$\frac{dy}{dx} = nx^{n-1}$
$y = e^x$	$\frac{dy}{dx} = e^x$
$y = b^x$	$\frac{dy}{dx} = (\ln b)b^x$
$y = \ln x$	$\frac{dy}{dx} = \frac{1}{x}$
$y = kf(x)$	$\frac{dy}{dx} = kf'(x)$

Perhaps the most important formula is the one that describes the derivative of an arbitrary function $f(x)$ in terms of the limiting value of slopes of nearby secant lines.

If $y = f(x)$, then $\frac{dy}{dx} = \lim\limits_{h \to 0} \frac{f(x + h) - f(x)}{h}$

This formula summarizes our conceptual development of a rate of change in terms of tangent lines and secant lines and provides the basis for the development of all the other formulas.

The Sum Rule

The Sum Rule tells us how to calculate rates of change of functions that are built from function additions. To calculate a rate of change for a function constructed by adding two functions, simply calculate the rates of change for the two component functions and add.

If $h(x) = f(x) + g(x)$, then $\frac{dh}{dx} = \frac{df}{dx} + \frac{dg}{dx}$

This rule also pertains to functions that are the sum of three or more functions. Simply calculate the individual rates of change and add them. Of course, the rule also applies to the subtraction of functions.

The Chain Rule

The Chain Rule is conceptually more difficult. It tells us how to calculate rates of change for a composite function $C(p(t))$. We presented it in two forms. Form 1 of the Chain Rule is most useful when we are given the two functions to compose: If C is a function of p and p is a function of t, then C can be regarded as a function of t, and

$$\frac{dC}{dt} = \left(\frac{dC}{dp}\right)\left(\frac{dp}{dt}\right)$$

If you keep track of the units on each of the two component rates of change, then this form of the Chain Rule is easy to use.

Form 2 of the Chain Rule is most useful when you are presented a function, $C(p(t))$, that you can recognize as the composition of an inside function p and an outside function C.

$$\frac{dC}{dt} = \begin{pmatrix} \text{the rate of} \\ \text{change of the} \\ \text{outside function} \\ \text{with the inside} \\ \text{function untouched} \end{pmatrix}\begin{pmatrix} \text{the rate of} \\ \text{change of the} \\ \text{inside function} \end{pmatrix}$$

$$= C'(p(t))\, p'(t)$$

The Product Rule

The Product Rule tells us how to calculate rates of change for a product function. If $f(x) = g(x) \cdot h(x)$, then

$$\frac{df}{dx} = \left(\frac{dg}{dx}\right)h(x) + g(x)\left(\frac{dh}{dx}\right)$$

If you need to calculate a rate of change for a general quotient function, say $h(x) = \frac{f(x)}{g(x)}$, simply view the quotient as a product $h(x) = f(x)\left(\frac{1}{g(x)}\right) = f(x)[g(x)]^{-1}$ and apply the Product Rule.

There are other formulas for derivatives that we have not given. However, we are providing the formulas that are most useful for the functions encountered in everyday situations associated with business, economics, finance, management, and the social and life sciences. You can look up other formulas (if you should ever need them) in a calculus book that emphasizes applications in science or engineering.

Chapter 4 Review Test

1. The number of in-hospital midwife-attended births for selected years between 1975 and 1993 is shown in Table 4.48.[52]

TABLE 4.48

Year	Number of births (thousands)
1975	19.7
1981	55.5
1987	98.4
1989	122.9
1990	139.2
1993	196.2

a. Find an exponential model for the data.

b. Numerically investigate the rate of change of the number of births in 1990. Choose at least three increasingly close points. In Table 4.49, record the close points, the slopes with four decimal places, and the limiting value with two decimal places.

TABLE 4.49

Close point	Slope
Limiting value =	

c. Interpret the limiting value in part b.

d. Give the formula for the derivative of your model in part a. Evaluate the derivative in 1990.

2. The total amount of long-term new mortgages (for 1–4-unit family homes) each year from 1985 through 1993 is shown in Table 4.50.[53] Also shown are a model for the data and a graph of the model. (See Figure 4.37).

TABLE 4.50

Year	Amount (billions of dollars)
1985	289.8
1986	499.4
1987	507.2
1988	446.3
1989	452.9
1990	458.4
1991	526.1
1992	893.7
1993	1019.9

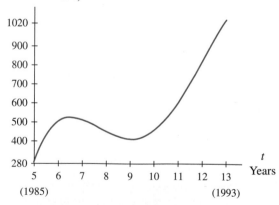

FIGURE 4.37

$A(t) = -1.74566t^4 + 68.8258t^3 - 966.380t^2 + 5759.1455t - 11,863.1488$ billion dollars t years after 1980

52. As reported in the Reno *Gazette-Journal*, August 5, 1996, page 1A

53. Cocheo, Steve, "Give Me Your Delinquents, Your Former Bankrupts, Yearning to Borrow . . ." *ABA Banking Journal*, Vol. 88 (August 1996), pp. 31–63.

a. Use only the data to estimate the rate of change in the yearly new-mortgage amounts in 1992.

b. Use the graph to estimate how quickly the new-mortgage amount was changing in 1992.

c. Find the derivative of the model in 1992. Interpret your answer.

3. Describe, in as much detail as possible, the slopes of the graph shown in Figure 4.38. On the basis of your description, sketch a slope graph on the axes given. Label the horizontal and vertical axes.

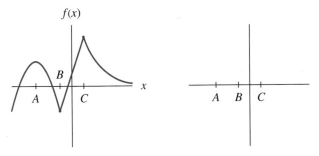

FIGURE 4.38

4. The total outstanding mortgage debt in the United States between 1980 and 1993 can be modeled[54] by

$$D(t) = \frac{3175}{1 + 12.38e^{-0.3902t}} + 1200 \text{ billion dollars}$$

t years after 1980

a. Find a formula for $\frac{dD}{dt}$.

b. Find the value of $D'(10)$. Interpret your answer.

c. How quickly was the total outstanding mortgage debt growing in 1993?

5. To find the percentage of outstanding mortgage debt represented by new mortgages each year, we divide the model in Question 2 by the model in Question 4 and multiply by 100.

$$P(t) = \frac{A(t)}{D(t)} \times 100 \text{ percent } t \text{ years after 1980}$$

a. Discuss how you would find a formula for the rate of change of the percentage of total mortgage debt represented by new mortgages each year.

b. Draw an input/output diagram for $P'(t)$, labeling units on input and output.

6. In your own words, outline the four-step method for calculating derivatives. Illustrate the method for the function $f(x) = 7x + 3$.

---

54. Based on data from *Statistical Abstract*, 1994.

Project 4.1

Fertility Rates

Setting

The *Statistical Abstract of the United States* (1995 Edition) reports the following fertility rates for whites in the United States.

TABLE 4.51

Year	Fertility rate
1970	2385
1972	1907
1974	1749
1976	1652
1978	1668
1980	1773
1982	1767
1984	1749
1986	1776
1988	1857
1990	2003
1992	1994
1994	1976

Tasks

1. Find a table of fertility rate data in a current edition of the *Statistical Abstract*, and summarize in your own words the meaning of "fertility rate." Assign units to the fertility rate column in Table 4.51.

2. Find a piecewise model that fits the given fertility rate data. Write the derivative function for this model. Construct a table of fertility rates and the rate of change of fertility rates for all years from 1970 through 1994. Discuss the points at which the derivative of your model does not exist (and why it does not exist), and explain how you estimated the rates of change at these points.

3. Using the data in a current edition of the *Statistical Abstract*, complete the same analysis as in Task 2 for the fertility rates for blacks in the United States.

4. Add 1996 and any other recent data for the fertility rate of whites. (*Note:* Recent editions of the *Statistical Abstract* occasionally update older data points. Therefore, you should check the data for pre-1996 as well and change any updated values so that your data agree with the most recent *Statistical Abstract*.) Find a piecewise model for the updated data. Use this new model to calculate the rates of change that occurred in the years since 1995.

Reporting

Write a report discussing your findings and their demographical impact. Include your mathematical computations as an appendix.

Project 4.2

Superhighway

Setting

The European Communities have decided to build a new superhighway that will run from Berlin through Paris and Madrid and end in Lisbon. This superhighway, like some others in Europe, will have no posted maximum speed so that motorists may drive as fast as they wish. There will be three toll stations installed, one at each border. Because these stations will be so far apart and motorists may not anticipate their need to stop, there has been widespread concern about the possibility of high speed collisions at these stations. There have been editorials protesting the installation of the toll stations, and a private-interest group has lobbied to delay the building of the new superhighway. In response to the concern for safety, the Committee on Transportation has determined that flashing warning lights should be installed at an appropriate distance before each toll station. Your firm has been contracted to study known stopping distances and to develop a model for predicting where the warning lights should be installed.

Tasks

1. *Getting Started* Find data that give stopping distances as a function of speed and cite the source for the data. (*Hint:* Drivers' handbooks and Department of Transportation documents are possible sources of data on stopping distance.) Present the data in a table and as a graph. Find a model to fit the data. Justify your choice. Before using your model to extrapolate, consult someone who could be considered an authority to determine whether the model holds outside of the data range. Consult a reliable source to determine probable speeds driven on such a highway. On the basis of your model, make a recommendation about where the warning lights should be placed. Justify your recommendation. Keep in mind that the Committee on Transportation does not wish to post any speed-limit signs. Suggest what other precautions could be taken to avoid accidents at the toll stations.

2. *Proceeding* Find rates of change of your model for at least three speeds, one of which should be the speed that you believe to be most likely. Interpret these rates of change in this context. Would underestimating the most likely speed have a serious adverse affect? Support your answer.

Reporting

(Bear in mind that you are reporting to Europeans who wish to see all results in the metric system. However, because you work for an American-based company, you must also have the English equivalent.)

1. Prepare a written report of your study for the Committee on Transportation.

2. Prepare a press release for the Committee on Transportation to use when it announces the implementation of your safety precautions. The press release should be succinct and should answer the questions Who, What, When, Where, and Why.

3. (Optional) Prepare a brief (15-minute) presentation of your results. You will be presenting it to members of the Committee on Transportation.

Analyzing Change: Extrema and Points of Inflection

Analyzing change is important to almost every institution in our society. Businesses regularly concern themselves with maximizing profits and minimizing costs, and most businesses also determine when sales and profits are increasing most rapidly. Industries need to maximize their production and identify points of diminishing returns—that is, points beyond which additional resources do not generate an increased rate of production. Our government must consider not only when the Gross Domestic Product will peak, but also under what conditions it will grow most rapidly.

This chapter is devoted to analyzing change. As we shall see, our geometric understanding of rates of change as slopes of tangent lines pays big dividends, for we are essentially concerned with locating horizontal tangents. Sometimes the horizontal tangents will be for the original graph of interest; other times, they will be for the rate-of-change graph of the original function. In either case, the location of these horizontal tangents provides the critical information that we seek.

Finally, we touch briefly on the topic of marginal analysis: the use of rates of change to approximate the actual change in economic quantities such as cost, revenue, production, and profit.

If we had a function modeling the altitude of a plane *m* minutes after takeoff and another function modeling the distance the plane has traveled *m* minutes after takeoff, then we could determine the time at which important changes in altitude or distance occur. (1) How long does it take to reach 10,000 feet, the altitude at which the flight attendants normally start their cabin activities? (2) At what times is the cabin pressure changing the fastest; that is, when is the plane descending or ascending most rapidly? (3) When is the plane's acceleration greatest? (4) When is the plane's speed decreasing most rapidly?

Visualize the functions modeling altitude and distance. Describe the slope of the tangent line at each of the times in the preceding questions.

■ 5.1 Optimization

In this section, we turn our attention to finding high points (maxima) and low points (minima) on the graph of a function. Points at which maximum or minimum outputs occur are called **extreme points**, and the process of **optimization** involves techniques for finding them. Maxima and minima often can be found using derivatives, and they have important applications to the world in which we live.

We begin by examining a model for the population of Kentucky[1] from 1980 through 1993:

$$\text{Population} = p(x) = 0.395x^3 - 6.674x^2 + 30.257x + 3661.147 \text{ thousand people}$$

where x is the number of years since the end of 1980.

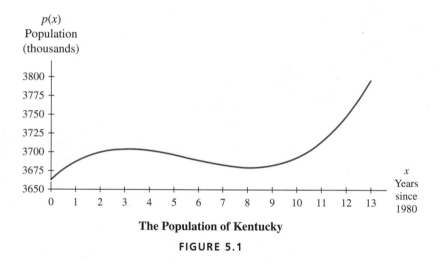

The Population of Kentucky

FIGURE 5.1

It is evident from the graph in Figure 5.1 that between 1980 and 1993, the population indicated by the model was smallest in 1980 (3661 thousand people) and greatest in 1993 (3794 thousand people). However, there are two other points of interest on the graph. Sometime near 1983, the population reached a peak. We call the peak a **relative** (or **local**) **maximum**. It does not represent the highest overall point, but it is a point to which the population model rises and after which it declines. Similarly, near 1988, the population reached a **relative** (or **local**) **minimum**. There are lower points on the graph, but in the region around this relative minimum, the population decreases and then increases as time increases.

It should be intuitively clear from the discussions in Chapter 3 that at a point where a smooth, continuous function reaches a relative maximum or minimum, the tangent line is horizontal and the slope is 0. We can consider this important link between such a function and its derivative in more detail by examining the relationship between the Kentucky population function and its slope graph. Horizontal tan-

---

1. Based on data from *Statistical Abstract*, 1994.

gent lines on the population function correspond to the points at which the slope graph crosses the horizontal axis. Figure 5.2 shows the graphs of the population function and its derivative.

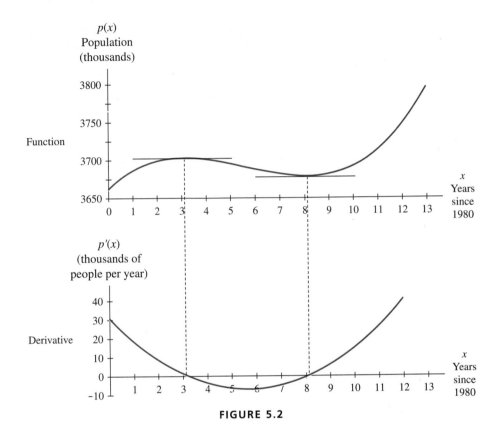

FIGURE 5.2

Note that near $x = 3$, the slope graph $p'(x)$ crosses the x-axis. This is the x-value for which $p(x)$ is at a local maximum. Likewise, $p'(x)$ crosses the x-axis at the x-value near 8 for which $p(x)$ is at a local minimum. The connection between the graphical and symbolic views of this situation is a key feature of optimization techniques.

> Finding where the slope graph crosses or touches the input axis is the same as finding *where the derivative is zero*.

The derivative of the population function is

$$\frac{dp}{dx} = 1.185x^2 - 13.348x + 30.257 \text{ thousand people per year}$$

where x is the number of years since 1980. Setting this expression equal to zero and solving for x results in two solutions: $x \approx 3.14$ and $x \approx 8.12$. This information, together with the graph of $p(x)$ shown in Figure 5.2, tells us that according to the model the population peaked in early 1984 at approximately $p(3.14) = 3703$ thousand people. We also conclude that the population declined to a local minimum in early 1989. The population at that time was approximately $p(8.12) = 3678$ thousand people.

Recall the statement at the beginning of this discussion: "It is evident from the graph in Figure 5.1 that the population from 1980 through 1993 was smallest in 1980 (3661 thousand people)." We call this point on the population graph an **absolute minimum** and refer to 3661 thousand people as the approximate **minimum value**. Also, because the 1993 population (3794 thousand people) is more than the population at the local maximum (3703 thousand people), the **absolute maximum** occurs in 1993. The **maximum value** of the population function is approximately 3794 thousand people. A function can have several different local maxima (or minima) in a given interval. However there can be only one absolute maximum value and one absolute minimum value for that interval. Any absolute maximum (or minimum) occurs either at a local maximum (or minimum) or at a point corresponding to an endpoint of the given input interval.

EXAMPLE 1   *Unemployment*

The unemployment rate in Somerset County, New Jersey, from 1983 through 1992, can be modeled[2] by

$$U(x) = 0.15603x^2 - 27.272x + 1193.493 \text{ percent}$$

where x is the number of years since the end of 1900.

a. Find the time between 1983 and 1992 at which the unemployment rate was lowest, and find the unemployment rate at that time.

b. Graph the function and its derivative. On each graph, clearly mark the time at which the unemployment rate was lowest between 1983 and 1992.

c. Is the relative minimum identified in part *a* an absolute minimum of $U(x)$ for the years 1983 through 1992?

Solution:

a. Because $U(x)$ is a concave-up parabola, we know that its minimum occurs where the derivative is zero. We consider input values where the derivative is zero:

$$U'(x) = 0.31206x - 27.272 = 0$$
$$x \approx 87.39$$

Because $x = 87$ corresponds to the end of 1987, we see that the unemployment rate was lowest sometime in the first half of 1988, with unemployment at that time near $U(87.39) \approx 1.8\%$.

---

2. Based on data from the New Jersey Department of Labor.

b.

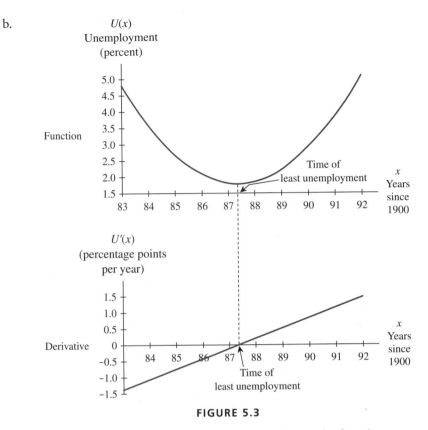

FIGURE 5.3

We see in Figure 5.3 that a relative minimum point on the function corresponds to the point at which the derivative crosses the *x*-axis.

c. Because there is no point on the graph of $U(x)$ in Figure 5.3 whose output is less than 1.8%, the absolute minimum value corresponds to the local minimum value of 1.8% found in part *a*.   ■

We have just seen how derivatives can be used to locate relative maxima and minima. *You should use caution*, however, and not automatically assume that just because the derivative is zero at a point, there is a relative maximum or relative minimum at that point. This is evident from the graphs in Figures 5.4c and 5.4d.

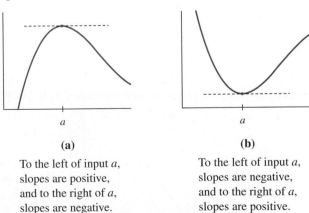

(a)

To the left of input *a*, slopes are positive, and to the right of *a*, slopes are negative.

(b)

To the left of input *a*, slopes are negative, and to the right of *a*, slopes are positive.

FIGURE 5.4

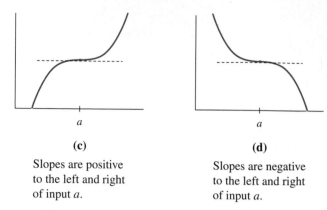

(c)

Slopes are positive
to the left and right
of input a.

(d)

Slopes are negative
to the left and right
of input a.

FIGURE 5.4

When there is a horizontal tangent line on the graph of a function (that is, when the derivative of the function is zero), one of the four situations depicted in Figure 5.4 occurs. It is therefore very important that you graph the function when using derivatives to locate extreme points.

> Always begin the process of finding extreme points by graphing the function to see whether there are any relative maxima or minima before proceeding to work with derivatives.

Let us further investigate the graph shown in Figure 5.4d. Figure 5.5 shows the graph of the function $f(x)$ that appears in Figure 5.4d and its slope (derivative) graph $f'(x)$.

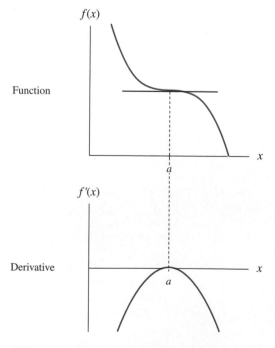

$f'(a) = 0$ but the graph of $f(x)$ has no relative maximum or minimum at a.

FIGURE 5.5

If you carefully examine the slope graph, you see that it touches the *x*-axis but does not cross it. Thus $f(x)$ does not have a relative maximum or minimum at *a*. You may notice that the derivative graph reaches its maximum at *a* as it touches the *x*-axis. Do not confuse maxima and minima on the derivative graph with maxima and minima of the original function. In the next section, we will see that maxima and minima of the derivative graph have other important interpretations in terms of the original function.

EXAMPLE 2   *Cable Company Revenue*

A cable company actively promoted sales in a town that previously had no cable service. Once the company saturated the market, it introduced a new 50-channel system and raised rates. As this company began to offer its expanded system, a different company began offering satellite service with more channels than the cable company and at a lower price. A model for the cable company's revenue for the 26 weeks after it began its sales campaign follows, and the graph of the model is shown in Figure 5.6.

$$R(x) = -3x^4 + 160x^3 - 3000x^2 + 24{,}000x \text{ dollars}$$

where *x* is the number of weeks since the cable company began sales.

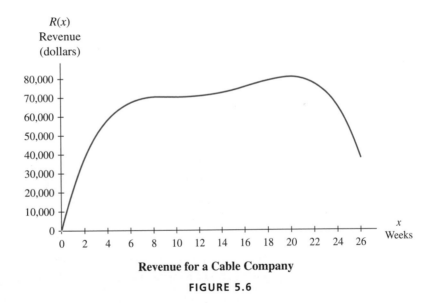

Revenue for a Cable Company

FIGURE 5.6

Determine when the company's revenue was greatest.

Solution: We know that revenue is greatest at the absolute maximum. An examination of the graph of $R(x)$ locates this point near 20 weeks. Solving the equation

$$R'(x) = -12x^3 + 480x^2 - 6000x + 24{,}000 = 0$$

gives two solutions, $x = 10$ weeks and $x = 20$ weeks. Revenue is greatest at 20 weeks, with a value of $R(20) = \$80{,}000$. This appears to correspond to the time immediately prior to when the satellite company's sales began negatively affecting the cable company.

The fact that the rate-of-change equation is zero at two places indicates that there are two places on the graph with horizontal tangent lines. Indeed, at $x = 10$ weeks, the line tangent to the curve is horizontal because the curve has leveled off.

This corresponds to the time when the cable company had saturated the market. However, no local maximum occurs at this point, because the slope graph only touches—it does not cross—the input axis at $x = 10$. Note the relationships between the rate-of-change graph and the revenue graph shown in Figure 5.7 and how they connect to the slope descriptions given in Figure 5.4.

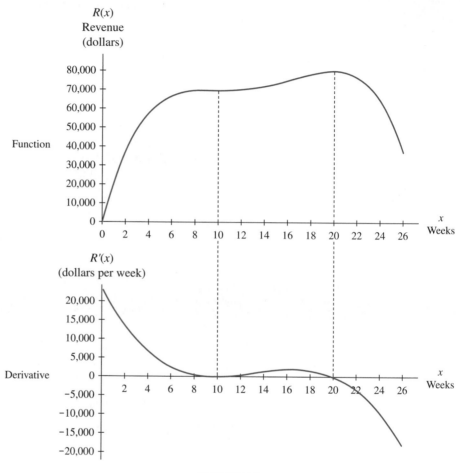

FIGURE 5.7

The maximum occurs where the derivative graph crosses the x-axis. The leveling-off point occurs where the derivative touches, but does not cross, the x-axis. ■

Is it true for every continuous function that relative maxima and minima occur only where the derivative crosses the input axis? Although this seems to be the case for most of the functions we use, consider the piecewise continuous function describing the average concentration (in nanograms per milliliter) of a 360-mg dose of a blood pressure drug in a patient's blood during the 24 hours after the drug is given:

$$C(h) = \begin{cases} -0.51h^3 + 7.65h^2 + 125 \text{ ng/mL} & \text{when } 0 \le h \le 10 \\ -16.07143h + 540.71430 \text{ ng/mL} & \text{when } 10 < h \le 24 \end{cases}$$

where h is the number of hours since the drug was given.

A close examination of $C(h)$ and its graph (see Figure 5.8) shows that $C(h)$ can be drawn without lifting your pencil from the paper; that is, $C(h)$ is continuous for all h between 0 and 24.

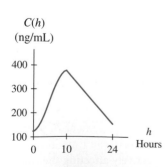

A Continuous Piecewise Function

FIGURE 5.8

It is also evident from the graph that the highest concentration of the drug occurs 10 hours after the patient receives the initial dose. Thus $C(10) = 380$ ng/ml is the maximum concentration. However, is $C'(10) = 0$? Does the slope graph for $C(h)$ cross the horizontal axis at $h = 10$? The answer is no, because $C'(10)$ does not exist. There is no line tangent to $C(h)$ at $h = 10$. That $C'(10)$ does not exist is illustrated in the graph of $C'(h)$ shown in Figure 5.9.

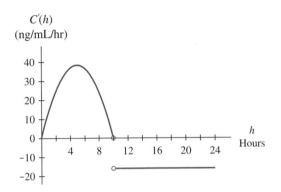

The Derivative Graph of Figure 5.8

FIGURE 5.9

It is therefore possible for an extreme point to occur where the derivative of the function does not exist as long as the function has a value at that point.

The results of these investigations are summarized in the following statements:

Conditions When Extreme Points Exist

1. If a function has a relative maximum or relative minimum at an input value c, then either $f'(c) = 0$ or $f'(c)$ does not exist but $f(c)$ *does* exist.

2. If $f'(c) = 0$ and the derivative crosses (not just touches) the input axis at c, then a relative maximum or relative minimum exists at c.

3. An absolute maximum or absolute minimum occurs at either a relative maximum or relative minimum or at a point whose input is an endpoint of a specified interval.

5.1   Concept Inventory

- Relative (local) maximum
- Relative (local) minimum
- Absolute maximum
- Absolute minimum
- Extreme points occur at an input value.
- The extreme value is an output value.
- Conditions under which extreme points exist

5.1   Activities

1. Which of the models discussed in this book could have relative maxima or minima?

2. Discuss in detail all of the options you have available for finding the relative maxima and relative minima of a function.

In Activities 3 through 9, mark the location of all relative maxima and minima with an X and all absolute maxima and minima with an O. For each extreme point that is not an endpoint, indicate whether the derivative at that point is zero or does not exist.

3.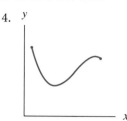

4.

5.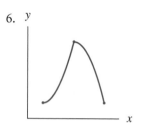

6.

7.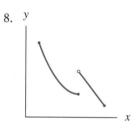

8.

9.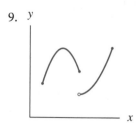

10. Sketch the graph of a function with a relative minimum at a point at which the derivative does not exist.

11. Sketch a graph of a function whose derivative is zero at a point but that does not have a relative maximum or minimum at that point.

12. Consider the function $f(x) = -0.6x^2 + 4.2x + 12.8$.

  a. Find the highest and lowest points of the function between $x = 0$ and $x = 8$.

  b. Graph the function and its derivative. Indicate the relationship between the highest point on

the function and the corresponding point on the derivative.

13. Consider the function
$$g(x) = 0.04x^3 - 0.88x^2 + 4.81x + 12.11$$

  a. Find the relative maximum and relative minimum of $g(x)$.

  b. Find the absolute maximum and the absolute minimum of $g(x)$ between $x = 0$ and $x = 14.5$.

  c. Graph the function and its derivative. Indicate the relationship between the relative maximum and minimum of $g(x)$ and the corresponding points on the derivative.

14. The U.S. Bureau of the Census' prediction for the percentage of the population 65 to 74 years old from 2000 through 2040 can be modeled[3] by

$$p(x) = (-4.474 \cdot 10^{-4}) x^3 + 0.023x^2 - 0.147x + 6.691 \text{ percent}$$

where x is the number of years after 2000.

Find the absolute maximum and the absolute minimum for the model between 2000 and 2040. Give the years and the corresponding percentages.

15. The percentage of southern Australian grasshopper eggs that hatch as a function of temperature (for temperatures between 7 °C and 25 °C) can be modeled[4] by

$$P(t) = -0.00645t^4 + 0.488t^3 - 12.991t^2 + 136.560t - 395.154 \text{ percent}$$

where t is the temperature in °C.

  a. Find the temperature between 7 °C and 25 °C that corresponds to the greatest percentage of eggs hatching.

  b. Use the equation $°F = \frac{9}{5}(°C) + 32$ to convert your answer to °F.

16. Lake Tahoe lies on the California/Nevada border, and its level is regulated by a 17-gate concrete dam at the lake's outlet. By Federal Court decree, the lake level must never be higher than 6229.1 feet above sea level. The lake level is monitored every midnight. On the basis of this lake level data,[5] the level of the lake from October 1, 1995, through July 31, 1996, can be modeled by

---

3. Based on data from *Statistical Abstract*, 1994.
4. Based on information in George L. Clarke, *Elements of Ecology* (New York: Wiley, 1954), p. 170.
5. From the Federal Watermaster, U.S. Department of the Interior.

$L(d) = (-5.345 \cdot 10^{-7})d^3 + (2.543 \cdot 10^{-4})d^2 - 0.0192d + 6226.192$ feet above sea level d days after September 30, 1995

According to the model, did the lake remain below the federally mandated level between October 1, 1995, and July 31, 1996?

17. The number of medical school students (in thousands) from 1980 through 1992, reported each fall, can be modeled[6] by

$m(t) = 0.0205t^3 - 0.396t^2 + 2.080t + 70.103$ thousand students

   t years after 1980.

   a. Find any relative maximum or minimum points on the model between 1980 and 1992.

   b. Although $m(t)$ is continuous, it must be discretely interpreted because only integer-valued inputs make sense in the context. With this restriction in mind, determine when, according to the model, the number of medical students was greatest and when it was least from 1980 through 1992.

18. The amount of fish produced for human food by fisheries in the United States from 1970 through 1992 can be modeled[7] by

$$F(t) = \begin{cases} 22.204t^2 - 108.431t + \\ \quad 2538.603 \text{ million} \\ \quad \text{pounds} \\ 85.622t^2 - 2253.951t + \\ \quad 17{,}772.387 \text{ million} \\ \quad \text{pounds} \end{cases} \begin{array}{l} \text{when} \\ 0 \le t < 10 \\ \\ \text{when} \\ 10 \le t \le 22 \\ \\ \end{array}$$

   where t is the number of years since 1970.

   a. Determine any local maxima or local minima between $t = 0$ and $t = 22$.

   b. $F(t)$ must be interpreted discretely because it gives yearly totals. With this restriction in mind, determine when production was greatest and when it was least from 1970 through 1992.

19. The flow rate (in cubic feet per second, cfs) of a river in the 24 hours after the beginning of a severe thunderstorm can be modeled by

$$C(h) = \begin{cases} -0.865h^3 + 12.045h^2 - \\ \quad 8.952h + 123.02 \text{ cfs} \\ -16.643h + 539.429 \text{ cfs} \end{cases} \begin{array}{l} \text{when} \\ 0 \le h \le 10 \\ \text{when} \\ 10 < h \le 24 \end{array}$$

   where h is the number of hours after the storm began.

   a. What were the flow rates for $h = 0$ and $h = 24$?

   b. Determine the absolute maximum and minimum flow rates between $h = 0$ and $h = 24$.

20. The average price (per 1000 cubic feet) of natural gas for residential use[8] from 1980 through 1991 is modeled by the equation

$p(x) = 0.01244x^3 - 0.2421x^2 + 1.407x + 3.4445$ dollars

   where x is the number of years since 1980.

   a. Graph the natural gas price equation, and find the local maximum and the local minimum.

   b. Graph the derivative $p'(x)$, and find the x-intercepts of the derivative. Interpret your answers.

   c. When (according to the model) was natural gas the most and least expensive between 1980 and 1991? (The average-price model must be interpreted discretely because it gives a yearly average.)

   d. Repeat part c for the interval 1983 through 1990.

21. *Swimming World* (August 1992) lists the time in seconds that an average athlete takes to swim 100 meters free style at age x years. The data are given in Table 5.1.

TABLE 5.1

Age, x (years)	Time (seconds)	Age, x (years)	Time (seconds)
8	92	22	50
10	84	24	49
12	70	26	51
14	60	28	53
16	58	30	57
18	54	32	60
20	51		

---

6. Based on data from *Statistical Abstract*, 1994.
7. Based on data from *Statistical Abstract*, 1994.
8. Based on data from *Statistical Abstract*, 1992.

a. Find the best model to fit the data.

b. Using the model, find the age at which the minimum swim time occurs. Also find the minimum swim time.

c. Compare the table values with the values in part b.

22. Table 5.2[9] lists the number of live births in the United States between the years 1950 and 1988 to women 45 years of age and older.

TABLE 5.2

Year	Number of births	Year	Number of births
1950	5322	1980	1200
1955	5430	1985	1162
1960	5182	1986	1251
1965	4614	1987	1375
1970	3146	1988	1427
1975	1628		

a. Use the data values to estimate when the number of births was greatest.

b. Use the data values to estimate when the number of births was least.

c. Find a model to fit the data, and use the model to answer parts a and b. (The model must be discretely interpreted because it describes yearly totals.)

d. How do the model answers differ from the table values?

23. *Consumer expenditure* and *revenue* are terms for the same thing from two perspectives. Consumer expenditure is the amount of money that consumers spend on a product, and revenue is the amount of money that businesses take in by selling the product. A street vendor constructs Table 5.3 on the basis of sales data.

a. Find a model for quantity sold.

b. Convert the demand equation to an equation for consumer expenditure.

c. What price should the street vendor charge to maximize consumer expenditure?

TABLE 5.3

Price of a dozen roses (dollars)	Number of dozens sold per week
10	190
15	145
20	110
25	86
30	65
35	52

d. If each dozen roses costs $6, what price should the street vendor charge to maximize profit?

24. An apartment complex has an exercise room and sauna, and tenants will be charged a yearly fee for the use of these facilities. A survey of tenants results in the demand/price data shown in Table 5.4.

TABLE 5.4

Quantity demanded	Price (dollars)
5	250
15	170
25	100
35	50
45	20
55	5

a. Find a model for price as a function of demand.

b. On the basis of the price model, give the equation for revenue.

c. Find the maximum point on the revenue model. What price and what demand give the highest revenue?

25. The yearly amount of garbage (in millions of tons) taken to a landfill outside a city during selected years from 1960 through 1990 is given in Table 5.5.

a. Find a cubic model to fit the data.

b. Give the slope formula for the model.

---

9. *The 1992 Information Please Almanac* (Boston: Houghton Mifflin Co.).

TABLE 5.5

Year	Amount (millions of tons)
1960	81
1965	99
1970	117
1975	122
1980	132
1985	145
1990	180

c. How rapidly was the amount of garbage taken to the landfill increasing in 1990?

d. Graph the derivative of your model, and determine whether your model has a relative maximum and/or minimum. Explain how you reached your conclusion.

26. A company analyzes the production costs for one of its products and determines the hourly operating costs when x units are produced each hour. The results are given in Table 5.6.

TABLE 5.6

Production level, x (units per hour)	Hourly cost (dollars)
1	210
7	480
13	650
19	760
25	810
31	845
37	880
43	950
49	1070
55	1280
61	1590

a. Find a model for hourly cost in terms of production level.

b. On the basis of the model in part a, what is the equation for the average hourly cost per unit when x units are produced each hour?

c. Find the production level that minimizes average hourly cost. Give the average hourly cost and total cost at that level.

27. Table 5.7[10] gives the amount spent on national defense in billions of dollars (measured in constant 1987 dollars).

TABLE 5.7

Year	Amount (billions of dollars)
1970	262.9
1972	219.7
1974	185.3
1976	177.8
1978	177.2
1980	187.1
1982	214.3
1984	241.7
1986	276.4
1988	283.3
1990	272.5
1992	250.2
1994	220.1 (estimated)

a. Fit a model to the data.

b. Give the slope formula for the model.

c. Find the local maximum and minimum points on the model.

d. A model for defense spending must be discretely interpreted because it provides yearly totals. With this restriction in mind, determine when, according to the model, defense spending was greatest and when it was least between 1970 and 1994.

e. If you found the answers to part c by using a graphing calculator or computer, discuss how you could find the answers without using that technology.

10. *Statistical Abstract*, 1994.

28. Table 5.8[11] gives the yearly emissions in millions of metric tons of nitrogen oxides, NO_x, in the United States from 1940 through 1990.

TABLE 5.8

Year	NO_x (millions of metric tons)
1940	6.9
1950	9.4
1960	13.0
1970	18.5
1980	20.9
1990	19.6

a. Fit a model to the data.

b. Give the slope formula for the model.

c. Determine the year in which emissions of NO_x were greatest between 1940 and 1990. What was the amount of emissions that year? (The model must be discretely interpreted.)

29. Imagine that you have been hired as director of a performing arts center for a mid-sized community. The community orchestra gives monthly concerts in the 400-seat auditorium. To promote attendance, the former director lowered the ticket price every 2 months. The ticket prices and corresponding average attendances are given in Table 5.9.

TABLE 5.9

Price (dollars)	Average attendance
35	165
30	200
25	240
20	280
15	335
10	400

a. Fit quadratic and exponential models to the data. Which model better reflects the probable attendance beyond a $35 ticket price? Explain.

b. On the basis of the model that you believe is more appropriate, give the equation for revenue.

c. Find the maximum revenue and the corresponding ticket price and average attendance.

d. What other things besides the maximum revenue should you consider in setting price?

30. Table 5.10[12] shows the Gross Domestic Product (in billions of dollars) for the United States from 1969 through 1991.

TABLE 5.10

Year	GDP (billions of dollars)
1969	498.0
1971	551.7
1973	688.3
1975	788.0
1977	1008.5
1979	1270.7
1981	1541.5
1983	1694.0
1985	2035.5
1987	2265.9
1989	2596.5
1991	2732.6

a. Find a cubic model that fits the data.

b. Write the derivative of the model you found in part a.

c. Using the answer to part b, determine in what year your model predicts that GDP will peak. What does your model predict GDP to be in that year? (The model must be interpreted discretely.)

d. Do you believe the predictions in part c to be accurate? Support your answer.

11. *Statistical Abstract*, 1992.
12. *Economic Report of the President*, 1993.

■ 5.2   Inflection Points

In Section 5.1, we discussed some important points that may occur on a model. These were called maxima and minima (or extreme points). Another important point is an **inflection point**.

Recall from our earlier work with cubic and logistic models that an inflection point is a point where a graph changes concavity. On a smooth, continuous model, the inflection point can also be thought of as the point of greatest or least slope. In real-life applications, this point is interpreted as *the point of most rapid change or least rapid change.* (See Figure 5.10.)

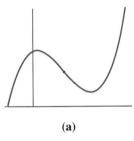

(a)

inflection point:
point of least slope,
point of most rapid decrease

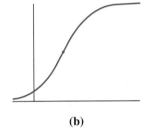

(b)

inflection point:
point of greatest slope,
point of most rapid increase

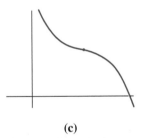

(c)

inflection point:
point of greatest slope,
point of least rapid decrease

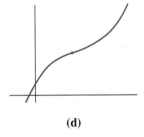

(d)

inflection point:
point of least slope,
point of least rapid increase

FIGURE 5.10

Relative maxima and minima on a smooth, continuous function can be found by locating the points at which the derivative graph crosses the horizontal axis. These points are among those where the original graph has horizontal tangent lines. Inflection points also can be found by examining the derivative graph and its relation to the function graph. To find the point of greatest or least slope on a function graph, we must find the point where the slope (derivative) graph is greatest or least. That is, we apply the method for finding maxima and minima to the *derivative graph.*

The Second Derivative

Consider, from the discussion in the previous section, the model for the population of Kentucky from 1980 through 1993:

$$p(x) = 0.395x^3 - 6.674x^2 + 30.257x + 3661.147 \text{ thousand people}$$

where x is the number of years since the end of 1980. Graphs of the model and its derivative are shown in Figure 5.11.

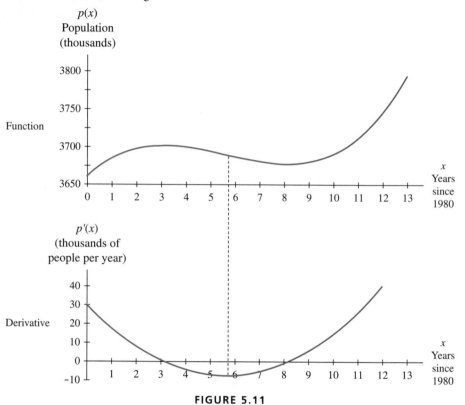

FIGURE 5.11

We wish to determine where the inflection point occurs—that is, where the population was declining most rapidly. It appears that $p'(x)$ has a minimum when $p(x)$ has an inflection point. In fact, this is exactly the case, so we can find the inflection point of $p(x)$ by finding the minimum of $p'(x)$. To find the minimum of $p'(x)$ for this smooth, continuous function $p(x)$, we must find where *its* derivative crosses the x-axis. The derivative of $p'(x)$ is called the **second derivative of $p(x)$**, because it is the derivative of a derivative. The second derivative of $p(x)$ is denoted $p''(x)$.[13] In this case, the second derivative is given by

$$p''(x) = 2.37x - 13.348 \text{ thousand people per year per year}$$

where $x = 0$ in 1980.

Because the second derivative represents the rate of change of the first derivative, the output units of $p''(x)$ are

$$\frac{\text{output units of } p'(x)}{\text{input units of } p'(x)}$$

The input/output diagram for this second derivative is shown in Figure 5.12.

---

13. Other acceptable notations for the second derivative of $p(x)$ include $\dfrac{d^2p}{dx^2}$ and $\dfrac{d^2}{dx^2}[p(x)]$.

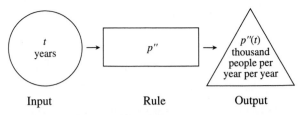

Input   Rule   Output

FIGURE 5.12

A graph of the second derivative (Figure 5.13) reveals that $p''(x)$ crosses the x-axis where $p(x)$ has an inflection point. Note that this identifies the minimum point on the graph of $p'(x)$ where the tangent line is horizontal.

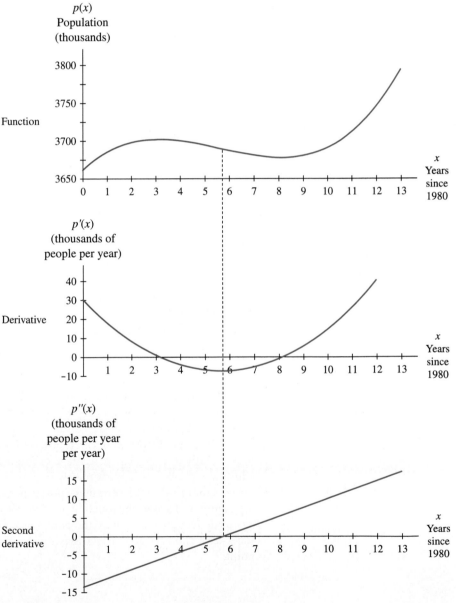

FIGURE 5.13

Setting the second derivative equal to zero and solving for x gives $x \approx 5.63$. According to the model, the population was declining most rapidly in mid-1986 at a rate of approximately $p'(5.63) \approx -7.3$ thousand people per year. At that time, the population was approximately $p(5.63) \approx 3690$ thousand people.

EXAMPLE 1   *Post-secondary Education*

Consider a model for the percentage[14] of students graduating from high school in South Carolina from 1982 through 1990 who entered post-secondary institutions.

$$f(x) = -0.1057x^3 + 1.355x^2 - 3.672x + 50.792 \text{ percent}$$

where $x = 0$ in 1982.

a. Find the inflection point of the model.

b. Determine the year in which the percentage was increasing most rapidly.

Solution:

a. Consider the point(s) at which the second derivative is zero. The rate-of-change formula for this function is

$$f'(x) = -0.3171x^2 + 2.71x - 3.672 \text{ percentage points per year}$$

where $x = 0$ in 1982.
The second derivative is

$$f''(x) = -0.6342x + 2.71 \text{ percentage points per year per year}$$

where $x = 0$ in 1982. The second derivative is zero when $x \approx 4.27$ years after 1982. Next, look at the graph of $f(x)$ shown in Figure 5.14. It does appear that $x = 4.27$ is the approximate input of the inflection point. The output is $f(4.27) \approx 51.6\%$, and the rate of change at that point is $f'(4.27) \approx 2.1$ percentage points per year.

b. Although $f(x)$ is a continuous function, it can be interpreted only at integer values of x because educational data such as these are reported for the fall of each year. Thus, to determine which valid input actually has the greatest rate of change, we evaluate $f'(x)$ at the integer values on either side of $x = 4.27$.

The rate of change of the model in 1986 is $f'(4) \approx 2.09$ percentage points per year. The rate of change in 1987 is $f'(5) \approx 1.95$ percentage points per year. Thus we can say that according to the model, the percentage of South Carolina high school graduates who enter post-secondary institutions was increasing most rapidly in 1986. The percentage of graduates going on for post-secondary education in 1986 was approximately $f(4) = 51.0\%$. The percentage was increasing by about $f'(4) = 2.1$ percentage points per year at that time.

Figure 5.14 shows the function, its derivative, and its second derivative. Note again the relationship among the points at which the second derivative crosses the x-axis, the derivative has a maximum, and the function has an inflection point.

14. Based on data in *South Carolina Statistical Abstract*, 1992.

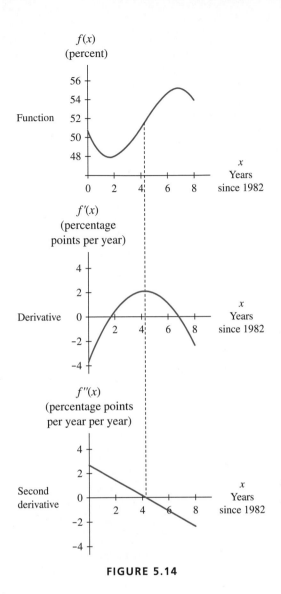

FIGURE 5.14

You have just seen two examples of how the second derivative of a function can be used to find an inflection point. It is important to use the second derivative whenever possible, because it gives an exact answer. Sometimes, however, finding the second derivative of a function can be tedious. In such cases, you will have to decide how important extreme accuracy is. If a close approximation will suffice (as is often the case in real-world modeling), then you may wish to find the first derivative only and use appropriate technology to estimate where its maximum (or minimum) occurs.

EXAMPLE 2   *The 1949 Polio Epidemic*

Consider the following model for the number of polio cases in the United States in 1949.

$$C(t) = \frac{42{,}183.911}{1 + 21{,}484.253e^{-1.248911t}} \text{ polio cases}$$

where $t = 1$ at the end of January, $t = 2$ at the end of February, and so forth. Find when the number of polio cases was increasing most rapidly, the rate of change of polio cases at that time, and the number of cases at that time.

Solution: The graphs of $C(t)$, $C'(t)$, and $C''(t)$ are shown in Figure 5.15. We seek the inflection point on $C(t)$ that corresponds to the maximum point on $C'(t)$ that corresponds to the point at which $C''(t)$ crosses the t-axis.

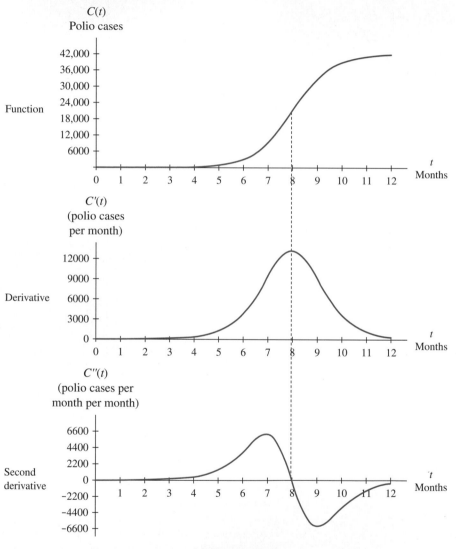

FIGURE 5.15

We choose to use technology to estimate the maximum point on the derivative graph. It occurs at $t \approx 8$, $C'(t) \approx 13{,}171$. It is important to understand what these numbers represent. The t-value tells us the month in which the greatest increase occurred: $t = 8$ corresponds to the end of August. The output value is a value on the *derivative* graph. *This is the slope of $C(t)$ at the inflection point.* We can therefore say that polio was spreading most rapidly at the end of August 1949 at a rate of approximately 13,171 cases per month. To find the number of polio cases at the inflection point, substitute the t-value of the point into the original function to obtain approximately 21,092 cases. Note that to the right of the inflection point, the *number* of

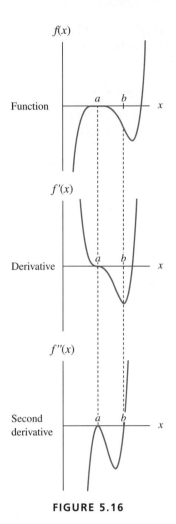

f(x)

Function

f'(x)

Derivative

f''(x)

Second
derivative

FIGURE 5.16

polio cases was increasing whereas the *rate* at which polio cases appeared was declining.   ■

We saw in Section 5.1 that for a smooth, continuous function, a relative maximum or minimum occurs where the derivative *crosses* the horizontal axis, but not where the derivative touches the horizontal axis without crossing it. A similar statement can be made about inflection points. If the second derivative *crosses* the horizontal axis, then an inflection point occurs on the graph of the function. The graphs in Figure 5.16 of a function, its derivative, and its second derivative illustrate this issue. Note that the point at which the second derivative touches, but does not cross, the horizontal axis actually corresponds to a relative maximum on the function, not to an inflection point.

Two other situations that could occur are illustrated in Figure 5.17.

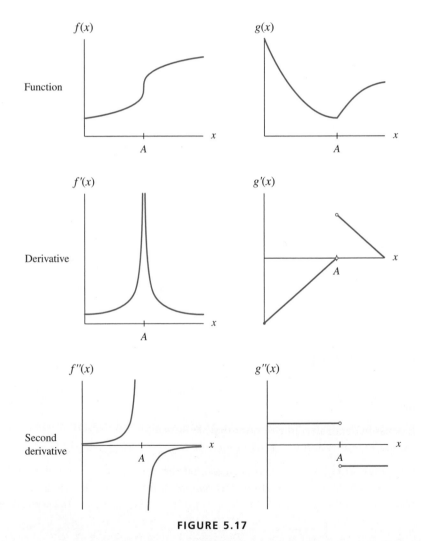

FIGURE 5.17

Note that both $f(x)$ and $g(x)$ have an inflection point at $x = A$ because they change concavity at that point. However, the second derivatives of $f(x)$ and $g(x)$ never cross the horizontal axis. In fact, in each case, the second derivative does not exist at

$x = A$, because the first derivative does not exist there. Even though such situations as this do not often occur in real-world applications, you should be aware that they could happen. Keep in mind the following result:

> At a point of inflection on the graph of a function, the second derivative is zero or does not exist. If the second derivative graph *crosses* the horizontal axis, then an inflection point of the function occurs at that input value.

In some applications, the inflection point can be regarded as the **point of diminishing returns**. Consider the college student who studies for 8 hours without a break before a major exam. The percentage of new material that the student will retain after studying for t hours can be modeled as

$$P(t) = \frac{45}{1 + 5.94e^{-0.969125t}} \text{ percent}$$

This function has an inflection point at $t \approx 1.8$. That is, after approximately 1 hour and 48 minutes, the rate at which the student is retaining new material begins to diminish. Studying beyond that point will improve the student's knowledge, but not as quickly. This is the idea behind diminishing returns: Beyond the inflection point, you gain fewer percentage points per hour than you gain at the inflection point. The existence of this point of diminishing returns is one factor that has led many educators and counselors to suggest studying in 2-hour increments with breaks in between.

5.2   Concept Inventory

- ■ Inflection point
- ■ Second derivative
- ■ Point of diminishing returns
- ■ Conditions under which inflection points exist

5.2   Activities

1. Which of the models discussed in this book could have inflection points?

2. Discuss in detail all of the options that are available for finding inflection points of a function.

3. The graph in Figure 5.2.1 shows an estimate of the ultimate crude oil production recoverable from Earth.[15]

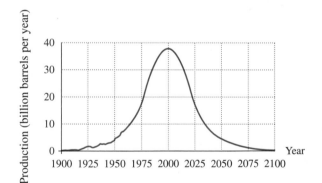

Source: Francois Ramade, *Ecology of Natural Resources* (New York: John Wiley and Sons, 1984).

FIGURE 5.2.1

a. Estimate the two inflection points on the graph.

b. Explain the meaning of the inflection points in the context of crude oil production.

15. Francois Ramade, *Ecology of Natural Resources* (New York: Wiley, 1984).

4. The graph in Figure 5.2.2 shows sales (in thousands of dollars) for a business as a function of the amount spent on advertising (in hundreds of dollars).

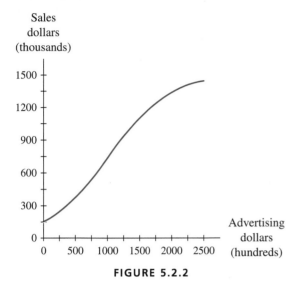

FIGURE 5.2.2

a. Mark the approximate location of the inflection point on the graph.

b. Explain the meaning of the inflection point in the context of this business.

c. Explain how knowledge of the inflection point might affect decisions made by the managers of this business.

5. Consider the function
$$g(x) = 0.04x^3 - 0.88x^2 + 4.81x + 12.11$$

a. Graph $g(x)$, $g'(x)$, and $g''(x)$ between $x = 0$ and $x = 15$. Indicate the relationships among points on the three graphs that correspond to maxima, minima, and inflection points.

b. Find the inflection point of $g(x)$. Is it a point of most rapid decline or least rapid decline?

6. Consider the function $f(x) = \dfrac{20}{1 + 19e^{-0.5x}}$

a. Graph $f(x)$, $f'(x)$, and $f''(x)$ between $x = 0$ and $x = 15$. Indicate the points on the graphs of $f'(x)$ and $f''(x)$ that correspond to the inflection point of $f(x)$.

b. Find the inflection point of $f(x)$. Is it a point of most rapid or least rapid increase?

7. The percentage of new material that a student will retain after studying t hours without a break can be modeled by

$$P(t) = \frac{45}{1 + 5.94e^{-0.969125t}} \text{ percent}$$

a. Find the inflection point on the graph, and interpret the answer.

b. Compare your answer with that given in the discussion at the end of this section.

8. The U.S. Bureau of the Census' prediction[16] for the percentage of the population that is 65 to 74 years old from 2000 through 2040 can be modeled by

$$p(x) = (-4.474 \cdot 10^{-4})x^3 + 0.023x^2 - 0.147x + 6.691 \text{ percent}$$

where x is the number of years after 2000. Determine the year in which the percentage is predicted to be increasing most rapidly, the percentage at that time, and the rate of change of the percentage at that time.

9. The percentage of southern Australian grasshopper eggs that hatch as a function of temperature (for temperatures between 7 °C and 25 °C) can be modeled[17] by

$$P(t) = -0.00645t^4 + 0.488t^3 - 12.991t^2 + 136.560t - 395.154 \text{ percent}$$

where t is the temperature in °C.

a. Graph $P(t)$, $P'(t)$, and $P''(t)$.

b. Find the point of most rapid decrease on $P(t)$. Interpret your answer.

10. The per capita consumption of chicken in the United States from 1975 through 1992 can be modeled[18] by

$$C(x) = 0.0059x^3 - 1.416123x^2 + 121.412x - 3347.707 \text{ pounds}$$

where x is the number of years since 1975.

a. Determine the time when per capita chicken consumption was increasing least rapidly. Find the consumption and rate of change of consumption at that time.

16. Based on data from the *Statistical Abstract*, 1994.
17. Based on information in George L. Clarke, *Elements of Ecology* (New York: Wiley, 1954), p. 170.
18. Based on data from *Statistical Abstract*, 1994.

b. Graph $C(x)$, $C'(x)$, and $C''(x)$, indicating the relationships between the inflection point on $C(x)$ and corresponding points on $C'(x)$ and $C''(x)$.

11. The average price (per 1000 cubic feet) of natural gas for residential use[19] from 1980 through 1991 is given by

$$p(x) = 0.01244x^3 - 0.2421x^2 + 1.407x + 3.4445 \text{ dollars}$$

where x is the number of years since 1980.

a. Sketch the graphs of $p(x)$ and its first and second derivatives. Label the vertical axes appropriately. Which point on the derivative graph corresponds to the inflection point of the original function? Which point on the second derivative graph corresponds to the inflection point of the original function?

b. Find the x-intercept of the second derivative, and interpret its meaning in the context of $p(x)$.

c. Although $p(x)$ is a continuous function, it must be interpreted discretely in this context. With that in mind, determine when, according to the model, average natural gas price was declining most rapidly between 1980 and 1991.

12. On the basis of data from the 1994 *Statistical Abstract* for years from 1978 through 1993, the percentage of households with TVs that also have VCRs can be modeled by

$$P(x) = \frac{77.3}{1 + 340.086e^{-0.699x}} \text{ percent}$$

where x is the number of years after 1978. When was the percentage of households with TVs that also have VCRs increasing the most rapidly? What were the percentage and the rate of change of the percentage at that time?

13. The number of people who donated to an organization supporting athletics at a certain university in the southeast from 1975 through 1992 can be modeled[20] by

$$D(t) = -10.247t^3 + 208.114t^2 - 168.805t + 9775.035 \text{ donors}$$

t years after 1975.

a. Find any relative maxima or minima that occur on the model.

b. Find the inflection point(s).

c. How do the following events in the history of football at that college correspond with the curvature of the model?

    i. In 1981, the team won the National Championship.

    ii. In 1988, Coach F was released and Coach H was hired.

14. The amount spent on cable television per person per year from 1984 through 1992 can be modeled[21] by

$$A(x) = -0.126x^3 + 1.596x^2 + 1.802x + 40.930 \text{ dollars}$$

where x is the number of years since 1984.

a. Find the inflection point on $A(x)$ and the corresponding points on $A'(x)$ and $A''(x)$.

b. Find the year in which the average amount spent per person per year on cable television was increasing most rapidly. What was the rate of change of the amount spent per person in that year?

15. A college student works for 8 hours without a break assembling mechanical components. The cumulative number of components she has assembled after h hours can be modeled by

$$N(h) = \frac{62}{1 + 11.49e^{-0.654h}} \text{ components}$$

a. Determine when the rate at which she was working was greatest.

b. How might her employer use the information in part *a* to increase the student's productivity?

16. The lake level of Lake Tahoe from October 1, 1995, through July 31, 1996, can be modeled[22] by

$$L(d) = (-5.345 \cdot 10^{-7})d^3 + (2.543 \cdot 10^{-4})d^2 - 0.0192d + 6226.192 \text{ feet above sea level}$$

d days after September 30, 1995.

a. Determine when the lake level was rising most rapidly between October 1, 1995, and July 31, 1996.

b. What factors may have caused the inflection point to occur at the time you found in part *a*?

19. Based on data from *Statistical Abstract*, 1992.
20. Based on data from the IPTAY Association at Clemson University.
21. Based on data from *Statistical Abstract*, 1994.
22. Based on data from the Federal Watermaster, U.S. Department of Interior.

c. Would you expect the most rapid rise to occur at approximately the same time each year? Explain.

17. The "life expectancy of a state" is the average age at the time of death of the residents of that state. Table 5.11 shows the number of states that have life expectancies below the given values of x (all values are for 1977).

TABLE 5.11

Life expectancy x years	Number of states below x
68	1
69	4
70	9
71	30
72	38
73	49
74	50

a. According to Table 5.11, for what life expectancy is the number of states increasing most rapidly?

b. Fit a model to the data, and find the inflection point of the model.

c. Use the model to find the age at which the number of states increases most rapidly. Find the age, the corresponding number of states, and the rate of change. (The model must be interpreted discretely.)

18. Table 5.12[23] lists the number of live births in the United States between the years 1950 and 1988 to women 45 years of age and older.

TABLE 5.12

Year	Births	Year	Births
1950	5322	1980	1200
1955	5430	1985	1162
1960	5182	1986	1251
1965	4614	1987	1375
1970	3146	1988	1427
1975	1628		

23. *The 1992 Information Please Almanac* (Boston: Houghton Mifflin).

a. Use the data to determine when the number of births was declining most rapidly.

b. Find a model to fit the data, and use the model to obtain the answer to part *a*. Note that the model must be interpreted discretely.

c. How does the answer to part *b* differ from the answer to part *a*?

d. Write a careful explanation of the procedure that you used with your model to find the point of most rapid decline.

19. The yearly amount of garbage (in millions of tons) taken to a landfill outside a city during selected years from 1960 through 1990 is given in Table 5.13.

TABLE 5.13

Year	Amount (millions of tons)
1960	81
1965	99
1970	117
1975	122
1980	132
1985	145
1990	180

a. Using the table values only, determine during which 5-year period the amount of garbage showed the slowest increase. What was the average rate of change during that 5-year period?

b. Fit a model to the data.

c. Give the second derivative formula for the model.

d. Use the second derivative to find the point of slowest increase on the model.

e. Graph the first and second derivatives, and explain how they support your answers to part *d*.

f. In what year was the rate of change of the yearly amount of garbage the smallest? What were the rate of increase and the amount of garbage in that year? Note that the model must be interpreted discretely.

20. A business owner's sole means of advertising is to put fliers on cars in a nearby shopping mall parking lot. Table 5.14 shows the number of labor-hours per month spent handing out fliers and the corresponding profit.

TABLE 5.14

Labor hours each month	Profit (dollars)
0	2000
10	3500
20	8500
30	19,000
40	32,000
50	43,000
60	48,500
70	55,500
80	56,500
90	57,000

a. Find a model for profit.

b. For what number of labor-hours is profit increasing most rapidly? Give the number of labor-hours, the profit, and the rate of change of profit at that number.

c. In this context, the inflection point can be thought of as the point of diminishing returns. Discuss how knowing the point of diminishing returns could help the business owner make decisions relative to employee tasks.

21. Table 5.15[24] gives the amount spent on national defense (in billions of dollars measured in constant 1987 dollars) between 1970 and 1994.

a. Fit a model to the data.

b. Give the second derivative formula for the model.

TABLE 5.15

Year	Amount (billions of dollars)
1970	262.9
1972	219.7
1974	185.3
1976	177.8
1978	177.2
1980	187.1
1982	214.3
1984	241.7
1986	276.4
1988	283.3
1990	272.5
1992	250.2
1994	220.1 (estimated)

c. Find the point between 1970 and 1994 at which spending on national defense showed the most rapid increase. Give the year, the amount spent, and how rapidly the amount was changing at that point. Note that the model must be interpreted discretely.

d. If you found the answer to part c by using a graphing calculator or computer, discuss how you could find the answer without using such a tool.

22. The net sales[25] (in thousands of dollars) of the Russell Corporation from 1984 through 1993 are given in Table 5.16.

a. Use the data to estimate the year in which sales were growing most rapidly.

b. Fit a model to the data.

c. Find the first and second derivatives of the model in part a.

d. Determine the year in which net sales were growing most rapidly. Find the net sales and the rate of change of net sales in that year. Note that the model must be interpreted discretely.

24. *Statistical Abstract*, 1994.
25. *1993 Russell Corporation Annual Report*.

TABLE 5.16

Year	Net sales (thousands of dollars)
1984	353,025
1985	385,433
1986	437,520
1987	479,880
1988	531,136
1989	687,954
1990	713,812
1991	804,585
1992	899,136
1993	930,787

23. The total labor force (in thousands) in the United States is modeled[26] by the equation

$$L(t) = \frac{22,000}{1 + 5.951e^{-0.3969t}} + 107,300 \text{ thousand workers}$$

where t = the number of years since the end 1981.

a. Give the slope formula for $L(t)$.

b. Find the rate of change of the labor force at the end of 1986.

c. Find the growth in the labor force during the first 6 months of 1987.

d. Find the time when the labor force grew most rapidly. Give the size of the labor force and how rapidly it was growing at that time.

24. Table 5.17[27] gives the total emissions in millions of metric tons of nitrogen oxides, NO_x, in the United States from 1940 through 1990.

TABLE 5.17

Year	NO_x (millions of metric tons)
1940	6.9
1950	9.4
1960	13.0
1970	18.5
1980	20.9
1990	19.6

a. Fit a cubic model to the data.

b. Give the slope formula for the model.

c. Determine when emissions were increasing most rapidly. Give the year, the amount of emissions, and how rapidly they were increasing. Note that the model must be interpreted discretely.

25. The personnel manager for a construction company keeps track of the total number of labor-hours spent on a construction job each week during the construction. Some of the weeks and the corresponding labor-hours are given in Table 5.18:

TABLE 5.18

Weeks after the start of a project	Cumulative labor-hours
1	25
4	158
7	1254
10	5633
13	9280
16	10,010
19	10,100

a. Fit a logistic model to the data.

b. Find the derivative of the model. What are the units on the derivative?

c. Graph the derivative, and discuss what information it gives the manager.

d. When is the maximum number of labor-hours per week needed? How many labor-hours are needed in that week?

e. Find the point of most rapid increase in the number of labor-hours per week. How many weeks into the job does this occur? How rapidly is the number of labor-hours per week increasing at this point?

f. Find the point of most rapid decrease in the number of labor-hours per week. How many weeks into the job does this occur? How rapidly is the number of labor-hours per week decreasing at this point?

26. Based on data from *Statistical Abstract*, 1992 and 1994.
27. *Statistical Abstract*, 1992.

g. Carefully explain how the exact values for the points in parts *e* and *f* can be obtained.

h. If the company has a second job requiring the same amount of time and labor-hours, a good manager will schedule the second job to begin so that the time when labor-hours per week for the first job are declining most rapidly corresponds to the time when labor-hours per week for the second job are increasing most rapidly. How many weeks into the first job should the second job begin?

26. Table 5.19[28] shows the Gross Domestic Product (in billions of dollars) for the United States from 1969 through 1991.

a. Find a cubic model that fits the data.

b. Write the second derivative of the model you found in part *a*.

c. In what year does your model exhibit a point of most rapid growth? What is the Gross Domestic Product for this year, and how rapidly is it growing? Note that the model must be interpreted discretely.

TABLE 5.19

Year	GDP (billions of dollars)	Year	GDP (billions of dollars)
1969	498.0	1981	1541.5
1970	511.6	1982	1573.4
1971	551.7	1983	1694.0
1972	612.4	1984	1918.3
1973	688.3	1985	2035.5
1974	737.1	1986	2114.9
1975	788.0	1987	2265.9
1976	890.8	1988	2467.3
1977	1008.5	1989	2596.5
1978	1147.2	1990	2707.2
1979	1270.7	1991	2732.6
1980	1371.7		

◼ **5.3**   **Approximating Change**

Recall from our discussion of linear models that the slope of a line $y = ax + b$ can be thought of as how much y changes when x changes by 1 unit. For example, the amount spent on pollution control in the United States during the 1980s can be described by the equation[29]

$$\text{Amount} = 3788.65t - 252{,}216.11 \text{ million dollars}$$

where $t = 80$ in 1980, 81 in 1981, and so on. The slope is $3788.65 million per year, so we can say that each year an *additional* $3788.65 million is spent on pollution control. It follows that every 2 years spending would increase by (2)($3788.65) million, every 3 years spending would increase by (3)($3788.65) million, and so on.

Because rates of change are slopes, we can apply a similar type of reasoning to functions other than lines. However, we must be careful in doing this, for although the slope of a line is constant, the slopes of other functions may change at every

28. *Economic Report of the President*, 1993.
29. Based on data from *Statistical Abstract*, 1992.

point. For example, if you are driving a car and know that your speed at a certain moment is 60 mph (1 mile per minute), then you can estimate that over the next minute you will travel approximately 1 mile. Over the next 30 seconds you will travel approximately half a mile, and over the next 2 minutes you will travel approximately 2 miles. Note that we are using the rate of change of distance to estimate by how much the distance traveled changes over a short time period.

The assumption underlying these approximations is that the speed remains constant during a short interval. If the speed deviated dramatically from 60 mph, then the approximations would be invalid. For instance, in the next 30 seconds, you could encounter a stop light and have to stop the car.

As another example, consider the average retail price (cost to the consumer) of hardback fiction books since 1980, which can be modeled by the equation[30]

$$p(t) = 0.0416t^2 - 6.3328t + 252.6486 \text{ dollars}$$

where $t = 80$ at the end of 1980, 81 at the end of 1981, on so on. We calculate how rapidly the average price was increasing at the end of 1993 as follows:

$$\frac{dp}{dt} = 2(0.0416)t - 6.3328 \text{ dollars per year}$$

When $t = 93$,

$$\frac{dp}{dt} = 2(0.0416)(93) - 6.3328 = \$1.4048 \text{ per year}$$

Thus, at the end of 1993, the price of hardback fiction was rising by approximately $1.40 per year. On the basis of this rate of change and the fact that the average price at the end of 1993 was approximately $p(93) = \$23.50$, we formulate the following approximations:

■ During the following year (1994), the price increased by approximately $1.40 to a price of $24.90.

■ During the first 6 months of 1994, the price increased by approximately $0.70 to a price of $24.20.

■ During the first quarter of 1994, the price increased by approximately $0.35 to a price of $23.85.

Again, we used the rate of change of price to estimate the change in price. Compare these approximations, including those for time periods of 9 months and 2 years, to the actual values given by the model, which are listed in Table 5.20.

TABLE 5.20

Time from end of 1993	Approximated price	Price from model	Difference between model value and approximation
2 years	$23.50 + 2($1.40) = $26.30	$p(95) \approx \$26.47$	$0.17
1 year	$23.50 + 1($1.40) = $24.90	$p(94) \approx \$24.94$	$0.04
9 months	$23.50 + (0.75)($1.40) = $24.55	$p(93.75) \approx \$24.57$	$0.02
6 months	$23.50 + (0.50)($1.40) = $24.20	$p(93.5) \approx \$24.21$	$0.01
3 months	$23.50 + (0.25)($1.40) = $23.85	$p(93.25) \approx \$23.85$	$0.00

30. Based on data from *Statistical Abstract*, 1992 and 1994.

Note that the shorter the time period, the closer the approximate price is to the price given by the model. This is no coincidence, because over a short time period, the rate of change is more likely to be nearly constant than over a longer period of time. Rates of change can often be used to approximate changes in a function, and they generally give good approximations over small intervals.

The justification for using rates of change to approximate changes in a function is best understood from a graphical point of view. We know that the rate of change of a function at a point is the slope of the line tangent to the function at that point. When we use the rate of change to make a prediction about the function, we are actually making a statement about what the tangent line will do and are assuming that the function will exhibit similar behavior. Indeed, over a small enough interval, the tangent line and the function are very much alike. (Recall the local linearity discussion in Section 3.3.) Consider the model for book prices from $t = 93$ through $t = 94$ and the line tangent to the model at $t = 93$. These graphs are shown in Figure 5.18.

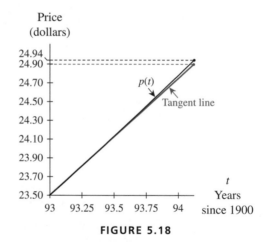

FIGURE 5.18

We used the slope of the line tangent to $p(t)$ at $t = 93$ to approximate values of $p(t)$ at $t = 95$, $t = 94$, $t = 93.75$, $t = 93.5$, and $t = 93.25$. The approximate values that we found are actually output values on the line tangent to $p(t)$ at these t-values. Note in Figure 5.18 that the closer we are to $t = 93$, the closer the tangent line is to $p(t)$. Similarly, the farther we move from $t = 93$, the more the tangent line deviates from the function. This is generally true, and it is the reason we restrict ourselves to small intervals when using a rate of change to approximate change in a function.

To summarize, consider the following statements:

- The change in a function from x to $x + h$ can be approximated by the change in the tangent line from x to $x + h$ when h is a small number.

- The change in the tangent line from x to $x + h$ is

$$\text{(slope of the tangent line at } x) \cdot (h)$$

The mathematical notation for the statement "the change in a function from x to $x + h$ is approximately the slope of the tangent line at x times h" is

$$f(x + h) - f(x) \approx f'(x) \cdot h$$

for small values of h.

Approximating Change

The approximate change in f is the rate of change of f times a small change in x. That is,

$$f(x + h) - f(x) \approx f'(x) \cdot h$$

where h represents the small change in x.

It follows from this formula for approximating change that we can approximate the function value $f(x + h)$ by adding the approximated change to $f(x)$.

Approximating the Result of Change

When x changes by a small amount to $x + h$, the output of f at $x + h$ is approximately the value of f at x plus the approximated change in f.

$$f(x + h) \approx f(x) + f'(x) \cdot h$$

It was this formula that we used to obtain the approximated price column in Table 5.20 in the book price example.

EXAMPLE 1   *Population of California*

At the end of 1993, the population of California was approximately 31,211,000 people and was increasing at a rate of about 449,000 people per year.

a. Estimate the change in the population of California during the first quarter of 1994.

b. Estimate the population of California in the middle of 1994.

c. What assumptions underlie these estimations?

Solution:

a. Change in population $\approx (449{,}000 \text{ people per year})\left(\tfrac{1}{4} \text{ year}\right)$
$$= 112{,}250 \text{ people}$$
The population increased by approximately 112,250 people during the first quarter of 1994.

b. Population in mid-1994 $\approx \left(\begin{array}{c}\text{population at end}\\ \text{of 1993}\end{array}\right) + \left(\begin{array}{c}\text{approximate change}\\ \text{in population}\end{array}\right)$
$$= 31{,}211{,}000 \text{ people} + (449{,}000 \text{ people per year})\left(\tfrac{1}{2} \text{ year}\right)$$
$$= 31{,}211{,}000 \text{ people} + 224{,}500 \text{ people}$$
$$= 31{,}435{,}500 \text{ people}$$
The population in the middle of 1994 was approximately 31,435,500 people.

c. In making these estimations, we assume that the growth rate remained essentially constant during 1994. ∎

EXAMPLE 2   *Temperature*

The temperature during and after a thunderstorm can be modeled by

$$T(h) = 2.37h^4 - 5.163h^3 + 8.69h^2 - 9.87h + 78 \text{ degrees Fahrenheit}$$

where h is the number of hours since the storm began.

a. Use the rate of change of $T(h)$ at $h = 0.25$ to estimate by how much the temperature changed between 15 and 20 minutes after the storm began.

b. Find the temperature and rate of change of temperature at $h = 1.5$ hours.

c. Using only the answers to part b, estimate the temperature 1 hour and 40 minutes after the storm began.

d. Sketch the graph of $T(h)$ from $h = 0$ to $h = 1.75$ with tangent lines drawn at $h = 0.25$ and $h = 1.5$. On the basis of the graph, determine whether the answers to parts a and c were overestimates or underestimates of the temperature given by the model.

Solution:

a. $T'(0.25) = 9.48(0.25)^3 - 15.489(0.25)^2 + 17.38(0.25) - 9.87$
$\approx -6.34 \text{ °F per hour}$

The change in the temperature between 15 and 20 minutes after the storm began is approximately $(-6.34 \text{ °F/hr})\left(\frac{1}{12} \text{ hr}\right) = -0.53$ °F. The temperature fell approximately half a degree.

b. $T(1.5) \approx 77.3$ °F and $T'(1.5) \approx 13.3$ °F per hour

c. First, note that 40 minutes $= \frac{2}{3}$ hour.

$$\begin{pmatrix}\text{Temperature at 1 hour} \\ \text{and 40 minutes}\end{pmatrix} \approx \begin{pmatrix}\text{temperature at} \\ 1\frac{1}{2} \text{ hours}\end{pmatrix} + \begin{pmatrix}\text{approximate change} \\ \text{in temperature}\end{pmatrix}$$

$$T\left(1 + \frac{2}{3}\right) \approx T(1.5) + T'(1.5)\left(\frac{1}{6}\right)$$

$$T\left(1 + \frac{2}{3}\right) \approx 77.3 \text{ °F} + (13.3 \text{ °F per hour})\left(\frac{1}{6} \text{ hour}\right)$$

$$\approx 77.3 \text{ °F} + 2.2 \text{ °F}$$

$$= 79.5 \text{ °F}$$

d. Temperature (°F)

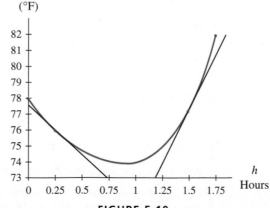

FIGURE 5.19

Because the tangent line at $h = 0.25$ hour is steeper than the graph is between $h = 0.25$ (15 minutes) and $h \approx 0.33$ (20 minutes), the approximate change in temperature overestimates the actual change (see Figure 5.19). Thus, using the approximate change to estimate the temperature at $h \approx 0.33$ gives a temperature that underestimates the temperature given by the model at $h \approx 0.33$. The tangent line at $h = 1.5$ is not as steep as the graph to the right of $h = 1.5$, so our temperature approximation for $h = 1.67$ is an underestimate of the temperature given by the model.   ■

You may be wondering why you would want to use a tangent line to estimate a function output value. Sometimes, as in Example 1, it is necessary because we do not have enough information available to develop a model. Other times, the tangent line approximation is used for short-term extrapolation.

This is illustrated in Example 3.

EXAMPLE 3   *Population of California*

TABLE 5.21

Year	Population (thousands)
1985	26,441
1986	27,102
1987	27,777
1988	28,464
1989	29,218
1990	29,760
1991	30,407
1992	30,895
1993	31,211

The population of California from 1985 through 1993 is shown in Table 5.21.

a. Use the last two data points given to estimate the population in 1994.

b. Find a model for the data. Use it to estimate the population in 1994.

c. Find the value of the model and the derivative of the model in 1993. Use these values to estimate the population in 1994.

d. Discuss the assumptions one makes when using each estimation technique in parts *a, b,* and *c.*

Solution:

a. The average rate of change between 1992 and 1993 is

$$\frac{31{,}211 - 30{,}895 \text{ thousand people}}{1 \text{ year}} = 316 \text{ thousand people per year}$$

Adding this to the population in 1993, we obtain an estimate of the 1994 population:

$$31{,}211{,}000 + 316{,}000 = 31{,}527{,}000 \text{ people}$$

b. The data are concave down and can be modeled by

$$P(t) = -21.001t^2 + 784.925t + 26{,}366.879 \text{ thousand people}$$

t years after 1985. The model estimates the population in 1994 as

$$P(9) \approx 31{,}730{,}000 \text{ people}$$

c. The derivative of $P(t)$ is

$$P'(t) = -42.002t + 784.925 \text{ thousand people per year}$$

t years after 1985. In 1993, the rate of change of the population was approximately

$$P'(8) \approx 448.9 \text{ thousand people per year}$$

and the population according to the model was 31,302,000. Thus

$$P(9) \approx P(8) + P'(8)$$
$$\approx 31{,}302{,}000 + 449{,}000$$
$$= 31{,}751{,}000 \text{ people}$$

d. To estimate using only the last two data points (part *a*) is to assume that the average rate of change in the population from 1993 through 1994 will be approximately the same as it was from 1992 through 1993. To estimate using only the model (part *b*) is to assume that future growth will continue in a concave-down, parabolic manner. To estimate using the derivative of the model (part *c*) is to assume that the rate of change at the end of 1993 (as estimated by the model) is a good predictor of the change in the population during 1994.

All three of these estimates are valid. If it were the beginning of 1993 and someone needed an estimate for the population at the end of the year, that person might use any one of these or many other techniques to make the prediction. You should be aware that because the data are concave down, the tangent line prediction is the largest of the three. ∎

Marginal Analysis

In economics, it is customary to refer to the rates of change of cost, revenue, and profit with respect to the number of units produced or sold as **marginal cost, marginal revenue,** and **marginal profit**. These rates are used to approximate actual change in cost, revenue, or profit when the number of units produced or sold is increased by one. The term **marginal analysis** is often applied to this type of approximation.

EXAMPLE 4 *Marginal Cost*

Suppose a manufacturer of toaster ovens currently produces 250 ovens per day with a total production cost of $12,000 and a marginal cost of $24 per oven. What information does the marginal cost value give the manufacturer?

Solution: The marginal cost is the rate of change of cost. It is the approximate increase in cost that will result if production is increased from 250 ovens per day to 251 ovens per day. ∎

The following example illustrates how marginal analysis can be used to help a business owner make decisions.

EXAMPLE 5 *Marginal Revenue*

A seafood restaurant has been keeping track of the price of its Monday night all-you-can-eat buffet and the corresponding number of nightly customers. These data are given in Table 5.22.

TABLE 5.22

Number of customers	86	83	80	78	76	73	70	68
Buffet price (dollars)	7.70	7.90	8.20	8.30	8.60	8.80	8.90	9.10

a. Find a model for the data, and convert it to a model for revenue.
b. Find the marginal revenue values for 60, 92, and 100 customers. Interpret the values in context.

Solution:

a. A linear model for these price data is

$$P(x) = -0.078702x + 14.477879 \text{ dollars}$$

where x is the number of customers. Because revenue is equal to (price)(number of customers), the revenue is given by the equation

$$R(x) = -0.078702x^2 + 14.477879x \text{ dollars}$$

where x is the number of customers.

b. The derivative of revenue is

$$R'(x) = -0.157404x + 14.477879 \text{ dollars per customer}$$

where x is the number of customers. Evaluating the derivative at 60, 92 and 100 customers gives the marginal revenue values shown in Table 5.23.

TABLE 5.23

Demand (number of customers)	Marginal revenue (dollars per customer)
60	5.03
92	0.00
100	-1.26

What do these marginals tell us? If the buffet price is set on the basis of 60 customers, then revenue is increasing by $5.03 per customer. In other words, increasing the number of customers to 61 (by lowering the price) will increase nightly revenue by approximately $5.03. It would benefit the restaurant to increase the number of customers by lowering price.

Similarly, we estimate that if the number of customers is increased from 92 to 93, then revenue will not change significantly. With 92 customers, stimulating sales by lowering the price will not benefit the restaurant.

Finally, note that when price is set so that the restaurant expects 100 customers, the marginal revenue is negative. Increasing the number of customers (by decreasing price) to 101 will result in an approximate decrease in nightly revenue of $1.26. That is clearly undesirable. ∎

Knowing marginal cost, revenue, or profit can be a valuable tool when making business decisions.

5.3   Concept Inventory

■ Change in function ≈ change in tangent line close
    to the point of tangency
    $f(x + h) - f(x) \approx f'(x) \cdot h$

■ $f(x + h) \approx f(x) + f'(x) \cdot h$ for small values of h

■ Marginal cost, marginal profit, marginal revenue

5.3   Activities

1. If the humidity is currently 32% and is falling at a rate of 4 percentage points per hour, estimate the humidity 20 minutes from now.

2. If an airplane is flying 300 mph and is accelerating at a rate of 200 mph per hour, estimate the airplane's speed in 5 minutes.

3. If $f(3) = 17$ and $f'(3) = 4.6$, estimate $f(3.5)$.

4. If $g(7) = 4$ and $g'(7) = -12.9$, estimate $g(7.25)$.

5. Interpret the following statements.

   a. At a production level of 500 units, marginal cost is $17 per unit.

   b. When weekly sales are 150 units, marginal profit is $4.75 per unit.

6. A fraternity currently realizes a profit of $400 selling T-shirts at the opening baseball game of the season. If its marginal profit is -$4 per shirt, what action should the fraternity consider taking to improve its profit?

7. A graph showing the annual premium for a one-million-dollar term life insurance policy as a function of the age of the insured person is given in Figure 5.3.1.

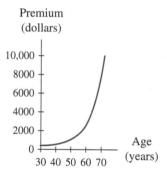

FIGURE 5.3.1

Sketch a tangent line at 70 years of age, and use it to predict the premium for a 72-year-old person.

8. A graph showing world life expectancy as a function of the number of decades since 1900 is given[31] in Figure 5.3.2.

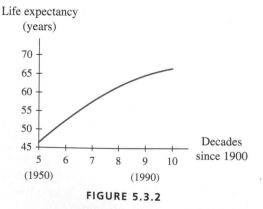

FIGURE 5.3.2

Sketch a tangent line at 1990, and use it to predict the world life expectancy in 2000.

9. A graph of the population of Cleveland, Ohio, as a function of the number of decades since 1900 is shown[32] in Figure 5.3.3.

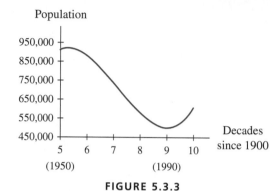

FIGURE 5.3.3

a. Sketch a tangent line at 1990, and use it to predict the population of Cleveland in 2000.

b. The model graphed in Figure 5.3.3 is

$$P(t) = 16,272t^3 - 349,625t^2 + 2,335,380t - 4,057,270 \text{ people}$$

   t decades after 1900. Note that the model applies for years after 1950. What does the model predict as the population in 2000?

c. Which prediction do you believe is the more valid one? Why?

10. In 1987, because of concern that CFCs have a detrimental effect on the stratospheric ozone layer, the Montreal Protocol calling for phasing out all chlorofluorocarbon (CFC) production was ratified. The graph in Figure 5.3.4 shows estimated releases of CFC-11, one of the two most prominent CFCs.[33]

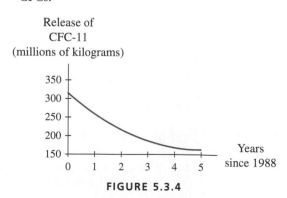

FIGURE 5.3.4

31. Based on data in *The True State of the Planet*, ed. Ronald Bailey (The Free Press for the Competitive Enterprise Institute, 1995).

32. Based on data from U.S. Department of Commerce, Bureau of the Census.

33. Based on data in *The True State of the Planet*, 1995.

a. Sketch a tangent line at 1992, and use it to estimate CFC-11 releases in 1993.

b. The model whose graph is shown in Figure 5.3.4 is

$$C(x) = 6.107x^2 - 60.799x + 315.994$$
million kilograms of CFC-11

where x is the number of years since 1988. What estimate does the model give for CFC-11 releases in 1993?

11. The population of South Carolina between 1790 and 1990 can be modeled[34] by

$$\text{Population} = 0.07767x^2 - 227.8377x + 248,720.649 \text{ thousand people in year } x$$

a. Find the rate of change of the population of South Carolina in 1990.

b. On the basis of your answer to part *a*, approximate how much the population changed between 1990 and 1992.

c. Write an explanation of the procedure you used to find the approximate change in the population between 1990 and 1992.

12. A pizza parlor has been lowering the price of a large one-topping pizza to promote sales. The average revenues from the sale of large one-topping pizzas on a Friday night (5 p.m. to midnight) are given in Table 5.24.

TABLE 5.24

Price	$9.25	$10.50	$11.75	$13.00	$14.25
Revenue	$1202.50	$1228.50	$1210.25	$1131	$1054.50

a. Find a model to fit the data.

b. Find and interpret the rate of change of revenue at a price of $9.25.

c. Find the change in revenue if the price is increased from $9.25 to $10.25.

d. Find and interpret the rate of change of revenue at a price of $11.50.

e. Find the change in revenue if the price is increased from $11.50 to $12.50.

f. Explain why the approximate change is an overestimate of the change from $9.25 to $10.25 but an underestimate of the change from $11.50 to $12.50.

13. The population of Mexico between 1921 and 1990 can be modeled[35] by

$$P(t) = 12.921e^{0.026578t} \text{ million persons}$$

where t is the number of years since 1921.

a. How rapidly was the population growing in 1985?

b. On the basis of your answer to part *a*, determine by approximately how much the population of Mexico should have increased between 1985 and 1986.

14. Suppose the graph in Figure 5.3.5 shows your grade out of 100 points as a function of the time that you spend studying.

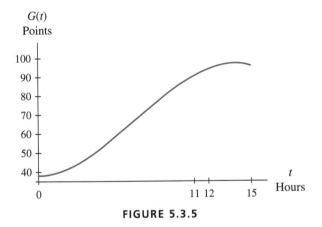

FIGURE 5.3.5

The equation for this graph is

$$G(t) = -0.044t^3 + 0.918t^2 + 38.001 \text{ points}$$

where t is the number of hours spent studying.

a. Confirm the following assertions:

i. After 11 hours of study, the slope is 4.2 points per hour, and the grade is 90.5 points.

ii. The grade after 12 hours of study is 94.2 points.

34. Based on data from *South Carolina Statistical Abstract*, 1994.

35. SPP and INEGI, Mexican Censuses of Population 1921 through 1990 as reported in Pick and Butler, *The Mexico Handbook* (Westview Press, 1994).

b. Use the information in the first part of *a* to estimate the grade after 12 hours. Is this an overestimate or an underestimate of the grade given by the model? Explain why.

15. A model[36] for the percentage of households with TVs that also have VCRs is

$$p(x) = \frac{75}{1 + 316.7509e^{-0.699060x}} \text{ percent}$$

where $x = 0$ in 1978. This model applies for the years 1978 through 1991.

a. How rapidly was the percentage of households with VCRs growing in 1986?

b. On the basis of the rate of change in 1986, what approximate increase would you expect between 1986 and 1987?

c. On the basis of the equation, what was the actual increase between 1986 and 1987?

d. According to the 1992 *Statistical Abstract*, the actual percentages for 1986 and 1987 were 36.0% and 48.7%, respectively. What was the actual 1986 through 1987 increase?

e. Compare the increases given by the derivative in part *b*, by the model in part *c*, and by the data in part *d*. Which of the three answers is most accurate? Explain.

16. Production costs for various hourly production levels of television sets are given in Table 5.25.

TABLE 5.25

Hourly production	Cost (dollars)
5	740
10	1060
15	1210
20	1320
25	1420
30	1580
35	1900

a. Fit a model to the data.

b. Find and interpret marginal cost at production levels of 5, 20, and 30 units.

c. Find the cost to produce the 6th unit, the 21st unit, and the 31st unit.

d. Why is the cost less than the marginal cost to produce the 6th unit but greater than the marginal cost to produce the 21st and 31st units?

e. Find a model for average cost.

f. Find and interpret the rate of change of average cost at production levels of 5, 20, and 30 units.

17. A concession stand owner finds that if he prices hot dogs so as to sell a certain number at each sporting event, then the corresponding revenues are those given in Table 5.26.

TABLE 5.26

Number of hot dogs sold	Revenue (dollars)
100	195
400	620
700	875
1000	1000
1200	1020
1500	975

a. Fit a model to the data.

b. Find and interpret the marginal revenue for sales levels of 200, 800, 1100, and 1400 hotdogs.

c. At what sales level will the marginal revenue be $1.10? $0.25?

18. A golf ball manufacturer knows that the cost associated with various hourly production levels are as shown in Table 5.27.

TABLE 5.27

Balls produced each hour (hundreds)	Cost (dollars)
2	248
5	356
8	432
11	499
14	532
17	567
20	625

36. Based on data from *Statistical Abstract*, 1992.

TABLE 5.28

Country	1980	1981	1982	1983	1984	1985	1986	1987
United States	100	110.4	117.2	120.9	126.1	130.5	133.1	137.9
Canada	100	112.4	124.6	131.8	137.5	143.0	148.9	155.4
Mexico	100	127.9	203.3	410.2	679.0	1071.2	1994.9	4624.7
Japan	100	104.9	107.8	109.9	112.3	114.6	115.3	115.4
Israel	100	217	478	1174	5560	22,498	33,330	39,937
Peru	100	175.4	288.4	609	1280	3372	5999	11,150
Brazil	100	206	407	984	2924	9556	23,436	77,258
Argentina	100	204	541	2403	17,462	134,833	256,308	592,900

a. Find a model to fit the data.

b. If 1000 balls are currently being produced each hour, find and interpret the marginal cost.

c. Repeat part *b* for 300 golf balls and for 2100 golf balls.

d. Convert the model in part *a* to one for average cost.

e. Find and interpret the rate of change of average cost for production levels of 300 and 1700 golf balls.

19. Rise in consumer prices is often used as a measure of inflation rate. Table 5.28[37] shows CPIs during the 1980s for several different countries.

a. Find the best models to fit the data for the United States, Canada, Peru, and Brazil.

b. How rapidly were consumer prices rising in each of those four countries in 1987?

c. On the basis of your answers to part *b*, what would you expect the CPI to have been in the four countries in 1988?

20. The distribution of wealth is an important issue in our society today. The percentages of all wealth held by the poorest fifth of the U.S. population in selected years from 1947 through 1984 are shown in Table 5.29.

TABLE 5.29

Year	Percentage	Year	Percentage
1947	5	1977	5.2
1957	5.1	1980	5.1
1967	5.5	1984	4.7

a. Fit a model to the data. Why did you choose this model?

b. What percentage of all wealth should the poorest fifth of Americans have in 1995 according to your model? Is this reasonable?

c. Graph your model. Is there something interesting happening in 1970?

d. How rapidly is the percentage of wealth held by the poorest fifth of Americans changing in 1970?

e. Use the derivative to approximate the change in percentage of wealth between 1984 and 1985.

21. Three hundred dollars is invested in an account that compounds 6.5% APR monthly.

a. Find an equation for the balance in the account after *t* years.

b. Rewrite the equation in part *a* to be of the form $A = P(b)^t$.

c. How much is in the account after 2 years?

d. How rapidly is the value of the account growing after 2 years?

e. Use the answer in part *d* to approximate how much the value of the account changes during the first quarter of the third year.

22. The amount in an investment after *t* years is given by

$$A(t) = 120,000 \, (1.12682503)^t \text{ dollars}$$

a. Give the rate-of-change formula for the amount.

b. Find the rate of change of the amount after 10 years. Write a sentence interpreting the answer.

37. *International Marketing Data and Statistics*, 1988/89.

c. On the basis of your answer to part *b*, determine by approximately how much the investment will grow during the first half of the 11th year.

d. Find the percentage rate of change after 10 years. Given that $A(t)$ is exponential, what is the significance of your answer?

23. A car dealership keeps track of how much it spends on advertising each month and of its monthly revenue. From this information, the list of advertising expenditures and probable associated revenues shown in Table 5.30 was compiled.

TABLE 5.30

Advertising (thousands of dollars)	Revenue (thousands of dollars)
5	150
7	200
9	250
11	325
13	400
15	450
17	500
19	525

a. Fit a model to the data.

b. Find and interpret the rate of change of revenue both as a rate of change and as an approximate change when $10,000 is spent on advertising.

c. Repeat part *b* for $18,000.

24. Write a brief essay that explains why, when rates of change are used to approximate change in a function, approximations over shorter time intervals generally give better answers than approximations over longer time intervals. Include graphical illustrations in your discussion.

Chapter 5   Summary

This chapter is devoted to analyzing change. The principal topics are optimization, inflection points, and approximating change.

Optimization

The word *optimization* (as we used it in Section 5.1) refers to locating extreme points on the graph of a function. By extreme points, we mean points where the graph reaches a relative (or local) maximum or a relative (or local) minimum. A relative maximum is simply a "peak" on the graph—a point to which the graph rises and after which the graph falls. Similarly, a relative minimum is a "valley" on the graph—a point to which the graph falls and after which the graph rises. There may be several relative maxima and relative minima on a graph. The highest and lowest points on a graph over an interval are called the absolute maximum and absolute minimum points. These points may coincide with a relative maximum or minimum, or they may occur at the endpoints of a given interval.

If a smooth, continuous graph has a relative maximum or relative minimum at a point, then the line tangent to the graph at that point is horizontal and has slope zero. In those cases, our search for relative maxima and minima narrows to points where the graph has horizontal tangent lines. However, it is not the case that every horizontal tangent occurs at a relative maximum or relative minimum, because horizontal tangents sometimes occur at points of inflection, as on the graph of $y = x^3$. Numerically, we find where horizontal tangent lines occur by determining input values where the derivative is zero. Analytically, the key to locating relative maxima and relative minima on smooth, continuous graphs is to locate the input values where the derivative actually *crosses* the horizontal axis. If the derivative merely *touches* the horizontal axis to produce a horizontal tangent, then we do not have a relative maximum or relative minimum.

Some functions may have extreme points where the function has a value, but the derivative does not exist; that is, there is no line tangent to the graph at those points. These functions may or may not be continuous. Some piecewise functions are examples of functions of this type.

It is always wise to begin any optimization investigation with a graph of the function before using derivatives.

Inflection Points

We first met inflection points in Chapter 2, where they appeared on the graphs of cubic and logistic models. Geometrically, inflection points are simply points where the concavity of the graph changes from concave up to concave down, or vice versa. Their importance, however, is that they identify the points of most rapid change or least rapid change.

Because the change along a graph is measured by its derivative, we often locate inflection points by finding the peaks and valleys of the derivative graph—that is, the relative maxima and relative minima of the derivative graph. Analytically, we look among the points where the second derivative (the derivative of the derivative) crosses the horizontal axis.

There are functions with inflection points where the second derivative fails to exist. These points can be found by determining where the second derivative is not defined and looking at the graph and its concavity on either side of these points. Again, we should always begin with a graph of the function and its derivative before attempting to use the second derivative to locate inflection points.

In certain applications, inflection points are referred to as points of diminishing returns. Most often, this occurs in situations where the graph is increasing and the concavity changes from concave up to concave down. The idea in these situations is that for input values beyond (to the right of) the inflection point, the rate of increase is decreasing; that is, output increases at a decreasing rate—hence the phrase *diminishing returns*.

Approximating Change

Few applications in business, economics, finance, management, and the social sciences demand the exact answers that you were taught to generate with mathematics in previous courses. Well-reasoned approximations are usually sufficient. Indeed, the very equations to which we apply some of the precise analytical techniques of mathematics are often approximating models constructed from data (which may be incomplete, inaccurate, or both). Nevertheless, we must still be careful in our approach and avoid bad reasoning, sloppy calculations, and improper interpretations. Precision in thought, calculation, and interpretation is vital in all mathematical endeavors.

One of the most useful approximations of change in a function is to use the behavior of a tangent line to approximate the behavior of the function. Because of the principle of local linearity, we know that tangent line approximations are quite accurate over small intervals. We approximate the output of $f(x + h)$ as $f(x) + f'(x) \cdot h$, where h represents the small change in input.

When cost, revenue, and profit are approximated by using these techniques, we call the approximation *marginal analysis*. The rates of change of cost, revenue, and profit with respect to the number of units produced or sold are commonly called marginal cost, marginal revenue, and marginal profit. In marginal analysis, the small change in input is one unit.

Chapter 5   Review Test

1. The number of tourists who visited Tahiti each year between 1988 and 1994 can be modeled[38] by

$$T(x) = -0.4804x^4 + 6.635x^3 - 26.126x^2 + 26.981x + 134.848 \text{ thousand tourists}$$

   x years after 1988.

   a. Give the rate-of-change formula for $T(x)$.

   b. Calculate the values of $T(3)$ and $T(5)$, and interpret your answers.

2. Find any relative maxima and minima of $T(x)$ between $x = 0$ and $x = 6$. Explain how you found the point(s).

3. Find any inflection points of $T(x)$ between $x = 0$ and $x = 6$. Explain how you found the point(s).

4. Graph $T(x)$, $T'(x)$, and $T''(x)$. Clearly label on each graph the points corresponding to your answers to Questions 2 and 3.

5. Although $T(x)$ is a continuous function, it must be interpreted discretely because it represents yearly totals. With this in mind, answer the following questions.

   a. Between 1988 and 1994, when was the number of tourists the greatest and when was it the least? What were the corresponding numbers of tourists in those years?

   b. Between 1988 and 1994, when was the number of tourists increasing the most rapidly, and when was it declining the most rapidly? Give the rates of change in each of those years.

6. a. Let $M(t)$ represent the population of French Polynesia (of which Tahiti is a part) at the end of year t. If $M(1996) = 225.2$ thousand people, and if $M'(1996) = 4.7$ thousand people per year, estimate the following:

   i. How much did the population of French Polynesia increase during the first quarter of 1997?

   ii. What was the population in the middle of 1997?

   b. The answer to question ii is an output value of what function?

---

38. Stephen J. Page, "The Pacific Islands," *EIU International Reports*, vol. 1 (1996), page 91.

Project 5.1

Hunting License Fees

Setting

In 1986, the state of California was trying to make a decision about raising the fee for a deer hunting license. Five hundred hunters were asked how much they would be willing to pay in excess of the current fee to hunt deer. The percentage of hunters to agree to a fee increase of $x is given by the logistic model[39]

$$\text{Percentage} = \frac{1.221}{1 + 0.221e^{0.0116x}}$$

Suppose that in 1986 the license fee was $100, and 75,000 licenses were sold. Suppose that you are part of the 1986 Natural Resources Team presenting a proposed increase in the hunting license fee to the head of the California Department of Natural Resources.

Tasks

1. Illustrate how the model can be used by answering the following questions.
   What was the hunting license revenue in 1986?
   Suppose that in 1987 the fee increased to $150.
   a. What percentage of the 1986 hunters would buy another license?
   b. How many hunters is that?
   c. What would be the 1987 revenue?
   Repeat the preceding analysis if the fee were increased to $300.

2. If the fee increase for 1987 were x dollars, find formulas for the following:
   a. the percentage of hunters willing to pay the new fee
   b. the number of the 75,000 hunters willing to pay the new fee
   c. the new fee
   d. the 1987 revenue

3. Use your formulas to determine the optimal license fee. Also, find the optimal fee increase, the number of hunters who will buy licenses at the new fee, and the optimal revenue.

Reporting

1. Prepare a letter to the Secretary of Natural Resources. Your letter should address the fee increase and expectations for revenue. You should not make it technical but should give some support to back up your conclusions.

2. Prepare a technical written report outlining your findings as well as the mathematical methods you used to arrive at your conclusions.

39. Based on information in *Journal of Environmental Economics and Management*, vol. 24, no. 1 (January 1993).

Project 5.2

Fund-Raising Campaign

Setting

In order to raise funds, the mathematics department in your college or university is planning to sell T-shirts before next year's football game against the school's biggest rival. Your team has volunteered to conduct the fund raiser. Because several other student groups have also volunteered to head this project, your team is to present its proposal for the fund drive, as well as your predictions about its outcome, to a panel of mathematics faculty.

Task A

Follow the tasks for Project 2.2 on pages 141–142. You will find a partial price listing for the T-shirt company in Table 2.73 on page 142.

Task B

1. *Getting Started:* Review your work for Task A. If you wish to make any changes in your marketing scheme, you should do so now. If you decide to make any changes, make sure that the polling that was done is still applicable (for example, you will not be able to change your target market.) Change (if necessary) any models from Task A to reflect any changes in your marketing scheme.

2. *Optimizing:* Use the models of demand, revenue, total cost, and profit developed in Task A to proceed with this section.

   Determine the selling price that generates maximum revenue. What is maximum revenue? Is the selling price that generates maximum revenue the same as the price that generates maximum profit? What is maximum profit? Which should you consider (maximum revenue or maximum profit) in order to get the best picture of the effectiveness of the drive? Re-evaluate the number of shirts you may wish to sell. Will this affect the cost you determined above? If so, change your revenue, total cost, and profit functions to reflect this adjustment and re-analyze optimal values. Show and explain the mathematics that underlies your reasoning.

   Discuss the sensitivity of the demand function to changes in price (check rates of change for $14, $10, and $6). Does the demand function have an inflection point? If so, find it. Find the rate of change of demand at this point, and interpret its meaning and impact in this context. How would the sensitivity of the demand curve affect your decisions about raising or lowering your selling price?

On the basis of your findings, predict the optimal selling price, the number of T-shirts you intend to print, the costs involved, the number of T-shirts you expect to sell before realizing a profit (that is, the break-even point), and the expected profit.

Reporting

1. Write a report for the mathematics department concerning your proposed campaign. They will be interested in the business interpretation as well as in an accurate description of the mathematics involved. Make sure that you include graphical as well as mathematical representations of your demand, revenue, cost, and profit functions. (Include graphs of any functions and derivatives that you use. Include your calculations and your survey as appendices.) Do not forget to cover Task A in this report as well.

2. Make your proposal and present your findings to a panel of mathematics professors in a 15-minute presentation. Your presentation should be restricted to the business interpretation, and you should use overhead transparencies of graphs and equations of all models and derivatives as well as any other visual aids that you consider appropriate.

Accumulating Change: Limits of Sums and the Definite Integral

Chapters 1 through 5 were directed toward helping you understand the derivative, one of the two fundamental concepts of calculus. We approached derivatives through their interpretation as rates of change, measured as slopes of tangent lines.

Now we begin a study of the second fundamental concept in calculus, the integral. As before, our approach will be through the mathematics of change. We start by analyzing the accumulated change in a quantity and quickly see the need to determine areas of regions between the rate-of-change function for that quantity and the horizontal axis. As we refine our thinking about area, we are led to consider limits of sums, which, in turn, show us how to account for the results of change in terms of integrals. Integrals are a new concept to us, but as we shall see, they are intimately connected to derivatives by the Fundamental Theorem of Calculus.

Heat loss through windows can be a headache for the managers at Dulles International Airport near Washington, D.C., unless they are prepared to take that heat loss into account when determining how much it costs to maintain heat in the building. In order to determine the amount of heat lost through a window, the managers must know the area of the window. The upper edge of the window at the end of the Dulles Airport terminal can be described by a quadratic model. Describe at least two different schemes for estimating the area of this window using geometric shapes. We can use calculus to find the exact area of the window.

■ 6.1   Results of Change

In our study of calculus so far, we have concentrated on finding rates of change. We now consider the results of change.

Velocity Examples

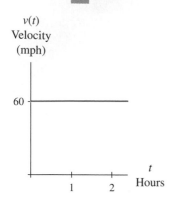

$v(t)$
Velocity
(mph)

60

1   2   Hours t

Velocity of a Vehicle

FIGURE 6.1

Suppose that you have been driving on an interstate highway for 2 hours at a constant speed of 60 miles per hour. Because velocity, $v(t)$, is the rate of change of distance traveled, $s(t)$, with respect to time, we write this mathematically as $v(t) = s'(t) = 60$ mph, where t is the time in hours. A graph of this rate function over a 2-hour period of time appears in Figure 6.1.

Recall that distance = (rate)(time), so at this constant rate, the distance traveled during any 1-hour period is (60 miles per hour)(1 hour) = 60 miles. Geometrically, we view this multiplication as giving the area of the region between the rate-of-change graph and the horizontal axis over any 1-hour period (see Figure 6.2a). Similarly, the distance traveled during the last 15-minute period of the 2-hour interval is (60 mph)(1/4 hour) = 15 miles. This gives the area of the region between the rate-of-change graph and the time axis from a time of 1.75 hours to a time of 2 hours. This region is shown in Figure 6.2b. In general, the distance traveled during t hours is $s(t) = 60t$ miles (see Figure 6.2c). In each case, the change in distance traveled during a time period is given by the area of a region between the rate-of-change graph and the horizontal axis from the beginning to the end of the time period.

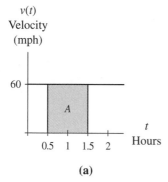

$v(t)$
Velocity
(mph)

60

A

0.5   1   1.5   2   Hours t

(a)

Area A is the distance traveled during 1 hour

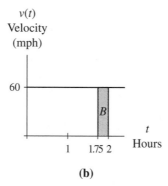

$v(t)$
Velocity
(mph)

60

B

1   1.75 2   Hours t

(b)

Area B is the distance traveled during last 15 minutes

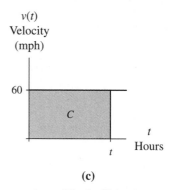

$v(t)$
Velocity
(mph)

60

C

t Hours t

(c)

Area C is the distance traveled during t hours

FIGURE 6.2

Now, imagine that after a 2-hour drive at a constant speed of 60 mph, you increase your speed at a constant rate to 75 mph over a 10-minute interval and then maintain that constant 75-mph speed during the next half-hour. A graph of your speed $v(t)$ appears in Figure 6.3a.

We know that the distance traveled during the times when speed is constant is the speed multiplied by the amount of time driven at that speed. But how can we calculate the distance driven between 2 hours and 2 hours 10 minutes when the speed is increasing linearly? If we knew the average speed over the 10-minute interval, then we could multiply that average speed by $\frac{1}{6}$ of an hour to obtain the distance

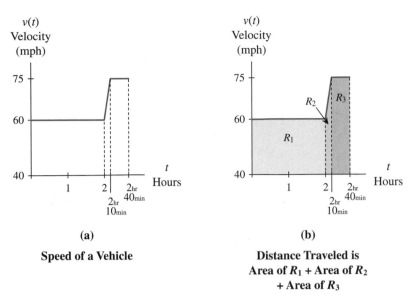

(a)

Speed of a Vehicle

(b)

**Distance Traveled is
Area of R_1 + Area of R_2
+ Area of R_3**

FIGURE 6.3

traveled. In this case, the average speed is simply the average of the beginning and ending speeds during the 10-minute interval. Thus we have

$$\begin{array}{l}\text{Distance traveled}\\ \text{between 2 hours}\\ \text{and 2 hours}\\ \text{10 minutes}\end{array} = \left(\begin{array}{l}\text{average}\\ \text{speed}\end{array}\right)(\text{time}) = \left(\frac{60 \text{ mph} + 75 \text{ mph}}{2}\right)\left(\frac{1}{6}\text{ hour}\right)$$

This distance is, in fact, the area of the trapezoid $\left[\text{Area} = (\text{base})\left(\frac{\text{side 1} + \text{side 2}}{2}\right)\right]$ labeled R_2 in Figure 6.3b. Thus, the distance traveled during the 2-hour 40-minute interval is the area of the region beneath the velocity graph and above the time axis between the specified inputs, calculated by dividing the region into three smaller regions as shown in Figure 6.3b.

$$\begin{array}{l}\text{Distance}\\ \text{traveled}\end{array} = \text{area of region } R_1 + \text{area of region } R_2 + \text{area of region } R_3$$

$$= (60 \text{ mph})(2 \text{ hr}) + \left(\frac{60 \text{ mph} + 75 \text{ mph}}{2}\right)\left(\frac{1}{6}\text{ hr}\right) + (75 \text{ mph})\left(\frac{1}{2}\text{ hr}\right)$$

$$= 120 \text{ miles} + 11.25 \text{ miles} + 37.5 \text{ miles}$$

$$= 168.75 \text{ miles}$$

Once more, the change in the distance traveled is given by the area of the region between the rate-of-change graph and the horizontal axis.

EXAMPLE 1   *Robot Speed*

A mechanical engineering graduate student designed a robot and is testing the ability of the robot to accelerate, decelerate, and maintain speed. The robot takes 1 minute to accelerate to 10 miles per hour (880 feet per minute). The robot maintains that speed for 2 minutes and then takes half a minute to come to a complete

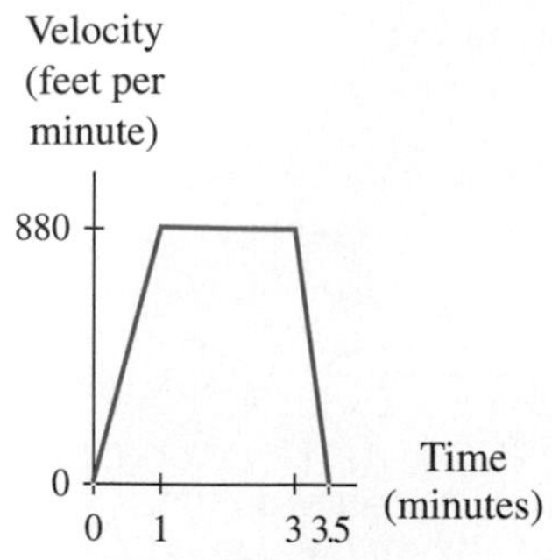

Velocity
(feet per
minute)

Speed of a Robot

FIGURE 6.4

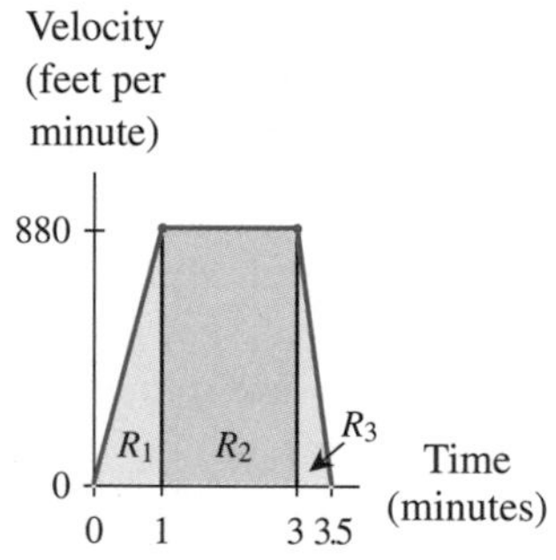

Velocity
(feet per
minute)

**Distance Traveled is
Area of R_1 + Area of R_2
+ Area of R_3**

FIGURE 6.5

stop. Assume that this robot's acceleration and deceleration are constant. A graph of the robot's speed during the experiment is shown in Figure 6.4.

a. Find the area of the region between the graph and the horizontal axis.

b. What is the practical interpretation of the area?

Solution:

a. The region can be divided into one rectangle and two triangles $\left[\text{Area of triangle} = \frac{1}{2}(\text{base})(\text{height})\right]$ as shown in Figure 6.5.

$$\text{Total area} = \text{area of } R_1 + \text{area of } R_2 + \text{area of } R_3$$

$$= \frac{1}{2}(1 \text{ min})(880 \text{ ft/min}) + (2 \text{ min})(880 \text{ ft/min}) +$$

$$\frac{1}{2}\left(\frac{1}{2} \text{ min}\right)(880 \text{ ft/min})$$

$$= 440 \text{ ft} + 1760 \text{ ft} + 220 \text{ ft}$$

$$= 2420 \text{ feet}$$

(*Note:* This area could also be calculated as the area of a single trapezoid.)

b. This area represents the distance the robot traveled during the $3\frac{1}{2}$-minute experiment.   ■

The previous velocity examples illustrate the following fundamental principle:

Results of Change

The accumulated change in a quantity is represented as the area of a region between the rate-of-change function for that quantity and the horizontal axis. In the case where the rate-of-change function is negative, the accumulated change in the quantity is the negative of the area of the region between the rate-of-change function and the horizontal axis.

We will examine a situation in which a portion of a rate of change function lies below the horizontal axis in Example 2. We will verify this principle later in the chapter by using the Fundamental Theorem of Calculus.

Approximating Accumulated Change

The preceding velocity examples were carefully chosen so that their rate-of-change graphs were easy to obtain and the areas of the desired regions were easy to calculate. But most real-life situations are not so simple. Indeed, there are two issues that we must face:

1. Obtaining the rate-of-change function for the quantity of interest

2. Calculating the area of the desired region between the rate-of-change function and the horizontal axis

In most cases, we must resort to approximating both the rate-of-change function and the desired area. Consider the example of a store manager of a large department store who wishes to estimate the number of customers who came to a Saturday sale from 9 a.m. to 9 p.m. The manager stands by the entrance for 1-minute intervals at different times throughout the day and counts the number of people entering the store. He uses these data as an estimate of the number of customers who enter the store each minute. The manager's data may look something like Table 6.1.

TABLE 6.1

Time	Number of customers per minute
9:00 a.m.	1
9:45 a.m.	2
10:15 a.m.	3
11:00 a.m.	4
11:45 a.m.	4
12:15 p.m.	5
1:15 p.m.	5
2:30 p.m.	5
3:10 p.m.	5
4:00 p.m.	4
5:15 p.m.	4
6:30 p.m.	3
7:30 p.m.	2
8:15 p.m.	2

The total number of customers who attended the sale can be calculated by summing the number of customers who entered the store during each hour for every hour of the 12-hour sale. We do not have enough information to determine the exact number of customers who entered the store each hour; however, we can estimate the number by using a continuous model for the customers-per-minute data. To build a model for the rate-of-change data, convert each of the above observation times to minutes after 9:00 a.m. Thus $m = 0$ at 9:00 a.m. and $m = 675$ at 8:15 p.m. A model for the customers-per-minute data is

$$c(m) = (4.58904 \cdot 10^{-8})m^3 - (7.78127 \cdot 10^{-5})m^2 + 0.03303m + 0.88763$$
$$\text{customers per minute}$$

where m is the number of minutes after 9:00 a.m. The graph of this model is shown in Figure 6.6. Note that $c(m)$ is a continuous function modeling a discrete situation.

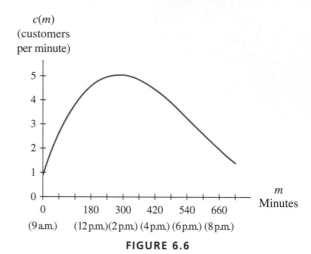

FIGURE 6.6

To estimate the total number of customers who attended the sale, we use the model to estimate the number of customers per minute entering the store at the beginning of each hour and multiply by 60 minutes to estimate the number of customers entering the store during that hour. Summing the estimates for each of the 12 hours results in the estimate we desire for the total number of customers. This process is the same as drawing a set of 12 rectangles under the graph of $c(m)$, one for each hour, and using the sum of their areas to estimate the total number of customers who came to the sale. (See Figure 6.7.)

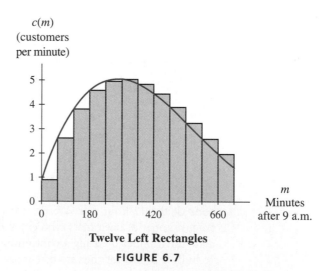

Twelve Left Rectangles

FIGURE 6.7

Note that the area of the rectangles is an estimate of the area of the region between $c(m)$ and the horizontal axis. This is precisely the point made in the statement after Example 1 concerning results of change. The total number of customers is approximately the area of the region beneath the rate-of-change model. Here we approximate the area using 12 rectangles. Later in this chapter, we will study other means of finding or approximating the area.

In Figure 6.7, the height of each rectangle is the function $c(m)$ evaluated at the left endpoint of the base of the rectangle. For this reason, we call these rectangles **left rectangles**.

Because we are using 12 rectangles of equal width to span the 12-hour (720-minute) sale, the width of each rectangle is $720 \div 12 = 60$ minutes. We use Table 6.2 to keep track of the areas that we are summing.

TABLE 6.2

Left endpoint of rectangle m	Height of rectangle $c(m)$ (customers per minute)	Width of rectangle (minutes)	Area of rectangle[1] = height·width [(customers/min)(min) → customers]
9 a.m.	$c(0) \approx 0.9$	60	53.3
10 a.m.	$c(60) \approx 2.6$	60	156.0
11 a.m.	$c(120) \approx 3.8$	60	228.6
noon	$c(180) \approx 4.6$	60	274.8
1 p.m.	$c(240) \approx 5.0$	60	298.1
2 p.m.	$c(300) \approx 5.0$	60	302.0
3 p.m.	$c(360) \approx 4.8$	60	290.2
4 p.m.	$c(420) \approx 4.4$	60	266.1
5 p.m.	$c(480) \approx 3.9$	60	233.4
6 p.m.	$c(540) \approx 3.3$	60	195.7
7 p.m.	$c(600) \approx 2.6$	60	156.4
8 p.m.	$c(660) \approx 2.0$	60	119.2
		Total area of rectangles $\approx$ 2574 customers	

Thus, using 12 rectangles, we estimate that 2574 customers came to the Saturday sale.

Note the importance in this example of measuring time in minutes. If time is measured in hours, then the area calculated by multiplying customers per minute by hours is not the number of customers. The output units for the function are the units of the heights of rectangles, and the input units are the units of the widths of rectangles. Make sure those units correspond so that their multiplication results in the desired units. Also note that our choice of using time intervals of 1 hour was arbitrary. In fact, using 30-minute or 15-minute intervals may give a more accurate estimate.

The previous discussion illustrated a way to approximate the number of customers attending a Saturday sale by using left-rectangle areas. In some situations, choosing rectangles whose heights are measured at the right endpoint of the base of each rectangle may give more reasonable area approximations. Such rectangles are called **right rectangles**. The use of such rectangles is illustrated in the following example.

1. These values were obtained using the unrounded model.

EXAMPLE 2   *Drug Absorption*

A pharmaceutical company has tested the absorption rate of a drug that is given in 20-milligram (mg) doses for 20 days. They have compiled data showing the rate of change of the concentration of the drug, measured in micrograms per milliliter per day (μg/mL/day), in the blood stream. The data and scatter plot are shown in Table 6.3 and Figure 6.8, respectively.

TABLE 6.3

Day	Concentration rate of change (μg/mL/day)
1	1.50
5	0.75
9	0.33
13	0.20
17	0.10
21	−1.10
25	−0.60
29	−0.15

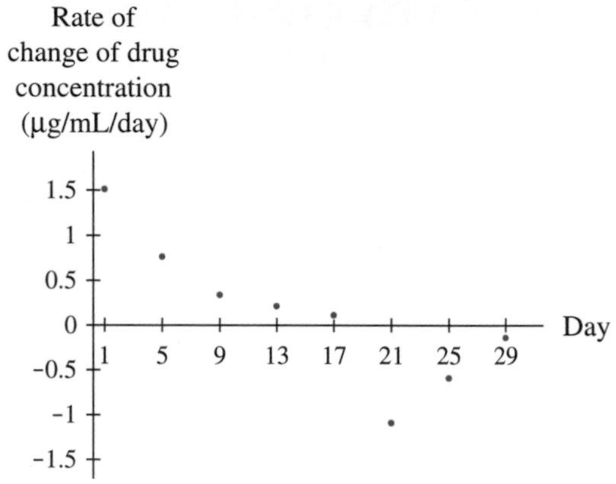

Rate of change of drug concentration (μg/mL/day)

FIGURE 6.8

a. Find a model for the data.

b. Use the model and right rectangles of width 2 days to estimate the change in the drug concentration from the beginning of day 1 through the end of day 20.

c. Use the model and right rectangles of width 1 day to estimate the change in the drug concentration from the beginning of day 21 through the end of day 29.

d. Combine your answers to parts *b* and *c* to estimate the change in the drug concentration from the beginning of day 1 through the end of day 29.

Solution:

a. It is clear that a piecewise model is appropriate, with one function modeling the data through the end of day 20 and a second function modeling the data after the drug is discontinued. A possible model is

$$r(x) = \begin{cases} 1.708(0.845)^x \ \mu\text{g/mL/day} & \text{when } 0 \le x \le 20 \\ 0.11875x - 3.5854 \ \mu\text{g/mL/day} & \text{when } 20 < x \le 29 \end{cases}$$

where x is the number of days after the drug is first administered.

b. To determine the change in the drug concentration from the beginning of day 1 ($x = 0$) through the end of day 20 ($x = 20$), we use the exponential portion of the model and 10 right rectangles as shown in Figure 6.9.

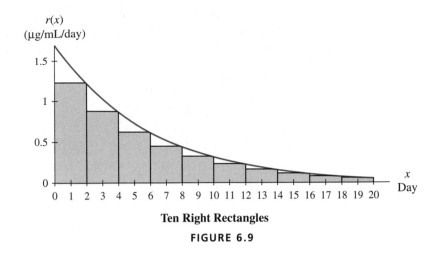

Ten Right Rectangles

FIGURE 6.9

TABLE 6.4

Right endpoint of rectangle x	Height of rectangle $r(x)$ (µg/mL/day)	Area of rectangle = height·width [(µg/mL/day)(days)→µg/mL]
2	1.220	2.439
4	0.871	1.742
6	0.622	1.244
8	0.444	0.888
10	0.317	0.634
12	0.226	0.453
14	0.162	0.323
16	0.115	0.231
18	0.082	0.165
20	0.059	0.118
Sum of areas of rectangles ≈ 8.235 µg/mL Change in the drug concentration ≈ 8.24 µg/mL		

From the beginning of day 1 through the end of day 20, the drug concentration increased by approximately 8.24 micrograms per milliliter.

c. To determine the change in the concentration from the beginning of day 21 through the end of day 29, we use the linear portion of the model and the 9 right rectangles shown in Figure 6.10.

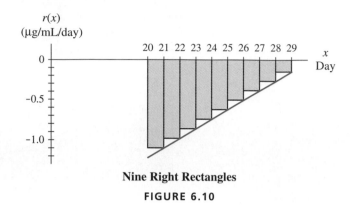

Nine Right Rectangles

FIGURE 6.10

There are two important things to note. First, the fact that this region lies below the axis indicates that the drug concentration is declining. Second, when we substitute x-values between 21 and 29 into $r(x)$ to determine the heights of the rectangles, we get negative numbers. Because height is always a positive measure, we use the absolute value of $r(x)$ for our heights.

TABLE 6.5

Right endpoint x	Height of rectangle $|r(x)|$ (μg/mL/day)	Area of rectangle = height·width [(μg/mL/day)(days)→μg/mL]
21	1.092	1.092
22	0.973	0.973
23	0.854	0.854
24	0.735	0.735
25	0.617	0.617
26	0.498	0.498
27	0.379	0.379
28	0.260	0.260
29	0.142	0.142
Sum of areas of rectangles $\approx$ 5.550 μg/mL Change in the drug concentration $\approx$ -5.55 μg/mL		

From the beginning of day 21 through the end of day 29, the concentration of the drug in the blood stream declined by approximately 5.55 μg/mL.

d. To determine the change in concentration from the beginning of day 1 through the end of day 29, we need only subtract the amount of decline from the amount of increase.

$$8.24 \ \mu g/m/L - 5.55 \ \mu gmL = 2.69 \ \mu g/mL$$

The drug concentration increased by approximately 2.69 μg/mL. ∎

TABLE 6.6

Year	Cans each year (billions)
1978	8.0
1979	8.5
1980	14.8
1981	24.9
1982	28.3
1983	29.4
1984	31.9
1985	33.1
1986	33.3
1987	36.6
1988	42.0

Frequently, rate-of-change data are not available for a quantity; instead, we have count data (that is, totals reported at the end of a period of time) pertaining to that quantity. Table 6.6[2] shows the number of aluminum cans that were recycled each year from 1978 through 1988.

A scatter plot of the data appears in Figure 6.11.

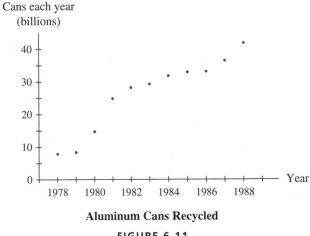

Aluminum Cans Recycled

FIGURE 6.11

The number of cans recycled during 1980 can be interpreted geometrically as the area of a rectangle. We use a right rectangle (one whose height is determined at the right side of the rectangle) because the data are reported at the end of each year. (See Figure 6.12.)

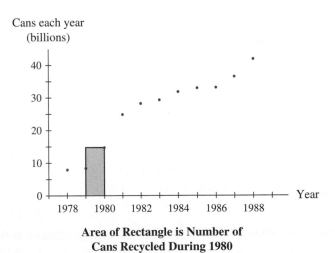

Area of Rectangle is Number of Cans Recycled During 1980

FIGURE 6.12

In other words, the number of cans recycled during 1980 is (height)(width) = (14.8 billion cans each year)(1 year) = 14.8 billion cans. If we consider the number of cans recycled during year t as the area of a rectangle, then the number of recycled

2. Data from The Aluminum Association, Inc.

aluminum cans from the beginning of 1978 through the end of 1988 is the sum of the areas of the rectangles shown in Figure 6.13.

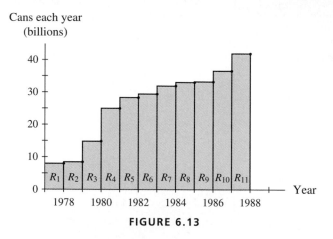

FIGURE 6.13

$$\text{Number of cans recycled} = \text{the sum of the areas of the rectangles}$$
$$= 8.0 + 8.5 + 14.8 + 24.9 + 28.3 + 29.4 +$$
$$31.9 + 33.1 + 33.3 + 36.6 + 42.0$$
$$= 290.8 \text{ billion cans}$$

Note that because each rectangle has width 1 year, the area is simply the sum of the data points.

What would we do if some of the data were missing? For instance, suppose that we had data for only even-numbered years from 1978 through 1988. We could use rectangles with width 2 years, or we could fit a model to the data and then construct estimates by using values determined from the model and rectangles with width 1 year.

6.1   Concept Inventory

- Area of region between rate-of-change function and horizontal axis between a and b = accumulated change in the amount function between a and b

- Left-rectangle approximation
- Right-rectangle approximation

6.1   Activities

1. The growth rate of bacteria (in thousands per hour) in milk at room temperature is modeled by $B(t)$, where t is the number of hours that the milk has been at room temperature. We wish to use rec-

tangles to estimate the area of the region between $B(t)$ and the t-axis. What are the units on

a. the heights of the rectangles?

b. the widths of the rectangles?

c. the areas of the rectangles?

d. the area of the region between $B(t)$ and the t-axis?

e. the accumulated change in the number of bacteria in the milk during the first hour that the milk is at room temperature?

2. The acceleration of a car (in feet per second per second) during a test conducted by a car manufacturer is given by $A(t)$, where t is the number of seconds since the beginning of the test.

a. What does the area of the region between the portion of $A(t)$ lying above the t-axis and the t-axis tell us about the car?

b. What are the units on
    i. the heights and widths of rectangles used to estimate area?
    ii. the area of the region between $A(t)$ and the t-axis?

3. The distance required for a car to stop is a function of the speed of the car when the brakes are applied. The rate of change of the stopping distance could be expressed in feet per mile per hour where the input is the speed of the car, in miles per hour, when brakes are applied.

    a. What does the area of the region between the rate-of-change graph and the input axis from 40 mph to 60 mph tell us about the car?

    b. What are the units on
        i. the heights and widths of rectangles used to estimate the area in part a?
        ii. the area in part a?

4. The atmospheric concentration of CO_2 is growing exponentially. If the growth rate in ppm per year is modeled by $C(t)$, where t is the number of years since 1980, what are the units on

    a. the area of the region between $C(t)$ and the t-axis from $t = 0$ to $t = 20$?

    b. the heights and widths of rectangles used to estimate the area in part a?

    c. the change in the CO_2 concentration from 1980 through 2000?

5. A real estate developer owns 160 lots in a newly opened subdivision for new homes.

    a. Write a function that gives the number of lots not yet sold in the subdivision if the developer sells lots at the rate of 24 lots per year. Note that this is a continuous model representing a discrete situation.

    b. Use your function to determine when there will be only 40 lots remaining to be sold.

    c. Graph the rate of change function for your continuous model for the number of lots not yet sold, and indicate the rectangles whose collective area represents the decline in the number of lots not yet sold during the first 5 years.

    d. Suppose that the developer sells 24 lots per year for the first 3 years, 12 lots per year for the next 4 years, and then 16 lots per year until all of the lots are sold. Graph the rate-of-change function for a continuous model of the number of lots not yet sold. Sketch the rectangles whose collective area represents the decline in the number of lots not yet sold from the beginning of year 3 through the end of year 9. Use these rectangles to find this decline.

6. Suppose that while driving your car at 45 mph, you apply the brakes in such a way that your speed decreases at a constant rate until you stop after 10 seconds.

    a. Find a formula for your speed as a function of time (in seconds).

    b. What was your speed 3 seconds after you first applied the brakes? 7 seconds after?

    c. Graph the speed equation, and find your stopping distance by calculating the area of the region between the graph and the horizontal axis from 0 seconds to 10 seconds.

    d. On the graph, indicate the region whose area gives the distance that you traveled during the 4-second interval that began 2 seconds after you first applied the brakes. How far did you travel during this 4-second interval?

7. In October of 1994, McDonnell Douglas[3] stock was worth $200 per share.

    a. Write a function for the value of x shares of stock. Graph this function. Note that this is a continuous model representing a discrete situation.

    b. Write the function for the rate of change of the continuous model for the value of McDonnell Douglas stock with respect to the number of shares held. Graph this rate-of-change function.

    c. Find the change in the value of stock held if the number of shares held is increased from 250 to 300 shares. Depict this change as the area of a region on the rate-of-change graph.

8. In 1996, a hospital reported that its paperwork was growing by 6 feet per day.

    a. Convert the rate of change from feet per day to feet per year.

    b. Assuming that this growth has remained constant since 1996, find a formula to represent the amount of paperwork accumulated by the xth year after 1996.

3. *Fortune Magazine*, October 3, 1994, p. 46.

c. Use the formula in part *b* to determine how much paperwork has accumulated since the end of 1996. Depict this accumulated amount as the area of a region on the rate-of-change graph.

9. On the basis of data reported in the 1994 *Statistical Abstract*, the rate of change of the population of Iowa between 1980 and 1993 can be modeled by

$$r(x) = \begin{cases} -3.864x - 7.841 \text{ thousand} & \text{when} \\ \quad \text{people per year} & 0 \le x < 7 \\ 2.714x - 18.929 \text{ thousand} & \text{when} \\ \quad \text{people per year} & 7 \le x \le 13 \end{cases}$$

where *x* is the number of years since 1980.

a. Sketch a graph of the rate-of-change function.

b. Find the area of the region between $r(x)$ and the horizontal axis from 0 to 7. Interpret your answer.

c. Find the area of the region between $r(x)$ and the horizontal axis from 7 to 13. Interpret your answer.

d. Was the population of Iowa in 1993 greater or less than the population in 1980? By how much did the population change between 1980 and 1993?

10. The rate of change of the population[4] of North Dakota from 1970 through 1993 can be modeled as

$$p(t) = \begin{cases} 3.87 \text{ thousand} & \text{when } 0 \le t < 15 \\ \quad \text{people per year} \\ -7.39 \text{ thousand} & \text{when } 15 \le t \le 21 \\ \quad \text{people per year} \\ 1.00 \text{ thousand} & \text{when } 21 < t \le 23 \\ \quad \text{people per year} \end{cases}$$

where *t* represents the number of years since 1970.

a. Sketch a graph of the rate-of-change function.

b. Find the area of the region between $p(t)$ and the horizontal axis from 0 to 15. Interpret your answer.

c. Find the area of the region between $p(t)$ and the horizontal axis from 15 to 21. Interpret your answer.

d. Find the area of the region between $p(t)$ and the horizontal axis from 21 to 23. Interpret your answer.

e. Was the population of North Dakota in 1993 greater or less than the population in 1970? By how much did the population change between 1970 and 1993?

11. The District of Columbia saw a decline in population from 1987 through 1993. The rate of change of population from 1987 through 1993 can be modeled[5] by

$$P(x) = 1.08333x^2 - 195.0238x + 8763.9365$$
$$\text{thousand people per year}$$

where *x* is the number of years since the end of 1900.

a. Graph $P(x)$ from the end of 1987 through 1993.

b. Determine when the population of the District of Columbia was decreasing most rapidly.

c. Use six right rectangles to approximate the area of the region between $P(x)$ and the *x*-axis from 87 through 93. Interpret your answer.

12. The data in Table 6.7 show the marginal cost for compact disc production at the indicated hourly production levels:

TABLE 6.7

Production (CDs per hour)	Marginal cost
100	$5
150	$3.50
200	$2.50
250	$2
300	$1.60

a. Use four left rectangles to approximate the change in cost when production is increased from 100 to 300 CDs per hour.

b. Fit a model to the data, and sketch a graph of the model for production levels from 100 to 300 CDs per hour.

c. Repeat part *a* using eight left rectangles. Why is a model necessary when you are using eight rectangles?

d. Sketch the rectangles used in part *c* on a graph of the model. Is the approximation larger or smaller than the area of the region between the model and the horizontal axis from 100 to 300?

4. Based on data from *Statistical Abstract*, 1994.
5. *Ibid.*

13. The number of live births each year from 1950 through 1988 in the United States to women 45 years of age and older is given in Table 6.8.[6]

TABLE 6.8

Year	Births
1950	5322
1955	5430
1960	5182
1965	4614
1970	3146
1975	1628
1980	1200
1985	1162
1986	1257
1987	1375
1988	1427

a. Draw a scatter plot of the data, and sketch right rectangles depicting the total number of live births from the beginning of 1985 through 1988. Use the data to find the total number of live births for this period. Is this number an approximation or exact? Explain.

b. Find a model to fit the data. Sketch a graph of this model.

c. Estimate the number of such births from the beginning of 1965 through the end of 1984 by using rectangles with a width of 1 year. How many rectangles were needed?

d. What would be required to find the exact total number of such births from the beginning of 1965 through 1984?

14. A major European tire manufacturer began international marketing of its radial tires in 1966, and sales soon skyrocketed. Table 6.9 shows the rate of change in sales revenue for selected years since 1970.

a. Fit a model to the data. Explain why you chose the model you did.

b. Sketch the graph of your model and six rectangles for approximating the area under the graph from 1970 through 1994. Estimate and interpret this area.

TABLE 6.9

Year	Rate of change in sales (thousands of dollars per year)
1970	4.92
1974	6.3
1978	13.7
1982	25.1
1986	38.3
1990	76.5
1994	106.85

c. Use 11 rectangles to approximate sales revenue from 1974 through 1985. Sketch the graph and the approximating rectangles that depict this revenue.

15. During a summer thunderstorm, the temperature drops and then rises again. The rate of change of the temperature during the hour and a half after the storm began is given by

$$T(h) = 9.48h^3 - 15.49h^2 + 17.38h - 9.87$$
°F per hour

where h is the number of hours since the storm began.

a. Graph the function $T(h)$ from $h = 0$ to $h = 1.5$. Find the point at which the graph crosses the horizontal axis.

b. Consider the portion of $T(h)$ lying below the horizontal axis. What does the area of the region between this portion of $T(h)$ and the horizontal axis represent?

c. What does the area of the region lying above the axis represent?

d. Use seven right rectangles to approximate the area of the region lying below the axis from 0 to 1.5 hours.

e. Use seven right rectangles to approximate the area of the region lying above the axis from 0 to 1.5 hours.

f. According to the model, one and a half hours after the storm began, was the temperature higher, lower, or the same as the temperature at the beginning of the storm? If it was higher or lower, by how many degrees?

6. *Statistical Abstract*, 1992.

16. The acceleration of a race car during the first 35 seconds of a road test is modeled by the equation

$$a(t) = 0.024t^2 - 1.72t + 22.58 \text{ ft/sec}^2$$

where t is the number of seconds since the test began.

a. Graph $a(t)$ from 0 to 35 seconds. Find the point at which the graph crosses the t-axis.

b. Use five left rectangles to approximate the area of the region between $a(t)$ and the t-axis that lies above the t-axis from 0 to 35 seconds. Interpret your answer.

c. Use five right rectangles to approximate the area of the region between $a(t)$ and the t-axis that lies below the axis from 0 to 35 seconds. Interpret your answer.

d. Do your answers in parts b and c overestimate or underestimate the actual areas?

e. Use your answers to parts b and c to estimate by how much the car's speed changed between the beginning of the test and 35 seconds later. Convert your answer to miles per hour.

17. Table 6.10[7] shows the number of new female Ph.D.s in computer science in selected years from 1970 through 1993.

a. Draw a scatter plot of the data, and sketch rectangles depicting the number of new female Ph.D.s in computer science from the beginning of 1985 through 1993. Find the total number of new female Ph.D.s from the beginning of 1985 through 1993. Is this number an approximation or exact?

b. Fit a model to the data.

TABLE 6.10

Year	New female Ph.D.s
1970	1
1973	7
1976	14
1979	24
1982	27
1985	32
1986	50
1987	51
1988	60
1989	87
1990	97
1991	113
1992	108
1993	126

c. Use the model to estimate the total number of new female Ph.D.s in computer science from the beginning of 1970 through 1993.

d. The total number of new female Ph.D.s in computer science from the beginning of 1970 through 1993 was 987. How does your estimate from part c compare to this number?

18. Many businesses spend large sums of money each year on advertising in order to stimulate sales of their products. The data given in Table 6.11 show the approximate increase in sales (in thousands of dollars) that an additional $100 spent on advertising, at various levels, can be expected to generate.

TABLE 6.11

Advertising expenditures (hundreds of dollars)	Revenue increase due to an extra $100 advertising (thousands of dollars)
25	5
50	60
75	95
100	105
125	104
150	79
175	34

7. *Computing Research News*, January 1994.

a. Fit a model to the data.

b. Use six rectangles of equal width to estimate the area under the rate-of-change-of-revenue function from 25 to 175 hundred advertising dollars. Sketch these rectangles on a graph of the model. This area is an approximation to what?

c. Repeat part *b* using 12 rectangles. Is the area of the 12 rectangles a better or worse approximation than the area of the 6 rectangles? Explain.

19. The 1992 *Statistical Abstract* gives the number of overseas telephone calls (in millions) in the United States as shown in Table 6.12.

a. Draw a scatter plot of the data, and sketch rectangles of width 1 year depicting the number of overseas calls from the beginning of 1984 through 1990. Find the total number of overseas calls from the beginning of 1984 through 1990. Is this an approximation or the exact number?

b. Fit a model to the data.

c. Use your model to estimate the number of overseas calls from the beginning of 1970 through 1990.

TABLE 6.12

Year	Calls (millions)
1970	23.4
1980	199.6
1984	427.6
1985	411.7
1986	477.6
1987	579.6
1988	705.7
1989	1007.8
1990	1200.7

20. In essay form (using pictures as necessary), explain why the approximation to the area of a region under a curve and above the horizontal axis generally improves when twice as many rectangles are used.

6.2   Trapezoid and Midpoint-Rectangle Approximations

In the last section, we saw that if a rate-of-change model is available for a certain quantity, then we can estimate changes in that quantity by finding areas of regions between the rate-of-change curve and the horizontal axis. In each application where the exact area could not easily be found, we used rectangles whose heights were calculated at the left or right endpoints of intervals to approximate the area. Let us review this process by approximating the area from 0 to 1 of the region between the graph of $f(x) = \sqrt{1 - x^2}$ and the horizontal axis using left rectangles (see Figure 6.14a) and right rectangles (see Figure 6.14b).

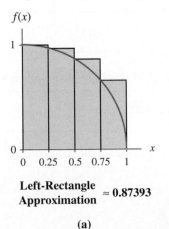

Left-Rectangle Approximation ≈ 0.87393

(a)

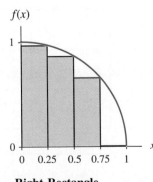

Right-Rectangle Approximation ≈ 0.62393

(b)

FIGURE 6.14

Because the graph of $f(x)$ is decreasing, any left-rectangle approximation of the area of the region between $f(x) = \sqrt{1 - x^2}$ and the x-axis exceeds the true area, and any right-rectangle approximation falls short of the true area value.

Trapezoid Approximation

What other kinds of area approximations can we use? It is often true that if a left-rectangle area is an over-approximation, then a right-rectangle area will be an under-approximation, and vice versa. For this reason, an average of the areas of left and right rectangles usually produces a more accurate area estimate. The average is calculated as follows:

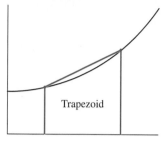

Trapezoid

FIGURE 6.15

$$\text{Average} = \frac{\left(\begin{array}{c}\text{height of left}\\\text{rectangle}\end{array}\right)(\text{width}) + \left(\begin{array}{c}\text{height of right}\\\text{rectangle}\end{array}\right)(\text{width})}{2}$$

$$= (\text{width})\frac{\left(\begin{array}{c}\text{height of left}\\\text{rectangle}\end{array} + \begin{array}{c}\text{height of right}\\\text{rectangle}\end{array}\right)}{2}$$

$$= (\text{width})(\text{average height})$$

This average is also the area of the trapezoid obtained by drawing a line segment between the two heights at the beginning and end of each subinterval (see Figure 6.15).

We call such an approximation a **trapezoid approximation**. Figure 6.16 shows a trapezoid approximation of the area of the region between $f(x) = \sqrt{1 - x^2}$ and the x-axis with four trapezoids. (Because the right height of the trapezoid on the far right is zero, that figure is actually a triangle.) We obtain the trapezoid approximation by summing the areas of the four approximating trapezoids, as shown in Table 6.13.

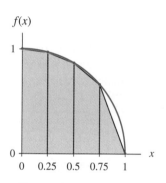

Trapezoid Approximation

FIGURE 6.16

TABLE 6.13

Trapezoid number	Interval	Average height	Width	Area = average height·width
1	0 to 0.25	$\dfrac{f(0) + f(0.25)}{2} \approx 0.98412$	0.25	0.24603
2	0.25 to 0.5	$\dfrac{f(0.25) + f(0.5)}{2} \approx 0.91714$	0.25	0.22928
3	0.5 to 0.75	$\dfrac{f(0.5) + f(0.75)}{2} \approx 0.76373$	0.25	0.19093
4	0.75 to 1	$\dfrac{f(0.75) + f(1)}{2} \approx 0.33072$	0.25	0.08268
				Total trapezoid area ≈ 0.74893

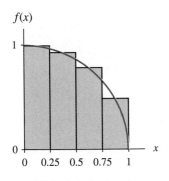

Midpoint-Rectangle Approximation

FIGURE 6.17

Note that if you already know the left- and right-rectangle approximations, the trapezoid approximation can be computed by averaging the two. In this example,

$$\text{Total trapezoid area} = \frac{\text{left-rectangle area} + \text{right-rectangle area}}{2}$$

$$\approx \frac{0.87393 + 0.62393}{2} = 0.74893$$

Midpoint-Rectangle Approximation

We next consider a third type of rectangle approximation. The **midpoint-rectangle approximation** uses rectangles whose heights are calculated at the midpoints of the subintervals (see Figure 6.17). Table 6.14 shows the calculations for the areas of the midpoint rectangles shown in Figure 6.17.

TABLE 6.14

Rectangle number	Midpoint of interval	Height of rectangle	Width of rectangle	Area = height·width
1	0.125	$f(0.125) \approx 0.99216$	0.25	0.24804
2	0.375	$f(0.375) \approx 0.92702$	0.25	0.23176
3	0.625	$f(0.625) \approx 0.78062$	0.25	0.19516
4	0.875	$f(0.875) \approx 0.48412$	0.25	0.12103
			Total midpoint area ≈ 0.79598	

Errors in Approximation

Which of the approximations we have considered is closest to the true area? The region whose area we are approximating is the interior of a quarter-circle with radius 1, so the true area is $\frac{1}{4}(\pi \cdot \text{radius}^2) = \frac{\pi(1)^2}{4} \approx 0.78540$. Examine the absolute errors in the approximations, shown in Table 6.15.

TABLE 6.15

	Approximation	|True value − approximation| = Absolute error
Left rectangles	0.87393	$\|0.785398 - 0.87393\| = 0.08853$
Right rectangles	0.62393	$\|0.785398 - 0.62393\| = 0.16147$
Trapezoids	0.74893	$\|0.785398 - 0.74893\| = 0.03647$
Midpoint rectangles	0.79598	$\|0.785398 - 0.79598\| = 0.01058$

Clearly, the midpoint-rectangle approximation is closest to the true value.

EXAMPLE 1 *Rising River*

For 20 hours following a heavy rain storm in the spring of 1996, the flow rate for the west fork of the Carson River was measured periodically. The flow rates in cubic feet per second (cfs) are shown in Table 6.16.[8]

TABLE 6.16

Time	Flow rate (cfs)
11:45 a.m. Wednesday	785
3:45 p.m. Wednesday	753
7:45 p.m. Wednesday	834
11:45 p.m. Wednesday	1070
3:45 a.m. Thursday	1470
7:45 a.m. Thursday	2090

a. If a model were found with input in hours since 11:45 a.m. Wednesday and output in cfs, what would be the units for the area of rectangles under the graph of the model?

b. Convert the flow rates to cubic feet per hour. How does this change the units on the area of the region under the graph of a model for the converted data?

c. Find a model for the converted data.

d. Use 10 intervals with left and right rectangles, trapezoids, and midpoint rectangles to estimate the area under the graph of the model from 11:45 a.m. Wednesday to 7:45 a.m. Thursday.

e. What does this area represent in the context of the Carson River?

Solution:

a. The heights of the rectangles will be measured in cfs and the widths of the rectangles will be in hours. Multiplying height by width results in units of (cubic feet per second)(hours). Without conversion, these are not meaningful units.

b. There are 3600 seconds each hour (60 minutes per hour times 60 seconds per minute). Multiplying cubic feet per second by seconds per hour gives cubic feet per hour. The data become

TABLE 6.17

Hours since 11:45 a.m. Wednesday	Flow rate (cubic feet per hour)
0	2,826,000
4	2,710,800
8	3,002,400
12	3,852,000
16	5,292,000
20	7,524,000

8. As reported in the *Reno Gazette Journal*, May 17, 1996, p. 4A.

The area of the region beneath a model of the data has units of (cubic feet per hour)(hour). The area is measured in cubic feet.

c. A quadratic model for the converted data is

$$f(h) = 18{,}225h^2 - 135{,}334.3h + 2{,}881{,}542.9 \text{ cubic feet per hour}$$

h hours after 11:45 a.m. Wednesday.

d. Using 10 intervals to span a 20-hour period requires that each interval be 2 hours long. Table 6.18 shows the calculation for left- and right-rectangle approximations.

TABLE 6.18

Interval endpoint h	Height at endpoint $f(h)$ (ft^3/hr)	Width of rectangle (hours)	Area of rectangle = height·width $[(\text{ft}^3/\text{hr})(\text{hour}) \rightarrow \text{ft}^3]$
0	$f(0) \approx 2{,}881{,}542.9$	2	5,763,085.7
2	$f(2) \approx 2{,}683{,}774.3$	2	5,367,548.6
4	$f(4) \approx 2{,}631{,}805.7$	2	5,263,611.4
6	$f(6) \approx 2{,}725{,}637.1$	2	5,451,274.3
8	$f(8) \approx 2{,}965{,}268.6$	2	5,930,537.1
10	$f(10) \approx 3{,}350{,}700.0$	2	6,701,400.0
12	$f(12) \approx 3{,}881{,}931.4$	2	7,763,862.9
14	$f(14) \approx 4{,}558{,}962.9$	2	9,117,925.7
16	$f(16) \approx 5{,}381{,}794.3$	2	10,763,588.6
18	$f(18) \approx 6{,}350{,}425.7$	2	12,700,851.4
20	$f(20) \approx 7{,}464{,}857.1$	2	14,929,714.3

The area values 5,763,085.7 through 12,700,851.4 form the sum for left rectangles; the values 5,367,548.6 through 14,929,714.3 form the sum for right rectangles.

Total left-rectangle area $\approx 74{,}823{,}686$ cubic feet
Total right-rectangle area $\approx 83{,}990{,}314$ cubic feet

The trapezoid approximation can be found by averaging the left- and right-rectangle approximations.

$$\frac{74{,}823{,}686 + 83{,}990{,}314}{2} = 79{,}407{,}000 \text{ cubic feet}$$

The midpoint approximation is calculated as shown in Table 6.19.

TABLE 6.19

Midpoint	Height of midpoint	Width of rectangle	Area of rectangle
1	$f(1) \approx 2{,}764{,}433.6$	2	5,528,867.1
3	$f(3) \approx 2{,}639{,}565.0$	2	5,279,130.0
5	$f(5) \approx 2{,}660{,}496.4$	2	5,320,992.9
7	$f(7) \approx 2{,}827{,}227.9$	2	5,654,455.7
9	$f(9) \approx 3{,}139{,}759.3$	2	6,279,518.6
11	$f(11) \approx 3{,}598{,}090.7$	2	7,196,181.4
13	$f(13) \approx 4{,}202{,}222.1$	2	8,404,444.3
15	$f(15) \approx 4{,}952{,}153.6$	2	9,904,307.1
17	$f(17) \approx 5{,}847{,}885.0$	2	11,695,770.0
19	$f(19) \approx 6{,}889{,}416.4$	2	13,778,832.9
		Total midpoint area $\approx$ 79,042,500 cubic feet	

e. The area (which appears to be approximately 79,000,000 cubic feet) represents the approximate amount of water that flowed through the Carson River (at the measuring point) from 11:45 a.m. Wednesday to 7:45 a.m. Thursday. ▪

6.2   Concept Inventory

- $\text{Trapezoid approximation} = \text{average of left- and right-rectangle approximations}$
- Midpoint-rectangle approximation
- Error in approximation

6.2   Activities

1. The graph of a function $f(x)$ is shown in Figure 6.2.1.

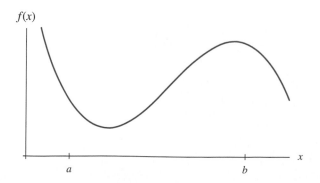

FIGURE 6.2.1

a. Discuss how to approximate the area of the region beneath the graph of $f(x)$ from $x = a$ to $x = b$ with four trapezoids that have the same width. Draw the trapezoids.

b. Discuss how to approximate this area with four midpoint rectangles that have the same width. Copy the graph, and draw the rectangles.

2. Look at the trapezoids drawn in part *a* of Activity 1. What generalization can you make about concavity and trapezoids underestimating or overestimating?

3. The graph in Figure 6.2.2 appears in *Ecology of Natural Resources* by François Ramade (New York: Wiley, 1984). It shows two estimates, labeled *A* and *B*, of oil production rates (in billions of barrels per year).

a. Use trapezoids of width 25 years to estimate the total amount of oil produced from 1900 through 2100 using graph *A*.

b. Repeat part *a* for graph *B*.

c. On page 31 of *Ecology of Natural Resources*, the total oil production is estimated from graph *A* to be 2100 billion barrels and from graph *B* to be 1350 billion barrels. How close were your estimates?

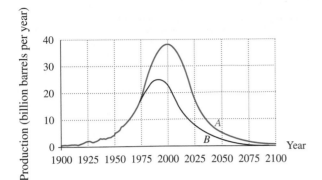

Source: François Ramade, *Ecology of Natural Resources*
(New York: John Wiley and Sons, 1984).

FIGURE 6.2.2

d. What can you conclude from the graph about the future of oil production?

4. Scientists have long been interested in studying global climatological changes and the effect of such changes on many aspects of the environment. From carefully controlled experiments, two scientists[9] constructed a model to simulate daily snow depth in a region of the Northwest Territories in Canada. Rates of change (in equivalent centimeters of water per day) estimated from their model are shown as a scatter plot in Figure 6.2.3. Appropriate models have been sketched on the scatter plot. Note that both vertical and horizontal scales change after June 9.

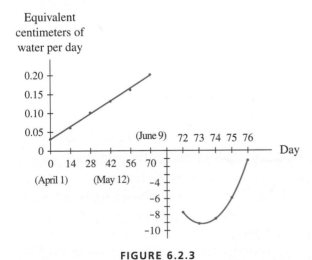

FIGURE 6.2.3

a. What does the figure indicate occurred between June 9 and June 11?

b. Estimate the area beneath the curve from April 1 through June 9. Interpret your answer.

c. Use four midpoint rectangles to estimate the area from June 11 through June 15. Interpret your answer.

5. Approximate the area of the region beneath the graph of $f(x) = \sqrt{1 - x^2}$ from $x = 0$ to $x = 1$ with eight left rectangles, right rectangles, trapezoids, and midpoint rectangles. In each case,

a. sketch the $f(x)$ curve from $x = 0$ to $x = 1$.

b. label the points on the x-axis, and draw the rectangles or trapezoids.

c. calculate the approximating areas.

d. calculate the absolute error. (Recall that the true area is $\frac{\pi}{4} \approx 0.7854$.)

6. Approximate the area of the region beneath the graph of $f(x) = e^{-x^2}$ from $x = -1$ to $x = 1$ using four left rectangles, right rectangles, trapezoids, and midpoint rectangles.

a. In each case,
   i. sketch the $f(x)$ curve from $x = -1$ to $x = 1$.
   ii. label the points on the x-axis, and draw the rectangles or trapezoids.
   iii. calculate the approximating areas.

b. Proceed as in part *a* to approximate the area of the region beneath the graph of $f(x) = e^{-x^2}$ from $x = -1$ to $x = 1$ using eight left rectangles, right rectangles, trapezoids, and midpoint rectangles.

c. The area, to nine decimal places, of the region beneath the graph of $f(x) = e^{-x^2}$ is 1.493648266. Complete Table 6.20 showing how the approximations to the area improve. Which approximation is the most accurate?

9. R.G. Gallimore and J.E. Kutzbach, "Role of Orbitally Induced Changes in Tundra Area in the Onset of Glaciation," *Nature*, vol. 381 (June 6, 1996), pp. 503–505.

TABLE 6.20

	Four intervals		Eight intervals	
	Area	**Absolute error**	**Area**	**Absolute error**
Left rectangles				
Right rectangles				
Trapezoids				
Midpoint rectangles				

7. Table 6.21 shows air speed recorded from a Cessna 172. Air speed is the speed at which the air flows into a Pitot tube located on the nose or wing of an aircraft and is measured in nautical miles per hour (knots). The data were recorded for 25 seconds after the plane began to taxi for takeoff.

TABLE 6.21

Time (seconds)	Air speed (knots)
0	0
10	40
15	60
20	70
25	80

a. Convert knots to miles per second, using the fact that 1 knot ≈ 1.15 mph.

b. Find a model for the converted data.

c. Estimate the area between the model and the horizontal axis from 0 seconds to 18 seconds.

d. If takeoff occurred at 18 seconds, interpret your estimate in part *c* in the context of the Cessna 172 data.

8. As the 76 million Americans born between 1946 and 1964 (the "baby boomers") begin to age, the United States will see an increasing proportion of Americans who are at the current "normal" retirement age of 65. Table 6.22 shows past and projected rates of change for the number of U.S. citizens 65 years of age or older. These are instantaneous rates of change measured at the end of each year based on data from the U.S. Census Bureau.

TABLE 6.22

Year	Rate of change of 65+ population (millions per year)
1940	0.21
1960	0.32
1980	0.50
2000	0.76
2010	0.94
2020	1.17
2040	1.45

a. Find a model to fit the rate-of-change data.

b. Use the model and ten midpoint rectangles to estimate the change in the population 65 years of age and older from the end of 2000 through the end of 2005.

c. Repeat part *b* using ten trapezoids.

9. Using data from the 1994 *Statistical Abstract*, we model the rate of change of the percentage of the U.S. population living around the Great Lakes from 1960 through 1990 as

$$P(t) = (9.371 \cdot 10^{-4})t^2 - 0.141t + 5.083$$
$$\text{percentage points per year}$$

where *t* is the number of years since 1900.

a. Sketch a graph of $P(t)$ from $t = 60$ to $t = 90$.

b. What does the fact that $P(t)$ lies below the *t*-axis from $t = 60$ to $t = 90$ tell you about the percentage of the population living in the Great Lakes region?

c. Use five trapezoids to estimate the area of the region between $P(t)$ and the *t*-axis from $t = 70$ to $t = 80$. Interpret your answer.

d. Repeat part *c* using ten midpoint rectangles.

10. The rate of change of per capita consumption of cottage cheese in the United States from 1970 through 1993 can be modeled[10] as

$$C(x) = -0.0011x^2 + 0.0200x - 0.1399$$
pounds per person per year

where x is the number of years since 1970.

a. Sketch a graph of $C(x)$ from $x = 0$ to $x = 30$.

b. What does the area of the region between $C(x)$ and the x-axis represent?

c. Use ten trapezoids to calculate the area from $x = 0$ to $x = 10$. Interpret your answer.

d. Use ten midpoint rectangles to calculate the area from $x = 10$ to $x = 20$. Interpret your answer.

11. Life expectancies in the United States are always rising because of advances in health care, increased education, and other factors. The rate of change (measured at the end of each year) of life expectancies for women in the United States for selected years since 1970 are shown in Table 6.23.[11]

TABLE 6.23

Year	Years of life expectancy per year
1970	0.51
1975	0.28
1980	0.14
1985	0.09
1990	0.13
1991	0.15
1992	0.17

a. Look at a scatter plot of the data. Does the fact that the data are declining from 1970 through 1985 contradict the statement that life expectancies are always rising? Explain.

b. Find a model for the data.

c. Use eight trapezoids to estimate the change in the life expectancy for women from 1970 through 1992.

d. Repeat part c using eight midpoint rectangles.

12. Accelerations for a vehicle during a road test are approximated in Table 6.24.

a. Find a model to fit the data.

b. Use ten midpoint rectangles to estimate the area of the region between your model and the input axis from 0 to 13.5 seconds. Interpret your answer.

c. Convert your answer in part b to miles per hour. The actual speed of the car after 13.5 seconds was 107 mph. How close is your answer to this speed?

TABLE 6.24

Time (seconds)	Acceleration (feet per second squared)
0	22.6
2	18.2
4	14.5
6	11.4
8	8.9
10	7.1
12	5.9

13. The rate of change of the level of Lake Tahoe d days after September 30, 1995, can be modeled[12] by

$$r(d) = (-1.6035 \cdot 10^{-6})d^2 + (5.086 \cdot 10^{-4})d - 0.0192$$
feet per day

The graph of the model is shown in Figure 6.2.4.

10. Based on data from *Statistical Abstract*, 1994.
11. *Ibid.*
12. Based on data from the Federal Watermaster, U.S. Department of the Interior.

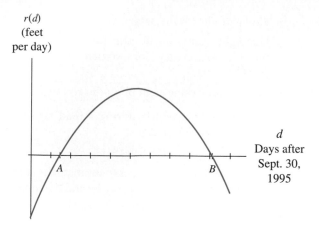

FIGURE 6.2.4

a. Find the points labeled A and B on the graph.

b. Use five midpoint rectangles to estimate the area between $r(d)$ and the horizontal axis from 0 to A. Interpret your answer.

c. Use ten midpoint rectangles to estimate the area between $r(d)$ and the horizontal axis from A to B. Interpret your answer.

d. Was the lake level higher or lower B days after September 30, 1995 than it was on that date? How much higher or lower was the lake level B days after September 30, 1995?

14. The rate of change in the percentage of southern Australian grasshopper eggs that hatch as a function of temperature (for temperatures between 7 °C and 25 °C) can be modeled[13] by

$$p(t) = -0.0258t^3 + 1.464t^2 - 25.982t + 136.560$$
percentage points per degree

where t is the temperature in °C. A graph of $p(t)$ is shown in Figure 6.2.5.

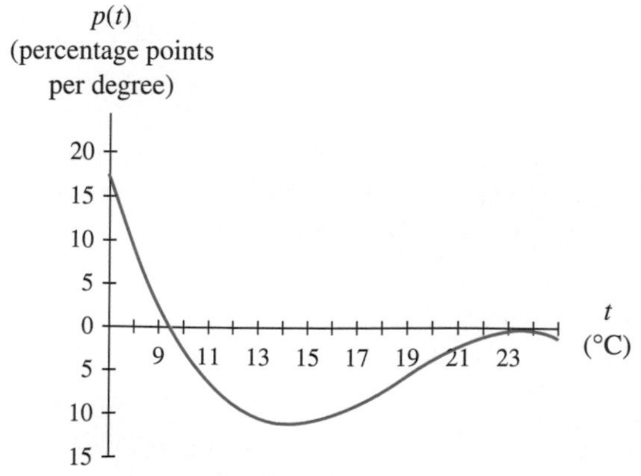

$p(t)$
(percentage points per degree)

FIGURE 6.2.5

a. Find the point at which the graph of $p(t)$ crosses the horizontal axis.

b. Use four trapezoids to estimate the area of the region between the $p(t)$ curve and the horizontal axis from $t = 7$ to the point at which the curve crosses the horizontal axis. Interpret your answer.

c. Use eight trapezoids to estimate the area of the region between the $p(t)$ curve and the horizontal axis from the point at which the curve crosses the horizontal axis to $t = 25$. Interpret your answer.

d. Estimate the difference between the percentage of grasshopper eggs that hatch at 25 °C and the percentage that hatch at 7 °C.

15. Another method of approximating the area of a region between a function and the horizontal axis is a weighted average of the midpoint-rectangle and trapezoid approximations. **Simpson's Rule** is a particular weighted average where the weight assigned to the midpoint-rectangle approximation is twice the weight given to the trapezoid approximation:

$$S = \frac{2}{3}\left(\begin{array}{c}\text{midpoint-rectangle} \\ \text{area}\end{array}\right) + \frac{1}{3}\left(\begin{array}{c}\text{trapezoid} \\ \text{area}\end{array}\right)$$

Simpson's Rule produces approximations to area that are usually much more accurate than any of the approximations we have considered so far.

a. Use four trapezoids to estimate the area of the region between $f(x) = -x^2 + 4$ from $x = -2$ to $x = 2$.

b. Repeat part a using four midpoint rectangles.

c. Calculate the Simpson's Rule weighted average of your answers to parts a and b to determine a better approximation to the area. (In fact, this gives the exact area.)

16. a. Approximate the area of the region beneath the graph of $f(x) = e^{-x^2}$ from $x = -1$ to $x = 1$ using four subintervals, once with trapezoids and once with midpoint rectangles. Calculate the Simpson's Rule weighted average.

b. Repeat part a using eight subintervals. Compare your results to the nine-digit approximation 1.493648266.

13. Based on information in George L. Clark, *Elements of Ecology* (New York: Wiley, 1954), p. 170.

■ **6.3**   **The Definite Integral as a Limit of Sums**

In Section 6.1 we saw that the accumulated change in a quantity can be interpreted in terms of areas of regions between the graph of a rate-of-change function for that quantity and the horizontal axis. Then, in Section 6.2, we approximated areas of regions between a curve and the horizontal axis by using midpoint rectangles. In the Activities, you should have noticed that the rectangle approximations became closer to the actual area of the region when twice as many rectangles were used. What would you expect to happen if you were to use four, eight, or even one hundred times as many rectangles?

We return to Example 2 in Section 6.1, where we considered the rate of change of the concentration of a drug in the blood stream. Recall that we modeled the rate of change of the drug concentration for the first 20 days as

$$r(x) = 1.708(0.845)^x \ \mu g/mL/day$$

where x is the number of days after the drug was first administered. We saw that we could estimate the change in the concentration of the drug between day 0 and day 20 as the area between $r(x)$ and the horizontal axis from $x = 0$ to $x = 20$.

We then used ten right rectangles to estimate this area (see Figure 6.18).

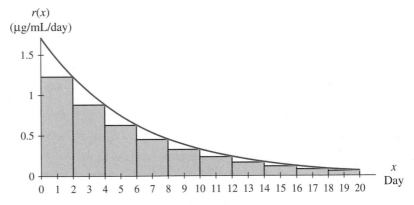

Sum of Areas of Ten Right Rectangles is 8.24 μg/mL

FIGURE 6.18

From our discussion in Section 6.2, we know that we obtain a better estimate of the area beneath $r(x)$ by using midpoint rectangles, as shown in Figure 6.19.

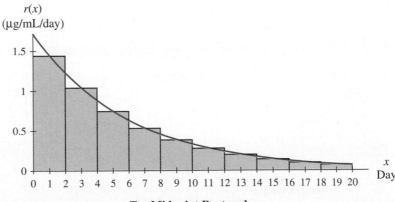

Ten Midpoint Rectangles

FIGURE 6.19

> From now on, whenever we speak of approximating areas using rectangles, we will use midpoint rectangles since they generally give the best approximations.

An estimate of the area beneath $r(x)$ from $x = 0$ to $x = 2$ derived by using ten midpoint rectangles is calculated in Table 6.25.

TABLE 6.25

Rectangle	Midpoint of interval	Height of rectangle (μg/mL/day)	Width of rectangle (days)	Area = (height)(width)
day 0–day 2	1	$r(1) \approx 1.443$	2	2.887
day 2–day 4	3	$r(3) \approx 1.031$	2	2.061
day 4–day 6	5	$r(5) \approx 0.736$	2	1.472
day 6–day 8	7	$r(7) \approx 0.525$	2	1.051
day 8–day 10	9	$r(9) \approx 0.375$	2	0.750
day 10–day 12	11	$r(11) \approx 0.268$	2	0.536
day 12–day 14	13	$r(13) \approx 0.191$	2	0.383
day 14–day 16	15	$r(15) \approx 0.137$	2	0.273
day 16–day 18	17	$r(17) \approx 0.098$	2	0.195
day 18–day 20	19	$r(19) \approx 0.070$	2	0.139
				Total area of rectangles ≈ 9.746 μg/mL

From this estimate of the area we can say that the drug concentration increased by approximately 9.75 μg/mL during the first 20 days of the test.

Finding a Trend

We can improve this estimate of the area beneath $r(x)$ by using 20 rectangles instead of 10. What do you think will happen to the accuracy of the approximations if we were to use even more rectangles? Figures 6.20 through 6.23 show approximations with 10, 20, 40, and 80 rectangles, respectively.

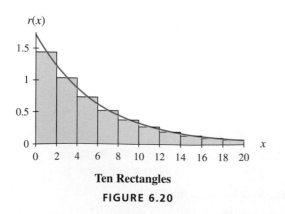

Ten Rectangles

FIGURE 6.20

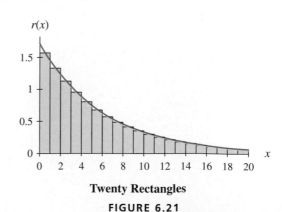

Twenty Rectangles

FIGURE 6.21

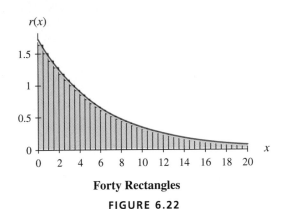

Forty Rectangles

FIGURE 6.22

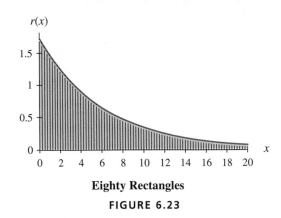

Eighty Rectangles

FIGURE 6.23

By carefully inspecting the graphs in these figures, you can see that as we use more rectangles, the shaded region accounted for by the rectangles more closely approximates the region of interest. This fact will help us approximate the area. If we make a table of the approximations, we may be able to recognize a *trend*—that is, a value to which the approximations seem to be getting closer and closer as the number of rectangles becomes larger and larger.

TABLE 6.26

Number of rectangles	Approximation of area
10	9.7459
20	9.7805
40	9.7892
80	9.7913
160	9.7919
320	9.7920
640	9.7921
1280	9.7921
Trend ≈ 9.79	

The trend evident from Table 6.26 indicates that the area under consideration is approximately 9.79. Interpreting this area in context tells us that the drug concentration increased by approximately 9.79 µg/mL during the first 20 days of the test.

EXAMPLE 1   *A More Accurate Area Approximation*

Consider the function $f(x) = \sqrt{1 - x^2}$ from $x = 0$ to $x = 1$. Approximate the area under this curve between $x = 0$ and $x = 1$, beginning with four midpoint rectangles and doubling the number each time until a trend is observed.

Solution: Figures 6.24 through 6.27 show the accuracy of the approximation improving when 4, 8, 16, and 32 midpoint rectangles are used.

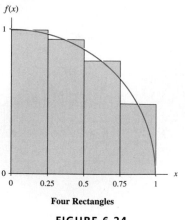

Four Rectangles

FIGURE 6.24

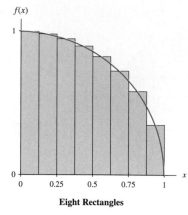

Eight Rectangles

FIGURE 6.25

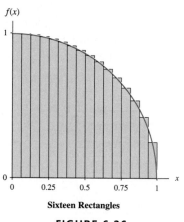

Sixteen Rectangles

FIGURE 6.26

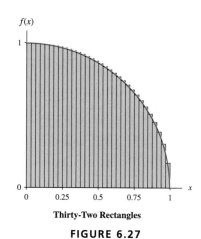

Thirty-Two Rectangles

FIGURE 6.27

Numerically, we observe the approximations approaching a trend in Table 6.27.

TABLE 6.27

Number of rectangles	Approximation of area
4	0.79598
8	0.78917
16	0.78674
32	0.78587
64	0.78557
128	0.78546
256	0.78542
512	0.78541
Trend ≈ 0.785	

We can be fairly confident that the trend to three decimal places is 0.785, because the value through the fourth decimal position has remained constant in several approximations.

The exact area of this region is $\frac{\pi}{4} \approx 0.7853982$. When $n = 512$, the absolute error is approximately 0.0000074.   ∎

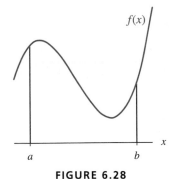

FIGURE 6.28

Area Beneath a Curve

Because the more rectangles we use to approximate area, the better we expect the approximation to be, we are led to consider area as the limiting value of the sums of areas of approximating midpoint rectangles as the number of rectangles increases without bound.

Let $f(x)$ be a function that is continuous and non-negative over the interval from a to b. (See Figure 6.28.) Partition the interval from a to b into n subintervals of equal length $\Delta x = \frac{b-a}{n}$, and on each subinterval construct a rectangle of width Δx whose height is given by the value of $f(x)$ at the midpoint of the subinterval. Figures 6.29 through 6.32 show the rectangles when $n = 4, 8, 16$, and 32.

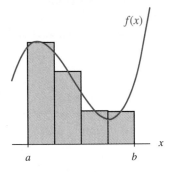

Four Rectangles

FIGURE 6.29

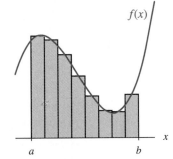

Eight Rectangles

FIGURE 6.30

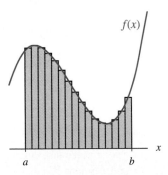

Sixteen Rectangles

FIGURE 6.31

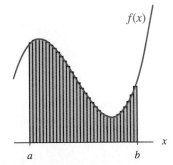

Thirty-Two Rectangles

FIGURE 6.32

The heights of the rectangles are given by the values

$$f(x_1), f(x_2), \ldots, f(x_n)$$

where $x_1, x_2, \ldots, x_n$ are the midpoints of the subintervals. Each rectangle has width Δx, so the areas of the rectangles are given by the values

$$f(x_1)\,\Delta x,\ f(x_2)\,\Delta x, \ldots,\ f(x_n)\,\Delta x$$

and the sum $[f(x_1) + f(x_2) + \ldots + f(x_n)]\Delta x$ is an approximation to the area of the region between the curve $f(x)$ and the x-axis from a to b. As our examples have shown, the approximations generally improve as n increases. In mathematical terms, the area of the region between the curve $f(x)$ and the x-axis from a to b is given by a limiting value of sums as n gets larger and larger.

$$\text{Area} = \lim_{n \to \infty} [f(x_1) + f(x_2) + \cdots + f(x_n)]\Delta x$$

Area Beneath a Curve

Let $f(x)$ be a continuous or a piecewise continuous non-negative function from a to b. The area of the region between the graph of $f(x)$ and the x-axis from a to b is given by the limit

$$\text{Area} = \lim_{n \to \infty} [f(x_1) + f(x_2) + \cdots + f(x_n)]\Delta x$$

where $x_1, x_2, \ldots, x_n$ are the midpoints of n subintervals of length $\Delta x = \frac{(b - a)}{n}$ between a and b.

The Definite Integral

The limit in the area definition is central to the development of calculus and is applied to more general functions as follows. Given an arbitrary function that is continuous or piecewise continuous over an interval from a to b, partition the interval into n subintervals of length $\Delta x = \frac{b - a}{n}$ and consider the sum

$$[f(x_1) + f(x_2) + \cdots + f(x_n)]\Delta x$$

Here, the values $x_1, x_2, \ldots, x_n$ are the midpoints of the n subintervals. The limit

$$\lim_{n \to \infty} [f(x_1) + f(x_2) + \cdots + f(x_n)]\Delta x$$

is called the **definite integral** of $f(x)$ from a to b and is represented by

$$\int_a^b f(x)\, dx$$

The Definite Integral

Let $f(x)$ be a continuous or piecewise continuous function from a to b. The definite integral of $f(x)$ from a to b is given by the limit

$$\int_a^b f(x)\, dx = \lim_{n \to \infty} [f(x_1) + f(x_2) + \cdots + f(x_n)]\Delta x$$

where $x_1, x_2, \ldots, x_n$ are the midpoints of n subintervals of length $\Delta x = \frac{(b - a)}{n}$ between a and b.

The integral sign $\int$ resembles an elongated S and reminds us that we are taking a limit of sums. The values a and b identify the input interval, $f(x)$ is the function, and the symbol dx reminds us of the width Δx of each subinterval. If $f(x)$ lies above the x-axis between a and b, then the definite integral is the area of the region between $f(x)$ and the x-axis from a to b.

EXAMPLE 2   *Rising River*

In Section 6.2, we saw flow rates of the Carson River modeled by

$$f(x) = 18{,}225x^2 - 135{,}334.3x + 2{,}881{,}542.9 \text{ ft}^3/\text{hr}$$

where x is the number of hours after 11:45 a.m. Wednesday.

   Use the idea of a limit of sums to estimate to the nearest ten thousand cubic feet the amount of water that flowed through the Carson River (past the point where the measurements were taken) from 11:45 a.m. Wednesday to 7:45 a.m. Thursday.

Solution: The amount of water that flowed through the Carson River from 11:45 a.m. Wednesday to 7:45 a.m. Thursday can be estimated by the area between the graph of $f(x)$ and the horizontal axis from $x = 0$ to $x = 20$. We begin with 10 mid-point rectangles, doubling the number each time until we are confident that we know the limiting value.

TABLE 6.28

Number of rectangles	Approximation of area
10	79,042,498
20	79,133,623
40	79,156,404
80	79,162,100
160	<u>79,163</u>,523
320	<u>79,163</u>,879
640	<u>79,163</u>,968
Trend $\approx$ 79,160,000	

It appears that, according to the model, to the nearest ten thousand cubic feet the amount of water flowing through the Carson River during the 20 hours after 11:45 a.m. Wednesday was 79,160,000 ft³. Using definite integral notation, we write

$$\int_0^{20} (18{,}225x^2 - 135{,}334.3x + 2{,}881{,}542.9)\, dx \approx 79{,}160{,}000 \text{ ft}^3 \qquad \blacksquare$$

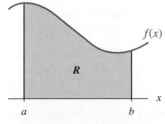

For $f(x) \geq 0$, $\int_a^b f(x)\, dx$ is the area of region R.

FIGURE 6.33

Interpretation of $\displaystyle\int_a^b f(x)\ dx$

When $f(x)$ is continuous or piecewise continuous and non-negative from a to b, then the integral $\displaystyle\int_a^b f(x)\, dx$ can be interpreted as the area of the region R between $f(x)$ and the x-axis from a to b. (See Figure 6.33.)

   What interpretation can we give to the integral $\displaystyle\int_a^b f(x)\, dx$ if $f(x)$ is negative from a to b? In this case, each of the terms $f(x_j)\Delta x$ for $1 \leq j \leq n$ appearing in the expression

$$\int_a^b f(x)\,dx = \lim_{n\to\infty}[f(x_1) + f(x_2) + \cdots + f(x_n)]\Delta x$$

will be negative, so the limit will also be negative. Thus the integral cannot represent the area of the region R between the graph $y = f(x)$ and the horizontal axis from a to b, because area is a measure of size and is thus non-negative.

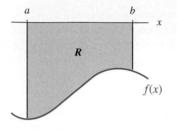

For $f(x) \le 0$, $\int_a^b f(x)\,dx$ is the negative of the area of region R.

FIGURE 6.34

> When $f(x)$ is negative from a to b, the integral is the *negative* of the area of region R. (See Figure 6.34.)

If the function $f(x)$ is sometimes positive and sometimes negative between a and b, then the integral $\int_a^b f(x)\,dx$ is equal to the area of the region lying under the graph of $f(x)$ and above the x-axis minus the area of the region lying above the graph of $f(x)$ and below the x-axis. (See Figure 6.35.)

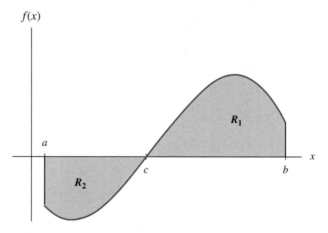

$\int_a^b f(x)\,dx$ = **Area lying above axis – area lying below axis**
= **Area of R_1 – area of R_2**

FIGURE 6.35

EXAMPLE 3   *Wine Consumption*

The rate of change of per capita consumption of wine in the United States from 1970 through 1990 can be modeled[14] as

$$W(x) = (1.243 \cdot 10^{-4})x^3 - 0.0314x^2 + 2.6174x - 71.977 \text{ gallons per}$$
$$\text{person per year}$$

where x is the number of years since the end of 1900. A graph of the model is shown in Figure 6.36.

a. Find the input value of the point labeled A.

b. From 1970 through 1990, according to the model, when was wine consumption increasing and when was it decreasing?

14. Based on data from *Statistical Abstract*, 1994.

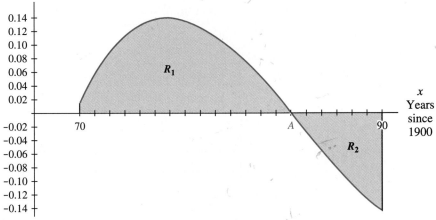

FIGURE 6.36

c. Use a limiting value of sums to estimate the areas, to two decimal places, of the regions labeled R_1 and R_2. Interpret your answers.

d. According to the model, what was the change in per capita consumption of wine from the end of 1970 through 1990?

e. Find the values of $\int_{70}^{83.97} W(x)\,dx$, $\int_{83.97}^{90} W(x)\,dx$, and $\int_{70}^{90} W(x)\,dx$.

Solution:

a. Solving $W(x) = 0$ gives $A \approx 83.97$, corresponding to the end of 1984.

b. Wine consumption was increasing where the rate-of-change graph is positive—from 1970 through 1984 ($x \approx 83.97$)—and was decreasing from 1984 through 1990, where the rate-of-change graph is negative.

c. In order to find the areas of the regions R_1 and R_2, we must know the lower limit and upper limit (that is, the endpoints) of each region. Because $W(x) = 0$ when $x \approx 83.97$, we use lower limit $x = 70$ and upper limit $x \approx 83.97$ for R_1 and lower limit $x \approx 83.97$ and upper limit $x = 90$ for R_2.

   The area of region R_1 is determined by examining sums of areas of midpoint rectangles for $n = 5, 10, 20, 40, 80$, and 160, as shown in Table 6.29.

TABLE 6.29

Number of rectangles	Approximation of Area
5	1.35964
10	1.34129
20	1.33671
40	1.33556
80	1.33527
160	1.33520
Trend ≈ 1.34	

The area of region R_1 is approximately 1.34 gallons per person. This indicates that wine consumption increased by approximately 1.34 gallons per person from the end of 1970 through 1984 ($x \approx 83.97$).

We calculate the area of region R_2 in a similar way, considering the absolute values of the sums.

TABLE 6.30

Number of rectangles	Approximation of Area
5	0.44927
10	0.44870
20	0.44856
40	0.44852
Trend ≈ 0.45	

The area of region R_2 is approximately 0.45 gallon per person. This is an estimate for the decrease in per capita wine consumption from 1984 ($x \approx 83.97$) through 1990.

d. To determine the net change in the per capita consumption of wine from the end of 1970 through 1990, we subtract the decrease (area of R_2) from the increase (area of R_1).

$$\text{Net change} = 1.34 - 0.45 = 0.89 \text{ gallon per person}$$

We see an approximate net increase of 0.89 gallons per person from the end of 1970 through 1990.

e. The area of region R_1 is the value of the definite integral of $W(x)$ from 70 to 83.97. The negative of the area of region R_2 is the value of the definite integral from 83.97 to 90, and the answer to part d is the value of the definite integral from 70 to 90.

$$\int_{70}^{83.97} W(x)\,dx \approx 1.34 \text{ gallons per person}$$

$$\int_{83.97}^{90} W(x)\,dx \approx -0.45 \text{ gallon per person}$$

$$\int_{70}^{90} W(x)\,dx \approx 0.89 \text{ gallon per person} \qquad \blacksquare$$

Although finding the limiting value of sums of areas of rectangles is an invaluable tool for finding accumulated change in a quantity, there are times when it is not an appropriate technique. For example, in Section 6.1 we saw data for the number of recycled aluminum cans each year. Because the data gave yearly totals, it would not be necessary to use rectangles of width any smaller than 1 year. A limiting value of sums is most useful in truly continuous or piecewise continuous situations in which a model is defined for all input values in a certain interval.

6.3   Concept Inventory

■ Area = limiting value of sums of areas of midpoint rectangles

■ $\int_a^b f(x)\,dx$ = definite integral

6.3   Activities

1. The rate of change of the population of a country, in thousands of people per year, is modeled by $P(t)$, where t is the number of years since 1995. What are the units on

   a. the area of the region between $P(t)$ and the t-axis from $t = 0$ to $t = 10$?

   b. $\int_{10}^{20} P(t)\,dt$?

   c. the change in the population from 1995 through 2000?

2. During the spring thaw a mountain lake rises by $L(d)$ feet per day, where d is the number of days since April 15. What are the units on

   a. the area of the region between $L(d)$ and the d-axis from $d = 0$ to $d = 15$?

   b. $\int_{16}^{31} L(d)\,dd$?

   c. the amount by which the lake rose from May 15 to May 31?

3. When warm water is released into a river from a source such as a power plant, the increased temperature of the water causes some algae to grow and other algae to die. In particular, blue-green algae that can be toxic to some aquatic life thrive. If $A(c)$ is the growth rate of blue-green algae (in organisms per °C) and c is the temperature of the water in °C, interpret the following in context:

   a. $\int_{25}^{35} A(c)\,dc$

   b. The area of the region between $A(c)$ and the c-axis from $c = 30\,°C$ to $c = 40\,°C$

4. The value of a stock portfolio is growing by $V(t)$ dollars per day, where t is the number of days since the beginning of the year. Interpret the following in context:

   a. The area of the region between $V(t)$ and the t-axis from $t = 0$ to $t = 120$

   b. $\int_{120}^{240} V(t)\,dt$

5. The graph in Figure 6.3.1 shows the rate of change of profit at various production levels for a pencil manufacturer. Fill in the blanks in the following discussion of the profit. If it is not possible to determine a value, write NA in the corresponding blank.

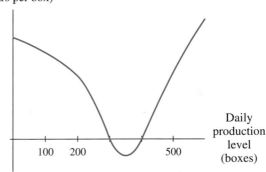

Rate of change
of profit
(dollars per box)

FIGURE 6.3.1

Profit is increasing when between (a)____ and (b)____ boxes of pencils are produced each day. The profit when 500 boxes of pencils are produced each day is (c)____ dollars. Profit is higher than nearby profits at a production level of (d)____ boxes each day, and it is lower than nearby profits at a production level of (e)____ boxes of pencils each day. The profit is decreasing most rapidly when (f)____ boxes are produced each day. The area between the rate-of-change-of-profit function and the production-level axis between production levels of 100 and 200 boxes each day has units (g)____. If $p'(b)$ represents the rate of change of profit (in dollars per box) at a daily production level of b boxes, would $\int_{300}^{400} p'(b)\,db$ be more than, less than, or the same value as $\int_{100}^{200} p'(b)\,db$? (h)____

6. The graph in Figure 6.3.2 shows the rate of change of cost for an orchard in Florida at various production levels during grapefruit season. Fill in the blanks in the following cost function discussion. If it is not possible to determine a value, write NA in the corresponding blank.

Rate of
change of cost
(dollars per carton)

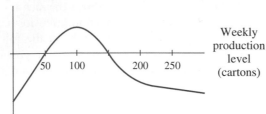

FIGURE 6.3.2

Cost is increasing when between (a)____ and (b)____ cartons of grapefruit are harvested each week. The cost to produce 100 cartons of grapefruit each week is (c)____ dollars. The cost is lower than nearby costs at a production level of (d)____ cartons, and it is higher than nearby costs at a production level of (e)____ cartons of grapefruit each week. The cost is increasing most rapidly when (f)____ cartons are produced each week. The area between the rate-of-change-of-cost function and the production-level axis between production levels of 50 and 150 cartons each week has units (g)____. If $c'(p)$ represents the rate of change of cost (in dollars per carton) at weekly production level of p cartons of grapefruit, would $\int_{200}^{250} c'(p)\, dp$ be greater than, less than, or the same value as $\int_{50}^{100} c'(p)\, dp$? (h)____

7. The rate of change of the weight of a laboratory mouse can be modeled by the equation

$$w(t) = \frac{13.785}{t} \text{ grams per week}$$

where t is the age of the mouse in weeks and $1 \le t \le 15$.

   a. Use the idea of a limit of sums to estimate the value of $\int_{3}^{11} w(t)\, dt$.

   b. Label units on the answer to part a. Interpret your answer.

8. The rate of change of annual U.S. factory sales (in billions of dollars per year) of electronics from 1986 through 1990 can be modeled[15] by the equation

$s(t) = 113.574e^{-0.3912t}$ billions of dollars per year

where t is the number of years since 1986.

   a. Use the idea of a limit of sums to estimate the change in factory sales from 1986 through 1990.

   b. Write the definite integral symbol for this limit of sums.

9. On the basis of data obtained from a preliminary report by a geological survey team, it is estimated that for the first ten years of production, a certain oil well can be expected to produce oil at the rate of $r(t) = 3.93546t^{3.55}e^{-1.35135t}$ thousand barrels per year t years after production begins.

   a. Use the idea of a limit of sums to estimate the yield from this oil field during the first 5 years of production.

   b. Use the idea of a limit of sums to estimate the yield during the first 10 years of production.

   c. Write the definite integral symbols representing the limits of sums in parts a and b.

   d. Estimate the percentage of the first 10 years' production that your answer to part a represents.

10. The rate of change of the temperature during the hour and a half after a thunderstorm began is modeled by

$T(h) = 9.48h^3 - 15.49h^2 + 17.38h - 9.87\ °F$ per hour

where h is the number of hours since the storm began. A graph of $T(h)$ is shown in Figure 6.3.3.

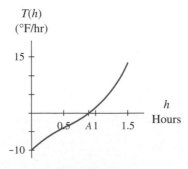

FIGURE 6.3.3

   a. Determine the value of A.

   b. Use a limit of sums to estimate $\int_{0}^{A} T(h)\, dh$. Interpret your answer.

15. Based on data from *Statistical Abstract*, 1992.

c. Use a limit of sums to estimate $\int_A^{1.5} T(h)\,dh$. Interpret your answer.

d. Estimate $\int_0^{1.5} T(h)\,dh$. Interpret your answer.

11. The acceleration of a race car during the first 35 seconds of a road test is modeled by

$$a(t) = 0.024t^2 - 1.72t + 22.58 \text{ ft/sec}^2$$

where t is the number of seconds since the test began. A graph of this acceleration function is shown in Figure 6.3.4.

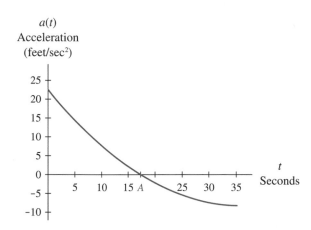

$a(t)$
Acceleration
(feet/sec²)

FIGURE 6.3.4

a. Find the time at which the acceleration curve becomes negative. This time is denoted A in the figure.

b. Use a limit of sums to estimate $\int_0^A a(t)\,dt$. Interpret your answer.

c. Use a limit of sums to estimate $\int_A^{35} a(t)\,dt$. Interpret your answer.

d. Estimate the change that occurred in the velocity of the car during the first 35 seconds of the road test.

12. Museums carefully monitor temperature and humidity. Suppose a museum has a gauge that monitors temperature as well as how rapidly temperature is changing. During a junior high school field trip to the museum, the rates of change of temperature shown in Table 6.31 are recorded.

a. Basing your explanation on the rate-of-change data, discuss what happens to the temperature from 8:30 a.m. to 10:15 a.m.

b. According to the data, when is the temperature greatest?

TABLE 6.31

Time of day	Rate of change (°F/hr)
8:30 a.m.	0
8:45 a.m.	2.2
9:00 a.m.	2.4
9:15 a.m.	1.1
9:30 a.m.	-0.8
9:45 a.m.	-2.3
10:00 a.m.	-2.7
10:15 a.m.	-1.0

c. Find a model for the data.

d. Use the model and a limit of sums to estimate the change in temperature between 8:30 a.m. and 9:30 a.m.

e. Use a limit of sums to estimate the change in temperature between 9:30 a.m. and 10:15 a.m.

f. What are the definite integral notations for your answers to parts d and e?

13. An article in the May 23, 1996, issue of *Nature* addresses the interest some physicists have in studying cracks in order to answer the question "How fast do things break, and why?" As stated in *Nature*, "In 1991, Fineberg, Gross, Swinney and others developed a method of looking at the motion of cracks, ... [making it] possible to measure the velocity of a crack on timescales much shorter than a millionth of a second, hundreds of thousands of times in succession."

   Data estimated from a graph in this article, showing the velocity of a crack during a 60-microsecond experiment, are given in Table 6.32.

TABLE 6.32

Time (microseconds)	Velocity (meters per second)
10	148.2
20	159.3
30	169.5
40	180.7
50	189.8
60	200

a. In order to determine the distance the crack traveled during the 60-microsecond experiment, we wish to determine the area beneath the velocity curve from 0 to 60. What are the units on the heights and widths of rectangles under this curve?

b. Convert the data to millimeters per microsecond (there are 1000 millimeters in a meter and 1,000,000 microseconds in a second). Using the converted data, identify the area units.

c. Find a quadratic model for the converted data.

d. Use a limit of sums to determine how far the crack traveled during the experiment.

e. Give the definite integral notation for your answer to part d.

14. Table 6.33 records the volume of sales (in thousands) of a popular movie for selected months the first 18 months after it was released on video cassette.

TABLE 6.33

Months after release	Number of cassettes sold each month (thousands)
2	565
4	467
5	321
7	204
10	61
11	31
12	17
16	3
18	2

a. Find a logistic model for the data.

b. Use 5, 10, and 15 right rectangles to estimate the number of cassettes sold during the first 15 months after release.

c. Which of the following would give the most accurate value of the number of cassettes sold during the first 15 months after release?
   i. The answer to part b for 15 rectangles.

   ii. The limiting value of the sums of midpoint rectangles using the model in part a.
   iii. The sum of actual sales figures for the first 15 months.

15. The personnel manager for a large construction company keeps records of the labor-hours per week spent on typical construction jobs handled by the company. He has developed the following model for a labor-power curve:

$$m(x) = \frac{6,608,830e^{-0.706x}}{(1 + 925e^{-0.706x})^2} \text{ labor-hours per week}$$

the xth week of the construction job

a. Use 5, 10, and 20 right rectangles to approximate the number of labor-hours spent during the first 20 weeks of a typical construction job.

b. If the number of labor-hours spent on a particular job exactly coincides with the model, which of the following would give the most accurate value of the number of labor-hours spent during the first 20 weeks of the job?
   i. The 20-right-rectangle sum found in part a.
   ii. The sum of 20 midpoint rectangles.
   iii. The limiting value of the sums of midpoint rectangles.

16. Though all companies receive revenue in a discrete fashion, if a company is large enough, then its revenue can be thought of as flowing in at a continuous rate. A utility company is a good example of this. Suppose that it is possible to measure the rate of flow of revenue for a company and that the flow rates measured at the end of the year can be modeled[16] by

$$r(x) = 9.907x^2 - 40.769x + 58.492$$
million dollars per year

x years after the end of 1987.

a. Use the idea of a limit of sums to estimate the value of $\int_0^6 r(x)\,dx$.

b. Interpret your answer to part a.

c. The company's revenue from 1987 through 1993 was reported to be 543 million dollars. How close is this value to your estimate from part a?

16. Based on data for Timberland in *Allen Financial Advisors*.

17. In Section 6.1 we considered a department store manager's effort to estimate the number of people who attended a Saturday sale. Recall that after taking samples of the number of customers who arrived during certain 1-minute times throughout the sale, the manager estimated the number of customers who entered the store each minute with the model

$$c(m) = (4.58904 \cdot 10^{-8})m^3 - (7.78127 \cdot 10^{-5})m^2 + 0.03303m + 0.88763 \text{ customers per minute}$$

where m is the number of minutes after 9:00 a.m.

a. Use a limiting value of sums to estimate the value of $\int_0^{720} c(m)\, dm$.

b. Interpret your answer to part a in context.

18. Table 6.34 gives rates of change of the amount in an interest-bearing account for which interest is compounded continuously.

TABLE 6.34

At end of year	Rate of change (dollars per day)
1	2.06
3	2.37
5	2.72
7	3.13
9	3.60

a. Convert the input to days. Disregard leap years. Why is a conversion important for a definite integral calculation?

b. Find an exponential model for the converted data.

c. Use a limiting value of sums to estimate the change in the balance of the account from the day the money was invested to the last day of the tenth year after the investment was made. Again, disregard leap years.

d. Give the definite integral notation for your answer to part c.

19. Blood pressure (BP) varies for individuals throughout the course of a day, typically being lowest at night and highest from late morning to early afternoon. The estimated rate of change in diastolic blood pressure for a patient with untreated hypertension is shown in Table 6.35.

TABLE 6.35

Time	Rate of change of diastolic BP (mm Hg per hour)
8 a.m.	3.0
10 a.m.	1.8
12 p.m.	0.7
2 p.m.	-0.1
4 p.m.	-0.7
6 p.m.	-1.1
8 p.m.	-1.3
10 p.m.	-1.1
12 a.m.	-0.7
2 a.m.	0.1
4 a.m.	0.8
6 a.m.	1.9

a. During which time intervals was the patient's diastolic blood pressure rising? falling?

b. Estimate the times when diastolic blood pressure was rising and falling most rapidly.

c. Find a model for the data.

d. Find the times at which the output of the model is zero. Of what significance are these times in the context of blood pressure?

e. Use the idea of a limiting value of sums to estimate by how much the diastolic blood pressure changed from 8 a.m. to 8 p.m.

f. Write the definite integral notation for your answer to part e.

6.4 Accumulation Functions

In the last three sections, we saw that when we have a rate-of-change function for a certain quantity, we approximate the accumulation of change in that quantity between two values of the input variable by calculating the area between the rate-of-change curve and the horizontal axis. We approximated area with sums of areas of rectangles and found that when we used more rectangles, the approximation usually improved. That discovery led us to view the area of a region between a rate-of-change function $f(x)$ and the x-axis as a limit of sums:

$$\lim_{n \to \infty} [f(x_1) + f(x_2) + \cdots + f(x_n)]\Delta x$$

This limit, when applied to an arbitrary, continuous function over an interval from a to b, is called the definite integral of $f(x)$ from a to b and is denoted by the symbol $\int_a^b f(x)\,dx$. In order to develop algebraic methods for determining the exact value of a definite integral $\int_a^b f(x)\,dx$, we now turn our attention to a more theoretical discussion of areas and integrals.

We begin by defining the accumulation function of a function $f(t)$, denoted $\int_a^x f(t)\,dt$ and read "the accumulation function of $f(t)$ from a to x." Unlike the definite integral, which is a number, the accumulation function is a formula in terms of the upper limit of the integral, x. The accumulation function is most easily thought of as an accumulation of area. For this reason, we make the following definition:

Accumulation Function

The accumulation function of a function $f(t)$, denoted by $\int_a^x f(t)\,dt$, gives the accumulation of the area between the horizontal axis and $f(t)$, from a to x, such that

1. if x is to the right of a, then
   - regions between $f(t)$ and the horizontal axis that lie above the axis contribute positive accumulation equal to the area of the region, and
   - regions that lie beneath the horizontal axis contribute negative accumulation equal to the negative of the area of the region.

2. if x is to the left of a, the opposite of the above statements is true. That is,
   - regions between $f(t)$ and the horizontal axis that lie above the axis contribute negative accumulation, and
   - regions that lie beneath the horizontal axis contribute positive accumulation.

3. if $x = a$, the accumulation is 0.

For example, consider the accumulation function $\int_0^x f(t)\,dt$, where $f(t) = 3$ for all values of t. We wish to find a formula for $\int_0^x 3\,dt$—that is, a formula for accumulated area between the horizontal line at 3 and the horizontal axis, with the beginning point being $t = 0$. We make a list of area values for specific values of x and use them to help us determine the accumulation function.

If $x = 0$, we know that $\int_0^0 3\,dt = 0$. It is true that every accumulation function will be zero at its starting value. If x is greater than zero, then the accumulation function values are determined by areas of rectangles. (See Figure 6.37.)

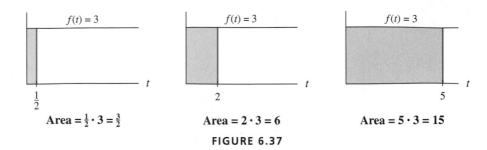

FIGURE 6.37

For values of x less than 0, we get similar areas, but the accumulation function values are the negative of those areas. (Refer to the first statement under Condition 2 in the definition of the accumulation function.) With this information, we construct Table 6.36, a table of values for $\int_0^x 3\,dt$.

TABLE 6.36

x	-5	-2	$-\dfrac{1}{2}$	0	$\dfrac{1}{2}$	2	5
$\displaystyle\int_0^x 3\,dt$	-15	-6	$-\dfrac{3}{2}$	0	$\dfrac{3}{2}$	6	15

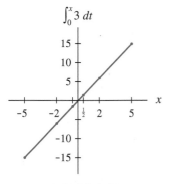

FIGURE 6.38

Plotting these values, we sketch the line in Figure 6.38. As you can easily verify, this line is $y = 3x$. Indeed, the accumulation function for $f(t) = 3$ beginning at 0 is

$$\int_0^x 3\,dt = 3x.$$

We could have determined the formula for $\int_0^x 3\,dt$ without using a table of values if we had noticed that for x to the right of 0, the region is a rectangle with height 3, width x, and area $3x$. This is our desired formula. For x to the left of 0, the argument is slightly more involved but results in the same conclusion: The accumulation formula for $f(t) = 3$ beginning at 0 is $\int_0^x 3\,dt = 3x$. For all cases in this section, the formula for x to the left of a will be the same as that for x to the right of a, so we consider only the second case when determining an accumulation formula.

EXAMPLE 1   *Finding Accumulation Functions*

Determine the following accumulation functions for $f(t) = 2t - 4$ and $x > 2$.

a. $\displaystyle\int_2^x f(t)\,dt$

b. $\displaystyle\int_{-1}^x f(t)\,dt$

c. $\displaystyle\int_6^x f(t)\,dt$

d. Use the formulas you obtained in parts a, b, and c to find the areas of the regions between $f(t) = 2t - 4$ and the horizontal axis from 2 to 5 and from 6 to 12.

Solution:

a. The graph in Figure 6.39 shows that the formula we desire is the area of a triangle with base $x - 2$ and height $2x - 4$.

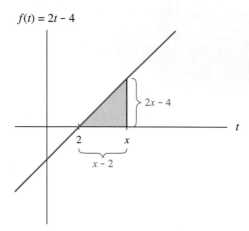

$f(t) = 2t - 4$

$2x - 4$

2 x

$x - 2$

t

FIGURE 6.39

Hence, $\displaystyle\int_2^x (2t - 4)\, dt = \frac{1}{2}(x - 2)(2x - 4) = x^2 - 4x + 4.$

b. If we change our starting point to -1, we now must consider two triangles, with the one above the horizontal axis contributing its area (which we found in part *a*) and the one beneath the axis contributing the negative of its area. (See Figure 6.40.)

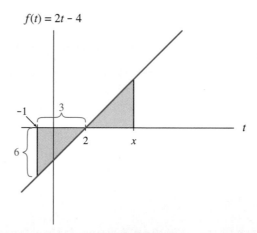

$f(t) = 2t - 4$

-1 3

2 x

6

t

FIGURE 6.40

The base of the triangle below the axis is 3 and the height is 6 (determined by substituting -1 for t in the formula $2t - 4$ and noting that heights are always positive measurements). Thus the area we must subtract is $\frac{1}{2}(3)(6) = 9.$

Therefore,

$$\int_{-1}^{x} (2t - 4)\, dt = \text{area of triangle above axis} - \text{area of triangle below axis}$$

$$= (x^2 - 4x + 4) - 9 = x^2 - 4x - 5$$

c. Using 6 as our starting value, we redraw Figure 6.40 with x to the right of 6 (for convenience) to obtain the graph in Figure 6.41.

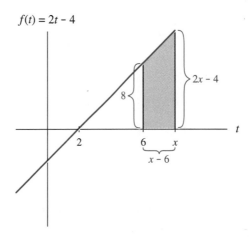

$f(t) = 2t - 4$

FIGURE 6.41

Now, instead of a triangle, we have a trapezoid. Its base is $x - 6$, and its heights are determined by evaluating $f(t) = 2t - 4$ at $t = 6$ and $t = x$. Thus the area accumulation function we desire is

$$\int_{6}^{x} (2t - 4)\, dt = (x - 6)\left(\frac{8 + (2x - 4)}{2}\right) = (x - 6)(x + 2) = x^2 - 4x - 12$$

d. The area from 2 to 5 (using the formula found in part a) is

$$\int_{2}^{5} (2t - 4)\, dt = 5^2 - 4(5) + 4 = 9$$

The area from 6 to 12 (using the formula found in part c) is

$$\int_{6}^{12} (2t - 4)\, dt = 12^2 - 4(12) - 12 = 84$$

Be sure you understand that we found formulas only for $x > 2$ and x to the right of the starting point. The formulas we found do hold for all values of x, but we have not proven that. ∎

Accumulation Function Graphs

Although it is fairly easy to find accumulation function formulas for lines, it is not always easy for other functions, so we proceed by limiting our focus to sketching accumulation function graphs. Let us begin with sketching an accumulation function for $f(t) = -t^2 + 4$ from -2 to x.

First, we know that when $x = -2$, the accumulation is 0. (There is no accumulation of area between $f(t)$ and the input axis from -2 to -2.) Next, note in Figure 6.42 that as x moves from -2 to 0, the accumulated area increases more and more rapidly.

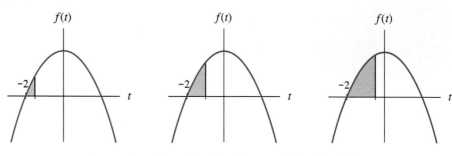

Accumulated Area Growing More and More Rapidly

FIGURE 6.42

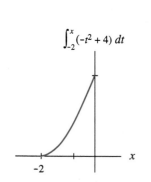

$$\int_{-2}^{x} (-t^2 + 4)\, dt$$

Accumulation Function Increasing and Concave Up

FIGURE 6.43

This indicates that from -2 to 0, the accumulation function is positive (because the accumulated area is above the axis), is increasing (because the accumulated area is getting larger), and is concave up (because the accumulated area is getting larger more and more rapidly). (See Figure 6.43.)

As x increases to the right of 0, the accumulated area continues to become larger (more positive), but it does so more and more slowly. (See Figure 6.44.)

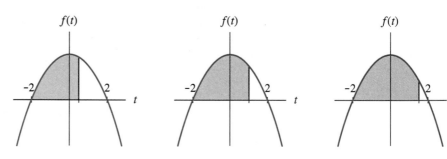

Accumulated Area Growing More and More Slowly

FIGURE 6.44

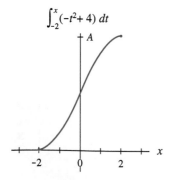

$$\int_{-2}^{x} (-t^2 + 4)\, dt$$

Accumulation Function From 0 to 2 is Increasing and Concave Down

FIGURE 6.45

That means that from 0 to 2, our accumulation function continues to increase but is concave down, as shown in Figure 6.45. Note that the output value labeled A is the area of the region beneath $f(t) = -t^2 + 4$ from -2 to 2.

As x moves to the right of 2, $f(t)$ is negative. Because a portion of the area is below the t-axis, the accumulation function begins to decline as the area below the axis is subtracted from the area that has accumulated above. (Refer to the second statement under Condition 1 in the definition of the accumulated area.) Also, the decline occurs more and more rapidly (resulting in the accumulation function being concave down) until the area below the axis is equal to the area above the axis. At this point, the accumulated area is again 0. This is illustrated in Figure 6.46.

$\int_{-2}^{x} f(t)\,dt$

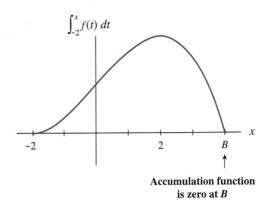

Accumulation function
is zero at B

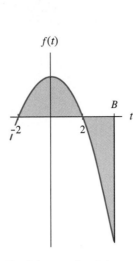

**Total Accumulated Area
From –2 to B is Zero**

FIGURE 6.46

$\int_{-2}^{x} f(t)\,dt$

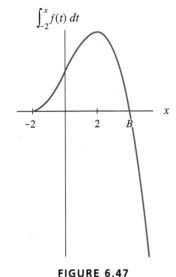

FIGURE 6.47

As x increases even further to the right of B, the area below the axis continues to grow more and more quickly. This results in the accumulation function being more and more negative, and this continues for all values of x greater than B, as shown in Figure 6.47.

We have only one more portion of the accumulation function graph to consider, and that is for values of x to the left of –2. We observe first that the accumulated area is below the axis and, therefore, is assigned a negative value. As we move from right to left, the accumulation is becoming more negative faster and faster and continues in that pattern, as shown in Figure 6.48.

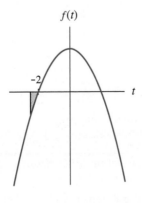
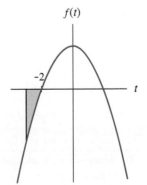
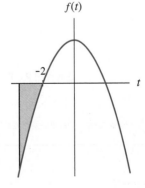

FIGURE 6.48

We would draw this accumulation as seen in Figure 6.49a, but remember that for values of x to the left of the starting point (in this case, –2), the accumulation is the opposite of what we would expect it to be. (Refer to Condition 2 in the definition of the accumulation function.) Thus we reflect the curve across the horizontal axis to see the correct portion of the accumulation graph in Figure 6.49b.

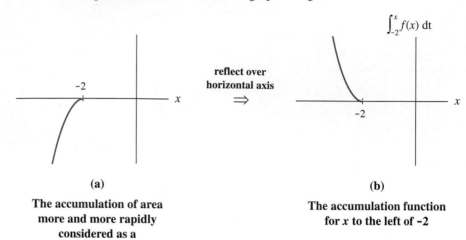

(a)
**The accumulation of area
more and more rapidly
considered as a
negative value**

**reflect over
horizontal axis
⇒**

(b)
**The accumulation function
for x to the left of –2**

FIGURE 6.49

Adding this final portion of the accumulation graph gives us the complete graph shown in Figure 6.50.

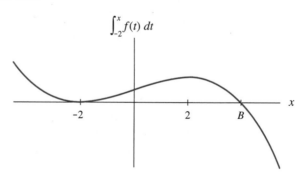

The accumulation function of $f(t) = -t^2 + 4$ from –2 to x

FIGURE 6.50

EXAMPLE 2   *Sketching Accumulation Graphs*

Consider the graph of $f(t)$ shown in Figure 6.51.

a. Give the notation for the accumulation function of $f(t)$ from A to x.

b. Sketch the accumulation function of $f(t)$ from A to x for values of x to the right of A.

c. Sketch the accumulation function of $f(t)$ from A to x for values of x to the left of A.

d. Combine parts b and c to sketch the complete accumulation function of $f(t)$ from A to x.

e. If $f(t)$ is the rate of change in the growth of a plant (in millimeters per day) and A is the Ath day after the plant germinates, what does the graph in part d represent?

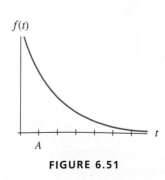

FIGURE 6.51

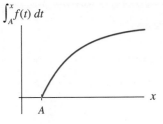

$\int_A^x f(t)\, dt$

FIGURE 6.52

Solution:

a. The accumulation function notation is $\int_A^x f(t)\, dt$.

b. As x moves to the right of A, the accumulated area is positive and is getting bigger more and more slowly. So for values of x to the right of A, the accumulation function is above the axis, is increasing, and is concave down as illustrated in Figure 6.52.

c. As x moves to the left of A, the accumulated area is positive and is getting bigger more and more quickly. Positive area getting bigger more and more quickly as x moves from A toward 0 is sketched as in Figure 6.53a, but the accumulation function graph is the "opposite"; that is, we reflect this portion of the graph across the horizontal axis as shown in Figure 6.53b.

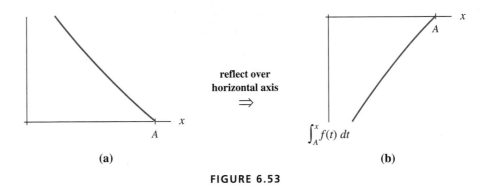

reflect over
horizontal axis
$\Rightarrow$

(a)

$\int_A^x f(t)\, dt$

(b)

FIGURE 6.53

d. The complete accumulation function graph is shown in Figure 6.54.

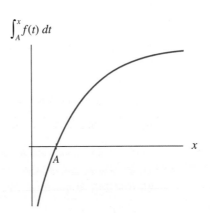

$\int_A^x f(t)\, dt$

FIGURE 6.54

e. The accumulation graph to the right of A represents how many millimeters the plant has grown since the Ath day.   ∎

To summarize, keep in mind the following guidelines for drawing an accumulation graph.

Sketching Accumulation Graphs

To sketch the graph of $\int_a^x f(t)\,dt$, consider the accumulated area between the horizontal axis and the graph of $f(t)$ as x moves to the right and then to the left of a on the horizontal axis.

1. Place the point $(a, 0)$ on the graph of $\int_a^x f(t)\,dt$ because the accumulation function value at the starting point is 0.

2. As x moves to the right of a,
   ■ accumulating area above the axis means the graph of $\int_a^x f(t)\,dt$ is increasing.
   ■ accumulating area below the axis means the graph of $\int_a^x f(t)\,dt$ is decreasing.

3. As x moves to the right of a,
   ■ accumulating area more and more quickly for positive accumulation means the graph of $\int_a^x f(t)\,dt$ is concave up.
   ■ accumulating area more and more quickly for negative accumulation means the graph of $\int_a^x f(t)\,dt$ is concave down.
   ■ accumulating area more and more slowly for positive accumulation means the graph of $\int_a^x f(t)\,dt$ is concave down.
   ■ accumulating area more and more slowly for negative accumulation means the graph of $\int_a^x f(t)\,dt$ is concave up.

4. As x moves to the left of a,
   ■ determine whether the graph would be increasing or decreasing from *right to left*, and concave up or concave down according to Steps 2 and 3.
   ■ reflect across the horizontal axis the portion of the graph you would have drawn to obtain the graph of $\int_a^x f(t)\,dt$.

Estimating Accumulation

How can we more accurately sketch accumulation function graphs? If we are given a graph with labeled tick marks on both the horizontal and the vertical axes, we can sketch an accumulation function fairly accurately by actually estimating areas. Consider the graph in Figure 6.55. This is the same graph that appears in Example 2, but it is drawn on a grid with labeled values on the vertical and horizontal axes. To sketch a more accurate graph of $\int_5^x f(t)\,dt$, we count boxes to estimate areas beneath the curve from 5 to various values of x.

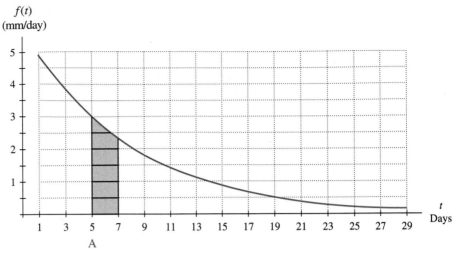

FIGURE 6.55

Begin by noting that each box on the grid has height 0.5 mm/day and width 2 days. Thus, each box has area (0.5mm/day)(2 days) = 1 mm. The shaded area in Figure 6.55 represents by how much the plant grew from day 5 to day 7. The shaded area is approximately $5\frac{1}{3}$ boxes, representing $5\frac{1}{3}$ mm of growth. This means that the point $\left(7, 5\frac{1}{3}\right)$ is a point on the accumulation function $\int_{5}^{x} f(t)\, dt$.

Table 6.37 shows other estimated areas (in this case, the same as the heights of the accumulation function), obtained by counting boxes from 5 to x. You should confirm these numbers by finding your own estimate of area by counting boxes. Your estimates probably won't agree exactly with these, but they should be close.

TABLE 6.37

x	7	11	15	19	23	27
Accumulated number of boxes = millimeters of growth	$5\frac{1}{3}$	$12\frac{2}{3}$	17	20	$21\frac{1}{2}$	$22\frac{1}{2}$

For values of x to the left of 5, we still count boxes, but because we are looking back in time (the 5th day was our starting point), we use the negative of the number of boxes. These values, from 5 to 3 and 5 to 1, are given in Table 6.38.

TABLE 6.38

x	3	1
Negative of the accumulated number of boxes = -(millimeters of growth)	$-6\frac{3}{4}$	$-15\frac{1}{2}$

Again, these represent points on the accumulation function. Plotting these eight points together with (5,0) (remember that every accumulation function is zero at its starting point) yields the scatter plot in Figure 6.56. The curve sketched through the points in the scatter plot confirms that the accumulation function we sketched in Figure 6.54 has the correct shape and gives us a more accurate sketch.

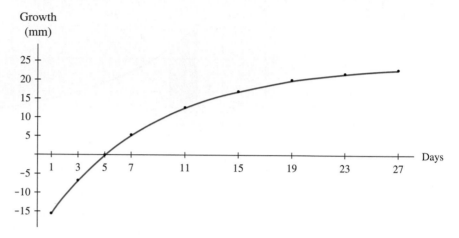

FIGURE 6.56

EXAMPLE 3 *Sketching an Accumulation Function Graph*

Consider the graph of $f(t)$ shown in Figure 6.57. We wish to sketch an accurate graph of the accumulation function of $f(t)$ beginning at 0.

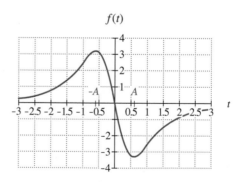

FIGURE 6.57

a. Construct a table of accumulation function values for $x = -3, -2.5, \ldots, 2.5, 3$.

b. Sketch a scatter plot and continuous graph of the accumulation function $\int_0^x f(t)\,dt$.

Solution:

a. By counting the boxes between $f(t)$ and the horizontal axis from 0 to x, we obtain the accumulation function values in Table 6.39. (*Note:* The area of each box is (0.5 units wide)(1 unit high) = 0.5. Also, the accumulation function value is negative for $x > 0$ because $f(t)$ is below the horizontal axis. For $x < 0$, $f(t)$ is above the horizontal axis; however, the accumulation function values are negative because x is to the left of the starting value.)

TABLE 6.39

x	Accumulated number of boxes	Accumulation function value	x	Accumulated number of boxes	Accumulation function value
0.5	2	−1	−0.5	2	−1
1	5	−2.5	−1	5	−2.5
1.5	7	−3.5	−1.5	7	−3.5
2	8	−4	−2	8	−4
2.5	8.5	−4.25	−2.5	8.5	−4.25
3	9	−4.5	−3	9	−4.5

b. A scatter plot and continuous graph for $\int_0^x f(t)\,dt$ is shown in Figure 6.58.

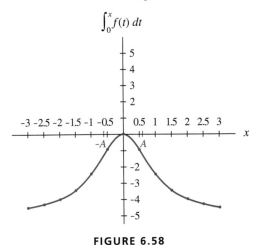

FIGURE 6.58

You may be wondering why we are interested in finding formulas for simple accumulation functions and sketching more complex ones. The reason is two-fold. First, because the accumulation of area beneath a rate-of-change function for a quantity represents an accumulation of change in that quantity, it is important that you have a thorough understanding of the accumulation process in order to understand accumulated change. Second, as we shall see in the next section, accumulation functions are an important step in developing an algebraic method for finding accumulated change.

e.

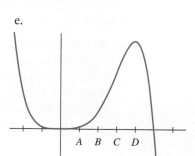

f.

6.4   Concept Inventory

■ Accumulation function:
Finding simple formulas
Sketching graphs
Interpreting

6.4   Activities

In each of Activities 1 through 4, a graph is given. Identify, from graphs *a* through *f*, the derivative graph and the accumulation graph (with 0 as the starting point) of the given graph. Graphs *a* through *f* may be used more than once.

1.

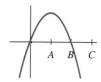

a.

2.

b.

3.

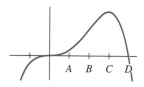

c.

4.

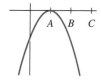

d.

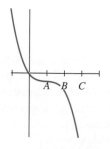

In each of Activities 5 and 6, a table of selected values for a function is given. Also shown are tables of values for the derivative and the accumulation function with 0 as the starting point. Determine which table is the derivative and which is the accumulation function, and justify your choice.

5.

t	$f(t)$	Input	Output	Input	Output
0	4	0	0	0	0
1	3	1	-2	1	3.667
2	0	2	-4	2	5.333
3	-5	3	-6	3	3
4	-12	4	-8	4	-5.333

6.

m	$p(m)$	Input	Output	Input	Output
0	0	0	0	0	0
1	-8	1	-12	1	-3
2	-16	2	0	2	-16
3	0	3	36	3	-27
4	64	4	96	4	0
5	200	5	180	5	125

Find formulas for the accumulation functions in Activities 7 through 12. Draw a graph, shading a region whose area the formula represents.

7. a. $\int_0^x 6\,dt$ b. $\int_0^x -2\,dt$ c. $\int_0^x \frac{1}{3}\,dt$

8. a. $\int_{-4}^x 6\,dt$ b. $\int_2^x 6\,dt$ c. $\int_{2000}^x 6\,dt$

9. a. $\int_{-4}^x -2\,dt$ b. $\int_2^x -2\,dt$ c. $\int_{2000}^x -2\,dt$

10. a. $\displaystyle\int_0^x 3t\,dt$ b. $\displaystyle\int_0^x -\tfrac{1}{2}t\,dt$ c. $\displaystyle\int_0^x 100t\,dt$

11. a. $\displaystyle\int_5^x 3t\,dt$ b. $\displaystyle\int_5^x -\tfrac{1}{2}t\,dt$ c. $\displaystyle\int_5^x 100t\,dt$

12. a. $\displaystyle\int_{-1}^x (2t+2)\,dt$ b. $\displaystyle\int_0^x (2t+2)\,dt$

    c. $\displaystyle\int_1^x (2t+2)\,dt$ d. $\displaystyle\int_{-2}^x (2t+2)\,dt$

On the basis of your answers to Activities 7 through 12, speculate about formulas for the accumulation functions in Activities 13 through 18.

13. $\displaystyle\int_0^x k\,dt$ 14. $\displaystyle\int_a^x k\,dt \quad k>0$

15. $\displaystyle\int_a^x k\,dt \quad k<0$ 16. $\displaystyle\int_0^x kt\,dt$

17. $\displaystyle\int_5^x kt\,dt$ 18. $\displaystyle\int_a^x (2t+2)\,dt$

19. a. Use the formula you found in Activity 17 to find the area of the region between $f(t)=3t$ and the horizontal axis from 5 to 19.

    b. If $f(t)$ is the rate of change of the revenue of an airline, in hundreds of dollars per day, and t is the number of days since December 31 of last year, what does the answer to part *a* represent?

20. a. Use the formula you found in Activity 16 to find the area of the region between $f(t)=7t$ and the horizontal axis from 0 to 9.

    b. If $f(t)$ is the rate of change in the temperature of an oven, in degrees per minute, and t is the number of minutes since the oven was turned on, what does your answer to part *a* represent?

In Activities 21 through 26, sketch the indicated accumulation function graphs.

21.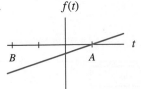

a. $\displaystyle\int_A^x f(t)\,dt$

b. $\displaystyle\int_B^x f(t)\,dt$

22.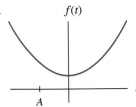

a. $\displaystyle\int_0^x f(t)\,dt$

b. $\displaystyle\int_A^x f(t)\,dt$

23.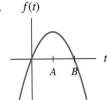

a. $\displaystyle\int_0^x f(t)\,dt$

b. $\displaystyle\int_A^x f(t)\,dt$

c. $\displaystyle\int_B^x f(t)\,dt$

24.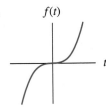

$\displaystyle\int_0^x f(t)\,dt$

25.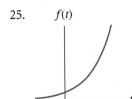

$\displaystyle\int_0^x f(t)\,dt$

26.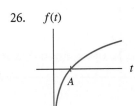

$\displaystyle\int_A^x f(t)\,dt$

27. The graph shown in Figure 6.4.1 is a model of the rate of change of profit for a new business during its first year. The input is weeks since the business opened, and the output units are thousands of dollars per week.

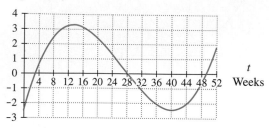

FIGURE 6.4.1

a. What does the area of each box in the grid represent?

b. What is the interpretation, in context, of the accumulation function $\int_0^x p(t)\,dt$?

c. Count boxes to estimate accumulation function values from 0 to x for the values of x given in Table 6.40. (Regions beneath the horizontal axis contribute negative accumulation.)

TABLE 6.40

x	Accumulation function value
0	
4	
8	
12	
16	
20	
24	
28	
32	
36	
40	
44	
48	
52	

d. Use the data in part c to sketch an accurate graph of the accumulation function $\int_0^x p(t)\,dt$. Label units and values on the horizontal and vertical axes.

28. After observing the growth of a certain bacterium, a microbiologist models the growth rate $b(t)$, measured in bacteria per hour, as a function of the number of hours since 8 a.m. on the day she began observing. A graph of the model is shown in Figure 6.4.2.

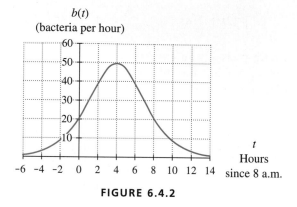

FIGURE 6.4.2

a. What does the area of each box on the grid represent?

b. Estimate the area under the curve by counting boxes from 0 to x for the values of x given in Table 6.41. For x to the left of zero, record the accumulation as a negative value.

TABLE 6.41

x	Accumulation function value
0	
2	
4	
6	
8	
10	
12	
14	
-2	
-4	
-6	

c. What is the interpretation of the accumulation function $\int_0^x b(t)\,dt$ in the context of bacteria growth?

d. Use the data in part b to sketch an accurate graph of the accumulation function of $b(t)$

from 0 to x. Label units and values on the horizontal and vertical axes.

29. a. Construct a table of values for areas of regions beneath $f(t) = t^2$ and the horizontal axis from 0 to x.

x	0	1	2	3	4	5
$\int_0^x t^2\, dt$						

   b. From the values in part a, suggest a formula for $\int_0^x t^2\, dt$.

30. a. Construct a table of values for areas of regions beneath $f(t) = t^2$ and the horizontal axis from 2 to x.

x	0	1	2	3	4	5
$\int_2^x t^2\, dt$						

   b. From the values in part a, suggest a formula for $\int_2^x t^2\, dt$.

31. a. Construct a table of values for $f(x) = e^x$ and for areas of regions beneath $f(t) = e^t$ and the horizontal axis from 0 to x.

x	0	1	2	3	4	5
e^x						
$\int_0^x e^t\, dt$						

   b. From the values in part a, suggest a formula for $\int_0^x e^t\, dt$.

32. a. Construct a table of values for $f(x) = e^x$ and for areas of regions beneath $f(t) = e^t$ and the horizontal axis from 1 to x.

x	0	1	2	3	4	5
e^x						
$\int_1^x e^t\, dt$						

   b. From the values in part a, suggest a formula for $\int_1^x e^t\, dt$.

33. Explain what effect changing the starting point has on a formula of an accumulation function and on its graph.

6.5   The Fundamental Theorem

In Section 6.4 we developed accumulation formulas for linear functions and found that these formulas vary slightly depending on the location at which we begin accumulating. Some of the formulas from the examples and activities in the last section are shown below:

$$f(t) = 3 \qquad\qquad \int_0^x 3\, dt = 3x$$

$$f(t) = -2 \qquad\qquad \int_0^x -2\, dt = -2x$$

$$\int_{2000}^x -2\, dt = -2x + 4000$$

$$f(t) = 2t - 4 \qquad\qquad \int_2^x (2t - 4)\, dt = x^2 - 4x + 4$$

$$\int_{-1}^x (2t - 4)\, dt = x^2 - 4x - 5$$

Note that if you take the derivative with respect to x of each of the accumulation formulas, you obtain the original function in terms of x. For example,

$$\frac{d}{dx}(x^2 - 4x + 4) = 2x - 4$$

which differs from $2t - 4$ only in the input variable. You may think this relationship is a special case for lines, but consider some of the more complicated accumulation functions that we sketched. In Example 2 in Section 6.4, we began with the function in Figure 6.59a and drew the accumulation function graph shown in Figure 6.59b.

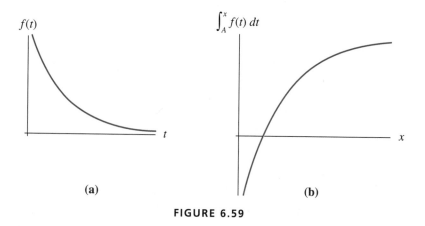

(a)   **(b)**

FIGURE 6.59

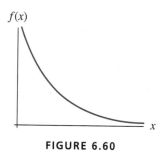

FIGURE 6.60

Now let us sketch the slope graph (or derivative graph) of the accumulation function in Figure 6.59b. Note that the slopes are positive everywhere but seem to be getting smaller and smaller as x gets larger. Also, close to the vertical axis, the graph is very steep, so the slopes are very large. The slope graph appears in Figure 6.60. The slope graph is exactly the graph with which we began in Figure 6.59a (with the input variable labeled x instead of t).

In Example 3 in Section 6.4, we began with the graph in Figure 6.61a and sketched the accumulation function shown in Figure 6.61b.

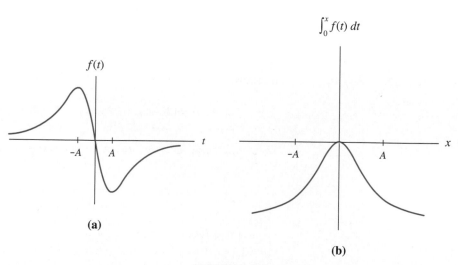

(a)

(b)

FIGURE 6.61

Again, let us sketch the slope graph of the accumulation function. To the left of zero, the graph has positive slopes. The slopes are near zero to the far left, and the graph becomes steeper until $-A$. Between $-A$ and 0, the slopes are still positive but are approaching 0 as the accumulation function approaches its maximum. At $x = 0$, the slope is zero. To the right of $x = 0$, the slopes are negative. They become more and more negative until A. To the right of A, the slopes are still negative but are getting closer and closer to zero as the graph levels off. The slope graph is shown in Figure 6.62.

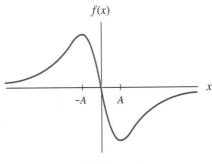

FIGURE 6.62

Again, this is exactly the graph with which we began in Figure 6.61a.

You are probably beginning to see that if we begin with a function $f(t)$, find an accumulation formula or accumulation graph, $\displaystyle\int_a^x f(t)\,dt$, and then take the derivative or draw the slope graph, we get $f(x)$, the function with which we began but in terms of x. This is actually a surprising result—accumulation functions and areas are related to derivatives! In fact, this is such an important result that it is called the **Fundamental Theorem of Calculus**.

The Fundamental Theorem of Calculus

For any continuous function $f(t)$, the derivative of an accumulation function of $f(t)$ is the function f in terms of x.

In symbols, we write

$$\frac{d}{dx}\left(\int_a^x f(t)\,dt\right) = f(x)$$

We can infer from the Fundamental Theorem that to find an accumulation formula, we need only reverse the process of finding a derivative. For this reason, we call the accumulation function $\displaystyle\int_a^x f(t)\,dt$ an **antiderivative** of the function $f(x)$.

Antiderivative

Let $f(x)$ be a function. A function $F(x)$ is called an antiderivative of $f(x)$ if $F'(x) = f(x)$; that is, the derivative of $F(x)$ is $f(x)$.

Our motivation for developing accumulation functions (antiderivatives) is not only to have a formula for the accumulated area between the graph of a function and the horizontal axis but also (and more importantly) to develop a function for a quantity if we know a function for that quantity's rate of change. In Example 1 of Section 6.2, we were interested in finding the total volume of water that ran through the Carson River in the 20 hours after 11:45 a.m. on a Wednesday but were given values only for the flow rate in cubic feet per hour. If we could find an antiderivative of the flow rate model, we would have a model for the volume of the water flowing through the river and could use the model to answer questions about the volume rather than summing areas of rectangles as we did in Section 6.2.

Similarly, in Example 2 of Section 6.1, we found a model for the rate of change of the concentration of a drug and summed areas of rectangles to find a change in the concentration. If we could find an antiderivative of the rate-of-change function, we would then have a function modeling the concentration of the drug and could use it to answer the questions posed in Section 6.1.

Recovering a Function

Recovering a function is the phrase we use for the process of beginning with a rate-of-change function for a quantity and finding its antiderivative to obtain a function for the quantity. An important part of recovering a function from its rate of change is recovering the units of the function from the units of its rate of change. If $\frac{dM}{dt}$ is the rate-of-change function of the amount of insulin in a patient's body t hours after an injection, and if it is reported in milliliters per hour, then we can recover the units of the amount function $M(t)$ by recalling that the rate of change of a function is a slope of a tangent line. Slope is calculated as $\frac{\text{rise}}{\text{run}}$, where the units are $\frac{\text{output units}}{\text{input units}}$. In the case of $\frac{dM}{dt}$, the units milliliters per hour can be rewritten as $\frac{\text{milliliters}}{\text{hour}}$. Now we can see that the output units of $M(t)$ are milliliters and the input units are hours. Figure 6.63 shows input/output diagrams for $\frac{dM}{dt}$ and $M(t)$.

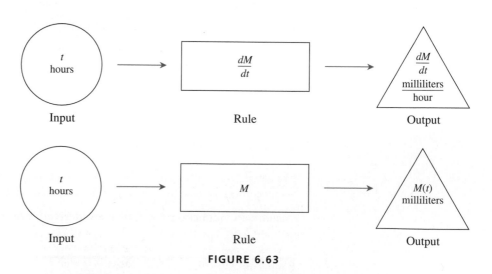

FIGURE 6.63

Occasionally, units of a rate of change are expressed in terms of a squared unit. For example, acceleration is often expressed in feet per second squared. Suppose $A(t)$ is acceleration of a vehicle in feet per second squared, where t is the number of seconds since the vehicle began accelerating. Again, this is obviously the rate of change of some function. In fact, it is the rate of change of velocity. However, the input and output units of that velocity function may not be immediately apparent because of the use of the word *squared*. Rewrite feet per second squared as

$$\frac{\text{feet}}{(\text{second})^2} = \frac{\text{feet}}{(\text{second})(\text{second})} = \frac{\text{feet}}{\text{second}} \div (\text{second})$$

$$= (\text{feet per second}) \text{ per second}$$

The output units of the velocity function are now identifiable as feet per second. Input/output diagrams for these functions are shown in Figure 6.64.

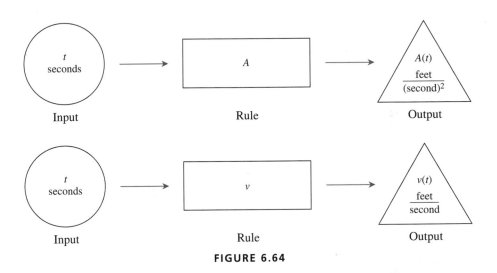

Input   Rule   Output

Input   Rule   Output

FIGURE 6.64

We have seen how to recover the quantity units from rate-of-change units, and we have noted that the Fundamental Theorem of Calculus implies that we have a powerful algebraic tool for recovering the accumulated-quantity formula from its rate-of-change formula. However, we have yet to develop the specific algebraic rules that will enable us actually to use this tool. We now turn our attention to developing those rules.

Antiderivative Formulas

Given a function, how do we find an antiderivative? As we have seen, we must reverse the process of differentiation. Antidifferentiation starts with the known derivative and finds the unknown function. For example, consider the constant function $f(x) = 3$. To find an antiderivative of $f(x)$, we need to find a function whose derivative is 3. One such function is $F(x) = 3x$. Other functions whose derivatives are 3 include $F(x) = 3x + 7$ and $F(x) = 3x - 24.9$.

In fact, having found one antiderivative $F(x)$ for a given function $f(x)$, we can obtain infinitely many antiderivatives for that function by adding an arbitrary constant C to $F(x)$. Thus we call $F(x) + C$ the **general antiderivative of $f(x)$**. We use the notation

$$\int f(x)\, dx = F(x) + C$$

for the general antiderivative. The general antiderivative is a group of infinitely many functions. (An accumulation function is one specific function from that group.) Note that the integral sign has no upper and lower limits for general antiderivative notation. The dx in this notation is to remind us that we are finding the general antiderivative with respect to x, so our antiderivative formula will be in terms of x. For example, we say that the general antiderivative for $f(x) = 3$ is $F(x) = 3x + C$ and we write $\int 3\,dx = 3x + C$.

As we saw in Section 6.3, an antiderivative (accumulation formula) of any constant function will be a line because lines have constant derivatives. We can write this general rule as

$$\int k\,dx = kx + C$$

where k and C are any constants.

Now consider finding the general antiderivative of $f(x) = 2x$. We are seeking a function whose derivative is $2x$. The function is $F(x) = x^2 + C$, and we write

$$\int 2x\,dx = x^2 + C$$

It is more difficult to reverse the derivative process for $f(x) = x^2$. Recall that the power rule for derivatives $\left[\frac{d}{dx}(x^n) = nx^{n-1}\right]$ instructs us to

■ multiply by the power

■ and then subtract 1 from the power to get the new power.

To reverse the process for antiderivatives, we

■ *add* 1 to the power to get the new power

■ and then *divide* by that new power.

This formula is known as the **Simple Power Rule** for antiderivatives.

Simple Power Rule for Antiderivatives

$$\int x^n\,dx = \frac{x^{n+1}}{n+1} + C, \text{ for any } n \neq -1$$

This rule requires that $n \neq -1$, because otherwise, we would be dividing by zero.

In the case of $f(x) = x^2$, the general antiderivative is $F(x) = \frac{x^{2+1}}{2+1} + C = \frac{x^3}{3} + C$:

$$\int x^2\,dx = \frac{x^3}{3} + C$$

EXAMPLE 1   *Finding Antiderivatives*

Find the following general antiderivatives and their appropriate units.

a. $\int -7\,dx$ degrees per hour squared where x is in hours

b. $\int h^{0.5}\,dh$ parts per million per day where h is in days

Solution:

a. $\int -7\,dx = -7x + C$ degrees per hour

b. $\int h^{0.5}\,dh = \frac{h^{1.5}}{1.5} + C$ parts per million ∎

Recall the Constant Multiplier Rule for derivatives:

If $g(x) = kf(x)$ then $g'(x) = kf'(x)$ where k is a constant

A similar rule applies for antiderivatives:

Constant Multiplier Rule for Antiderivatives

$$\int kf(x)\,dx = k\int f(x)\,dx$$

Thus $\int 12x^6\,dx = 12\int x^6\,dx = 12\left(\frac{x^7}{7}\right) + C = \left(\frac{12x^7}{7}\right) + C.$

Another property of antiderivatives that can be easily deduced from a similar property for derivatives is the Sum Rule.

Sum Rule

$$\int [f(x) \pm g(x)]\,dx = \int f(x)\,dx \pm \int g(x)\,dx$$

The Sum Rule lets us find an antiderivative for a sum (or difference) of functions by operating on each function independently. For example,

$$\int (7x^3 + x)\,dx = \int 7x^3\,dx + \int x\,dx$$

$$= \left(\frac{7x^4}{4} + C_1\right) + \left(\frac{x^2}{2} + C_2\right)$$

$$= \frac{7x^4}{4} + \frac{x^2}{2} + C \quad \text{(Combine } C_1 \text{ and } C_2 \text{ into one constant } C)$$

Repeated applications of the Simple Power Rule, the Constant Multiplier Rule, and the Sum Rule enable us to find an antiderivative of any polynomial function.

We now have the tools we need to begin with a simple polynomial rate-of-change function for a quantity and recover an amount function for that quantity.

EXAMPLE 2   *Birth Rate*

An African country has an increasing population but a declining birth rate, a situation that results in the number of babies born each year increasing but at a slower rate. The rate of change in the number of babies born each year is given by

$$b(t) = 87{,}000 - 1600t \text{ births per year}$$

t years from the end of this year. Also, the number of babies born in the current year is 1,185,800.

  a. Find a function describing the number of births each year t years from now.

  b. Use the function in part a to estimate the number of babies born next year.

Solution:

  a. A function $B(t)$ describing the number of births each year is found as

$$B(t) = \int b(t)\,dt = 87{,}000t - 800t^2 + C \text{ births}$$

    t years from now.

       We also know that $B(0) = 1{,}185{,}800$, so $C = 1{,}185{,}800$. Thus we have

$$B(t) = 87{,}000t - 800t^2 + 1{,}185{,}800 \text{ births each year}$$

    t years from now.

  b. The number of babies born next year is estimated as $B(1) = 1{,}272{,}000$ babies. ∎

We have just presented and applied three antiderivative rules: the Simple Power Rule, the Constant Multiplier Rule, and the Sum Rule. Now let us look at three more rules for finding antiderivatives.

Refer to the Simple Power Rule, and note that it did not apply for $n = -1$. The case where $n = -1$ is special. This is the antiderivative $\int x^{-1}\,dx = \int \frac{1}{x}\,dx$. Recall that $\frac{d}{dx}(\ln x) = \frac{1}{x}$. This is valid only for $x > 0$, because $\ln x$ is not defined for $x < 0$. When $x < 0$, we use $\ln(-x)$ because $\frac{d}{dx}[\ln(-x)] = \frac{-1}{-x} = \frac{1}{x}$. Both cases ($x > 0$ and $x < 0$) can be handled very simply by using $\ln|x|$.

$$\int \frac{1}{x}\,dx = \ln|x| + C$$

The two final antiderivative formulas that we consider are for exponential functions. Recall that the derivative of $f(x) = e^x$ is e^x. Similarly, the general antiderivative of $f(x) = e^x$ is also e^x plus a constant.

$$\int e^x\,dx = e^x + C$$

The other exponential function that we have encountered is $f(x) = b^x$. Its derivative was found by multiplying b^x by $\ln b$. To find the general antiderivative, we divide b^x by $\ln b$ and add a constant:

$$\int b^x\,dx = \frac{b^x}{\ln b} + C$$

EXAMPLE 3   *Marginal Cost*

Suppose that a manufacturer of small toaster ovens has collected the data given in Table 6.42, which shows, at various production levels, the approximate cost to produce one more oven. Recall from Section 5.3 that this is marginal cost and can be interpreted as the rate of change of cost.

TABLE 6.42

Production level (ovens per day)	200	300	400	500	600	700
Cost to produce an additional oven	$29	$20	$15	$11	$9	$7

The manufacturer also knows that the total cost to produce 250 ovens is $12,000.

a. Find a model for the marginal cost data.

b. Recover a cost model from the model you found in part *a*.

c. Estimate the cost of producing 500 ovens.

Solution:

a. Either a quadratic or an exponential model is a good fit to the data. We choose an exponential model:

$$C'(x) = 47.638(0.9972)^x \text{ dollars per oven}$$

where x is the number of ovens produced each day.

b. To recover a model for cost, we need an antiderivative of $C(x)$ satisfying the known condition that the cost to produce 250 ovens is $12,000. The general antiderivative is

$$C(x) = \frac{47.638(0.9972)^x}{\ln(0.9972)} + K \approx -16{,}991.852(0.9972)^x + K$$

where K is a constant.

    Using the fact that $C(250) = \$12{,}000$, we substitute 250 for x, set the antiderivative equal to $12,000, and solve for K.

$$-16{,}991.852(0.9972)^{250} + K = 12{,}000$$

$$-8430.309 + K = 12{,}000$$

$$K \approx 20{,}430.309$$

Thus the approximate cost of producing x toaster ovens is

$$C(x) = -16{,}991.852(0.9972)^x + 20{,}430.309 \text{ dollars}$$

c. You can readily verify that the total cost of producing 500 toaster ovens is estimated by $C(500) \approx \$16{,}248$.   ■

In summary, we now have the following antiderivative formulas:

Antiderivative Formulas		
	Function: $f(x)$	**General antiderivative:** $\int f(x)\,dx$
Constant Rule	k	$kx + C$
Simple Power Rule	$x^n, n \neq -1$	$\left(\dfrac{1}{n+1}\right) x^{n+1} + C$
Exception to Simple Power Rule	$\dfrac{1}{x}$	$\ln|x| + C$
Exponential Rule	b^x	$\left(\dfrac{1}{\ln b}\right) b^x + C$
e^x Rule	e^x	$e^x + C$
Constant Multiplier Rule	$kg(x)$	$k \int g(x)\,dx$
Sum Rule	$g(x) \pm h(x)$	$\int g(x)\,dx \pm \int h(x)\,dx$

EXAMPLE 4   *More Antiderivatives*

Find the following general antiderivatives.

a. $\int \left(3^x - 7e^x + \frac{5}{x}\right) dx$ quarts per hour where x is measured in hours

b. $\int \left(4\sqrt{x} + 100e^{0.06x} + 0.46\right) dx$ mpg per mph where x is measured in mph

Solution:

a. $\displaystyle \int \left(3^x - 7e^x + \frac{5}{x}\right) dx = \int 3^x\,dx - 7\int e^x\,dx + 5\int \frac{1}{x}\,dx$

$\displaystyle \qquad\qquad = \frac{3^x}{\ln 3} - 7e^x + 5\ln|x| + C \text{ quarts}$

b. We must first rewrite $\sqrt{x}$ as $x^{\frac{1}{2}}$ and $e^{0.06x}$ in the form b^x with $b = e^{0.06}$

$\displaystyle \int \left(4\sqrt{x} + 100e^{0.06x} + 0.46\right) dx = \int \left[4x^{\frac{1}{2}} + 100(e^{0.06})^x + 0.46\right] dx$

$\displaystyle \qquad\qquad = 4\int x^{\frac{1}{2}}\,dx + 100\int (e^{0.06})^x\,dx + \int 0.46\,dx$

$\displaystyle \qquad\qquad = 4\frac{x^{\frac{3}{2}}}{\frac{3}{2}} + 100\frac{(e^{0.06})^x}{\ln(e^{0.06})} + 0.46x + C$

$\displaystyle \qquad\qquad = \frac{8}{3}x^{\frac{3}{2}} + \frac{100}{0.06}e^{0.06x} + 0.46x + C \text{ mpg}$

It is sometimes necessary to find an antiderivative twice in order to obtain the appropriate accumulation formula. For example, to obtain distance from acceleration, you must determine the antiderivative of the acceleration function to obtain a velocity function and then determine the antiderivative of the velocity function to obtain a function for distance traveled.

Let us summarize what we have learned thus far about integrals. The definite integral $\int_a^b f(x)\,dx$ is a limiting value of sums of areas of rectangles and gives us the area of the region between $f(x)$ and the x-axis if $f(x)$ lies above the horizontal axis from a to b. When $f(x)$ is a function lying below the horizontal axis from a to b, the definite integral is the negative of the area inscribed. If $f(x)$ is a rate of change of some quantity, then $\int_a^b f(x)\,dx$ is the change in the quantity from a to b. The accumulation function $\int_a^x f(t)\,dt$ is a formula in terms of x for the accumulated area from a to x. We use the integral symbol without the upper and lower limits, $\int f(x)\,dx$, to represent the general antiderivative of $f(x)$. Although these three symbols are similar, it is important that you have a clear understanding of what each one represents. Their interpretations are summarized in Table 6.43.

TABLE 6.43

Symbol	Name	Interpretation
$\int_a^b f(x)\,dx$	definite integral	a number that can be thought of in terms of area
$\int_a^x f(t)\,dt$	accumulation function	a formula for an accumulated amount
$\int f(x)\,dx$	general antiderivative	a formula whose derivative is $f(x)$

The Fundamental Theorem of Calculus tells us that accumulation formulas are specific antiderivatives. As we shall see in Section 6.6, antiderivatives enable us to find areas algebraically by using accumulation formulas rather than numerically as limiting values of sums of areas of rectangles.

Apart from helping us find areas, antiderivatives are useful in allowing us to recover functions from rates of change. We have seen several examples of that in this section. It may seem difficult to reverse your thinking from finding derivatives to finding antiderivatives, but with practice you will soon be proficient at both.

6.5   Concept Inventory

- Fundamental Theorem of Calculus
- Antiderivative
- Recovering a function from its rate of change
- $\int f(x)\,dx$ = general antiderivative
- Antiderivative formulas

6.5   Activities

In Activities 1 through 12, a and b are constants and x and t are variables. In these activities, label each notation as always representing

a. a function of x, or

b. a function of t, or

c. a number.

1. $f'(t)$

2. $\dfrac{df}{dx}$

3. $f'(3)$

4. $\displaystyle\int f(t)\,dt$

5. $\displaystyle\int f(x)\,dx$

6. $\displaystyle\int_a^b f(t)\,dt$

7. $\displaystyle\int_a^b f(x)\,dx$

8. $\displaystyle\int_a^x f(t)\,dt$

9. $\displaystyle\int_b^t f(x)\,dx$

10. $\dfrac{d}{dx}\displaystyle\int_a^x f(t)\,dt$

11. $\dfrac{d}{dt}\displaystyle\int_a^t f(x)\,dx$

12. $\dfrac{d}{dx}\displaystyle\int_a^a f(t)\,dt$

For each of the rates of change in Activities 13 through 16:

  a. Write the units of the rate of change as a fraction.

  b. Draw an input/output diagram for the recovered function.

  c. Interpret the recovered function in a sentence.

13. When m thousand dollars are being spent on advertising, the annual revenue of a corporation is changing by $\frac{dR}{dm}$ million dollars per thousand dollars spent on advertising.

14. The percentage of households with washing machines was changing by $\frac{dW}{dt}$ percentage points per year, where t is the number of years since 1950.

15. The concentration of a drug in the blood stream of a patient is changing by $\frac{dc}{dh}$ milligrams per liter per hour h hours after the drug was given.

16. The level of production at a tire manufacturer h hours after production began is increasing by $P'(h)$ tires per hour squared.

Find the general antiderivative as indicated in Activities 17 through 22. Check each of your antiderivatives by taking its derivative.

17. $\displaystyle\int 19.436(1.07)^x\,dx$

18. $\displaystyle\int (32.685\,x^3 + 3.296x - 15.067)\,dx$

19. $\displaystyle\int [6e^x + 4(2^x)]\,dx$

20. $\displaystyle\int 39.24e^{3.9x}\,dx$

21. $\displaystyle\int \left(10^x + 4\sqrt{x} + 8\right)dx$

22. $\displaystyle\int \left[\tfrac{1}{2}x + \tfrac{1}{2x} + \left(\tfrac{1}{2}\right)^x\right]dx$

For each of the rate-of-change functions in Activities 23 through 27, find the general antiderivative, and label the units on the antiderivative.

23. $s(m) = 6250(0.92985)^m$ CDs per month m months since the beginning of the year

24. $p(x) = 0.03731x^2 - 0.4841x + 1.4069$ dollars per 1000 cubic feet per year x years since 1989

25. $j(x) = \frac{15.29}{x} + 7.95$ units per dollar where x is the price (in dollars)

26. $c(x) = \frac{0.7925}{x} + 0.3292(0.009324)^x$ dollars per unit squared when x units are produced

27. $p(t) = 1.724928e^{0.0256t}$ millions of people per year t years after 1990

In Activities 28 through 31, find F, the antiderivative of f.

28. $f(x) = 5x + 3$; $F(1) = 3$

29. $f(t) = t^2 + 2t$; $F(12) = 700$

30. $f(u) = \frac{2}{u} + u$; $F(1) = 5$

31. $f(z) = \frac{1}{z^2} + e^z$; $F(2) = 1$

32. The civilian labor force in the Trident region of South Carolina has been expanding[17] at a rate of $p(x) = -2915x + 19{,}433$ people each year x years after 1989. The civilian labor force was 227,120 people in 1989.

  a. Find a model for the size of the civilian labor force.

  b. Determine, according to the model, how many people were in the labor force by 1992.

  c. According to the model, how many people were added to the labor force between 1989 and 1991?

33. For the first 2 months after sweet corn is planted (given optimal growing conditions) the rate of

17. Based on information in *1993 Trident Economic Forecast*.

change of its growth rate can be modeled by

$g'(d) = -0.00036d - 0.018$ inches per day squared

where d is the number of days after sprouting.

a. Find a function for the growth rate of sweet corn d days after it sprouts. The corn was growing by 1.95 inches per day on the day it sprouted.

b. Find a function for the height of sweet corn d days after it sprouts.

c. How tall is the sweet corn 60 days after sprouting?

34. An investment worth $1 million in 1990 has been growing at a rate of $f(t) = 0.1397619(1.15)^t$ million dollars per year t years after 1990.

a. Recover the amount function, and determine the current value of the investment and its projected value next year.

b. Determine how much the investment has grown since 1990 and how much it is projected to grow over the next year.

35. The amount[18] of poultry produced in the United States annually from 1960 through 1993 can be modeled as $p(x) = 6.178983e^{0.047509x}$ million pounds per year x years after 1960.

a. Write the general antiderivative of $p(x)$.

b. According to the model, how much poultry was produced in the United States from 1960 through 1993?

c. Find a model for the rate of change in annual poultry production.

d. How do $p(x)$, the answer to part a, and the answer to part c differ?

36. As a person's age increases, the amount she or he would have to pay as a first-year premium on a new life insurance policy also increases. According to a recent report from an insurance company, the first-year premium (per $1000) on an annual renewable term life insurance policy for nonsmokers between the ages of 30 and 60 increases at a rate of $p'(x) = 0.017714(1.158887)^x$ dollars per year of age, where $x + 30$ is the age of the insured.

a. What value is x if the person is 45 years old?

b. Find the function for the premium (per $1000). The premium for a 40-year-old is $1.49.

c. Determine the first-year premium for a 30-year-old who is purchasing a $500,000 policy.

37. The Washington Monument, located at one end of the Federal Mall in Washington, D.C., is the world's tallest obelisk at 555 feet. Suppose that a tourist drops a penny from the observation deck atop the monument. Let us assume that the penny falls from a height of 540 feet.

a. Recover the velocity function for the penny using the facts that
   i. acceleration due to gravity near the surface of the earth is -32 feet per second squared.
   ii. because the penny is dropped, velocity is 0 when time is 0.

b. Recover the distance function for the penny using the velocity function from part a and the fact that distance is 540 feet when time is 0.

c. When will the penny hit the ground?

d. What is the impact velocity of the penny in miles per hour?

38. According to the *Guinness Book of Records*, the world's record high dive from a diving board is 176 feet, 10 inches. It was made by Olivier Favre (Switzerland) in 1987. Ignoring air resistance, approximate Favre's impact velocity in miles per hour from a height of 176 feet, 10 inches.

39. a. What is the impact velocity (in feet per second and miles per hour) of a cat that accidentally falls off a building from a height of 66 feet $\left(5\frac{1}{2} \text{ stories}\right)$?

b. In the 1960s, Donald McDonald claimed in an article in *The New Scientist* that plummeting cats never fall faster than 40 mph. What accounts for the difference between your answer to part a and McDonald's claim (assuming McDonald's claim is accurate)?

40. Table 6.44[19] gives the increase or decrease in the number of donors to a college athletics support organization for selected years.

18. Based on information in *USA Today*, August 12, 1994. *Source*: National Agricultural Statistics Service.
19. Based on data from IPTAY, Clemson University.

TABLE 6.44

Year	Rate of change in donors (donors per year)
1975	-169
1978	803
1981	1222
1984	1087
1987	399
1990	-842

a. Find a model for the rate of change in the number of donors.

b. Find a model for the number of donors. Use the fact that in 1980 there were 10,706 donors.

c. Estimate the number of donors in 1992.

41. From 1982 through 1992, a microchip manufacturer in Silicon Valley was hiring new employees at a rate of $n(x) = \frac{593}{x} + 138$ new employees per year, where x represents the number of years since the company was founded in 1978. By 1982 the company had hired 896 employees.

a. Write the function that gives the number of employees who had been hired by the xth year after the company was founded in 1978.

b. For what years will the function in part a apply?

c. Find the total number of employees the company had hired by the end of 1992. Would this figure necessarily be the same as the number of employees the company had at the end of 1992? Explain.

42. Table 6.45 shows the rate of change of the purchasing power of one dollar at various times since 1978.

TABLE 6.45

Year	Rate of change of purchasing power of $1 (dollars per year)
1979	-0.14
1981	-0.07
1983	-0.02
1985	-0.003
1987	-0.006
1989	-0.03
1991	-0.09

a. Examine a scatter plot of the data, and find a model for the rate of change over the period from 1978 through 1991. (Align the data so that time t is 0 in 1978.)

b. Use your model to derive a function that models the purchasing power of one dollar as a function of time. Write the function in terms of constant 1990 dollars. (That is, one dollar was worth one dollar in 1990.)

c. Rewrite the function in terms of constant 1980 dollars.

d. Rewrite the function in terms of the current value of a dollar.

e. What is the difference in the three functions from parts b, c, and d?

■ 6.6   The Definite Integral

So far, we have been finding general and specific antiderivatives to recover functions from their rate-of-change functions. The Fundamental Theorem tells us that accumulation functions are antiderivatives. Consider finding the accumulation function for $f(t) = 6t^2 - 2t + 3$ from 4 to x, $\int_4^x (6t^2 - 2t + 3)\, dt$. We know that the answer is an antiderivative of the form $F(x) = 2x^3 - x^2 + 3x + C$, and we also know that when $x = 4$, the accumulation must be zero: $\int_4^4 (6t^2 - 2t + 3)\, dt = 0$. Substituting $x = 4$

into the general antiderivative, setting it equal to zero, and solving for C gives the value of C in the accumulation function with starting point $x = 4$.

$$F(4) = 2(4)^3 - (4)^2 + 3(4) + C = 0$$
$$128 - 16 + 12 + C = 0$$
$$124 + C = 0$$
$$C = -124$$

Thus we have

$$\int_4^x (6t^2 - 2t + 3)\,dt = 2x^3 - x^2 + 3x - 124$$

We use this formula to find the area of the region between $f(t) = 6t^2 - 2t + 3$ and the horizontal axis from 4 to 7 by substituting 7 into the accumulation formula.

$$2(7)^3 - (7)^2 + 3(7) - 124 = 534$$

EXAMPLE 1   *Finding Accumulation Formulas*

a. Find a formula for $\displaystyle\int 69.7966(1.07229)^t\,dt$.

b. Find a formula for the accumulation function $\displaystyle\int_1^x 69.7966(1.07229)^t\,dt$.

c. Use the accumulation function to find the area of the region between the function $f(t) = 69.7966(1.07229)^t$ and the horizontal axis from 1 to 4.

d. If $f(t) = 69.7966(1.07229)^t$ is the rate of change of the balance in a savings account given in dollars per year, and t is the number of years since the savings account was opened, what does your answer to the question in part c represent?

Solution:

a. $\displaystyle\int 69.7966(1.07229)^t\,dt = \frac{69.7966(1.07229)^t}{\ln 1.07229} + C$

b. The accumulation function is the specific antiderivative (in terms of x) for which the antiderivative is zero when $x = 1$.

$$\frac{69.7966(1.07229)^1}{\ln 1.07229} + C = 0$$
$$C \approx -1072.2908$$

Thus we have

$$\int_1^x 69.7966(1.07229)^t\,dt \approx \frac{69.7966(1.07229)^x}{\ln 1.07229} - 1072.2908$$

c. Substituting 4 for x in the formula above gives the area as approximately 249.7637.

d. The answer to part c represents the change in the amount in the savings account between 1 and 4 years. The amount grew by \$249.76.   ∎

The Fundamental Theorem of Calculus gives us a method, as illustrated in Example 1, for finding an accumulation function and using it to determine the area of the region between a function and the horizontal axis. In general, we know that $\int_a^x f(t)\,dt = F(x) + C$, where $F(x)$ is an antiderivative of $f(x)$. We also know that when $x = a$, the accumulation function is zero.

$$\int_a^a f(t)\,dt = F(a) + C = 0$$

This tells us that $C = -F(a)$. Thus we have

$$\int_a^x f(t)\,dt = F(x) - F(a)$$

To find the value of the accumulation function from a to b, we simply substitute b for x.

$$\int_a^b f(t)\,dt = F(b) - F(a)$$

We now have an efficient algebraic method for evaluating definite integrals. We summarize the discussion as follows:

Evaluating a Definite Integral

If $f(x)$ is a continuous function from a to b and $F(x)$ is any antiderivative of $f(x)$, then

$$\int_a^b f(x)\,dx = F(b) - F(a)$$

Recall from Section 6.3 that we define the definite integral, $\int_a^b f(x)\,dx$, as the limiting value of sums of areas of rectangles. That is,

$$\int_a^b f(x)\,dx = \lim_{n \to \infty} [f(x_1) + f(x_2) + \cdots + f(x_n)]\Delta x$$

The antiderivative definition for a definite integral, $\int_a^b f(x)\,dx = F(b) - F(a)$, gives us another method for evaluating a definite integral, so for many functions, we no longer have to rely on the sometimes tedious process of finding limiting values of sums of areas of rectangles. The fact that we can find areas using a reverse derivative process should continue to surprise you. The connection between areas and derivatives is a remarkable one.

To illustrate the ease with which we can calculate areas of regions between functions and the horizontal axis, consider the region between $f(x) = x^2 + 2$ and the x-axis from -2 to 4. (See Figure 6.65.)

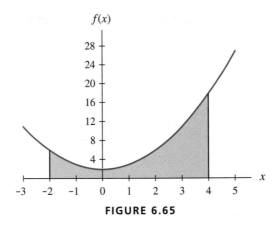

FIGURE 6.65

To find the area of this region in Sections 6.1, 6.2, and 6.3, we would have used sums of areas of rectangles. Now all we need to do is calculate the value of the definite integral, $\int_{-2}^{4}(x^2 + 2)\,dx$, by simply finding an antiderivative and subtracting the value of the antiderivative at -2 from the value at 4.

$$\int_{-2}^{4}(x^2 + 2)\,dx = F(4) - F(-2)$$

where $F(x)$ is an antiderivative of $x^2 + 2$. Here are the details of this process:

1. Find an antiderivative:

$$\int_{-2}^{4}(x^2 + 2)\,dx = \left(\frac{x^3}{3} + 2x\right)\Bigg|_{-2}^{4}$$

   This notation is used to indicate that we have found an antiderivative and now must evaluate it at 4 and -2 and then subtract the results.

2. Evaluate at upper and lower limits and then subtract:

$$= \left[\frac{4^3}{3} + 2(4)\right] - \left[\frac{(-2)^3}{3} + 2(-2)\right]$$

$$= 29\tfrac{1}{3} - \left(-6\tfrac{2}{3}\right) \approx 29.333 - (-6.667)$$

$$= 36$$

Thus the area of the region depicted in Figure 6.65 is exactly 36.

EXAMPLE 2   *Plant Growth*

Suppose a plant is 9 mm tall at the beginning of an experiment and is growing at a rate of $f(t) = 6t$ mm per day t days after the beginning of the experiment.

a. Find the specific antiderivative of $f(t)$ that gives the plant's height t days after the experiment began.

b. Find $\int_{0}^{2} f(t)\,dt$. Interpret your answer.

Solution:

a. The general antiderivative we seek is

$$F(t) = \frac{6t^2}{2} + C = 3t^2 + C$$

Using the fact that when $t = 0$ the plant is 9 mm tall, we have

$$F(0) = 9$$
$$3(0)^2 + C = 9$$
$$C = 9$$

Thus the specific antiderivative giving the plant's height is

$$F(t) = 3t^2 + 9 \text{ mm}$$

t days after the beginning of the experiment.

b. $\displaystyle\int_0^2 f(t)\, dt = F(t)\big|_0^2$

$$= \left(3t^2 + 9\right)\bigg|_0^2$$
$$= [3(2)^2 + 9] - [3(0)^2 + 9]$$
$$= (12 + 9) - 9$$
$$= 12 \text{ mm}$$

This is the area of the region between $f(t) = 6t$ and the horizontal axis from 0 to 2. It is the change in the plant's height during the first 2 days of the experiment. That is, the plant has grown 12 mm during the first 2 days of the experiment. ∎

Now consider the result of Example 1 if the plant had been 0.5 mm tall at the beginning of the experiment. The function giving the plant's height would then be $F(t) = 3t^2 + 0.5$ mm t days into the experiment. The definite integral calculation would be

$$\int_0^2 f(t)\, dt = F(t)\big|_0^2$$
$$= (3t^2 + 0.5)\big|_0^2$$
$$= [3(2)^2 + 0.5] - [3(0)^2 + 0.5]$$
$$= (12 + 0.5) - 0.5$$
$$= 12 \text{ mm}$$

Note that the initial height of the plant made no difference in the definite integral calculation. This is not a coincidence. If $F_1(x)$ and $F_2(x)$ are any two antiderivatives of $f(x)$, they differ only by a constant: $F_1(x) = F_2(x) + C$. Thus the change in $F_1(x)$ from a to b is

$$F_1(b) - F_1(a) = [F_2(b) + C] - [F_2(a) + C] = F_2(b) - F_2(a)$$

the same as the change in $F_2(x)$ from a to b. The constant term C always cancels out during the calculation.

> The constant term in an antiderivative does not affect definite integral calculations. If you are concerned only with finding change in a quantity, finding the constant in the antiderivative is not necessary.

We return to Example 4 in Section 6.5. Recall that we modeled the marginal cost function for toaster ovens as the exponential function

$$C'(x) = 47.638(0.9972)^x \text{ dollars per oven}$$

where x is the number of ovens produced per day.

Suppose that the current production level is 300 ovens per day and that the manufacturer wishes to increase production to 500 ovens per day. The definite integral $\int_{300}^{500} C'(x)\,dx = C(500) - C(300)$ gives the change in cost as a result of this increase. Finding the change requires two steps. First, find an antiderivative of $C'(x)$ (that is, find the cost function). Then evaluate the cost function at the two production levels and subtract the value at the lower limit from the value at the upper limit of the integral.

In Section 6.5 (using an unrounded model), we found the general antiderivative for $C'(x)$ to be

$$C(x) = \int C'(x)\,dx = -16{,}991.852(0.9972)^x + K$$

The constant K will not affect our calculations of change, so we set K to be 0. (Note that if we need to answer other questions about cost, we should use the information that the cost to produce 250 ovens is $12,000 to find the proper value of K, as was illustrated in Example 4.)

The definite integral is

$$\int_{300}^{500} C'(x)\,dx = C(x)\Big|_{300}^{500}$$

$$= C(500) - C(300)$$
$$= -16{,}991.852(0.9972)^{500} - [-16{,}991.852(0.9972)^{300}]$$
$$\approx \$3145$$

We conclude that when production is increased from 300 to 500 ovens per day, cost increases by approximately $3145.

Sums of Definite Integrals

Definite integral calculations for piecewise continuous functions require special care. In Example 1 of Section 6.2, we saw flow rates for a river during heavy rains; however, only some of the data were given. The complete available data[20] and a scatter plot are shown in Table 6.46 and Figure 6.66, respectively.

---

20. As reported in the *Reno Gazette Journal*, May 17, 1996, p. 4a.

TABLE 6.46

Hours since 11:45 a.m. Wednesday	Flow rate (cubic feet per hour)
0	2,826,000
4	2,710,800
8	3,002,400
12	3,852,000
16	5,292,000
20	7,524,000
23	6,624,000
27	5,760,000

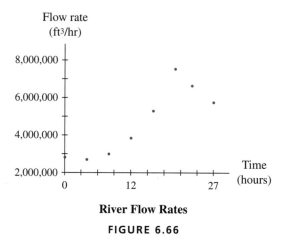

River Flow Rates

FIGURE 6.66

A piecewise continuous function is appropriate. We choose to divide the data at 20 hours and obtain the model[21]

$$f(h) = \begin{cases} 18{,}225h^2 - 135{,}334.3h + 2{,}881{,}542.9 \text{ ft}^3/\text{hr} & \text{when } 0 \le h \le 20 \\ 12{,}000h^2 - 816{,}000h + 19{,}044{,}000 \text{ ft}^3/\text{hr} & \text{when } 20 < h \le 27 \end{cases}$$

where h is the number of hours since 11:45 a.m. Wednesday.

To estimate the amount of water that flowed through the river from 11:45 a.m. Wednesday to 2:45 p.m. Thursday, 27 hours later, we calculate the value of the definite integral, $\int_0^{27} f(h)\,dh$. Note that the point of division for the model occurs in the interval from 0 to 27. For this reason, we cannot calculate the value of the definite integral simply by evaluating an antiderivative of $f(h)$ at 27 and 0 and subtracting.

Note that the area of the region shaded in Figure 6.67 is equal to the sum of the area of R_1 and the area of R_2.

---

21. When the entire data set is viewed, the three points on the right appear to follow a linear pattern. However, when viewed in isolation, the points are distinctively concave up. For this reason, we chose quadratic models for both portions of the data.

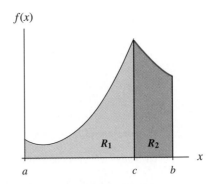

The Area of the Region from
a to b = Area of R_1 + Area of R_2

FIGURE 6.67

This figure illustrates the following property of integrals:

$$\int_a^b f(x)\,dx = \int_a^c f(x)\,dx + \int_c^b f(x)\,dx$$

It is this property that enables us to calculate definite integrals for piecewise continuous functions.

Returning to the river flow function, in order to calculate $\int_0^{27} f(h)\,dh$, we divide the integral into two pieces at the point where the model changes and sum the results.

$$\int_0^{27} f(h)\,dh = \int_0^{20} f(h)\,dh + \int_{20}^{27} f(h)\,dh$$

$$= \int_0^{20} (18{,}225h^2 - 135{,}334.3h + 2{,}881{,}542.9)\,dh +$$

$$\int_{20}^{27} (12{,}000h^2 - 816{,}000h + 19{,}044{,}000)\,dh$$

$$= \left(\frac{18{,}225}{3}h^3 - \frac{135{,}344.3}{2}h^2 + 2{,}881{,}542.9h\right)\Big|_0^{20} +$$

$$\left(\frac{12{,}000}{3}h^3 - \frac{816{,}000}{2}h^2 + 19{,}044{,}000h\right)\Big|_{20}^{27}$$

$$= (79{,}164{,}000 - 0) + (295{,}488{,}000 - 249{,}680{,}000)$$

$$= 79{,}164{,}000 + 45{,}808{,}000$$

$$= 124{,}972{,}000 \text{ cubic feet}$$

We estimate that during the first 20 hours, 79,164,000 ft³ of water flowed through the river. Between 20 and 27 hours, the volume of water was approximately 45,808,000 ft³. Summing these two values, we estimate that from 11:45 a.m. Wednesday to 2:45 p.m. Thursday, 124,972,000 ft³ of water flowed through the river.

In Example 3 in Section 6.3, we saw a rate-of-change function for per capita wine consumption that was both above and below the horizontal axis. In order to calculate the change in per capita wine consumption from the end of 1970 through 1990, we calculated the increase in wine consumption from the end of 1970 through

1984 (actually through $x \approx 83.97$) and subtracted the decrease in wine consumption from 1984 ($x \approx 83.97$) through 1990. In other words, we calculated the definite integral from 1970 ($x = 70$) through 1990 ($x = 90$) as

$$\int_{70}^{83.97} W(x)\,dx + \int_{83.97}^{90} W(x)\,dx \approx 1.34 - 0.45 = 0.89 \text{ gallon per person}$$

where the value of the integral from 83.97 to 90 is negative because $W(x)$ is below the horizontal axis. Because $W(x)$ is continuous from 70 to 90 (not piecewise continuous), we can calculate $\int_{70}^{90} W(x)\,dx$ by finding an antiderivative of $W(x)$, evaluating it at 90 and 70, and subtracting.

$$\int_{70}^{90} W(x)\,dx = \int_{70}^{90} [(1.243 \cdot 10^{-4})x^3 - 0.0314x^2 + 2.6174x - 71.977]\,dx$$

$$= [(3.1075 \cdot 10^{-5})x^4 - 0.01046667x^3 + 1.3087x^2 - 71.977x]\Big|_{70}^{90}$$

$$\approx -1468.83 - (-1469.72) = 0.89 \text{ gallon per person}$$

EXAMPLE 3   *Changing Sea Levels*

Scientists believe that the average sea level is dropping and has been for some 4000 years. They also believe that was not always the case. Estimated rates of change in the average sea level in meters per year during the past 7000 years are given in Table 6.47.[22]

TABLE 6.47

Time, t (thousands of years before present)	Rate of change of average sea level, $r(t)$ (meters/year)
-7	3.8
-6	2.6
-5	1.0
-4	0.1
-3	-0.6
-2	-0.9
-1	-1.0

a. Find a model $r(t)$ for the data.

b. Find the areas of the regions above and below the t-axis from $t = -7$ to $t = 0$. Interpret the areas in the context of sea level.

c. Find $\displaystyle\int_{-7}^{0} r(t)\,dt$, and interpret your answer.

22. Estimated from information in François Ramade, *Ecology of Natural Resources* (New York: Wiley, 1981).

Solution:

a. A quadratic model for the data is

$$r(t) = 0.14762t^2 + 0.35952t - 0.8 \text{ meters per year}$$

t thousand years from the present (past years are represented by negative numbers).

b. A graph of $r(t)$ is shown in Figure 6.68. The function crosses the t-axis at $t \approx -3.845$ thousand years.

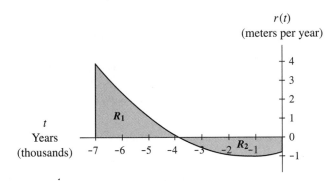

Rate of Change of Sea Level

FIGURE 6.68

The area of region R_1 (above the t-axis) is

$$\int_{-7}^{-3.845} (0.14762t^2 + 0.35952t - 0.8)\,dt = \left(\frac{0.14762}{3}t^3 + \frac{0.35952}{2}t^2 - 0.8t\right)\Bigg|_{-7}^{-3.845}$$

$$\approx 2.9365 - (-2.4694) \approx 5.4 \text{ meters}$$

The area of region R_2 (below the t-axis) is

$$-\int_{-3.845}^{0} (0.14762t^2 + 0.35952t - 0.8)\,dt = -\left(\frac{0.14762}{3}t^3 + \frac{0.35952}{2}t^2 - 0.8t\right)\Bigg|_{-3.845}^{0}$$

$$\approx -(0 - 2.9365) \approx 2.9 \text{ meters}$$

From 7000 years ago through 3845 years ago, average sea level rose by approximately 5.4 meters. From 3845 years ago to the present, average sea level fell by approximately 2.9 meters.

c. $\displaystyle\int_{-7}^{0} r(t)\,dt = \int_{-7}^{0} (0.14762t^2 + 0.35952t - 0.8)\,dt$

$$= \left(\frac{0.14762}{3}t^3 + \frac{0.35952}{2}t^2 - 0.8t\right)\Bigg|_{-7}^{0}$$

$$\approx 0 - (-2.4694) \approx 2.5 \text{ meters}$$

From 7000 years ago to the present, average sea level has risen approximately 2.5 meters. This result is the same as that obtained by subtracting the amount that sea level has fallen from the amount that it has risen:

$$5.4 \text{ meters} - 2.9 \text{ meters} = 2.5 \text{ meters} \qquad \blacksquare$$

The Fundamental Theorem gives us a technique for evaluating definite integrals using antiderivatives. However, we have seen rules for finding antiderivatives for very few classifications of functions. There are many other rules for antiderivatives, and there are some functions to which no antiderivative rule applies.

Consider the function that models the rate of worldwide[23] oil production:

$$p(x) = 0.018e^{-0.001x^2} \text{ billion barrels per year}$$

where $x = 0$ in the year 2000. Consider using this model to predict how much oil will be produced from the year 2000 through the year 2010.

To make this prediction, we find the value of $\int_0^{10} 0.018e^{-0.001x^2} dx$. This is not a function for which there is an antiderivative rule. However, recall that the definite integral was first defined as a limiting value of sums. We rely on that definition whenever an algebraic formula for the antiderivative of a function is not known.

Let us investigate the limiting value of sums beginning with 10 midpoint rectangles:

TABLE 6.48

n	Approximation of area
10	0.174189
20	0.174179
40	0.174177
80	0.174176
160	0.174176
320	0.174176
Trend $\approx$ 0.17418	

Approximately 174,180,000 barrels of oil will be produced from the year 2000 through the year 2010.

It is important for you to understand that if you are ever unable to find an algebraic formula for the antiderivative of a function, you can still estimate the value of a definite integral of that function by using the limiting value of sums of areas of rectangles.

6.6 Concept Inventory

- ■ $\int_a^b f(x) dx = F(b) - F(a)$
- ■ Finding a specific antiderivative by evaluating the constant C in the general antiderivative $\int f(x) dx = F(x) + C$.
- ■ $\int_a^b f(x) dx = \int_a^c f(x) dx + \int_c^b f(x) dx$

6.6 Activities

For each of Activities 1 through 7, determine which of the following is necessary to answer the question posed.

a. Find a general antiderivative (with unknown constant).

b. Find a specific antiderivative (solve for the constant).

c. Find a derivative.

23. Estimated from information in François Ramade, *Ecology of Natural Resources* (New York: Wiley, 1981).

1. Given a rate-of-change function for population and the population in a given year, find the population in year t.

2. Given a velocity function, determine the distance traveled from time a to time b.

3. Given a function, find its accumulation function from a to x.

4. Given a velocity function, determine acceleration at time t.

5. Given a rate-of-change function for population, find the change in population from year a to year b.

6. Given a function, find the area of the region between the function and the horizontal axis from a to b.

7. Given a function, find the slope of the tangent line at input a.

8. a. Find the accumulation function $\int_0^x 2e^t \, dt$.

   b. Use part a to find the value of $\int_0^3 2e^t \, dt$.

9. a. Find the accumulation function
   $$\int_1^x (t^3 - 2t + 3) \, dt.$$

   b. Use part a to find the value of
   $$\int_1^2 (t^3 - 2t + 3) \, dt.$$

10. a. Find a formula for $\int_2^x (4.63t^2 - 9.2t + 6) \, dt$.

    b. Use the formula from part a to find the area of the region between $f(t) = 4.63t^2 - 9.2t + 6$ and the horizontal axis from 2 to 3.

    c. If $f(t)$ is the rate of change of the population of a state, in thousands of people per year, and t is the number of years since 1975, what is the interpretation of the answer to part b?

11. a. Find the accumulation function for
    $f(t) = 18,000(0.974)^t + 1500$ from 0 to x.

    b. Use the formula in part a to find the area of the region between $f(t) = 18,000(0.974)^t + 1500$ and the horizontal axis from 0 to 10.

    c. If $f(t)$ is the rate of increase in a piece of real estate in dollars per year, where t is the number of years since the current owner purchased the property, what is the meaning of the answer to part b?

Evaluate the definite integrals in Activities 12 to 17.

12. $\int_{-1}^4 3 \, dx$

13. $\int_1^2 (3x - 7) \, dx$

14. $\int_3^{10} 4.298(1.036)^x \, dx$

15. $\int_1^2 \left(1 + \frac{1}{x} + \frac{1}{x^3}\right) dx$

16. $\int_0^{4.5} (6.27e^{-1.3296x} - 0.324x^3 + 1.59) \, dx$

17. $\int_{10}^{40} [427.705(1.043)^x - 413.226e^{-0.4132x}] \, dx$

In Activities 18 through 21:
   a. Graph the function $f(x)$ from a to b.
   b. Find the area of region between the graph and the x-axis from a to b.

18. $f(x) = -4x^{-2}; \quad a = 1, b = 4$

19. $f(x) = -1.3x^3 + 0.93x^2 + 0.49; \quad a = -1, b = 2$

20. $f(x) = \dfrac{9.295}{x} - 1.472; \quad a = 5, b = 10$

21. $f(x) = -965.27(1.079)^x; \quad a = 0.5, b = 3.5$

22. The air speed of a small airplane during the first 25 seconds of takeoff and flight can be modeled by
    $$v(t) = -940,602t^2 + 19,269.3t - 0.3 \text{ mph}$$
    t hours after takeoff.

    a. Find the value of $\int_0^{0.005} v(t) \, dt$.

    b. Interpret your answer in context.

23. From data taken from the 1994 *Statistical Abstract*, we model the rate of change of the percentage of the population of the United States living in the Great Lakes region from 1960 through 1990 as
    $$P(t) = (9.371 \cdot 10^{-4})t^2 - 0.141t + 5.083$$
    $$\text{percentage points per year}$$
    where t is the number of years since 1900. Find and interpret the value of $\int_{80}^{90} P(t) \, dt$.

24. The rate of change of the weight of a laboratory mouse t weeks (for $1 \le t \le 15$) after the beginning of an experiment can be modeled by the equation
    $$w(t) = \frac{13.785}{t} \text{ grams per week}$$
    Evaluate $\int_3^9 w(t) \, dt$, and interpret your answer.

25. A corporation's revenue flow rate can be modeled by

$$r(x) = 9.907x^2 - 40.769x + 58.492$$
million dollars per year

x years after the end of 1987. Evaluate $\int_0^5 r(x)\,dx$, and interpret your answer.

26. A dam develops a leak that ultimately results in a major fissure. The rate at which water is flowing out of the leak is given by

$$l(m) = e^{0.0062m^2} \text{ cubic feet per minute}$$

m minutes after the leak develops.
Evaluate $\int_0^{30} e^{0.0062m^2}\,dm$, and interpret your answer.

27. The production rate of a certain oil well can be modeled by

$$r(t) = 3.93546t^{3.55}e^{-1.35135t} \text{ thousand barrels per year}$$

t years after production begins.
Evaluate $\int_0^{10} r(t)\,dt$, and interpret your answer.

28. In Section 6.1 we saw the rate of change in the concentration of a drug modeled by

$$r(x) = \begin{cases} 1.708(0.845)^x & \text{when } 0 \le x \le 20 \\ \mu g/mL/day \\ 0.11875x - 3.5854 & \text{when } 20 < x \le 29 \\ \mu g/mL/day \end{cases}$$

where x is the number of days after the drug was administered. Determine the values of the following definite integrals, and interpret your answers.

a. $\int_0^{20} r(x)\,dx$ b. $\int_{20}^{29} r(x)\,dx$

c. $\int_0^{29} r(x)\,dx$

29. The rate of change of the snow pack in an area in the Northwest Territories in Canada can be modeled by

$$s(t) = \begin{cases} 0.00241t + 0.02905 \text{ cm} & \text{when} \\ \quad \text{of water per day} & 0 \le t \le 70 \\ 1.011t^2 - 147.971t + & \text{when} \\ \quad 5406.578 \text{ cm of} & 72 \le t \le 76 \\ \quad \text{water per day} \end{cases}$$

where t is the number of days since April 1.

a. Evaluate $\int_0^{70} s(t)\,dt$, and interpret your answer.

b. Evaluate $\int_{72}^{76} s(t)\,dt$, and interpret your answer.

c. Explain why it is not possible to find the value of $\int_0^{76} s(t)\,dt$.

30. The rate of change of the temperature during the hour and a half after the beginning of a thunderstorm is given by

$$T(h) = 9.48h^3 - 15.49h^2 + 17.38h - 9.87$$
°F per hour

where h is the number of hours since the storm began.

a. Graph the function $T(h)$ from $h = 0$ to $h = 1.5$.

b. Calculate the value of $\int_0^{1.5} T(h)\,dh$. Interpret your answer.

31. The rate of change of the temperature in a museum during a junior high school field trip can be modeled by

$$T(h) = 9.07h^3 - 24.69h^2 + 14.87h - 0.03$$
°F per hour

h hours after 8:30 a.m.

a. Find the area of the region between $T(h)$ and the h-axis that lies above the axis between 8:30 a.m. and 10:15 a.m. Interpret the answer.

b. Find the area of the region between $T(h)$ and the h-axis that lies below the axis between 8:30 a.m. and 10:15 a.m. Interpret the answer.

c. There are items in the museum that should not be exposed to temperatures greater than 73 °F. If the temperature at 8:30 a.m. was 71 °F, did the temperature exceed 73 °F between 8:30 a.m. and 10:15 a.m.?

32. The acceleration of a race car during the first 35 seconds of a road test is modeled by

$$a(t) = 0.024t^2 - 1.72t + 22.58 \text{ ft/sec}^2$$

where t is the number of seconds since the test began.

a. Graph $a(t)$ from $t = 0$ to $t = 35$.

b. Write the definite integral notation representing the amount by which the car's speed increased during the road test. Calculate the value of the definite integral.

33. The estimated production rate of marketed natural gas, in trillion cubic feet per year, in the United

States (excluding Alaska) from 1900 through 1960 is shown in Table 6.49.[24]

TABLE 6.49

Year	Estimated production rate (trillions of cubic feet per year)
1900	0.1
1910	0.5
1920	0.8
1930	2.0
1940	2.3
1950	6.0
1960	12.7

a. Find a model for the data in Table 6.49.

b. Use the model to estimate the total production of natural gas from 1940 through 1960.

c. Give the definite integral notation for your answer to part *b*.

34. Many businesses spend large sums of money each year on advertising in order to stimulate sales of their products. The data given in Table 6.50 show the approximate increase in sales (in thousands of dollars) that an additional $100 spent on advertising, at various levels, can be expected to generate.

a. Fit a model to these data.

b. Use the model in part *a* to determine a model for the total sales revenue $R(x)$ as a function of the amount x spent on advertising. Use the fact

that revenue is approximately 877 thousand dollars when $5000 is spent on advertising.

c. Find the point where returns begin to diminish for sales revenue.

d. The managers of the business are considering an increase in advertising expenditures from the current level of $8000 to $13,000. What effect could this decision have on sales revenue?

35. Table 6.51 shows the marginal cost to produce one more compact disc, given various hourly production levels.

TABLE 6.51

Production (CDs per hour)	Cost of an additional CD
100	$5
150	$3.50
200	$2.50
250	$2
300	$1.60

a. Fit an appropriate model to the data.

b. Use your model from part *a* to derive an equation that specifies production cost $C(x)$ as a function of the number x of CDs produced. Use the fact that it costs approximately $750 to produce 150 CDs in a 1-hour period.

c. Calculate the value of $\int_{200}^{300} C'(x)\,dx$. Interpret your answer.

TABLE 6.50

Advertising expenditures (hundreds of dollars)	Revenue increase due to an extra $100 advertising (thousands of dollars)
25	5
50	60
75	95
100	105
125	104
150	79
175	34

24. From information in *Resources and Man*, National Academy of Sciences, 1969, p. 165.

Chapter 6   Summary

Results of Change

Derivatives enable us to describe, determine, and analyze change, but sooner or later we must confront the accumulated results of change. The accumulated results of change are best understood in geometric terms, as areas of regions between the rate-of-change function (the derivative) and the horizontal axis. We can approximate the areas of the regions of interest by summing areas of rectangular regions.

Approximating Area

There are several ways to approximate areas of regions between graphs of functions and the horizontal axis: left rectangles, right rectangles, and midpoint rectangles. Of these three, midpoint rectangles usually produce the most accurate approximations. Approximating area with trapezoids is also an option, but although trapezoid approximations are usually better than left- or right-rectangle approximations, they are generally less accurate than approximations with midpoint rectangles.

Limits of Sums

If we agree to estimate the accumulated change in a quantity by calculating areas of regions with sums of midpoint-rectangle areas, then it is evident that the estimates generally improve as we increase the number of approximating rectangles. If we imagine that there is no bound on the number of rectangles, then we see that the area of a region between the graph of a continuous, non-negative function $f(x)$ and the horizontal axis from a to b is given by a limit of sums:

$$\text{Area} = \lim_{n \to \infty} [f(x_1) + f(x_2) + \cdots + f(x_n)] \, \Delta x$$

Here, the points $x_1, x_2, \ldots, x_n$ are the midpoints of n rectangles of width $\Delta x = \frac{b-a}{n}$ between a and b.

More generally, we consider the limit applied to an arbitrary continuous function $f(x)$ over the interval from a to b and call this limit the definite integral of $f(x)$ from a to b. In symbols, we write

$$\int_a^b f(x) \, dx = \lim_{n \to \infty} [f(x_1) + f(x_2) + \cdots + f(x_n)] \, \Delta x$$

When $f(x)$ is a negative function, the definite integral $\int_a^b f(x) \, dx$ is the negative of the area of the region between the graph of $f(x)$ and the x-axis from a to b.

When $f(x)$ is sometimes positive and sometimes negative, the integral $\int_a^b f(x) \, dx$ is the area of the region lying under the graph of $f(x)$ and above the x-axis from a to b minus the area of the region lying above the graph of $f(x)$ and below the x-axis from a to b.

Accumulation Functions

What is an accumulation function? It is simply an integral of the form $\int_a^x f(t) \, dt$ where the upper limit x is regarded as being a variable. The value of this integral depends on the variable x, so the accumulation function $\int_a^x f(t) \, dt$ is actually a function of x: $F(x) = \int_a^x f(t) \, dt$. This function gives us a formula for calculating accumulated change in a quantity. By conducting geometric investigations, we were able to uncover accumulation formulas for lines. We also sketched accumulation function graphs for more complex functions.

The Fundamental Theorem of Calculus

The Fundamental Theorem sets forth the fundamental connection between the two main concepts of calculus, the derivative and the integral. It tells us that for any continuous function $f(x)$,

$$\frac{d}{dx} \int_a^x f(t) \, dt = f(x)$$

In other words, the derivative of an accumulation function of $f(t)$ is precisely $f(x)$. If we imagine differentiation and integration as processes, then the process of integration (from a fixed value a to a variable x) followed by the process of differentiation takes us back to where we began, namely $f(x)$.

If we reverse the order of these two processes and begin by differentiating first, then we obtain the starting function $f(x)$ plus a constant.

$$\int_a^x f'(t) \, dt = f(x) + C$$

Antiderivatives

A function $G(x)$ is an antiderivative of $f(x)$ if $G'(x) = f(x)$. Because the derivative of $\int_a^x f(t)\,dt$ is $f(x)$, we see that $\int_a^x f(t)\,dt$ is an antiderivative of $f(x)$. Each continuous function $f(x)$ has infinitely many antiderivatives, but any two differ by only a constant. If $F(x)$ is an antiderivative of $f(x)$, then any other antiderivative $G(x)$ can be obtained from $F(x)$ by adding a constant: $G(x) = F(x) + C$. Although the constant C is usually unspecified, it represents the vertical shift required to move the graph of $F(x)$ to the graph of $G(x)$.

The Symbol $\int$

We used the symbol $\int$ in three ways. How are these uses related and how do they differ? The definite integral $\int_a^b f(t)\,dt$ is a number that can be interpreted as the area of the region between the graph of $f(t)$ and the horizontal axis from a to b. The accumulation function $\int_a^x f(t)\,dt$ is a function of x; for each input value of x, it gives the accumulated change in the quantity for which $f(t)$ is a rate of change as t changes from a to x. Finally, the general antiderivative of a function $f(x)$ is written $\int f(x)\,dx$; it is a set of functions whose derivatives are $f(x)$ and involves an arbitrary constant C. By the Fundamental Theorem of Calculus, the accumulation function $\int_a^x f(t)\,dt$ is an antiderivative of $f(x)$.

Recovering Functions from Rates of Change

The formulas for antiderivatives that were developed in Section 6.5 enable us to recover an unknown function $f(x)$ quickly from its rate of change $f'(x)$ because

$$\int f'(x)\,dx = f(x) + C$$

In real-life situations, the constant C must be determined from information about a known value of the function $f(x)$.

An important aspect of recovering a function $f(x)$ from its rate of change $f'(x)$ is to recover the units used to describe the function from the units associated with its rate of change. Ordinarily, this is not difficult; $f'(x)$ is usually expressed in terms of (output units) per (input unit), where the input and output units are those of $f(x)$.

The Definite Integral

The definite integral was defined as a limit of sums.

$$\int_a^b f(x)\,dx = \lim_{n \to \infty} [f(x_1) + f(x_2) + \cdots + f(x_n)]\,\Delta x$$

where $x_1, x_2, \ldots, x_n$ are the midpoints of n equally spaced subintervals of length $\Delta x = \frac{b-a}{n}$ between a and b.

We initially found the value of a definite integral by examining the trend in a sequence of sums of the form $f(x_1)\,\Delta x + \ldots + f(x_n)\,\Delta x$ as n increases. But the Fundamental Theorem of Calculus enabled us to show that when $f(x)$ is a smooth, continuous function the definite integral $\int_a^b f(x)\,dx$ can be evaluated by

$$\int_a^b f(x)\,dx = F(b) - F(a)$$

where $F(x)$ is any antiderivative of $f(x)$.

The Fundamental Theorem of Calculus ensures that each continuous function $f(x)$ does indeed have an antiderivative. Thus, to the extent that we can actually obtain a closed-form algebraic expression[25] in x for an antiderivative $F(x)$, we can easily evaluate a definite integral. In situations where an antiderivative cannot be found, we use one of the approximation techniques discussed in Sections 6.1 through 6.3. In most real-life situations that are not directly associated with science or engineering, it is possible to use reasonably accurate approximations to the value of a definite integral. These approximations can usually be obtained with a calculator or computer.

---

25. Indeed, there is a large number of antiderivative formulas for highly specialized classes of functions as well as a sizable collection of techniques for finding antiderivatives. However, it is easy to give simple examples of functions, such as $f(x) = e^{-x^2}$ and $g(x) = \sqrt{1 + x^4}$, for which it is known that no closed-form algebraic expression for an antiderivative exists. In such cases, we must resort to numerical approximations.

Chapter 6   Review Test

1. The rate at which crude oil flows through a pipe into a holding tank can be modeled by

   $r(t) = 10(-3.2t^2 + 93.3t + 50.7)$ cubic feet per minute

   where t is the number of minutes the oil has been flowing into the tank.

   a. Sketch a graph of $r(t)$ for t between 0 and 25 minutes.

   b. Use five trapezoids to estimate the area of the region between $r(t)$ and the t-axis from 0 to 25 minutes. Sketch the trapezoids on the graph you drew in part a.

   c. Interpret your answer to part b.

2. A hurricane is 300 miles off the east coast of Florida at 1 a.m. The speed at which the hurricane is moving toward Florida is measured each hour. Speeds between 1 a.m. and 5 a.m. are recorded in Table 6.52.

   TABLE 6.52

Time	Speed (mph)
1 a.m.	15
2 a.m.	25
3 a.m.	35
4 a.m.	38
5 a.m.	40

   a. Find a model for the data.

   b. Use a limiting value of sums of areas of midpoint rectangles to estimate how far (to the nearest tenth of a mile) the hurricane traveled between 1 a.m. and 5 a.m. Begin with five rectangles, doubling the number each time until you are confident that you know the limiting value.

3. The graph shown in Figure 6.69 depicts the rate of change in the weight of someone who diets for 20 weeks.

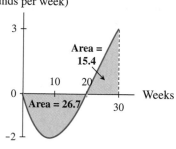

FIGURE 6.69

   a. What does the area of the shaded region beneath the horizontal axis represent?

   b. What does the area of the shaded region above the horizontal axis represent?

   c. Is this person's weight at 30 weeks more or less than it was at 0 weeks? How much more or less?

   d. If $w(t)$ is the function shown in Figure 6.69, sketch a graph of $\int_0^x w(t)\,dt$ on the axes provided. Label units on both axes as well as values on the vertical axis.

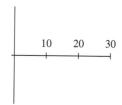

FIGURE 6.70

   e. What does the graph in part d represent?

4. Consider again the model for the flow rate of crude oil into a holding tank.

   $r(t) = 10(-3.2t^2 + 93.3t + 50.7)$ ft^3/minute

   after t minutes.

a. If the holding tank had 5000 ft^3 of oil in it at $t = 0$, find a model for the amount of oil in the tank after t minutes.

b. Use your model in part a to find how much oil flowed into the tank during the first 10 minutes.

c. If the capacity of the tank is 150,000 ft^3, according to the model, how long can the oil flow into the tank before the tank is full?

5. Ten thousand dollars invested in a mutual fund is growing at a rate of

$$a(x) = 840(1.08763)^x \text{ dollars per year}$$

x years after it was invested.

a. Determine the value of $\int_0^{2.75} a(x)\,dx$.

b. Interpret your answer to part a.

Project 6.1

Acceleration, Velocity, and Distance

Setting

According to tests conducted by *Road and Track*, a 1993 Toyota Supra Turbo accelerates from 0 to 30 mph in 2.2 seconds and travels 1.4 miles (1320 feet) in 13.5 seconds, reaching a speed of 107 mph. *Road and Track* reported the data given in Table 6.53.

TABLE 6.53

Time (seconds)	Speed reached from rest (mph)
0	0
2.2	30
2.9	40
4.0	50
5.0	60
6.5	70
8.0	80
9.0	90
11.8	100

Tasks

1. Convert the speed data to feet per second, and find a quadratic model for velocity (in feet per second) as a function of time (in seconds). Discuss how close your model comes to predicting the 107 mph reached after 13.5 seconds.

2. Add the data point for 13.5 seconds, and find a quadratic model for velocity $v(t)$.

3. Use four rectangles and your model from Task 2 to estimate the distance traveled during acceleration from rest to a speed of 50 mph and the distance traveled during acceleration from a speed of 50 mph to a speed of 100 mph. Repeat the estimate using twice as many rectangles.

4. Use nine rectangles to approximate the distance traveled during the first 13.5 seconds. How close is your estimate to the reported value?

5. Find the distances traveled during

   a. acceleration from rest to a speed of 50 mph

   b. acceleration from a speed of 50 mph to a speed of 100 mph

   c. the first 13.5 seconds of acceleration

   Compare these answers to your estimates in Tasks 3 and 4. Explain how estimating with areas of rectangles is related to calculating the definite integral.

Reporting

Prepare a written report of your work. Include scatter plots, models, graphs, and discussions of each of the above tasks.

Project 6.2

Estimating Growth

Setting

Table 6.54[26] lists the rate of growth of a typical male from birth to 18 years.

TABLE 6.54

Age (years)	Rate of growth (centimeters per year)
0	18.0
1	16.0
2	12.0
3	6.5
4	6.0
5	6.25
6	6.5
7	6.0
8	5.75
9	5.25
10	5.0
11	4.5
12	5.0
13	6.75
14	9.0
15	7.0
16	2.0
17	0.75
18	0.5

26. Based on data collected in the Berkeley Growth Study.

Tasks

1. Use the data and right rectangles to approximate the height of a typical 18-year-old male.

2. Sketch a smooth, continuous curve over a scatter plot of the data. Remember that such a curve should exhibit only the curvature implied by the data. Find a piecewise model for the data. Use no more than three pieces.

3. Use your piecewise model and limits of sums to approximate the height of a typical 18-year-old male. Convert centimeters to feet and inches, and compare your answer to the estimate you obtained using right rectangles. Which is likely to be the more accurate approximation? Why?

4. Use your piecewise model and what you know about definite integrals to find the height of a typical 18-year-old male in feet and inches. Compare your answer with the better of the approximations you obtained in Task 3.

5. Randomly choose ten 18-year-old male students, and determine their heights. (Include your data—names are not necessary, only the heights.) Discuss your selection process and why you feel that it is random. Find the average height of the 18-year-old males in your sample. Compare this average height with your answer to Task 4. Discuss your results.

6. Refer to your sketch of the rate-of-growth graph in Task 2, and draw a possible graph of the height of a typical 18-year-old male from birth to age 18.

Reporting

Prepare a report that presents your findings in Tasks 1 through 6. Explain the different methods that you used, and discuss why these methods should all give similar results. Attach your mathematical work as an appendix to your report.

Analyzing Accumulated Change: More Applications of Integrals

Chapter 6 established that the accumulated results of change are limiting values of approximating sums. These are known to us as definite integrals and can be expressed in geometric terms as areas of regions between a rate-of-change function and the horizontal axis. Our work with accumulation functions led us to develop the Fundamental Theorem of Calculus, which in turn provided a simple method for evaluating definite integrals using antiderivatives.

In Chapter 7 we consider the accumulation of change that is expected to occur perpetually, as well as the difference of two accumulated changes. In doing so, we examine several applications of integrals in economics, business, and biology. We will also use integrals to calculate averages.

In business and economics there are many examples of accumulated change. For instance, a CEO of a clothing corporation is interested in the accumulation of sales and concerned about the accumulation of costs. The difference between these two accumulations is the accumulation of profit.

If the corporation were up for sale, a prospective buyer would consider the value of the corporation to be the present value of the expected future profits. An economist might be interested in the total social gain that can be expected from selling clothing at a certain market price.

In what other fields might there exist applications where the accumulation of change could be estimated by using the area of regions that lie between two curves? In what fields might there be questions about the accumulation of change as the input variable grows infinitely large?

■ 7.1   Differences of Accumulated Changes

The idea of accumulated change was introduced in Chapter 6. We saw that when given a continuous or piecewise continuous rate-of-change function, we obtain the accumulation of change over an interval by finding the limiting value of sums of areas of rectangles. In Section 6.3, we defined this limiting value as the definite integral

$$\int_a^b f(x)\,dx = \lim_{n\to\infty}\left[f(x_1) + f(x_2) + \cdots + f(x_n)\right]\Delta x$$

We used accumulation functions to present the Fundamental Theorem of Calculus, which introduced the concept of an antiderivative. We saw that if $F(x)$ is an antiderivative of a smooth, continuous function $f(x)$, then

$$\int_a^b f(x)\,dx = F(b) - F(a)$$

As we proceeded through Chapter 6, we considered many applications of the definite integral. All of these applications had two characteristics in common: (1) the accumulation could be represented as the area (or negative of the area) of a region between the graph of a function and the horizontal axis, and (2) the accumulation was considered to take place between two finite input values.

Now we turn our attention to the difference of two accumulated changes. This difference can often be thought of as the area of a region between two curves. For example, suppose the number of patients admitted to a large inner-city hospital is changing by

$$a(h) = 0.0145h^3 - 0.549h^2 + 4.85h + 8.00 \text{ patients per hour}$$

h hours after 3 a.m. We find the approximate number of patients admitted between 7 a.m. ($h = 4$) and 10 a.m. ($h = 7$) as

$$\int_4^7 a(h)\,dh = \int_4^7 (0.0145h^3 - 0.549h^2 + 4.85h + 8.00)\,dh$$

$$= (0.003625h^4 - 0.183h^3 + 2.425h^2 + 8.00h)\Big|_4^7$$

$$\approx 120.760 - 60.016 \approx 61 \text{ patients}$$

Graphically, this value is the area of the region between the $a(h)$ curve and the horizontal axis from 4 to 7. (See Figure 7.1.)

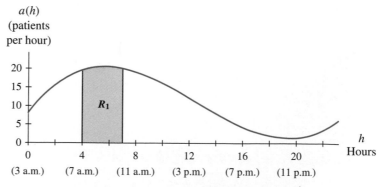

Area of R_1 is the number of patients admitted
between 7 a.m. and 10 a.m.

FIGURE 7.1

Now suppose that the rate at which patients are discharged is modeled by

$$y(h) = \begin{cases} -0.028h^3 + 0.528h^2 + 0.056h - 1.5 & \text{when } 4 \leq h \leq 17 \\ \text{patients per hour} \\ 0 \text{ patients per hour} & \text{when } 0 \leq h < 4 \text{ and } 17 < h \leq 24 \end{cases}$$

where h is the number of hours after 3 a.m. The approximate number of patients discharged between 7 a.m. and 10 a.m. is calculated as

$$\int_4^7 y(h)\,dh = \int_4^7 (-0.028h^3 + 0.528h^2 + 0.056h - 1.5)\,dh$$

$$= (-0.007h^4 + 0.176h^3 + 0.028h^2 - 1.5h)\Big|_4^7$$

$$= 34.433 - 3.92 \approx 31 \text{ patients}$$

Graphically, this value is the area of the region between the $y(h)$ curve and the horizontal axis from 4 to 7. (See Figure 7.2.)

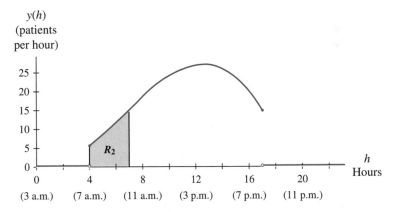

Area of R_2 is the number of patients discharged
between 7 a.m. and 10 a.m.

FIGURE 7.2

The net change in the number of patients at the hospital from 7 a.m. to 10 a.m. is the difference between the number of patients admitted and the number discharged between 7 a.m. and 10 a.m. That is,

$$\text{Change in the number of patients from 7 a.m. to 10 a.m.} = \int_4^7 a(h)\,dh - \int_4^7 y(h)\,dh$$

$$\approx 60.744 - 30.513$$

$$\approx 30 \text{ patients}$$

Geometrically, we represent this value as the area of the region below the graph of $a(h)$ and above the graph of $y(h)$ from 4 to 7. (See Figure 7.3.)

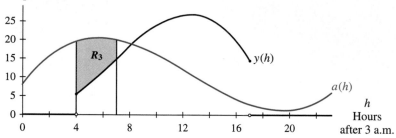

Patients admitted
and discharged
(patients per hour)

Area of R_3 is the net change in hospital patients
between 7 a.m. and 10 a.m.

FIGURE 7.3

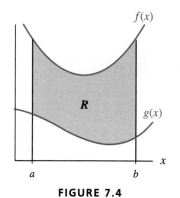

FIGURE 7.4

In general, when we want to find the area of a region that lies below one curve $f(x)$ and above another curve $g(x)$ from a to b (as in Figure 7.4), we calculate it as

$$\text{Area of the region between } f(x) \text{ and } g(x) = \text{area beneath } f(x) - \text{area beneath } g(x)$$

$$= \int_a^b f(x)\,dx - \int_a^b g(x)\,dx$$

Using the Constant Multiplier Rule and the Sum Rule for antiderivatives, we obtain

$$\text{Area of the region between } f(x) \text{ and } g(x) = \int_a^b [f(x) - g(x)]\,dx$$

Note that when $f(x)$ and $g(x)$ are obtained by fitting curves to data, the input variable of each function must represent the same quantity measured in the same units.

Area Between Two Curves

If the graph of $f(x)$ lies above the graph of $g(x)$ from a to b, then the area of the region between the two curves from a to b is given by

$$\int_a^b [f(x) - g(x)]\,dx$$

EXAMPLE 1   *Tire Manufacturers*

A major European tire manufacturer has seen its sales from tires skyrocket since 1974. A model for the rate of change of sales (in U.S. dollars) accumulated since 1974 is

$$s(t) = 3.7(1.19376)^t \text{ million dollars per year}$$

where t is the number of years since the end of 1974.

At the same time, an American tire manufacturer's rate of change of sales accumulated since 1974 can be modeled by

$$a(t) = 0.04t^3 - 0.54t^2 + 2.5t + 4.47 \text{ million dollars per year}$$

where t is the number of years since the end of 1974.

a. According to the models given, by how much did the amount of accumulated sales differ for these two companies from the end of 1980 through 1988?

b. Interpret the answer to part a geometrically.

Solution:

a. First, determine whether one graph lies above the other on the interval in the question. The two rate-of-change functions are graphed in Figure 7.5.

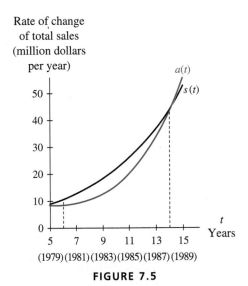

FIGURE 7.5

The graph of $s(t)$ is above the graph of $a(t)$ for all input values between 6 and 14, so the difference in the two companies' accumulated sales is

$$\begin{pmatrix} \text{Difference} \\ \text{in total sales} \end{pmatrix} = \begin{pmatrix} \text{accumulated sales for} \\ \text{the European company} \end{pmatrix} - \begin{pmatrix} \text{accumulated sales for} \\ \text{the American company} \end{pmatrix}$$

$$= \int_6^{14} s(t)\, dt - \int_6^{14} a(t)\, dt$$

$$= \int_6^{14} [s(t) - a(t)]\, dt$$

$$= \int_6^{14} [3.7(1.19376)^t - (0.04t^3 - 0.54t^2 + 2.5t + 4.47)]\, dt$$

$$= \left[\frac{3.7(1.19376)^t}{\ln 1.19376} - 0.01t^4 + 0.18t^3 - 1.25t^2 - 4.47t \right]\Bigg|_6^{14}$$

$$\approx 36.9660 \text{ million dollars}$$

The European company's accumulated sales were approximately $37,000,000 more than the American company's sales between 1980 and 1988.

b. The difference in the accumulated sales for the two companies is represented as the area of the region between the $s(t)$ curve and the $a(t)$ curve from $t = 6$ to $t = 14$. (See Figure 7.6.) As we saw in part *a*, this area is approximately $37 million.

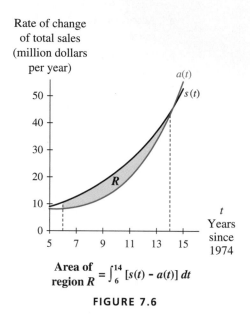

$$\text{Area of region } R = \int_6^{14} [s(t) - a(t)]\, dt$$

FIGURE 7.6   ■

In both the hospital example and the tire example, we saw two functions that intersect. So far we have considered intervals that do not include the point or points of intersection of the curves. What would happen if the interval in which we were interested included the intersection point?

Let us consider the two tire companies from Example 1. What is the difference in the accumulated sales for the two companies from 1984 through 1994? As shown in Figure 7.7, the two rate-of-change functions $s(t)$ and $a(t)$ cross close to the middle of 1979 and cross again near the beginning of 1989.

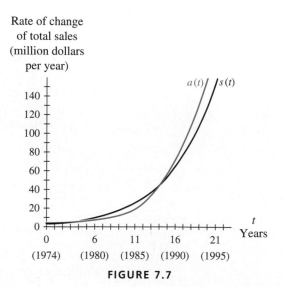

FIGURE 7.7

If we set $s(t) = 3.7(1.19376)^t$ equal to $a(t) = 0.04t^3 - 0.54t^2 + 2.5t + 4.47$ and solve for t, we find that the two functions intersect when $t \approx 4.657$ (in 1979) and when $t \approx 14.242$ (in 1989), as well as when $t \approx -0.372$. Accumulated sales were greater for the European company than for the American company from $t \approx 4.657$ to $t \approx 14.242$. After $t \approx 14.242$, the American company saw greater accumulated sales than the European company.

From the beginning of 1985 ($t = 10$) through most of the first quarter of 1989 ($t \approx 14.242$), the European company accumulated approximately

$$\int_{10}^{14.242} [s(t) - a(t)] \, dt = 18.5383 \text{ million dollars}$$

more in sales than the American company. This is the area of region R_1 in Figure 7.8.

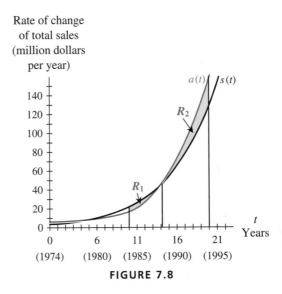

FIGURE 7.8

From close to the end of the first quarter of 1989 ($t \approx 14.242$) through 1994 ($t = 20$), the American company accumulated approximately

$$\int_{14.242}^{20} [a(t) - s(t)] \, dt = 79.4076 \text{ million dollars}$$

more in sales than the European company. This is the area of region R_2 in Figure 7.8.

In order to calculate the estimated difference in accumulated sales between the two companies from 1984 through 1994, we subtract the portion where the American company's accumulated sales were greater from the portion where the European company's accumulated sales were greater. That is,

$$\begin{aligned} \text{Difference in} \atop \text{accumulated sales} &= \int_{10}^{14.242} [s(t) - a(t)] \, dt - \int_{14.242}^{20} [a(t) - s(t)] \, dt \\ &\approx 18.5383 \text{ million dollars} - 79.4076 \text{ million dollars} \\ &\approx -60.869 \text{ million dollars} \end{aligned}$$

The European company's accumulated sales were nearly 61 million dollars less than the American company's accumulated sales over the years considered.

If we use the Constant Multiplier Rule and the Sum Rule for antiderivatives, we see that we did not need to split the interval from 10 to 20 into two intervals.

$$\begin{aligned}
\text{Difference in} \atop \text{accumulated sales} &= \int_{10}^{14.242} [s(t) - a(t)]\,dt - \int_{14.242}^{20} [a(t) - s(t)]\,dt \\
&= \int_{10}^{14.242} [s(t) - a(t)]\,dt + \int_{14.242}^{20} [-a(t) + s(t)]\,dt \\
&= \int_{10}^{14.242} [s(t) - a(t)]\,dt + \int_{14.242}^{20} [s(t) - a(t)]\,dt \\
&= \int_{10}^{20} [s(t) - a(t)]\,dt
\end{aligned}$$

This is true in general—whenever you wish to find the difference between accumulated change for two continuous rate-of-change functions, you can calculate the definite integral of the difference of the functions, regardless of where the functions intersect.

Difference of Two Accumulated Changes

If $f(x)$ and $g(x)$ are two continuous rate-of-change functions, then the difference between the accumulated change of $f(x)$ from a to b and the accumulated change of $g(x)$ from a to b is

$$\int_a^b [f(x) - g(x)]\,dx$$

It is important to note, however, that the integral $\int_a^b [f(x) - g(x)]\,dx$ may not represent the total area between $f(x)$ and $g(x)$ from a to b. Note that in the tire sales example, the area of the regions between the two curves from 10 to 20 is $18.538 + 79.408 = 97.946$, whereas the difference in accumulated sales is $18.538 - 79.408 = -60.869$.

If the two rate-of-change functions intersect somewhere in the interval from a to b, then the difference between their accumulated changes is *not* the same as the total area of the regions between the two curves.

EXAMPLE 2   *Area Between Two Curves*

The functions $f(x) = 0.3x^3 - 3.3x^2 + 9.6x + 3.3$ and $g(x) = -0.15x^2 + 2.03x + 3.33$ are shown in Figure 7.9.

a. Determine the input values marked A and B.

b. Find the areas of regions R_1, R_2, and R_3.

c. Find the value of $\int_1^7 [f(x) - g(x)]\,dx$.

d. What is the total area of the regions between the two functions from 1 to 7?

Solution:

a. By solving the equation $f(x) = g(x)$ for x, we find that $A \approx 3.715$ and $B \approx 6.781$. (The functions also intersect at $x \approx 0.004$).

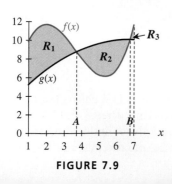

FIGURE 7.9

b. Area $R_1 = \int_1^A [f(x) - g(x)]\, dx \approx 9.797$.

Area $R_2 = \int_A^B [g(x) - f(x)]\, dx \approx 7.557$.

Area $R_3 = \int_B^7 [f(x) - g(x)]\, dx \approx 0.160$.

c. $\int_1^7 [f(x) - g(x)]\, dx = 2.4$

d. Using unrounded values, we find the area to be approximately

$$9.797 + 7.557 + 0.160 \approx 17.513$$ ■

7.1 Concept Inventory

■ Area(s) of region(s) between two curves
■ Differences of accumulated changes

7.1 Activities

For Activities 1 through 4:

a. Sketch the functions $f(x)$ and $g(x)$ on the same axes.

b. Shade the region between $f(x)$ and $g(x)$ from a to b.

c. Calculate the area of the shaded region.

1. $f(x) = 0.25x^2 + 2$ $a = 0$
    $g(x) = -2x + 24$ $b = 6$

2. $f(x) = 10(0.85)^x$ $a = 2$
    $g(x) = 6(0.75)^x$ $b = 10$

3. $f(x) = x^2 - 4x + 10$ $a = 1$
    $g(x) = 2x^2 - 12x + 14$ $b = 7$

4. $f(x) = e^x$ $a = 0$
    $g(x) = e^{-x}$ $b = 2$

For Activities 5 through 8:

a. Sketch the functions $f(x)$ and $g(x)$ on the same axes.

b. Find the input value(s) at which $f(x)$ and $g(x)$ intersect.

c. Shade the region(s) between $f(x)$ and $g(x)$ from a to b.

d. Calculate the difference in the area of the region between $f(x)$ and the horizontal axis and the area of the region between $g(x)$ and the horizontal axis from a to b.

e. Calculate the total area of the shaded region(s).

5. $f(x) = e^{0.5x}$ $a = 0.5$
    $g(x) = \dfrac{2}{x}$ $b = 3$

6. $f(x) = 0.25x - 3$ $a = 15$
    $g(x) = 14(0.93)^x$ $b = 50$

7. $f(x) = x + 1.5$ $a = -5$
    $g(x) = 2(1.25)^x$ $b = 8$

8. $f(x) = -0.25x^2 + 1.9x + 1.8$ $a = 0$
    $g(x) = -0.08x^3 + 0.5x^2 + 0.3x + 2.0$ $b = 10$

9. Figure 7.1.1 shows $r'(x)$, the rate of change of revenue, and $c'(x)$, the rate of change of costs (both in thousands of dollars per thousand dollars of capital investment) associated with the production of solid wood furniture as functions of the amount (in thousands of dollars) invested in capital. The area of the shaded region is 1329.

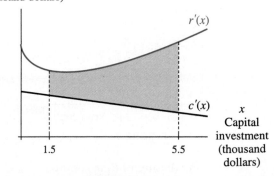

Rates of change of cost and revenue (thousand dollars per thousand dollars)

FIGURE 7.1.1

a. Interpret the area in the context of furniture manufacturing.

b. Write a mathematical equation for the area of the shaded region.

10. Figure 7.1.2 depicts the rate of change of total revenue $R'(x)$ (in billions of dollars per year) and the rate of change of total cost $C'(x)$ (in billions of dollars per year) of a company. The area of the shaded region is 126.5.

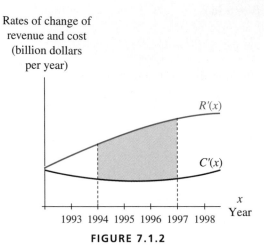

Rates of change of
revenue and cost
(billion dollars
per year)

FIGURE 7.1.2

a. Interpret the area in context.

b. Write a mathematical equation for the area of the shaded region.

11. The rate of change of total revenue $R'(x)$ (in dollars per month) and the rate of change of total cost $C'(x)$ (in dollars per month) for a company are shown in Figure 7.1.3. Region R_1 has area 369,000, and region R_2 has area 972,000.

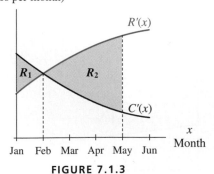

Rates of change of
revenue and cost
(dollars per month)

FIGURE 7.1.3

a. Interpret the area of R_1 in context.

b. Interpret the area of R_2 in context.

c. Find the profit realized by the company from the beginning of January through May.

d. Explain why the answer to part c is not the sum of the areas of the two regions.

12. Figure 7.1.4 depicts the functions $c(t)$, the rate at which people contract a virus during an epidemic, and $r(t)$, the rate at which people recover from the virus, where t is the number of days after the epidemic begins.

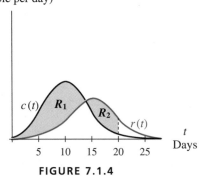

Rates of change of
contraction and recovery
(people per day)

FIGURE 7.1.4

a. Interpret the area of region R_1 in the context of the epidemic.

b. Interpret the area of region R_2 in the context of the epidemic.

c. Explain how you could use a definite integral to find the number of people who contracted the virus since day 0 but have not recovered by day 20.

13. Table 7.1 shows the time it takes for a Toyota Supra and a Porsche 911 Carerra to accelerate from 0 mph to the speeds given.[1]

a. Find models for the speed of each car, given the number of seconds after starting from 0 mph. (*Hint*: Add the point (0, 0), and convert miles per hour to feet per second before modeling.)

b. How much farther than a Porsche 911 Carerra does a Toyota Supra travel during the first 10 seconds, assuming that both cars begin from a standing start?

1. *Road and Track.*

TABLE 7.1

Toyota Supra		Porsche 911 Carerra	
Time (seconds)	Speed (mph)	Time (seconds)	Speed (mph)
2.2	30	1.9	30
2.9	40	3.0	40
4.0	50	4.1	50
5.0	60	5.2	60
6.5	70	6.8	70
8.0	80	8.6	80
9.9	90	10.7	90
11.8	100	13.3	100

  c. How much farther than a Porsche 911 does a Toyota Supra travel during the last 5 seconds of the 10-second run?

14. The rate of change of the value of goods exported from the United States can be modeled[2] as

$$E'(t) = 394.14(1.0645)^t \text{ billion dollars per year}$$

   t years after the end of 1990. Likewise, the rate of change of the value of goods imported into the United States can be modeled[3] as

$$I'(t) = 13.2t^2 - 9.94t + 491.97 \text{ billion dollars per year}$$

   t years after the end of 1990.

   a. Which was greater between the end of 1990 and the end of 1997, the value of imported goods or the value of exported goods?

   b. Find the difference in the accumulated value of imports and the accumulated value of exports from the end of 1990 through 1997.

   c. Is your answer from part b the same as the area of the region(s) between the graphs of $E'(t)$ and $I'(t)$? Explain.

7.2   Perpetual Accumulation

Limits as Input Increases or Decreases Without Bound

Definite integrals have specific numbers a and b for both their lower and their upper limits. We now consider what happens to the accumulation of change when one or both of the limits of the integral, a or b, is infinite. That is, we wish to evaluate integrals of the form $\int_a^{\infty} f(x)\,dx$, $\int_{-\infty}^b f(x)\,dx$, or $\int_{-\infty}^{\infty} f(x)\,dx$. For example, archeologists, geologists, and environmentalists might want to know how much of a radioactive isotope will ultimately decay.

We call integrals of the form $\int_a^{\infty} f(x)\,dx$, $\int_{-\infty}^b f(x)\,dx$, or $\int_{-\infty}^{\infty} f(x)\,dx$ **improper integrals**. Before we proceed with a discussion of these integrals, let us first consider what happens to a function $f(x)$ as x becomes infinitely large. Mathematically, we write $\lim_{x\to\infty} f(x)$ to express the idea that we want to know what the output of a function $f(x)$ approaches as x increases without bound. Consider the four graphs in Figure 7.10.

---

2. Based on data from *Statistical Abstract*, 1995.
3. Based on data from *Statistical Abstract*, 1995.

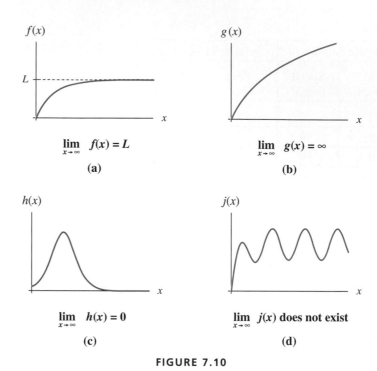

$$\lim_{x \to \infty} f(x) = L$$

(a)

$$\lim_{x \to \infty} g(x) = \infty$$

(b)

$$\lim_{x \to \infty} h(x) = 0$$

(c)

$$\lim_{x \to \infty} j(x) \text{ does not exist}$$

(d)

FIGURE 7.10

It is possible for a function, such as $f(x)$ in Figure 7.10a, to approach some horizontal limiting value L and remain close to L as x grows large. A logistic model is an example of this behavior. A function can continue to increase in height as x grows large, in which case we say the output is growing without bound. See the function $g(x)$ in Figure 7.10b. We express this type of behavior by writing $\lim_{x \to \infty} g(x) = \infty$. A function can approach and remain close to the horizontal axis as x grows large, as is the case with $h(x)$ in Figure 7.10c. Or, a function can oscillate and not approach any specific output value as x increases without bound. Such is the case with the function $j(x)$ in Figure 7.10d. In this case, we say the limit as x approaches infinity does not exist.

Of course, if we can consider x approaching infinity, we can also consider x approaching negative infinity—that is, x decreasing without bound. In this case, we write $\lim_{x \to -\infty} f(x)$. If the function $f(x)$ has the same limit, L, as x approaches negative infinity as it does as x approaches infinity, then we write $\lim_{x \to \pm\infty} f(x) = L$.

For the applications we consider, we are interested in those functions whose limit is a number as the input increases without bound. Two such functions are $f(x)$ and $h(x)$ in Figure 7.10.

We discuss limits of three specific functions: $f(x) = c$, $f(x) = b^x$, and $f(x) = x^n$, where n is negative. In the first case, $f(x) = c$, the function is a horizontal line whose output is constant. In this case, we say that the limit as x approaches ∞ or $-\infty$ is c, and we write $\lim_{x \to \infty} f(x) = c$ and $\lim_{x \to -\infty} f(x) = c$ or $\lim_{x \to \pm\infty} f(x) = c$. This rule, called the *Constant Rule*, is illustrated in Figure 7.11.

A function that plays an important role in radioactive decay and finance applications is the exponential function $f(x) = b^x$. If $b > 1$, then this function has the graph shown in Figure 7.12a, rising infinitely as x approaches ∞ and getting closer and closer to zero as x approaches $-\infty$. If $0 < b < 1$, then the function exhibits the opposite behavior, as shown in Figure 7.12b.

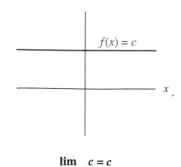

$f(x) = c$

$$\lim_{x \to \infty} c = c$$

$$\lim_{x \to -\infty} c = c$$

FIGURE 7.11

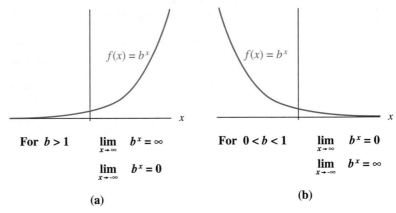

For $b > 1$ $\displaystyle\lim_{x \to \infty} b^x = \infty$ For $0 < b < 1$ $\displaystyle\lim_{x \to \infty} b^x = 0$

$\displaystyle\lim_{x \to -\infty} b^x = 0$ $\displaystyle\lim_{x \to -\infty} b^x = \infty$

(a)   (b)

FIGURE 7.12

We summarize these two cases of the *Exponential Rule* by writing

$$\lim_{x \to \infty} b^x = \begin{cases} \infty & \text{if } b > 1 \\ 0 & \text{if } 0 < b < 1 \end{cases} \qquad \lim_{x \to -\infty} b^x = \begin{cases} 0 & \text{if } b > 1 \\ \infty & \text{if } 0 < b < 1 \end{cases}$$

A third important function is $f(x) = x^n$, where n is negative. Three functions of this form are shown in Figure 7.13. In each case, $\displaystyle\lim_{x \to \infty} f(x) = 0$, and for those functions defined for negative x-values, $\displaystyle\lim_{x \to -\infty} f(x) = 0$.

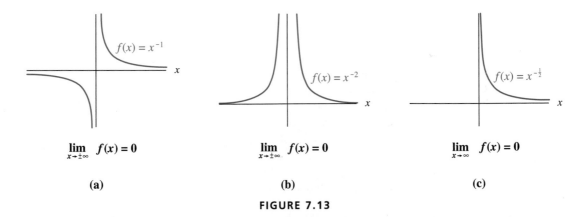

$\displaystyle\lim_{x \to \pm\infty} f(x) = 0$ $\displaystyle\lim_{x \to \pm\infty} f(x) = 0$ $\displaystyle\lim_{x \to \infty} f(x) = 0$

(a)   (b)   (c)

FIGURE 7.13

In general, it is true that

$$\lim_{x \to \infty} x^n = 0 \text{ if } n \text{ is negative} \qquad \lim_{x \to -\infty} x^n = 0 \text{ if } n \text{ is negative and } x^n \text{ is defined}$$
$$\text{for negative input values}$$

We call these statements the *Negative Power Rule*.

We now establish two rules for limits that we have already seen in the context of definite integrals. (Recall that the definite integral is defined as a limit of sums of areas of rectangles as the number of approximating rectangles approaches infinity.)

Constant Multiplier Rule $\displaystyle\lim_{x \to \pm\infty} cf(x) = c\left[\lim_{x \to \pm\infty} f(x)\right]$

Sum Rule $\displaystyle\lim_{x \to \pm\infty} \left[f(x) + g(x)\right] = \lim_{x \to \pm\infty} f(x) + \lim_{x \to \pm\infty} g(x)$

(provided these limits exist.)

For convenience, we summarize the limit rules given in this section:

Limit Rules

Constant Rule $\lim\limits_{x \to \pm\infty} c = c$

Exponential Rule $\lim\limits_{x \to \infty} b^x = \begin{cases} \infty & \text{if } b > 1 \\ 0 & \text{if } 0 < b < 1 \end{cases}$

$\lim\limits_{x \to -\infty} b^x = \begin{cases} 0 & \text{if } b > 1 \\ \infty & \text{if } 0 < b < 1 \end{cases}$

Negative Power Rule $\lim\limits_{x \to \infty} x^n = 0$ if n is negative

$\lim\limits_{x \to -\infty} x^n = 0$ if n is negative and x^n is defined
for negative input value

Constant Multiplier Rule $\lim\limits_{x \to \pm\infty} cf(x) = c\left[\lim\limits_{x \to \pm\infty} f(x)\right]$

Sum Rule $\lim\limits_{x \to \pm\infty} [f(x) + g(x)] = \lim\limits_{x \to \pm\infty} f(x) + \lim\limits_{x \to \pm\infty} g(x)$

EXAMPLE 1   *Evaluating Limits*

If they exist, determine the values of the following limits:

a. $\lim\limits_{x \to -\infty} 14.7(2.369)^x$ b. $\lim\limits_{x \to \infty} (12.905 + 88.421e^{-1.3x})$

c. $\lim\limits_{x \to \infty} \left(\dfrac{3}{x^2} + 12x^{-0.5} + 36\right)$

Solution:

a. $\lim\limits_{x \to -\infty} 14.7(2.369)^x = 14.7 \lim\limits_{x \to -\infty} (2.369)^x$ Constant Multiplier Rule

$= 14.7(0) = 0$ Exponential Rule

b. $\lim\limits_{x \to \infty} (12.905 + 88.421e^{-1.3x})$

$= \lim\limits_{x \to \infty} 12.905 + 88.421 \lim\limits_{x \to \infty} (0.27253)^x$ Sum and Constant Multiplier Rules

$= 12.905 + 88.421(0) = 12.905$ Constant and Exponential Rules

c. $\lim\limits_{x \to \infty} \left(\dfrac{3}{x^2} + 12x^{-0.5} + 36\right)$

$= 3\lim\limits_{x \to \infty} x^{-2} + 12\lim\limits_{x \to \infty} x^{-0.5} + \lim\limits_{x \to \infty} 36$ Sum and Constant Multiplier Rules

$= 3(0) + 12(0) + 36 = 36$ Constant and Negative Power Rules   ∎

Evaluating Improper Integrals

Now that we have laid the groundwork for finding limits as x increases and/or decreases without bound, we can return to the question of how to evaluate improper integrals. An improper integral $\int_a^\infty f(x)\,dx$ is evaluated by replacing infinity with a variable, say N, and evaluating the integral $\int_a^N f(x)\,dx$ as N approaches infinity. That is, provided the limits exist,

$$\int_a^\infty f(x)\,dx = \lim_{N\to\infty} \int_a^N f(x)\,dx = \left[\lim_{N\to\infty} F(N)\right] - F(a)$$

or

$$\int_{-\infty}^b f(x)\,dx = \lim_{N\to-\infty} \int_N^b f(x)\,dx = F(b) - \lim_{N\to-\infty} F(N)$$

where $F(x)$ is an antiderivative of $f(x)$.

We now have the tools we need to apply improper integrals to some real-world problems.

EXAMPLE 2   *Carbon-14*

Carbon-14 dating methods are sometimes used by archeologists to determine the age of an artifact. The rate at which 100 milligrams of ^{14}C is decaying can be modeled by

$$r(t) = -0.01209(0.999879)^t \text{ milligrams per year}$$

where t is the number of years since the 100 milligrams began to decay.

a. How much of the ^{14}C will have decayed after 1000 years?

b. How much of the ^{14}C will eventually decay?

Solution:

a. The amount of ^{14}C to decay during the first 1000 years is

$$\int_0^{1000} r(t)\,dt = \int_0^{1000} -0.01209(0.999879)^t\,dt \approx -11.4 \text{ milligrams}$$

Thus approximately 11.4 milligrams will decay during the first 1000 years. Note that, -11.4 milligrams is the negative of the shaded area in Figure 7.14.

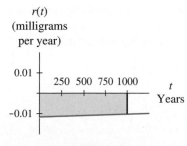

r(t)
(milligrams
per year)

0.01

250 500 750 1000

t
Years

-0.01

FIGURE 7.14

b. In the long run, the amount that will decay is

$$\int_0^\infty r(t)\,dt = \lim_{N\to\infty}\int_0^N -0.01209(0.999879)^t\,dt$$

$$= \lim_{N\to\infty}\left[\frac{-0.01209(0.999879)^t}{\ln 0.999879}\right]\Bigg|_0^N$$

$$= \lim_{N\to\infty}\left(\frac{-0.01209(0.999879)^N}{\ln 0.999879} - \frac{-0.01209(0.999879)^0}{\ln 0.999879}\right)$$

$$\approx -99.91131\left[\lim_{N\to\infty}(0.999879)^N\right] - 99.91131$$

$$= -99.91131(0) - 99.91131 \approx -100 \text{ milligrams}$$

Eventually all of the ^{14}C will ultimately decay. In terms of the graph shown in Figure 7.15, the area between $r(t)$ and the horizontal axis gets closer and closer to 99.91131 as t gets larger and larger. Because the parameters in the model $r(t)$ were rounded, the area of the region between $r(t)$ and the t-axis is getting closer and closer to 99.91131 rather than to 100 which is the amount that must ultimately decay.

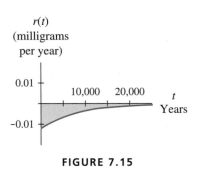

FIGURE 7.15　■

<table>
<tr><td>

7.2 Concept Inventory

■ Limits of functions as input increases or decreases without bound

■ Improper integrals

7.2 Activities

For Activities 1 through 8, evaluate the indicated limit.

1. $\lim_{x\to\infty} 7$

2. $\lim_{x\to-\infty} 13$

3. $\lim_{x\to\infty} e^x$

4. $\lim_{x\to\infty} 0.95^x$

5. $\lim_{x\to-\infty} 5(3^x)$

6. $\lim_{x\to-\infty} \frac{1}{x}$

7. $\lim_{x\to\infty}\left(e^{-0.1x} + \frac{2}{x}\right)$

8. $\lim_{x\to\infty}[5x^{-2} + 3(0.9)^x + 2]$

</td><td>

For Activities 9 through 13, evaluate the indicated improper integral.

9. $\int_0^\infty 3e^{-0.2t}\,dt$

10. $\int_{15}^\infty 5(0.36^t)\,dt$

11. $\int_{10}^\infty 3x^{-2}\,dx$

12. $\int_{-\infty}^3 7e^{7x}\,dx$

13. $\int_{-\infty}^{-10} 4x^{-3}\,dx$

14. The rate at which 15 grams of ^{14}C is decaying can be modeled by

$r(t) = -0.001814(0.998188)^t$ grams per year

where t is the number of years since the 15 grams began decaying.

a. How much of the ^{14}C will decay during the first 1000 years? during the fourth 1000 years?

b. How much of the ^{14}C will eventually decay?

</td></tr>
</table>

15. An isotope of uranium, ^{238}U, is commonly used in atomic weapons and nuclear power generators. Because of its radioactive nature, the United States government is concerned with safe ways of storing used uranium. The rate at which 100 milligrams of ^{238}U is decaying can be modeled by

$$r(t) = -[1.55(0.9999999845)^t] \cdot 10^{-6} \text{ milligrams per year}$$

where t is the number of years since the 100 milligrams began decaying.

a. How much of the ^{238}U will decay during the first 100 years? during the first 1000 years?

b. How much of the ^{238}U will eventually decay?

16. The glasshouse red spider (*Tetranychus urticae Koch*) was tested for susceptibility to arachnicides.[4] Originally 40 female spiders were tested. Each spider laid approximately 5.1 eggs per day, of which only 10% hatched. Because glasshouse red spiders live at most 15 days, a model for the population of the test spiders and their offspring b days after the experiment began is

$$P(b) = 40(0.7684)^b + \int_0^b 0.51(0.7684)^{b-x} dx \text{ spiders}$$

a. What was the spider population after 30 days?

b. What is the spider population in the long run?

17. The king penguin has a 25% mortality rate.[4] If 1000 new king penguins are added to the penguin population each year, the change in the king penguin population over b years can be calculated as

$$P(b) = \int_0^b 1000(0.75)^{b-t} dt \text{ penguins}$$

a. Find the change in the population of king penguins over 10 years. over 20 years.

b. Find the change in the population of king penguins in the long run.

■ **7.3   Streams in Business and Biology**

Picture a stream flowing into a pond. You have probably just created a mental picture of water that is flowing continuously into the pond. We can also imagine moneys that are "flowing" continuously into an investment or new individuals that are "flowing" continuously into an existing population.

When you make periodic payments to a bank or to some other financial institution for the purpose of investing money or repaying a loan, your payments are usually for the same fixed amount and are made at regular times. However, we can think of the income of large financial institutions and major corporations as being received continuously over time in varying amounts. Such a flow of money is called an **income stream** and is usually described as a rate $R(t)$ that varies with time t.

We saw how to calculate future and present values of single payments in Section 2.2. Future and present values of a stream of payments also have meaningful applications.

Future Value

The **future value** is the total accumulated value of the income stream and its earned interest. Suppose that an income stream flows continuously into an interest-bearing account at the rate of $R(t)$ dollars per year and the account earns interest at the

4. J. Meltzer and N.V. Philips-Roxane, "Arachnicidal Properties of 2,4,5,4'-Tetrachoro-Diphenyl Sulphone (Tedion)," *Proceedings of the Tenth International Congress of Entomology*, Montreal, August 17–25, 1956.

5. Bryan Nelson, *Seabirds: Their Biology and Ecology* (New York: Hamlyn Publishing Group, 1979).

annual rate of $r\%$ compounded continuously (where r is a decimal number). What is the future value of the account at the end of T years?

To answer this question, we begin by imagining the time interval from 0 to T years as being divided into n subintervals, each of length Δt.

$$\begin{array}{c|c|c|c} \vphantom{} & & & \\ \hline 0 & t \quad t+\Delta t & & T = (\Delta t)(n) \end{array}$$

We regard Δt as being very small—so small that over a typical subinterval $[t, t + \Delta t]$, the rate $R(t)$ can be considered constant. Then the amount paid into the account during this subinterval can be approximated by

$$\text{Amount paid in} \approx (R(t) \text{ dollars per year})(\Delta t \text{ years})$$
$$\approx R(t)\Delta t \text{ dollars}$$

We consider this amount as being paid in at t, the beginning of the interval, and earning interest continuously for $(T - t)$ years. Using the continuously compounded interest formula $(A = Pe^{rt})$, we see that the amount grows to

$$R(t)\Delta t\, e^{r(T-t)} = R(t)e^{r(T-t)}\Delta t \text{ dollars}$$

at the end of T years. Summing over the n subintervals, we have the approximation

$$\text{Future value} \approx [R(t_1)e^{r(T-t_1)} + R(t_2)e^{r(T-t_2)} + \cdots + R(t_n)e^{r(T-t_n)}]\Delta t \text{ dollars}$$

where $t_1, t_2, \ldots, t_n$ are the left endpoints of the n subintervals. This sum should look familiar to you. If we simplify the expression by letting $f(t) = R(t)e^{r(T-t)}$ and rewrite the sum as

$$[f(t_1) + f(t_2) + \cdots + f(t_n)]\Delta t$$

then you should recognize it as the type of sum we used in Sections 6.1 through 6.3.

Because we are considering the income as a continuous stream and interest as being compounded continuously, we let the time interval Δt become extremely small ($\Delta t \to 0$). That is, we use an infinite number of intervals ($n \to \infty$). Thus

$$\text{Future value} = \lim_{n \to \infty} [f(t_1) + f(t_2) + \cdots + f(t_n)]\Delta t$$
$$= \int_0^T f(t)\, dt$$
$$= \int_0^T R(t)e^{r(T-t)}\, dt \text{ dollars}$$

Future Value of an Income Stream

Suppose that an income stream flows continuously into an interest-bearing account at the rate of $R(t)$ dollars per year and the account earns interest at the annual rate of $r\%$ compounded continuously (where r is a decimal number). The future value of the account at the end of T years is

$$\text{Future value} = \int_0^T R(t)e^{r(T-t)}\, dt \text{ dollars}$$

Using the Fundamental Theorem, we can find the rate-of-change function for future value.

$$\text{Rate of change of future value} = \frac{d}{dx} \int_0^x R(t)e^{r(T-t)} dt \quad \text{for } 0 \le x \le T$$

$$= R(x)e^{r(T-x)} \text{ dollars per year}$$

Thus the function $f(t) = R(t)e^{r(T-t)}$ gives the rate of change after t years of the future value (in T years) of an income stream whose income is flowing in at a rate of $R(t)$ dollars per year. It is the rate-of-change function $f(t) = R(t)e^{r(T-t)}$, not the flow rate of the income stream $R(t)$, that we graph when depicting future value as the area of a region beneath a rate-of-change function.

EXAMPLE 1   *Airline Expansion*

The owners of a small airline are making big plans. They hope to be able to buy out a larger airline 10 years from now. They have determined that they can afford to invest $3.3 million each year in investments that return a 19.4% APR.

a. Assuming a continuous income stream and continuous compounding, how much will these investments be worth 10 years from now?

b. Illustrate the result to part *a* using a graph.

Solution:

a. The flow rate of the income stream is $R(t) = 3.3$ million per year with $r = 0.194$ and $T = 10$ years. The value of these investments in 10 years is calculated as

$$\text{Future value} = \int_0^{10} 3.3e^{0.194(10-t)} dt$$

$$= \int_0^{10} 3.3e^{1.94}e^{-0.194t} dt$$

$$= \frac{3.3e^{1.94}}{-0.194} e^{-0.194(10)} - \frac{3.3e^{1.94}}{-0.194} e^{-0.194(0)}$$

$$\approx -17.010 + 118.371$$

$$\approx \$101.4 \text{ million}$$

b. Even though the income stream is flowing at the constant rate $R(t) = \$3.3$ million per year, the future value is graphically interpreted as the area of the region between the rate-of-change function $f(t) = 3.3e^{0.194(10-t)}$ and the horizontal axis between 0 and 10 years. The future value is the area of the shaded region.

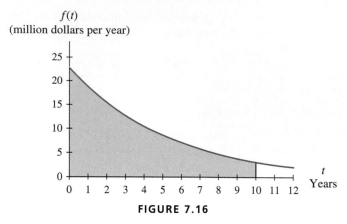

$f(t)$
(million dollars per year)

FIGURE 7.16   ■

Present Value

As before, the **present value** of an income stream is the amount P that would have to be invested now in an interest-bearing account now in order for the amount to grow to a given future value. Because P dollars earning continuously compounded interest would grow to Pe^{rT} dollars in T years, we have

$$Pe^{rT} = \int_0^T R(t)e^{r(T-t)}\,dt = \int_0^T R(t)e^{rT}e^{-rt}\,dt = e^{rT}\int_0^T R(t)e^{-rt}\,dt$$

Solving for P, we obtain

$$\text{Present value} = \int_0^T R(t)e^{-rt}\,dt$$

Present Value of an Income Stream

Suppose that an income stream flows continuously into an interest-bearing account at the rate of $R(t)$ dollars per year and that the account earns interest at the annual rate of $r\%$ compounded continuously (where r is a decimal number). The present value of the account is

$$\text{Present value} = \int_0^T R(t)e^{-rt}\,dt \text{ dollars}$$

EXAMPLE 2   *Revenue Expansion*

Last year, profit for the HiTech Corporation was $1.3 million. Assuming that HiTech's profits increase continuously for the next 5 years at a rate of $1.3 million per year what are the future and present values of the corporation's 5-year profits? Assume an interest rate of 12% compounded continuously.

Solution: We note that the rate of the stream is $R(t) = 1.3t + 1.3$ million dollars per year in year t. In order to calculate the future value of this stream, we evaluate $\int_0^5 (1.3t + 1.3)e^{0.12(5-t)}\,dt$. We have not developed a method for finding the antiderivative of $f(t) = (1.3t + 1.3)e^{0.12(5-t)}$, so we numerically estimate the definite integral using a limiting value of sums.

$$\text{Future value} = \int_0^5 (1.3t + 1.3)e^{0.12(5-t)}\, dt \approx \$29.96 \text{ million}$$

The invested revenue will be worth approximately \$30 million in 5 years. We also numerically estimate a limiting value of sums to find the present value.

$$\text{Present value} = \int_0^5 (1.3t + 1.3)e^{-0.12t}\, dt \approx \$15.89 \text{ million}$$

This is the lump sum (\$15.9 million) that would have to be invested in order to earn \$30 million (the future value) in 5 years.

It is worth noting that once you have found a future value, it is easy to calculate the associated present value by solving for P in the equation $Pe^{rt} =$ future value. In this case,

$$Pe^{(0.12)(5)} \approx \$29.96 \text{ million, so } P \approx \$16 \text{ million}$$

The integral definition of the present value is most useful in situations in which you do not know the future value. ∎

Discrete Income Streams

The assumptions that income is flowing continuously and that interest is compounded continuously make it possible to use mathematical ideas and are often imposed by economists. Unfortunately, they do not generally hold in the real world of business. It is much more realistic to consider an income stream that flows monthly into an account with monthly compounding of interest or a stream flowing quarterly with quarterly compoundings.

EXAMPLE 3   *Saving for the Future*

When you graduate from college (say, in 3 years), you would like to purchase a car. You have a job and can put \$75 into savings each month for this purchase. The best savings plan you can find offers an APR of 6.2% compounded monthly.

a. How much money will you have deposited in 3 years?

b. What will be the value of your savings in 3 years?

c. How much money would you have to deposit now (in one lump sum) to achieve the same future value in 3 years?

Solution:

a. The total amount deposited is $(36)(\$75) = \$2700$.

b. The \$75 deposit made at time t will earn interest for $3 - t$ years, so the change in the future value that occurs because of the deposit at time t is

$$f(t) = 75\left(1 + \frac{0.062}{12}\right)^{12(3-t)} \approx 75(1.06379)^{(3-t)} \text{ dollars}$$

where t is the number of years since the first deposit was made.

Even though $f(t)$ is a continuous model, the amount in the account changes only when $t = 0, \frac{1}{12}, \frac{2}{12}, \frac{3}{12}$, and so on, because the activity in the account takes place only once a month, not continuously. To compute the value of this account in 3 years, we sum the values of $f(t)$—that is, the change in the future value—each month using $t = 0, \frac{1}{12}, \frac{2}{12}, \ldots \frac{35}{12}$. See Table 7.2.

TABLE 7.2

Time of monthly deposit	Accumulated value of monthly deposit
0	$f(0) \approx 90.288$
$\frac{1}{12}$	$f\left(\frac{1}{12}\right) \approx 89.824$
$\frac{2}{12}$	$f\left(\frac{2}{12}\right) \approx 89.363$
⋮	⋮
$\frac{33}{12}$	$f\left(\frac{33}{12}\right) \approx 76.169$
$\frac{34}{12}$	$f\left(\frac{34}{12}\right) \approx 75.777$
$\frac{35}{12}$	$f\left(\frac{35}{12}\right) \approx 75.388$
	Sum ≈ 2974.338

$f(t)$
Change in
future value
(dollars)

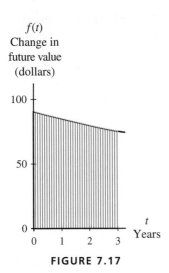

FIGURE 7.17

$p(t)$
Change in
future value
(dollars)

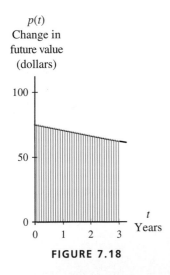

FIGURE 7.18

Thus the value of this account after 3 years will be $2974.34. Note that the value of the account can be interpreted as 12 times the area of 36 left rectangles under the $f(t)$ curve. (See Figure 7.17.)

c. Because we know the future value, we can solve $P\left(1 + \frac{0.062}{12}\right)^{12(3)} \approx 2974.338$ for the present value P to obtain $P \approx \$2470.70$. This is the amount that you would need to deposit now to have $2974.34 in 3 years.

   If we do not know future value, then we can calculate the present value as follows: The change in the present value that occurs because of the deposit at time t is

$$p(t) = 75\left(1 + \frac{0.062}{12}\right)^{-12t} \approx 75(0.94003)^t \text{ dollars per month}$$

where t is the number of years after the initial deposit. The present value is the accumulated change in the amount from 0 to 3. See Table 7.3.

Again, we see that the present value of this income stream over 3 years is $2470.70. This can be interpreted as 12 times the sum of the areas of the 36 rectangles under the $p(t)$ curve. (See Figure 7.19.) ■

Streams in Biology

Biology and other fields involve situations very similar to income streams. An example of this is the growth of populations of animals. As of 1978, there were approximately 1.5 million sperm whales[6] in the world's oceans. Each year approximately 0.06 million sperm whales are added to the population. Also each year, 4% of the sperm whale population either die of natural causes or are killed by hunters. Assuming that these rates (and percentage rates) have remained constant since 1978, we

---

6. Delphine Haley, *Marine Mammals* (Seattle: Pacific Search Press, 1978).

TABLE 7.3

Time of monthly deposit	Accumulated value of monthly deposit
0	$p(0) \approx 75.000$
$\dfrac{1}{12}$	$p\left(\dfrac{1}{12}\right) \approx 74.614$
$\dfrac{2}{12}$	$p\left(\dfrac{2}{12}\right) \approx 74.231$
$\vdots$	$\vdots$
$\dfrac{33}{12}$	$p\left(\dfrac{33}{12}\right) \approx 63.271$
$\dfrac{34}{12}$	$p\left(\dfrac{34}{12}\right) \approx 62.946$
$\dfrac{35}{12}$	$p\left(\dfrac{35}{12}\right) \approx 62.622$
Sum ≈ 2470.697	

estimate the sperm whale population in 1998 using the same procedure as when determining future value of income streams.

There are two aspects of the population that we must consider when estimating the population of sperm whales in 1998. First, we must determine the number of whales that were living in 1978 that will still be living in 1998. Because 4% of the sperm whales die each year, we calculate the number of whales that have survived the entire 20 years as $1.5(0.96)^{20} \approx 0.663$ million whales.

The second aspect that we must consider is the impact on the population made by the birth of new whales. We are told that 0.06 million whales per year are added to the population and that 96% of those survive each year. Therefore, the rate at which the population of sperm whales associated with those that were born t years after 1978 is growing is

$$f(t) = 0.06(0.96)^{(20-t)} \text{ million whales per year}$$

Thus the sperm whale population in 1998 is calculated as

$$\text{Whale population} = 1.5(0.96)^{20} + \int_0^{20} 0.06(0.96)^{(20-t)}\, dt$$

$$\approx 0.663 + \int_0^{20} 0.06(0.96)^{20}(0.96)^{-t}\, dt$$

$$= 0.663 + 0.06(0.96)^{20}\int_0^{20} (0.96^{-1})^t\, dt$$

$$= 0.663 + \left.\frac{0.06(0.96)^{20}(0.96^{-1})^t}{\ln(0.96^{-1})}\right|_0^{20}$$

$$\approx 1.48 \text{ million sperm whales}$$

Functions that model such biological streams, in which new individuals are added to the population and the rate of survival of the individuals is known, are referred to as *survival* and *renewal functions*.

In general, the formula for the future value (in b years) of a biological stream with initial population size P, survival rate s (in decimals), and renewal rate $r(t)$, where t is the number of years, is

$$\text{Future value} \approx Ps^b + \int_0^b r(t)s^{(b-t)}\, dt$$

In the sperm whale example, the initial population is $P = 1.5$ million. The survival rate is 96% per year, so $s = 0.96$, and the renewal rate is $r(t) = 0.06$ million whales per year.

EXAMPLE 4   *Flea Population*

An example of a stream in entomology is the growth of a flea population. In cooler areas of the country, adult fleas die before winter, but flea eggs survive and hatch the following spring when temperatures again reach 70 °F. Not all the eggs hatch at the same time, so part of the growth in flea population is due to the hatching of the original eggs. Another part of the growth in flea population is due to propagation. Suppose fleas propagate at the rate of 134% per day and that the original set of fleas (from the dormant eggs) become reproducing adults at the rate of 600 fleas per day. What will the flea population be 10 days after the first 600 fleas begin reproducing? Assume that none of the fleas die during the 10-day period.

Solution: We first note that because we begin counting when the first 600 fleas have become mature adults, we consider the initial population to be $P = 600$ fleas. The renewal rate is also 600 fleas per day, so $r(t) = 600$.

Because, in this case, the renewal rate function $r(t)$ does not account for renewal due to propagation, we must incorporate the propagation rate of 134% into the survival rate of 100%. Thus the survival/propagation rate is $s = 2.34$.

Because the renewal rate and survival/propagation rate are given in days, we let t be the input variable measured in days. The flea population will grow over 10 days to

$$\text{Flea population} \approx Ps^{10} + \int_0^{10} r(t)s^{(10-t)}\, dt$$

$$= 600(2.34)^{10} + \int_0^{10} 600(2.34)^{(10-t)}\, dt$$

$$\approx 2{,}953{,}315 + 3{,}473{,}166 \approx 6.4 \text{ million fleas} \qquad \blacksquare$$

7.3   Concept Inventory

- Income stream
- Flow rate of a stream
- Future value
- Present value
- Biological stream
- Survival and renewal functions

7.3   Activities

1. To prepare for your future retirement (in 40 years), suppose you begin investing $500 per month in an annuity with a fixed rate of return of 8.34%.

   a. Assuming a continuous stream, what will the annuity be worth at the end of 40 years?

   b. Assuming monthly activity (deposits and interest compounding), what will the annuity be worth at the end of 40 years?

c. Is the answer to part *a* or part *b* more likely to be the actual future value of the annuity? Explain.

2. In preparing for your retirement (in 40 years), suppose you plan to invest 14% of your salary each month in an annuity with a fixed rate of return of 9.2%. You currently make $2800 per month and expect your income to increase by 4% per year.

   a. Assuming a continuous stream, what will the annuity be worth at the end of 40 years?

   b. Assuming monthly activity (deposits and interest compounding), what will the annuity be worth at the end of 40 years?

   c. Discuss whether parts *a* and *b* overestimate or underestimate the future value of the annuity.

3. Compaq Computer Corporation's 1996 third-quarter profits[7] were $350 million. Assume that these profits will increase by 5% per quarter and that Compaq will invest 15% of its quarterly profits in an investment with a quarterly return of 13%.

   a. Write a function for the rate at which money flows into this investment each quarter.

   b. Write a function for the rate at which the 4-year future value of this investment is changing.

   c. Find the value of this investment at the end of the year 2000. (Assume a quarterly stream beginning on January 1, 1997 with the investment of 4th quarter 1996 profits.)

4. For the year ending June 30, 1994, Creative Technology posted an annual net income of $97,941,000.[8] Assume the income can be reinvested at an annual rate of return of 10% compounded continuously. Also assume that Creative Technology will maintain this annual net income continuously for the next 5 years.

   a. What is the future value of its 5-year net income?

   b. What is the present value of its 5-year net income?

   c. Graphically illustrate the results of parts *a* and *b* as areas of regions under a curve.

5. In 1993, PepsiCo installed a new soccer scoreboard for Alma College in Alma, MI. The terms of the installation were that Pepsi would have sole vending rights at Alma College for the next 7 years. It is estimated that in the 3 years after the scoreboard was installed, Pepsi sold 36.4 thousand liters of Pepsi products to Alma College students, faculty, staff, and visitors. Suppose that the average yearly sales and associated revenue remain constant and that the revenue from Alma College sales is reinvested at 4.5% APR. Also assume that PepsiCo makes a revenue of $0.80 per liter of Pepsi.

   a. The vending of Pepsi products on campus can be considered a continuous process. Assuming that the revenue is invested in a continuous stream and that interest on that investment is compounded continuously, how much will Pepsi make from its 7 years of sales at Alma College?

   b. Still assuming a continuous stream, find how much Pepsi would have had to invest in 1993 to create the same 7-year future value.

6. In preparing for your future retirement (in 40 years), you begin investing $500 per month in an annuity with a fixed rate of return of 8.34%.

   a. Write a function for the rate at which you will invest money into the annuity.

   b. How much would you have to invest now, in one lump sum instead of in a continuous stream, in order to build to the same future (40-year) value? Assume that interest is compounded continuously.

   c. How much would you have to invest now, in one lump sum instead of in a monthly stream, in order to build to the same future (40-year) value? Assume monthly compounding of interest.

   d. Is the answer to part *b* or to part *c* more likely to be the actual present value of the annuity? Explain.

7. In preparing for your retirement (in 40 years), you plan to invest 14% of your salary each month in an annuity with a fixed rate of return of 9.2%. You currently make $2800 per month, and you expect your income to increase by 4% per year. How much would you have to invest now, in one lump sum instead of in a continuous stream, in order to build to the same future (40-year) value?

7. "Compaq Profit Jumps 43% as Sales Strengthen, Component Costs Decrease," *Wall Street Journal*, October 5, 1996.
8. *Wall Street Journal*, August 9, 1994.

8. The revenue of Sears, the second largest retailer in the United States, can be modeled[9] as

$$R(t) = 33.6(1.08)^t \text{ billion dollars per year}$$

t years after 1995. Assume that the revenue can be reinvested at 12% compounded continuously.

   a. How much will the revenue invested since 1995 of Sears be worth in 2000?

   b. How much would this accumulated investment have been worth in 1995?

9. In 1956, AT&T laid its first underwater phone line. By 1996, AT&T Submarine Systems, the division of AT&T that installs and maintains undersea communication lines, had seven cable ships and 1000 workers. On October 5, 1996, AT&T[10] announced that it was seeking a buyer for its Submarine Systems division. The Submarine Systems division of AT&T was posting a profit of $850 million per year.

   a. If AT&T assumed that the Submarine Systems division's annual profit would remain constant and could be reinvested at an annual return of 15%, what would AT&T have considered to be the 20-year present value of its Submarine Systems division? (Assume a continuous stream.)

   b. If prospective bidder A considered that the annual profits of this division would remain constant and could be reinvested at an annual return of 13%, what would bidder A consider to be the 20-year present value of AT&T's Submarine Systems? (Assume a continuous stream.)

   c. If prospective bidder B considered that over a 20-year period, profits of the division would grow by 10% per year (after which it would be obsolete) and that profits could be reinvested at an annual return of 14%, what would bidder B consider to be the 20-year present value of AT&T's Submarine Systems? (Assume a continuous stream.)

10. On October 4, 1996, Tenet Healthcare Corporation, the second-largest hospital company in the

United States at that time, announced that it would buy Ornda Healthcorp.[11]

   a. If Tenet Healthcare Corporation assumed that Ornda's annual revenue of $0.273 billion would increase by 10% per year and that the revenues could be continuously reinvested at an annual return of 13%, what would Tenet Healthcare Corporation consider to be the 15-year present value of Ornda Healthcorp at the time of the buyout?

   b. If Ornda Healthcorp's forecast for its financial future was that its $0.273 billion annual revenue would remain constant and that revenues could be continuously reinvested at an annual return of 15%, what would Ornda Healthcorp consider its 15-year present value to be at the time of the buyout?

   c. Tenet Healthcare Corporation bought Ornda Healthcorp for $1.82 billion in stock. If the sale price was the 15-year present value, did either of the companies have to compromise on what it believed to be the value of Ornda Healthcorp?

11. CSX Corporation, a railway company, announced in October of 1996, its intention to buy Conrail Inc. for $8.1 billion.[12] The combined company, CSX-Conrail, would control 29,000 miles of track and have an annual revenue of $14 billion the first year after the merger, making it one of the largest railway companies in the country.

   a. If Conrail assumed that its $2 billion annual revenue would decrease by 5% each year for the next 10 years but that the annual revenue could be reinvested at an annual return of 20%, what would Conrail consider to be its 10-year present value at the time of CSX's offer? Is this more or less than the amount CSX offered?

   b. CSX Corporation forecast that its Conrail acquisition would add $1.2 billion to its annual revenue the first year and that this added annual revenue would increase by 2% each year. Suppose CSX is able to reinvest that revenue at an annual return of 20%. What would CSX

9. Based on information from "Sears Posts 22% Increase in Profit for 3rd Period, Beating Estimates," *Wall Street Journal*, October 5, 1996.

10. "AT&T Seeking a Buyer for Cable-Ship Business," *Wall Street Journal*, October 5, 1996.

11. "Tenet to Acquire Ornda," *Wall Street Journal*, October 5, 1996.

12. "Seeking Concessions from CSX-Conrail Is Seen as Most Likely Move by Norfolk," *Wall Street Journal*, October 5, 1996.

Corporation have considered to be the 10-year present value of the Conrail acquisition in October of 1996?

c. Why might CSX Corporation have forecast an increase in annual revenue when Conrail forecast a decrease?

12. Company A is attempting to negotiate a buyout of Company B. Company B accountants project an annual income of 2.8 million dollars per year. Accountants for Company A project that with Company B's assets, Company A could produce an income starting at 1.4 million dollars per year and growing at a rate of 5% per year. The discount rate (the rate at which income can be reinvested) is 8% for both companies. Suppose that both companies consider their incomes over a 10-year period. Company A's top offer is equal to the present value of its projected income, and Company B's bottom price is equal to the present value of its projected income. Will the two companies come to an agreement for the buyout? Explain.

13. A company involved in videotape reproduction has just reported $1.2 million net income during its first year of operation. Projections are that net income will grow over the next 5 years at the rate of 6% per year. The *capital value* (present sales value) of the company has been set as its present value over the next 5 years. If the rate of return on reinvested income can be compounded continuously for the next 5 years at 12% per year, what is the capital value of this company?

14. There were once more than 1 million elephants in West Africa.[13] Now, however, the elephant population has dwindled to 19,000. Each year 17.8% of West Africa elephants die or are killed by hunters. At the same time, elephant births are decreasing by 13% per year.

a. How many of the current population of 19,000 elephants will still be alive 30 years from now?

b. Considering that 47 elephants were born in the wild this year, write a function for the number

of elephants that will be born t years from now and will still be alive 30 years from now.

c. Estimate the elephant population of West Africa 30 years from now.

15. In 1979 there were 12 million sooty terns (a bird) in the world.[14] Assume that the percentage of terns that survive from year to year has stayed constant at 83% and that approximately 2.04 million terns hatch each year.

a. How many of the terns that were alive in 1979 are still alive?

b. Write a function for the number of terns that hatched t years after 1979 and are still alive.

c. Estimate the present population of sooty terns.

16. From 1936 through 1957, a population of 15,000 muskrats in Iowa[15] bred at a rate of 468 new muskrats per year and had a survival rate of 75%.

a. How many of the muskrats alive in 1936 were still alive in 1957?

b. Write a function for the number of muskrats that were born t years after 1936 and were still alive in 1957.

c. Estimate the muskrat population in 1957.

17. There are approximately 200 thousand northern fur seals.[16] Suppose the population is being renewed at a rate of $r(t) = 60 - 0.5t$ thousand seals per year and that the survival rate is 67%.

a. How many of the current population of 200 thousand seals will still be alive 50 years from now?

b. Write a function for the number of seals that will be born t years from now and will still be alive 50 years from now.

c. Estimate the northern fur seal population 50 years from now.

18. Explain, using related examples, the difference between a continuous income stream and a discrete income stream.

---

13. Douglas Chawick, *The Fate of the Elephant* (Sierra Club Books, 1992).
14. Bryan Nelson, *Seabirds: Their Biology and Ecology* (New York: Hamlyn Publishing Group, 1979).
15. Paul L. Errington, *Muskrat Population* (Ames, IA: Iowa State University Press, 1963).
16. Delphine Haley, *Marine Mammals* (Seattle Pacific Search Press, 1978).

7.4   Integrals in Economics

q
Quantity

D(p)

p
Price
per unit

FIGURE 7.19

When you purchase an item in a store, you ordinarily have no control over the price that you pay. Your only choice is whether to buy or not to buy the item at the current price. In general, consumers hold to the view that price is a variable to which they can only respond. As the price per unit increases, consumers usually respond by purchasing (demanding) less. The typical relation between the price per unit (as input) and the quantity in demand (as output) is shown in Figure 7.19.

The traditional approach to graphing in economic theory is to put the price per unit (input) along the vertical axis and the quantity in demand (output) along the horizontal axis. (See Figure 7.20.) We choose not to graph the economists' way but instead to continue to follow the mathematical convention for graphing. This will help us visualize the definite integrals used later in this section.

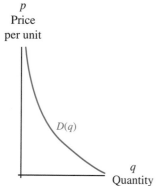

p
Price
per unit

D(q)

q
Quantity

FIGURE 7.20

Demand Curves

The curve relating quantity in demand *q* to price per unit *p* is called a **demand curve**. In economic theory, demand is actually a function that has several input variables, such as price per unit, consumers' ability to buy, consumers' need, and so on. The demand curve we consider here is a simplified version. We assume that all the possible input variables are constant except price. We denote this demand function as $D(p)$.

Even though the demand function $D(p)$ is not a rate-of-change function, there are economic interpretations for the areas of certain regions lying beneath the $D(p)$ curve. In order to interpret the area of these regions, you must understand how to interpret the information the demand curve represents.

For instance, suppose the graph in Figure 7.21 represents the annual demand for parsley (a fairly common herb used in cooking). A point on the demand curve indicates the quantity that consumers will purchase at a given price. At $33.86 per pound, consumers will purchase 1 million pounds of parsley. At $20.00 per pound, consumers will purchase 2 million pounds of parsley. Even though points on the demand curve tell us how much consumers will actually purchase at certain prices, consumers are willing and able to pay more than that for the quantity they purchase. For instance, consumers are willing and able to spend approximately $33.86 per pound for the first million pounds, but they are willing and able to spend only approximately $20.00 per pound for the second million pounds. Thus, in total, consumers are willing and able to spend approximately $53.86 for two million pounds of parsley.

If the price of parsley is $11.89, consumers are willing and able to buy the third million pounds of parsley. That is, consumers are willing and able to spend approximately

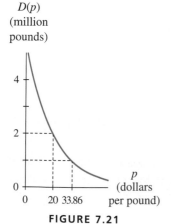

D(p)
(million
pounds)

4

2

0

0   20 33.86

p
(dollars
per pound)

FIGURE 7.21

(1 million pounds)($33.86 per pound) + (1 million pounds)($20.00 per pound)
  + (1 million pounds)($11.89 per pound) = $65.75 million

for 3 million pounds of parsley, even though in actuality they spend only

(3 million pounds)($11.89 per pound) = $35.67 million

See Figures 7.22a and b.

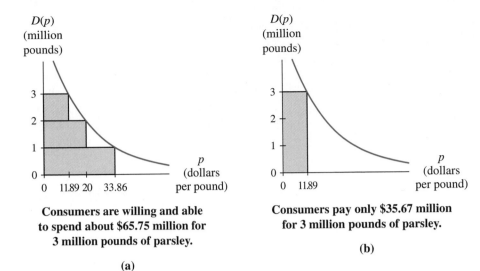

Consumers are willing and able to spend about $65.75 million for 3 million pounds of parsley.

(a)

Consumers pay only $35.67 million for 3 million pounds of parsley.

(b)

FIGURE 7.22

Consumers' Willingness and Ability to Spend

You should have noticed that the amount that consumers are willing and able to spend for 3 million pounds was given as *approximately* $65.75 million. We can make this approximation better by considering smaller increments for price. Given the following table of prices and their relative demands, let us calculate the quantities in demand at various prices and then interpret them as consumers' willingness and ability to spend. We use Table 7.4 to present certain prices and their relative demands.

TABLE 7.4

p (dollars per pound)	61.89	51.89	41.89	31.89	21.89	11.89
$D(p)$ (million pounds)	0.2	0.4	0.7	1.1	1.8	3.0

At $61.89 per pound, consumers are willing and able to purchase 0.2 million pounds of parsley. At $51.89 per pound, consumers are willing and able to purchase 0.2 million pounds more. At $41.89 per pound, consumers are willing and able to purchase another 0.3 million pounds of parsley, and so on.

At $11.89 per pound, consumers are willing and able to purchase 1.2 million pounds in excess of the 1.8 million pounds they were willing and able to purchase at prices exceeding $21.89. These estimates of the consumers' willingness and ability to purchase are represented as stacked rectangles in Figure 7.23.

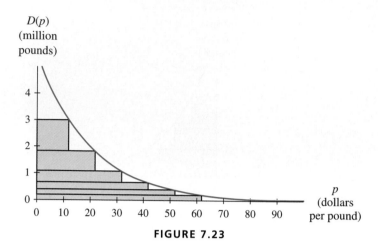

FIGURE 7.23

How much are consumers willing and able to spend in order to purchase 3 million pounds of parsley?

$$(0.2 \text{ million pounds})(\$61.89 \text{ per pound})$$
$$+ (0.2 \text{ million pounds})(\$51.89 \text{ per pound})$$
$$+ (0.3 \text{ million pounds})(\$41.89 \text{ per pound})$$
$$+ (0.4 \text{ million pounds})(\$31.89 \text{ per pound})$$
$$+ (0.7 \text{ million pounds})(\$21.89 \text{ per pound})$$
$$+ (1.2 \text{ million pounds})(\$11.89 \text{ per pound})$$
$$= \$77.67 \text{ million}$$

Consumers are willing and able to spend approximately $77.67 million to purchase 3 million pounds of parsley.

If we were to approximate consumers' willingness and ability to spend using price increments of $5 per pound, or $2.5 per pound, or $1.25 per pound, etc., we would see that the areas of the stacked rectangles representing these approximations would become closer to being the true area depicted in Figure 7.24.

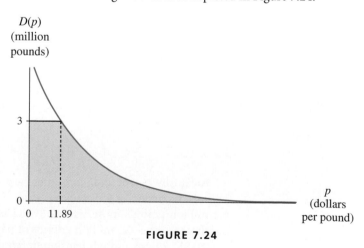

FIGURE 7.24

Thus consumers' willingness and ability to purchase 3 million pounds of parsley can be visually represented by the region under the horizontal line $D(p) = 3$ and under the $D(p)$ curve from 0 to P, where P is the price above which consumers cannot and will not purchase any parsley. We calculate the consumers' willingness and ability to purchase as

$$\int_0^{11.89} 3\,dp + \int_{11.89}^{P} D(p)\,dp \text{ million dollars}$$

Suppose the demand for parsley can be modeled by

$$D(p) = 5.4366(0.95123)^p \text{ million pounds}$$

where p dollars is the price per pound. The only piece of information we still need is P, the price above which no parsley will be purchased. You should notice that our demand function $D(p) = 5.4366(0.95123)^p$ approaches 0 as p becomes large; however, $D(p) = 5.4366(0.95123)^p$ will never be exactly 0 for any p. Hence, we let P approach ∞. This is true for most demand functions in economics. In this case, we consider the area under the $D(p)$ curve as P becomes infinitely large. That is,

$$\int_0^{11.89} 3\,dp + \int_{11.89}^{\infty} 5.4366(0.95123)^p\,dp$$

$$= \int_0^{11.89} 3\,dp + \lim_{P\to\infty} \int_{11.89}^{P} 5.4366(0.95123)^p\,dp$$

$$= 35.67 + \lim_{P\to\infty} \left. \left(\frac{5.4366(0.95123)^p}{\ln 0.95123} \right) \right|_{11.89}^{P}$$

$$= 35.67 + \left(\lim_{P\to\infty} \frac{5.4366(0.95123)^P}{\ln 0.95123} \right) - \left(\frac{5.4366(0.95123)^{11.89}}{\ln 0.95123} \right)$$

$$\approx 35.67 + 0 - (-60.00365)$$

$$\approx \$95.67 \text{ million}$$

Thus consumers are willing and able to spend \$95.67 million in order to purchase 3 million pounds of parsley.

In general, we make the following definition:

Consumers' Willingness and Ability to Spend

For a continuous demand curve $D(p)$, the maximum amount that consumers are willing and able to spend for a certain quantity q_0 of goods or services is

$$\int_0^{p_0} q_0\,dp + \int_{p_0}^{P} D(p)\,dp$$

where p_0 is the market price at which q_0 units are in demand and P is the price above which consumers will purchase none of the goods or services. (See Figure 7.25).

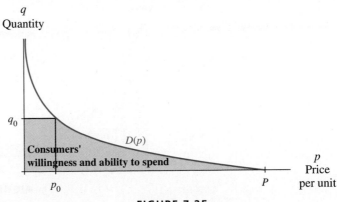

FIGURE 7.25

EXAMPLE 1   *Cellular Phones*

Suppose the average weekly demand of a certain brand of cellular phone can be modeled by the equation $D(p) = 1952(0.959)^p$ phones, where p is the wholesale price per phone in dollars. How much money are consumers willing and able to spend each week on the wholesale market for 300 such cellular phones?

Solution:

The amount that consumers are willing and able to spend on the wholesale market for 300 cellular telephones each week is given by the area of the region below the line $q_0 = 300$ and the $D(p)$ curve.

We use $D(p) = 300$ to solve for $p_0 \approx 44.73577$ and note that $D(p)$ is never 0. However, it approaches 0 as p increases, so consumers are willing and able to spend

$$\int_0^{44.73577} 300\, dp + \int_{44.73577}^{\infty} 1952(0.959)^p\, dp$$

$$= (300)(44.73577) + \lim_{P \to \infty} \left(\frac{1952(0.959)^P}{\ln 0.959} \right) \Big|_{44.73577}^{P}$$

$$\approx 13{,}420.73019 + \left(\lim_{P \to \infty} \frac{1952(0.959)^P}{\ln 0.959} \right) - (-7166.026596)$$

$$= 13{,}420.73019 + 0 - (-7166.026596)$$

$$\approx \$20{,}587$$

According to the demand model, $D(p)$, consumers are willing and able to spend an average of \$20,587 for 300 cellular phones each week.   ∎

Consumers' Expenditure and Surplus

Now that we have considered what consumers are willing and able to spend for a certain quantity of a product, let us turn our attention to calculating what consumers actually spend for that quantity. We return to the discussion of parsley demand. As was previously mentioned, if the market price for parsley is \$11.89 per pound, consumers will purchase 3 million pounds. The actual amount spent by consumers is (3 million pounds)(\$11.89 per pound) = \$35.67 million, even though

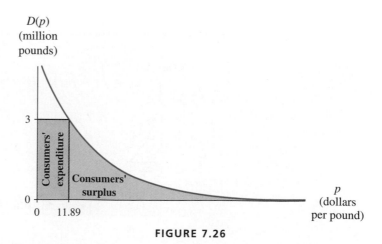

FIGURE 7.26

they are willing and able to spend much more. This actual amount spent is the area of the rectangular region from the vertical axis to $p = 11.89$ with height 3 as shown in Figure 7.26. This amount is known as the **consumers' expenditure**. The amount that consumers are willing and able to spend but do not actually spend is known as the **consumers' surplus**. (See Figure 7.26.)

Earlier we found that consumers are willing and able to spend $95.67 million to purchase 3 million pounds of parsley, so the consumers' surplus from buying 3 million pounds of parsley at $11.89 per pound is

$$\$95.67 \text{ million} - \$35.67 \text{ million} = \$60.00 \text{ million}$$

Consumers' surplus can also be computed directly from the demand function as

$$\text{Consumers' surplus} = \int_{11.89}^{\infty} 5.4366(0.95123)^p \, dp \approx \$60.00 \text{ million}$$

In general, we make the following definitions:

Consumers' Expenditure and Surplus

For a continuous demand curve $D(p)$, the amount that consumers spend at a certain market price p_0 is

$$\text{Consumers' expenditure} = \int_{0}^{p_0} q_0 \, dp = q_0 p_0$$

where $q_0 = D(p_0)$ is the quantity in demand at the market price p_0.

Furthermore, the amount consumers are willing and able to spend but do not spend for q_0 items at market price p_0 is

$$\text{Consumers' surplus} = \int_{p_0}^{P} D(p) \, dp$$

where P is the price above which consumers will purchase none of the goods or services. (See Figure 7.27).

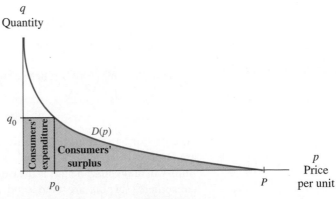

FIGURE 7.27

EXAMPLE 2   *Mini-vans*

Suppose the demand for mini-vans in the United States can be modeled as

$$D(p) = 14.12(0.933)^p - 0.25 \text{ million mini-vans}$$

when the market price is p thousand dollars per mini-van.

a. At what price per mini-van will consumers purchase 2.5 million mini-vans?

b. What is the consumers' expenditure when purchasing 2.5 million mini-vans?

c. Does the model indicate a possible price above which consumers will purchase no mini-vans? If so, what is this price?

d. When 2.5 million mini-vans are purchased, what is the consumers' surplus?

e. What is the total amount consumers are willing and able to spend on 2.5 million mini-vans?

Solution:

a. We solve $D(p) = 2.5$ to find the market price at which consumers will purchase 2.5 million mini-vans. The equation

$$14.12(0.933)^p - 0.25 = 2.5$$

is satisfied when $p \approx 23.59033$. That is, at a market price p_0 of approximately $23,600 per mini-van, consumers will purchase $q_0 = 2.5$ million mini-vans.

b. When they purchase 2.5 million mini-vans, consumers' expenditure will be

$$\int_0^{p_0} q_0 \, dp = p_0 q_0$$

$$\approx (23.59033 \text{ thousand dollars per mini-van})(2.5 \text{ million mini-vans})$$

$$\approx 59.0 \text{ billion dollars}$$

c. If the demand function approaches but does not cross the horizontal axis as price per unit increases without bound, then there is no price above which consumers will not purchase mini-vans. However, in this case, the demand function crosses the horizontal axis near $p = 58.16701$ (found by solving $D(p) = 0$). Even though in real life, we cannot be certain at which price consumers will refuse to purchase any mini-vans, we use the price indicated by the function to find the consumers' surplus (and willingness and ability to spend). According to the model, the price above which consumers will purchase no mini-vans is approximately $p = \$58.2$ thousand per mini-van.

d. Consumers' surplus is calculated as

$$\int_{p_0}^{P} D(p) \, dp \approx \int_{23.59033}^{58.16701} [14.12(0.933)^p - 0.25] \, dp$$

$$\approx 27.40482$$

To determine the appropriate units for consumers' surplus, remember that we are finding the area of a region whose width is measured in thousand dollars per mini-van and whose height is measured in million mini-vans. Thus the units on consumers' surplus are (thousand dollars per mini-van)(million mini-vans) which simplify to billion dollars.

Therefore, we estimate the consumers' surplus when purchasing 2.5 million mini-vans to be 27.4 billion dollars.

e. Consumers are willing and able to spend approximately $59.0 + 27.4 = \$86.4$ billion on 2.5 million mini-vans.   ■

Supply Curves

We have seen that when prices go up, consumers usually respond by demanding less. However, manufacturers and producers respond to higher prices by supplying more. Thus a typical curve that relates the quantity supplied to price per unit is usually increasing and appears as shown in Figure 7.28.

The curve $S(p)$ that expresses the quantity supplied in terms of the price per unit is called a **supply curve**. You should note from Figure 7.28 that there is a price p_1 under which producers are not willing or able to supply any quantity of the product. The point $(p_1, S(p_1))$ is known in economics as the **shutdown point**. If the market price (and the corresponding quantity) fall below this point, producers will shut down their production.

The supply function $S(p)$ has an interpretation very similar to that of the demand function $D(p)$. Suppose the quantity of parsley that producers will supply is modeled as

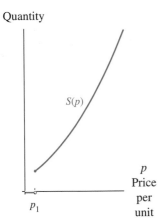

Quantity

$S(p)$

p
Price per unit

p_1

**Supply Curve $S(p)$
with Shutdown Price p_1**

FIGURE 7.28

$$S(p) = \begin{cases} 0 \text{ million pounds} & \text{when } p < 10 \\ 0.0003p^2 + 0.015p + 0.5 \text{ million pounds} & \text{when } p \geq 10 \end{cases}$$

where the market price of parsley is p dollars per pound. This function is graphed in Figure 7.29. At \$22.88 per pound, producers will supply 1 million pounds of parsley. At \$50.00 per pound, producers will supply 2 million pounds of parsley. That is, producers are willing and able to supply the first million pounds for \$22.88 per pound, but they are willing and able to supply the second million pounds for no less than \$50.00 per pound.

We approximate how much producers are willing and able to receive (at the very minimum) for 3 million pounds of parsley as follows:

First, we must know what market price corresponds to 3 million pounds of parsley, so we solve $0.0003p^2 + 0.015p + 0.5 = 3$ for p. The solution is $p_0 \approx \$69.64847$. (Recall that we do not round numbers until we have finished all of our calculations.) Another solution is $p \approx -119.65$, but negative price has no meaning in this context.

Next, we approximate producers' minimum willingness and ability to receive by using six intervals from $p = 10$ (the shutdown price for parsley production) to $p \approx 69.64847$. We construct Table 7.5.

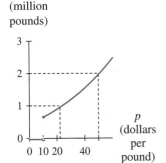

$S(p)$
(million pounds)

3

2

1

0

0 10 20   40

p
(dollars per pound)

FIGURE 7.29

TABLE 7.5

p (dollars per pound)	$S(p)$ (million pounds)
10.00000	0.68000
19.94141	0.91842
29.88282	1.21614
39.82424	1.57315
49.76565	1.98947
59.70706	2.46509
69.64847	3.00000

Using these values, we calculate the areas of stacked rectangles as we did to find consumers' willingness and ability to purchase. Refer to Figure 7.30 and Table 7.6.

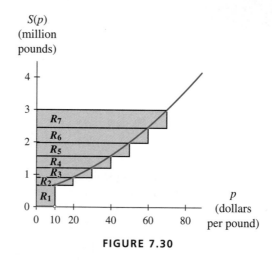

FIGURE 7.30

TABLE 7.6

Rectangle (numbered from bottom to top)	p Width of rectangle (dollars per pound)	$S(p)$ (million pounds)	Height of rectangle (million pounds)	Area of rectangle (million dollars)
1	10.00000	0.68000	0.68000	6.80000
2	19.94141	0.91842	0.23842	4.75441
3	29.88282	1.21614	0.29772	8.89666
4	39.82424	1.57315	0.35702	14.21794
5	49.76565	1.98947	0.41632	20.71824
6	59.70706	2.46509	0.47562	28.39758
7	69.64847	3.00000	0.53491	37.25596
			Total area $\approx$ 121.04	

Thus the minimum amount that suppliers are willing and able to receive is approximately $121.04 million for 3 million pounds of parsley. Note that the bottom rectangle used in this approximation has its upper right-hand corner at the shutdown point. This will always be the case.

As we use more intervals from $p = 10$ to $p \approx 69.64847$, the area of stacked rectangles comes closer to the true area of the region below the $q = 3$ line and above the $q = S(p)$ curve shown in Figure 7.31. Because a portion of $S(p)$ is zero, we find the total area by dividing the region into a rectangular region and the region below $q = 3$ and above $q = S(p)$ to the right of the shutdown point.

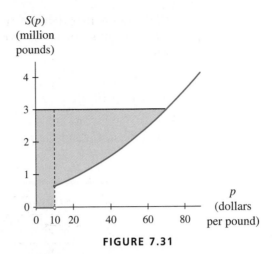

FIGURE 7.31

Therefore, the minimum amount that suppliers are willing and able to receive is calculated as

$$\int_0^{10} 3\,dp + \int_{10}^{69.64847} [3 - S(p)]\,dp$$

$$= \int_0^{10} 3\,dp + \int_{10}^{69.64847} (-0.0003p^2 - 0.015p + 2.5)\,dp$$

$$\approx \$109.80 \text{ million}$$

According to the supply model $S(p)$, suppliers are willing and able to receive no less than \$109.80 million for 3 million pounds of parsley.

The market price that will lead to the supply of 3 million pounds of parsley is $p \approx \$69.65$. At this market price, producers will receive a **total revenue** of \$208.95 million from the sale of 3 million pounds of parsley. (See Figure 7.32.) Producers will therefore receive \$99.14 million[17] in excess of the minimum they are willing to receive. This excess is known as the **producers' surplus**.

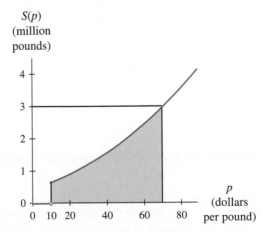

The area of the rectangle is producers' total revenue.
The area of the shaded region is producers' surplus.

FIGURE 7.32

17. Because values used in calculations should be unrounded, we use for this calculation (shown to 4 decimal places) 208.9451 − 109.80351 = 99.1419.

We calculate the producers' surplus from the sale of 3 million pounds of parsley at the market price of approximately \$69.65 directly from the supply function $S(p)$ as follows:

$$\int_{10}^{69.64847} S(p)\, dp = \int_{10}^{69.6487} (0.0003p^2 + 0.015p + 0.5)\, dp$$

$$\approx \$99.14 \text{ million}$$

In general, we find the producers' total revenue and the producers' surplus as follows:

Producers' Total Revenue and Surplus

For a piecewise continuous supply curve $S(p)$, the amount that producers receive at a certain market price p_0 is

$$\text{Total revenue} = \int_0^{p_0} q_0\, dp = q_0 p_0$$

where $q_0 = S(p_0)$ is the quantity supplied at the market price p_0.

Furthermore, the amount that producers receive above the minimum amount they are willing and able to receive for q_0 items at market price p_0 is

$$\text{Producers' surplus} = \int_{p_1}^{p_0} S(p)\, dp$$

where p_1 is the price below which production shuts down. (See Figure 7.33)

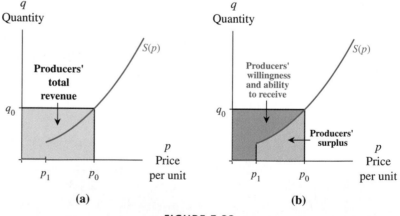

FIGURE 7.33

EXAMPLE 3   *Cellular Phone Supply*

Suppose the function for the average weekly supply of a certain brand of cellular phone can be modeled by the equation

$$S(p) = \begin{cases} 0 \text{ phones} & \text{when } p < 15 \\ 0.047p^2 + 9.38p + 150 \text{ phones} & \text{when } p \geq 15 \end{cases}$$

where p is the market price in dollars per phone.

a. How many phones (on average) will producers supply at a market price of $45.95?

b. What is the least amount that producers are willing and able to receive for the quantity of phones that corresponds to a market price of $45.95?

c. What is the producers' total revenue when the market price is $45.95?

d. What is the producers' surplus when the market price is $45.95?

Solution:

a. When the market price is $45.95, producers will supply an average of $S(45.95) \approx$ 680 phones each week.

b. Producers are willing and able to receive no less than

$$\int_0^{15} 680.247 \, dp + \int_{15}^{45.95} [680.247 - S(p)] \, dp$$

$$= 10{,}203.704 + (-0.0157p^3 - 4.69p^2 + 530.247p) \Big|_{15}^{45.95}$$

$$\approx \$16{,}300.53$$

c. When the market price is $45.95, the producers' total revenue is

(Quantity supplied at $45.95)($45.95 per phone)

$$\approx (680.247 \text{ phones})(\$45.95 \text{ per phone})$$

$$\approx \$31{,}257.35$$

d. When the market price is $45.95, the producers' surplus is

$$\int_{15}^{45.95} S(p) \, dp = (0.0157p^3 + 4.69p^2 + 150p) \Big|_{15}^{45.95} \approx \$14{,}956.82$$

Note that the producers' surplus plus the minimum amount that the producers are willing and able to receive is equal to the producers' total revenue. ∎

Social Gain

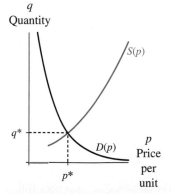

q
Quantity

$S(p)$

q^*

$D(p)$

p
Price
per
unit

p^*

Market equilibrium (p^*, q^*) occurs when demand is equal to supply.

FIGURE 7.34

Consider the economic market for a particular item for which the demand and supply curves are shown in Figure 7.34. The point (p^*, q^*) where the demand curve and supply curve cross is called the **equilibrium point**. At the equilibrium price p^*, the quantity demanded by consumers coincides with the quantity supplied by producers. This quantity is q^*.

Economists consider that society is benefited whenever consumers and/or producers have surplus funds. When the market price of a product is the equilibrium price for that product, the total benefit to society is the consumers' surplus plus the producers' surplus. This amount is known as the **total social gain**.

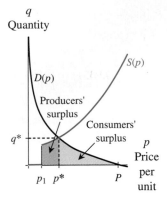

q
Quantity

FIGURE 7.35

Market Equilibrium and Social Gain

Market equilibrium occurs when the supply of a product is equal to the demand for that product. Thus the equilibrium point is the point (p^*, q^*), where p^* is the price that satisfies the equation $S(p) = D(p)$ and $q^* = D(p^*) = S(p^*)$, where $S(p)$ is the supply curve and $D(p)$ is the demand curve for a product.

The total social gain of a product when q^* units are produced and the market price is p^* is

$$\text{Total social gain} = \text{producers' surplus} + \text{consumers' surplus}$$

$$= \int_{p_1}^{p^*} S(p)\, dp + \int_{p^*}^{P} D(p)\, dp$$

where p_1 is the price below which production shuts down and P is the price above which consumers will not purchase. (See Figure 7.35.)

EXAMPLE 4   *Mini-vans*

Suppose the demand and supply curves for mini-vans in the United States are

$$\text{Demand} = D(p) = 14.12(0.933)^p - 0.25 \text{ million mini-vans}$$

and

$$\text{Supply} = S(p) = \begin{cases} 0 \text{ million mini-vans} & \text{when } p < 15 \\ 0.25p - 3.75 \text{ million mini-vans} & \text{when } p \geq 15 \end{cases}$$

where p is the market price in thousand dollars per mini-van.

a. Find the market equilibrium for mini-vans. Report both the price per unit and the quantity at market equilibrium.

b. Find the total social gain when mini-vans are sold at the market equilibrium price.

Solution:

a. Solving

$$14.12(0.933)^p - 0.25 = 0.25p - 3.75$$

for p yields $p^* \approx \$24.39963$ thousand. At this market price, $q^* \approx 2.34991$ million mini-vans will be purchased. [*Note*: q^* can be found as either $D(p^*)$ or $S(p^*)$.]

b. The total social gain at market equilibrium is

$$\text{Total social gain} = (\text{producers' surplus}) + (\text{consumers' surplus})$$

$$= \int_{p_1}^{p^*} S(p)\, dp + \int_{p^*}^{P} D(p)\, dp$$

where p_1 is the shutdown price and P is the price beyond which consumers will purchase no mini-vans. We must find these two prices before we can proceed.

The shutdown price is given in the statement of the supply function $S(p)$ as

$p_1 = 15$. The price beyond which consumers will not purchase can be found by solving $D(p) = 0$. In this case, $P \approx 58.16701$.

Now we proceed with our calculation of total social gain:

$$\text{Total social gain} \approx \int_{15}^{24.39963} (0.25p - 3.75)\,dp + \int_{24.39963}^{58.16701} [14.12(0.933)^p - 0.25]\,dp$$

$$\approx 11.04413 + 25.44287$$

$$\approx 36.487$$

Hence the total social gain is approximately \$36.5 billion. See Example 2, solution d for an explanation of these units. ∎

7.4   Concept Inventory

- Market price
- Demand curve
- Consumers' willingness and ability to spend
- Consumers' expenditure
- Consumers' surplus
- Supply curve
- Shutdown point
- Producers' total revenue
- Producers' surplus
- Producers' willingness and ability to receive
- Market equilibrium
- Total social gain

7.4   Activities

1. Give the economic name by which we call each of the following.

   a. The function relating the number of items the consumer will purchase at a certain price and the price per item

   b. The function relating the quantity of items the supplier of the items will sell at a certain price and the price per item

   c. The area of the region below the supply curve between the shutdown price and the market price

   d. The area of the region below the demand curve between the market price and the price above which consumers will cease to purchase

2. For each of the following amounts:

   i. Describe the region whose area gives the specified amount.

   ii. Illustrate that region by sketching an example.

   a. The maximum amount that consumers are willing and able to spend

   b. The minimum amount that producers are willing and able to receive

   c. The consumers' expenditure

   d. The total social gain at market equilibrium

3. Explain how to find each of the following.

   a. The price P above which consumers will purchase none of the goods or services

   b. The shutdown point

   c. The point of market equilibrium

4. The demand for wooden chairs can be modeled as

   $$D(p) = -0.01p + 5.55 \text{ million chairs}$$

   where p is the price (in dollars) of a chair.

   a. According to the model, at what price will consumers no longer purchase chairs? Is this price guaranteed to be the highest price any consumer will pay for a wooden chair? Explain.

   b. Find the quantity of wooden chairs that consumers will purchase if the market price is \$99.95.

   c. Determine the amount that consumers are willing and able to spend to purchase 3 million wooden chairs.

   d. Find the consumers' surplus when consumers purchase 3 million wooden chairs.

5. The demand for ceiling fans can be modeled as

$$D(p) = 125.92(0.996)^p \text{ thousand ceiling fans}$$

where p is the price (in dollars) of a ceiling fan.

a. According to the model, is there a price above which consumers will no longer purchase fans? If so, what is it? If not, explain why not.

b. Find the amount that consumers are willing and able to spend to purchase 10 thousand ceiling fans.

c. Find the quantity of fans consumers will purchase if the market price is $500.

d. Find the consumers' surplus when the market price is $500.

6. The demand for premium grade unleaded gasoline is given in Table 7.7.

TABLE 7.7

Price (dollars per gallon)	Demand (million gallons)
1.09	48
1.29	45
1.49	28
1.69	18
1.89	1.2
2.09	0.2

a. Find a model for demand as a function of price.

b. Does your model indicate a price above which consumers will purchase no premium grade gasoline? If so, what is it? If not, explain.

c. Find the quantity of premium grade gasoline that consumers will purchase if the market price is $1.59.

d. Find the amount that consumers are willing and able to spend to purchase the quantity you found in part c.

e. Find the consumers' surplus when the market price is $1.59.

7. The average daily demand for a new type of kerosene lantern in a certain hardware store is as shown in Table 7.8.

TABLE 7.8

Price (dollars per lantern)	Average quantity demanded (lanterns)
21.52	1
17.11	3
14.00	5
11.45	7
9.23	9
7.25	11

a. Find a model giving the average quantity demanded as a function of the price.

b. How much are consumers willing and able to spend each day for these lanterns if the market price is $12.34 per lantern?

c. Find the consumers' surplus if the equilibrium price for these lanterns is $12.34 per lantern.

8. The willingness of saddle producers to supply saddles can be modeled by the following function:

$$S(p) = \begin{cases} 0 \text{ thousand saddles} & \text{if } p < 5 \\ 2.194(1.295)^p \text{ thousand saddles} & \text{if } p \geq 5 \end{cases}$$

when saddles are sold for p thousand dollars.

a. How many saddles will producers supply if the market price is $4000? $8000?

b. At what price will producers supply 10 thousand saddles?

c. Find the producers' revenue if the market price is $7500.

d. Find the producers' surplus if the market price is $7500.

9. The willingness of answering machine producers to supply can be modeled by the following function:

$$S(p) = \begin{cases} 0 \text{ thousand answering machines} & \text{if } p < 20 \\ 0.024p^2 - 2p + 60 \text{ thousand} & \text{if } p \geq 20 \\ \quad \text{answering machines} \end{cases}$$

when answering machines are sold for p dollars.

a. How many answering machines will producers supply if the market price is $40? $150?

b. Find the producers' revenue and the producers' surplus if the market price is $99.95.

10. Table 7.9 shows the number of CDs that producers will supply at the given prices.

TABLE 7.9

Price per CD (dollars)	CDs supplied (millions)
5.00	1
7.50	1.5
10.00	2
15.00	3
20.00	4
25.00	5

a. Find a model giving the quantity supplied as a function of the price per CD. *Note*: Producers will not supply CDs if the market price falls below $4.99.

b. How many CDs will producers supply if the market price is $15.98?

c. At what price will producers supply 2.3 million CDs?

d. Find the producers' revenue and producers' surplus if the market price is $19.99.

11. Table 7.10 shows the average number of prints of a famous painting that producers will supply at the given prices.

TABLE 7.10

Price per print (hundred dollars)	Prints supplied (hundreds)
5	2
6	2.2
7	3
8	4.3
9	6.3
10	8.9

a. Find a model giving the quantity supplied as a function of the price per print. *Note*: Producers will not supply prints if the market price falls below $500.

b. At what price will producers supply 5 hundred prints?

c. Find the producers' revenue and producers' surplus if the market price is $630.

12. The daily demand for beef can be modeled by

$$D(p) = \frac{40.007}{1 + 0.033e^{0.35382p}} \text{ million pounds}$$

when the price for beef is p dollars per pound. Likewise, the supply for beef can be modeled by

$$S(p) = \begin{cases} 0 \text{ million pounds} & \text{if } p < 0.5 \\ \dfrac{51}{1 + 53.98e^{-0.3949p}} \text{ million pounds} & \text{if } p \geq 0.5 \end{cases}$$

when the price for beef is p dollars per pound.

a. How much beef is supplied when the price is $1.50 per pound? Will supply exceed demand at this quantity?

b. Find the point of market equilibrium.

13. The average quantity of sculptures that consumers will demand can be modeled as $D(p) = -1.003p^2 - 20.689p + 850.375$ sculptures, and the average quantity that producers will supply can be modeled as

$$S(p) = \begin{cases} 0 \text{ sculptures} & \text{when } p < 4.5 \\ 0.256p^2 + 8.132p & \text{when } p \geq 4.5 \\ \quad + 250.097 \text{ sculptures} \end{cases}$$

where the market price is p hundred dollars per sculpture.

a. How much are consumers willing and able to spend for 20 sculptures?

b. How many sculptures will producers supply at $500 per sculpture? Will supply exceed demand at this quantity?

c. Determine the total social gain when sculptures are sold at the equilibrium price.

14. A florist constructs Table 7.11 on the basis of sales data for roses.

TABLE 7.11

Price of 1 dozen roses (dollars)	Dozens sold per week
10	190
15	145
20	110
25	86
30	65
35	52

a. Find a model for quantity demanded.

b. Determine how much money consumers will be willing and able to spend for 80 dozen roses each week.

c. If the actual market price of the roses is $22 per dozen, find the consumers' surplus.

Suppose the suppliers of roses collect the data shown in Table 7.12.

TABLE 7.12

Price of 1 dozen roses (dollars)	Dozens supplied per week
20	200
18	150
14	100
11	80
8	60
5	50

d. Find an equation that models the supply data.

e. What is the producers' surplus when the market price is $17 per dozen?

f. For what price will roses be sold at the equilibrium point?

g. What is the total social gain from the sale of roses at market equilibrium?

15. Table 7.13 gives both the number of copies of a hardback science fiction novel in demand and the number supplied at certain prices.

a. Find an exponential model for demand given the price per book.

b. Find a model for supply given the price per book. *Note*: Producers are not willing to supply any books when the market price is less than $18.97.

c. At what price will market equilibrium occur? How many books will be supplied and demanded at this price?

d. Find the total social gain from the sale of a hardback science fiction novel at the market equilibrium price.

16. Table 7.14 shows both the number of a certain type of graphing calculator in demand and the number supplied at certain prices.

a. Find a model for demand given the price per calculator.

b. Find a model for supply given the price per calculator. *Note*: Producers are not willing to supply any of these graphing calculators when the market price is less than $45.95.

c. At what price will market equilibrium occur? How many calculators will be supplied and demanded at this price?

d. Find the producers' surplus at market equilibrium.

e. Estimate the consumers' surplus at market equilibrium.

f. Estimate the total social gain from the sale of this type of graphing calculator at the market equilibrium price.

TABLE 7.13

Price (dollars per book)	20	23	25	28	30	32
Books demanded (thousands)	214	186	170	150	138	128
Books supplied (thousands)	120	130	140	160	190	210

TABLE 7.14

Price (dollars per calculator)	60	90	120	150	180	210
Calculators demanded (millions)	35	31	15	5	3	0.1
Calculators supplied (millions)	10	32	50	80	100	120

7.5   Average Values and Average Rates of Change

In our study of calculus, we have been analyzing situations that are continuously undergoing change. We have seen that when we model a quantity that is constantly changing, we use derivatives to find instantaneous rates of change. Sometimes we can actually measure rates of change (such as velocity, flow rates, wind speed, etc.) and use definite integrals to find accumulated change.

It is often interesting or necessary to be able to find the average value of something that is in a constant state of flux. For instance, we may be interested in knowing the average temperature on a given day, or we may need to know the average water level on a certain beach. An investor would be more interested in the average rate of return on an investment than in its instantaneous rate of return. A meteorologist may be as interested in the average wind speed during a storm as in the instantaneous wind speeds (gusts and lulls).

As we saw in Chapter 3, if we have a model $f(x)$ for some changing quantity, then we calculate the average rate of change in the quantity over an interval from a to b as

$$\text{Average rate of change} \atop \text{of } f(x) \text{ from } a \text{ to } b = \frac{f(b) - f(a)}{b - a}$$

For instance, given a function for population, we can calculate the average rate of growth of population. Aurora, Nevada, was a booming mining town in the 1860s and early 1870s. The sites of the Winnemucca and Esmeralda lodes were discovered by three men on August 25, 1860.[18] As word of the discovery of silver spread, miners flocked to the region, and soon the town of Aurora was established. By August of 1865, the town had a population of 3286. Aurora reached its peak population of over 6000 people in 1870. By 1915, the last of the hangers-on had left, and Aurora became a ghost town. The population of Aurora can be modeled as

$$p(t) = \begin{cases} -7.91t^3 + 120.96t^2 + 193.92t - 123.\dot{2}1 \text{ people} & \text{when } 0.7 \leq t \leq 13 \\ 45{,}544(0.8474)^t \text{ people} & \text{when } 13 < t \leq 55 \end{cases}$$

where t is the number of years since the beginning of 1860.

Using the model $p(t)$ and its derivative function

$$p'(t) = \begin{cases} -23.73t^2 + 241.92t + 193.92 \text{ people per year} & \text{when } 0.7 \leq t < 13 \\ -7541.287(0.8474)^t \text{ people per year} & \text{when } 13 < t \leq 55 \end{cases}$$

we calculate the following: by January 1861, there were 184 people in Aurora, and by January 1871, there were 6118 people. The population of Aurora was growing by 412 people per year in January 1861 (that is a percentage rate of growth of 224% per year) and was declining by 16 people per year in January 1871.

Knowing these two instantaneous rates of change gives us a picture of the change occurring in 1861 and 1871 but does not reveal much about the change that occurred between these two times. To get an overall picture of the change that occurred in the population of Aurora between January 1861 and January 1871, we could calculate the instantaneous rates at several points between January 1861 and

18. Don Ashbaugh, *Nevada's Turbulent Yesterday: A Study in Ghost Towns* (Los Angeles: Westernlore Press, 1963).

January 1871, or we could calculate the average rate of change between January 1861 and January 1871.

The average rate of change that occurred in the population between January 1861 and January 1871 is

$$\text{Average rate of change from January 1861 through January 1871} = \frac{p(11) - p(1)}{11 - 1} \approx 593.41 \text{ people per year}$$

In other words, the population of Aurora increased by 593 people per year (on average) between January 1861 and January 1871.

Average Value of a Rate of Change

Now let us look at this change from a different perspective. If it is available, we can use the information about instantaneous rate of change given by this model to find an average rate of change. Suppose that we do not have data or a model for the population of Aurora but have only the data (and hence a model) for the rate at which Aurora's population was growing or declining:

$$p'(t) = \begin{cases} -23.73t^2 + 241.92t + 193.92 \text{ people per year} & \text{when } 0.7 \leq t < 13 \\ -7541.287(0.8474)^t \text{ people per year} & \text{when } 13 < t \leq 55 \end{cases}$$

where t is the number of years since the beginning of 1860.

We use the rate-of-change function $p'(t)$ to find the rate of change of population at the beginnings of 1861, 1863, 1865, 1867, and 1869. See Table 7.15.

TABLE 7.15

Beginning of year	t	$p'(t)$ (people per year)
1861	1	412.11
1863	3	706.11
1865	5	810.27
1867	7	724.59
1869	9	449.07

If we consider these five rate-of-change values representative of all the rates of change from 1861 to 1871, then the average of the rates of change in population is approximated as

$$\text{Average value of the rate of change from January 1861 through January 1871} \approx \frac{412.11 + 706.11 + 810.27 + 724.59 + 449.07}{5} \text{ people per year}$$

$$= 620.43 \text{ people per year}$$

We call this type of average the **average value of the rate of change.** The value 620.43 is a fairly rough approximation of the average value of the rate of change. If we use more readings, then we improve our approximation of the average of the rates of change. Yearly rates of change for 1861 through 1870 are shown in Table 7.16.

TABLE 7.16

Beginning of year	t	$p'(t)$ (people per year)
1861	1	412.11
1862	2	582.84
1863	3	706.11
1864	4	781.92
1865	5	810.27
1866	6	791.16
1867	7	724.59
1868	8	610.56
1869	9	449.07
1870	10	240.12

The average of these rates of change is

Average value of the
rate of change from $\approx \dfrac{412.11 + 582.84 + 706.11 + \cdots + 449.07 + 240.12}{10}$
January 1861
through January 1871

$$= 610.875 \text{ people per year}$$

In general, the average value of the rate of change of a quantity that is changing continuously can be approximated as

Average value of the rate of change $\approx \dfrac{f'(x_1) + f'(x_2) + \cdots + f'(x_n)}{n}$
of $f(x)$ from a to b

where the rate of change $f'(x)$ is given for n equally spaced input values.

The sum in the numerator of this approximation should look familiar. It should remind you of the sum of rectangles from Sections 6.1 through 6.3, except that these "heights" are not multiplied by the widths of the rectangles. We can rectify this by multiplying both numerator and denominator by the width of the rectangles (the distance between the equally spaced input values): $\Delta x = \frac{b - a}{n}$.

Thus

Average value of rate of change $\approx \dfrac{[f'(x_1) + f'(x_2) + \cdots + f'(x_n)]\Delta x}{n\Delta x}$
of $f(x)$ from a to b

$$= \dfrac{[f'(x_1) + f'(x_2) + \cdots + f'(x_n)]\,\Delta x}{b - a}$$

Because the population of Aurora could be considered as changing continuously from 1861 until 1871, we expect that we can improve the accuracy of our approximation of the average of the rates of change by using more frequent readings of the instantaneous rate of change $p'(t)$. That is, instead of readings every year, we use readings every quarter, or every month, or every week, and so on. As our readings become more frequent, the interval Δt between readings becomes smaller, and the number n of readings becomes larger.

In other words, we consider the average value of the rate of change as n approaches infinity:

$$\text{Average value of rate of change of } f(x) \text{ from } a \text{ to } b = \frac{\lim\limits_{n \to \infty} [f'(x_1) + f'(x_2) + \cdots + f'(x_n)]\Delta x}{b - a}$$

The numerator is the limit of sums that is defined as the definite integral from a to b. We therefore make the following definition:

The Average Value of the Rate of Change

If $f'(x)$ is a smooth, continuous rate-of-change function from a to b, then the average value of $f'(x)$ from a to b is

$$\text{Average value of rate of change of } f(x) \text{ from } a \text{ to } b = \frac{\displaystyle\int_a^b f'(x)\,dx}{b - a} = \frac{f(b) - f(a)}{b - a}$$

where $f(x)$ is an antiderivative of $f'(x)$.

In the case of the population of Aurora, we calculate the average value of the rate of change of population between January 1861 and January 1871 as

$$\text{Average value of the rate of change from January 1861 through January 1871} = \frac{\displaystyle\int_1^{11} p'(t)\,dt}{11 - 1}$$

$$= \frac{\displaystyle\int_1^{11} (-23.73t^2 + 241.92t + 193.92)\,dt}{11 - 1}$$

$$\approx \frac{6241.07 - 306.97}{11 - 1}$$

$$\approx 593.41 \text{ people per year}$$

You should note that the average value of the rate of change in population is the same as the average rate of change of population that we calculated earlier. This is true in general for any continuous function. Hence we will use the phrases *average rate of change* and *average value of the rate of change* interchangeably. Note that we have two methods for calculating the average rate of change: one using the rate-of-change function, the other using the quantity function. Also note that the average value of the rate of change is always measured using the same units as the rate-of-change function.

EXAMPLE 1   *Carbon-14*

Scientists estimate that 100 milligrams of the isotope ^{14}C used in carbon dating methods decays at a rate of

$$r(t) = -0.01209(0.999879)^t \text{ milligrams per year}$$

where t is the number of years since the 100 milligrams of isotope began to decay.

What is the average rate of decay during the first 1000 years? during the second 1000 years?

Solution: We find the average rate of decay during the first 1000 years as

$$\frac{\int_0^{1000} r(t)\,dt}{1000 - 0} \approx \text{-}0.0114 \text{ milligram per year}$$

and that the average rate of decay during the second 1000 years is

$$\frac{\int_{1000}^{2000} r(t)\,dt}{2000 - 1000} \approx \text{-}0.0101 \text{ milligram per year}$$

In other words, the amount of ^{14}C decreased by 0.0114 milligram per year (on average) during the first 1000 years and by 0.0101 milligram per year (on average) during the second 1000 years.   ■

Average Value of a Quantity

We can use definite integrals to calculate the average value of the rate of change if we are given an instantaneous rate-of-change function. Similarly, if we are given a function for quantity, we can use definite integrals to calculate the average value of that quantity.

Recall the function giving the population of Aurora t years after the beginnning of 1860:

$$p(t) = \begin{cases} \text{-}7.91t^3 + 120.96t^2 + 193.92t - 123.21 \text{ people} & \text{when } 0.7 \le t \le 13 \\ 45{,}544(0.8474)^t \text{ people} & \text{when } 13 < t \le 55 \end{cases}$$

What is the average population of Aurora between January 1861 and January 1871?

Earlier we accumulated the rates of change and divided by $(b - a)$, the width of the interval, to find average rate of change. Now we accumulate the quantity and divide by the width of the interval to find average value.

$$\begin{aligned} \text{Average population from January 1861 through January 1871} &= \frac{\int_1^{11} p(t)\,dt}{11 - 1} \\[2mm] &= \frac{(\text{-}1.9775t^4 + 40.32t^3 + 96.96t^2 - 123.21t)\big|_1^{11}}{11 - 1} \\[2mm] &\approx \frac{35{,}090.1925 - 12.0925}{11 - 1} \\[2mm] &\approx 3507.8 \text{ people} \end{aligned}$$

In general, we define the average value of a quantity function as follows:

Average Value

If $f(x)$ is a smooth, continuous function from a to b, then the average value of $f(x)$ from a to b is

$$\text{Average value of } f(x) \text{ from } a \text{ to } b = \frac{\int_a^b f(x)\,dx}{b - a}$$

$$= \frac{F(b) - F(a)}{b - a}$$

where $F(x)$ is an antiderivative of $f(x)$.

EXAMPLE 2   *Temperature*

Suppose that the hourly temperatures shown in Table 7.17 were recorded from 7 a.m. to 7 p.m. one day in September.

TABLE 7.17

Time	Temperature (°F)
7 a.m.	49
8 a.m.	54
9 a.m.	58
10 a.m.	66
11 a.m.	72
noon	76
1 p.m.	79
2 p.m.	80
3 p.m.	80
4 p.m.	78
5 p.m.	74
6 p.m.	69
7 p.m.	62

a. Fit a cubic model to this set of data.

b. Calculate the average temperature between 9 a.m. and 6 p.m.

c. Calculate the average rate of change of temperature from 9 a.m. to 6 p.m.

Solution:

a. The temperature on this particular day can be modeled as

$$t(h) = -0.03526h^3 + 0.71816h^2 + 1.584h + 13.689 \text{ degrees Fahrenheit}$$

h hours after midnight. This model applies only from $h = 7$ (7 a.m.) to $h = 19$ (7 p.m.).

b. The average temperature between 9 a.m. ($h = 9$) and 6 p.m. ($h = 18$) is

$$\text{Average temperature} = \frac{\displaystyle\int_9^{18} t(h)\,dh}{18 - 9} \approx 74.4\,°\text{F}$$

c. The average rate of change of temperature from 9 a.m. to 6 p.m. is

$$\begin{array}{l}\text{Average rate of change} \\ \text{of temperature}\end{array} = \frac{t(18) - t(9)}{18 - 9} \approx 0.98\,°\text{F per hour} \qquad ∎$$

If we have a function $f'(x)$ for the rate of change of a quantity and we know the value of that quantity for a certain input value, then we can find the average value of the quantity.

EXAMPLE 3   *Population Growth*

The growth rate of the population of South Carolina can be modeled[19] as

$$p'(t) = 0.1552t + 0.223 \text{ thousand people per year}$$

where t is the number of years since 1790. The population of South Carolina in 1990 was 3487 thousand people.

a. What was the average rate of change in population from 1990 through 1994?
b. What was the average size of the population from 1990 through 1994?

Solution:

a. The average rate of change is calculated directly from the rate-of-change function $p'(t)$ as

$$\begin{array}{l}\text{Average rate of change} \\ \text{from 1990 through 1994}\end{array} = \frac{\displaystyle\int_{200}^{204} p'(t)\,dt}{204 - 200}$$

$$\approx 31.6 \text{ thousand people per year}$$

b. In order to calculate average population, we must have a function for population. That is, we need an antiderivative of $p'(t)$:

$$p(t) = \int p'(t)\,dt$$

$$= 0.0776t^2 + 0.233t + C \text{ thousand people}$$

---

19. Based on data from *South Carolina Statistical Abstract*, 1994.

We know that the population in 1990 was 3487 thousand people. Using this fact, we solve for C, so the function for population is

$$p(t) = 0.0776t^2 + 0.233t + 336.4 \text{ thousand people}$$

where t is the number of years since 1790.

Now we calculate the average population as

$$\text{Average population} \atop \text{from 1990 through 1994} = \frac{\displaystyle\int_{200}^{204} p(t)\,dt}{204 - 200} \approx 3550 \text{ thousand people} \qquad \blacksquare$$

7.5   Concept Inventory

- ■ Average value of a function
- ■ Average rate of change of a function
- ■ Average value of a rate-of-change function

7.5   Activities

1. The Highway Department is concerned about the high speed of traffic during the weekday afternoon rush hours from 4 p.m. to 7 p.m. on a newly widened stretch of interstate highway that is just inside the city limits of a certain city. The Office of Traffic Studies has collected the data given in Table 7.18, which show typical weekday speeds during the 4 p.m. to 7 p.m. rush hours.

TABLE 7.18

Time	Speed (mph)
4:00	60
4:15	61
4:30	62.5
4:45	64
5:00	66.25
5:15	67.5
5:30	70
5:45	72.25
6:00	74
6:15	74.5
6:30	75
6:45	74.25
7:00	73

a. Fit a model to the data.

b. Use your model to approximate the average weekday rush-hour speed from 4 p.m. to 7 p.m.

c. Use your model to approximate the average weekday rush-hour speed from 5 p.m. to 7 p.m.

2. U.S. factory sales of electronics from 1986 through 1990 can be modeled[20] by the equation

$$\text{Sales} = 220 - 44.43e^{-0.3912t} \text{ billion dollars}$$

where t is the number of years since 1986.

a. Use a definite integral to approximate the average annual value of U.S. factory sales of electronics over the 4-year period from 1986 through 1990.

b. Sketch the graph of sales from 1986 through 1990, and draw the horizontal line representing the average value.

3. The most expensive rates (in dollars per minute) for a 2-minute telephone call using a long-distance carrier are listed in Table 7.19.

TABLE 7.19

Year	Rate (dollars per minute)
82	1.32
84	1.24
85	1.14
86	1.01
87	0.83
88	0.77
89	0.65
90	0.65

---

20. Based on data from *Statistical Abstract*, 1992.

a. Find a model to fit the data.

b. Use a definite integral to estimate the average of the most expensive rates from 1982 through 1990.

c. Use a definite integral to estimate the average of the most expensive rates from 1985 through 1990.

4. Table 7.20 gives the price (in dollars) of a round-trip flight from Denver to Chicago on a certain airline and the corresponding monthly profit (in millions of dollars) for that airline for that route.

TABLE 7.20

Ticket price (dollars)	Profit (millions of dollars)
200	3.08
250	3.52
300	3.76
350	3.82
400	3.70
450	3.38

a. Fit a model to the data.

b. Determine the average profit for ticket prices from $325 to $450.

c. Determine the average rate of change of profit when the ticket price rises from $325 to $450.

5. The population of Mexico between 1921 and 1990 is given by the model[21]

$$\text{Population} = 12.921e^{0.026578t} \text{ million people}$$

where t is number of years since the end of 1921.

a. What was the average population of Mexico from the beginning of 1980 through the end of 1989?

b. In what year was the population of Mexico equal to its 1980s average?

c. What was the average rate of change of the population of Mexico during the 1980s?

6. The number of AIDS cases diagnosed from 1988 through 1991 can be modeled[22] by

$$\frac{\text{Cases}}{\text{diagnosed}} = -1049.50x^2 + 5988.7x + 33,770.7 \text{ cases}$$

where x is the number of years since the end of 1988.

a. Use a definite integral to estimate the average number of cases diagnosed each year between the end of 1988 and the end of 1991.

b. Find the average rate of change in cases diagnosed from the end of 1988 through 1991.

c. In which year was the number of cases diagnosed closest to the average number of cases diagnosed from the end of 1988 through 1991?

7. The number of general-aviation aircraft accidents from 1975 through 1992 can be modeled[23] by

$$\frac{\text{Number of}}{\text{accidents}} = -123.7746x + 4057.6633 \text{ accidents}$$

where x is the number of years since 1975.

a. Calculate the average rate of change in the yearly number of accidents from 1976 through 1992.

b. Use a definite integral to estimate the average number of accidents that occurred each year from 1976 through 1992.

8. During a summer thunderstorm, the temperature drops and then rises again. The rate of change of the temperature during the hour and a half after the storm began is given by

$$T(h) = 9.48h^3 - 15.49h^2 + 17.38h - 9.87 \text{ °F per hour}$$

where h is the number of hours since the storm began.

a. Calculate the average rate of change of temperature from 0 to 1.5 hours after the storm began.

b. If the temperature was 85 °F at the time the storm began, find the average temperature during the first 1.5 hours of the storm.

21. Based on data from SPP and INEGI, Mexican Censuses of Population 1921 through 1990 as reported by Pick and Butler, *The Mexico Handbook* (Westview Press, 1994).

22. Based on information in the *HIV/AIDS Surveillance* 1992 Year End Edition.

23. Based on data from *Statistical Abstract*, 1994.

9. The acceleration of a race car during the first 35 seconds of a road test is modeled by

$$a(t) = 0.024t^2 - 1.72t + 22.58 \text{ ft/sec}^2$$

where t is the number of seconds since the test began. Assume that velocity and distance were both 0 at the beginning of the road test.

a. Calculate the average acceleration during the first 35 seconds of the road test.

b. Calculate the average velocity during the first 35 seconds of the road test.

c. Calculate the distance traveled during the first 35 seconds of the road test.

d. If the car had been traveling at its average velocity throughout the 35 seconds, how far would the car have traveled during that 35 seconds?

10. On the basis of data obtained from a preliminary report by a geological survey team, it is estimated that for the first 10 years of production, a certain oil well can be expected to produce oil at the rate of $r(t) = 3.93546t^{3.55}e^{-1.35135t}$ thousand barrels per year, t years after production begins. Estimate the average annual yield from this oil field during the first 10 years of production.

11. An article in the May 23, 1996, issue of *Nature* addresses the interest some physicists have in studying cracks in order to answer the question "How fast do things break, and why?" Data estimated from a graph in this article showing velocity of a crack during a 60-microsecond experiment are shown in Table 7.21.

TABLE 7.21

Time (microseconds)	Velocity (meters per second)
10	148.2
20	159.3
30	169.5
40	180.7
50	189.8
60	200

a. Find a model to fit the data.

b. Determine the average speed at which a crack travels between 10 and 60 microseconds.

12. The amount[24] of poultry produced in the United States annually from 1960 through 1993 can be modeled as $p(x) = 6.178983e^{0.047509x}$ million pounds per year, x years after 1960.

a. Estimate the average amount of poultry produced annually from 1960 through 1993.

b. In what year was poultry production closest to the average annual production from 1960 through 1993?

13. Blood pressure varies for individuals throughout the course of a day, typically being lowest at night and highest from late morning to early afternoon. The estimated rate of change in diastolic blood pressure for a patient with untreated hypertension is shown in Table 7.22.

TABLE 7.22

Time	Diastolic BP (mm Hg per hour)
8 a.m.	3.0
10 a.m.	1.8
12 p.m.	0.7
2 p.m.	-0.1
4 p.m.	-0.7
6 p.m.	-1.1
8 p.m.	-1.3
10 p.m.	-1.1
12 a.m.	-0.7
2 a.m.	0.1
4 a.m.	0.8
6 a.m.	1.9

a. Find a model for the data.

b. Estimate the average rate of change in diastolic blood pressure from 8 a.m. to 8 p.m.

c. Assuming that diastolic blood pressure was 95mm Hg at 12 p.m., estimate the average diastolic blood pressure between 8 a.m. and 8 p.m.

24. Based on information in *USA Today*, August 12, 1994. Source: National Agricultural Statistics Service.

14. The air speed of a small airplane during the first 25 seconds of takeoff and flight can be modeled by

$$v(t) = -940{,}602t^2 + 19{,}269.3t - 0.3 \text{ mph}$$

    t hours after takeoff.

    a. Find the average air speed during the first 25 seconds of takeoff and flight.

    b. Find the average acceleration during the first 25 seconds of takeoff and flight.

15. From data taken from the 1994 *Statistical Abstract*, we model the rate of change of the percentage of the population of the United States living in the Great Lakes region from 1960 through 1990 as

$$P(t) = (9.371 \cdot 10^{-4})t^2 - 0.141t + 5.083$$
$$\text{percentage points per year}$$

where t is the number of years since 1900. Find the average rate of change of the percentage of the population of the United States living in the Great Lakes region from 1960 through 1990.

Chapter 7   Summary

In this chapter we have discussed many uses for integrals in real-world situations. More applications have been given for definite integrals: finding the differences of accumulated changes, the future and present value of an income stream, the future value of a biological stream, consumers' and producers' surplus, the average value of a rate of change, and the average value of a function.

Because some applications involve processes that continue indefinitely, we introduced a different type of integral—an improper integral. This enabled us to expand many of the definite integral applications to cases where there is no finite upper limit on the input interval.

Differences of Accumulated Changes

As we saw in Chapter 6, many real-world problems can be solved by relating the situation in the problem to the area under a curve. In this chapter, we investigated situations that lend themselves to analysis by examination of the area between two curves. To compute the area between two curves, we used the fact that if the graph of $f(x)$ lies above the graph of $g(x)$ from a to b, then the integral $\int_a^b [f(x) - g(x)]\, dx$ is the area of the region between the two graphs from a to b.

We observed that when $f(x)$ and $g(x)$ are two continuous rate-of-change functions, then the difference between the accumulated change of $f(x)$ from a to b and the accumulated change of $g(x)$ from a to b is also given by $\int_a^b [f(x) - g(x)]\, dx$. However, if the two functions intersect between a and b, then the difference between the accumulated changes of the functions is *not* the same as the total area of the regions between the two rate-of-change curves.

Perpetual Accumulation

It is sometimes necessary to accumulate quantities over infinitely long intervals. The integrals involved in such accumulations are called improper integrals. To understand how to evaluate improper integrals, we first considered the behavior of several different functions as the input either increased or decreased without bound. This gave us a group of limit rules that are useful when we evaluate improper integrals.

Improper integrals differ from definite integrals in that either one or both of the limits on the integral are infinite. We considered only those improper integrals that equal specific numbers and found the perpetual accumulation of change by using the definitions

$$\int_a^\infty f(x)\, dx = \lim_{N \to \infty} \int_a^N f(x)\, dx = \left[\lim_{N \to \infty} F(N)\right] - F(a)$$

$$\int_{-\infty}^b f(x)\, dx = \lim_{N \to -\infty} \int_N^b f(x)\, dx = F(b) - \lim_{N \to -\infty} F(N)$$

where $F(x)$ is an antiderivative of $f(x)$.

Streams in Business and Biology

An income stream is a flow of money into an interest-bearing account over a period of time. If the stream flows continuously into an account at a rate of $R(t)$ dollars per year and the account earns annual interest at the rate of $r\%$ compounded continuously (where r is a decimal number), then the future value of the account at the end of T years is given by

$$\text{Future value} = \int_0^T R(t) e^{r(T-t)}\, dt \text{ dollars}$$

The present value of an income stream is the amount that would have to be invested now in order for the account to grow to a given future value. The present value of an income stream whose future value is given by the above equation is

$$\text{Present value} = \int_0^T R(t) e^{-rt}\, dt \text{ dollars}$$

Although it may be feasible to consider money flowing as a continuous stream in very large corporations, there are many business applications for income streams that have quarterly or monthly activity. We call such streams discrete income streams and determine their present and future values by summing rather than integrating.

Streams also have applications in biology and related fields. The future value (in b years) of a biological stream with initial population size P, survival rate s (in decimals), and renewal rate $r(t)$, where t is the number of years of the stream, is

$$\text{Future value} \approx Ps^b + \int_0^b r(t) s^{(b-t)}\, dt$$

Integrals in Economics

A demand curve for a commodity is determined by economic factors. Points on the demand curve tell us how much consumers will purchase at the associated prices. However, consumers are usually willing and able to pay more for the quantity they purchase. Once a market price (the price at which the item is sold) is set, the actual amount consumers spend for a certain quantity of items is the consumers' expenditure. The amount that consumers are willing and able to spend but do not actually spend is known as the consumers' surplus. We use areas of regions under the demand curve to represent these quantities.

In Figure 7.36 the area of region R_1 is the consumers' expenditure. The area of region R_2 is the consumers' surplus. The area of region R_1 plus the area of region R_2 is the maximum amount that consumers are willing and able to spend for q_0 items.

The interaction between supply and demand usually determines the quantity of an item that is available. The producers' supply curve $S(p)$ gives the number of units that a producer is willing and able to supply as a function of p, the price per unit of the item. When prices increase, consumers usually respond by demanding less, but producers ordinarily respond by supplying more. Commodities may have a price, called the shutdown price, under which producers are not willing or able to supply any of the product, so the supply curve has output 0 to the left of this price. Thus a supply curve is the graph of a piecewise continuous function.

Points on the supply curve tell us how much producers will supply at the associated prices. However, producers are often willing and able to supply more of the commodity they produce. The amount that producers receive for supplying a certain quantity of items at a certain market price is the producers' total revenue. The amount that producers receive above the minimum amount they are willing and able to receive is the producers' surplus.

Like demand curves, supply curves are graphs of nonnegative functions, and the integral formulas for areas of certain regions associated with the supply curve have economic interpretations. In Figure 7.37, the area of region R_1 is the least amount that producers are willing and able to receive for q_0 items. The area of region R_2 is the producers' surplus. The area of region R_1 plus the area of region R_2 (the area of the rectangle) is the producers' total revenue.

The point (p^*, q^*) where the supply and demand curves for a particular item cross is called the equilibrium point. At the equilibrium price p^*, the quantity demanded by consumers coincides with the quantity supplied by producers, which is q^*. When the market price of a commodity is the equilibrium price for that product, the total social gain is the sum of the consumers' surplus and the producers' surplus.

Average Values and Average Rates of Change

When we are given a rate-of-change function $f'(t)$, the average rate of change of $f(t)$ from $t = a$ to $t = b$ is

$$\text{Average rate of change of } f(t) \text{ from } a \text{ to } b = \frac{\displaystyle\int_a^b f'(t)\,dt}{b - a}$$

If $f(t)$ is an antiderivative of $f'(t)$, the average rate of change can also be calculated by using the quantity function $f(t)$:

$$\text{Average rate of change of } f(t) \text{ from } a \text{ to } b = \frac{f(b) - f(a)}{b - a}$$

Just as we can use definite integrals to calculate the average value of the rate of change when we are given an instantaneous rate of change function, we can use definite integrals to calculate the average value of a continuous function for a quantity:

$$\text{Average value of } f(x) \text{ from } a \text{ to } b = \frac{\displaystyle\int_a^b f(x)\,dx}{b - a}$$

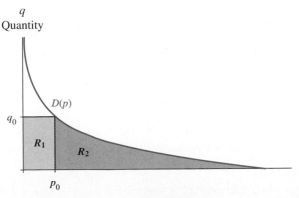

FIGURE 7.36

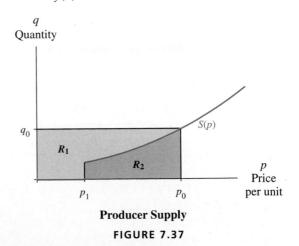

Producer Supply

FIGURE 7.37

Chapter 7 Review Test

1. In preparing for your retirement (in 40 years), you plan to invest 10% of your salary each month in an annuity with a fixed rate of return of 8.3%. You currently make $3000 per month and expect your income to increase by $500 per year.

   a. Find a function for the yearly rate at which you will invest money in the annuity.

   b. If you start investing now, to what amount will your annuity grow in 40 years? (Consider a continuous stream.)

   c. How much would you have to invest now in one lump sum, instead of in a continuous stream, in order to build to the same future (40-year) value?

2. Suppose a 1990 population of 10,000 foxes breeds at a rate of 500 pups per year and has a survival rate of 63%.

   a. Assuming that the survival and renewal rates remain constant, determine how many of the foxes alive in 1990 will still be alive in 2010.

   b. Write a function for the number of foxes that were born t years after 1990 and will still be alive in 2010.

   c. Estimate the fox population in the year 2010.

3. The demand for 20-inch color televisions can be modeled as

   $$D(p) = 230(0.993)^p \text{ thousand televisions}$$

   where p is the price (in dollars) of a television.

   a. According to the model, is there a price above which consumers will no longer purchase televisions? If so, what is it? If not, explain why not.

   b. Find the amount that consumers are willing and able to spend to purchase 10 thousand televisions.

   c. Find the quantity of televisions that consumers will purchase if the market price is $300.

   d. Find the consumers' surplus when the market price is $300.

4. The average quantity of marble fountains that consumers will demand can be modeled as

   $$D(p) = -1.0\,p^2 - 20.6p + 900 \text{ fountains}$$

   and the average quantity that producers will supply can be modeled as

   $$S(p) = \begin{cases} 0 \text{ fountains} & \text{if } p < 2 \\ 0.3p^2 + 8.1p + 300 \text{ fountains} & \text{if } p \geq 2 \end{cases}$$

   when the market price is p hundred dollars per fountain.

   a. How much are consumers willing to spend for 30 fountains?

   b. How many fountains will producers supply at $1000 per fountain? Will supply exceed demand at this quantity?

   c. Determine the total social gain when fountains are sold at the equilibrium price.

5. The national unemployment rates from July of 1995 through January of 1996 as supplied by the United States Department of Labor can be modeled by the function

   $$U(m) = 0.02476m^2 - 0.48095m + 7.85429$$
   percent unemployment

   where $m = 7$ in July of 1995, $m = 8$ in August of 1995, and so on.

   a. Find the average unemployment rate between July of 1995 and January of 1996.

   b. Find the average rate of change of unemployment for that same time period.

Project 7.1

Arch Art

Setting

A popular historical site in Missouri is the Gateway Arch. Designed by Eero Saarinen, it is located on the original riverfront town site of St. Louis and symbolizes the city's role as gateway to the West. The stainless steel Gateway Arch (also called the St. Louis Arch) is 630 feet (192 meters) high and has an equal span.

In honor of the 200th anniversary of the Louisiana Purchase, which made St. Louis a part of the United States, the city has commissioned an artist to design a work of art at the Jefferson National Expansion Memorial National Historic Site. The artist plans to construct a hill beneath the Gateway Arch, located at the Historic Site, and hang strips of mylar from the arch to the hill so as to completely fill the space. (See Figure 7.38.) The artist has asked for your help in determining the amount of mylar needed.

Tasks

1. If the hill is to be 30 feet tall at its highest point, find an equation for the height of the cross-section of the hill at its peak. Refer to Figure 7.38.

2. Estimate the height of the arch in at least ten different places. Use the estimated heights to construct a model for the height of the arch. (You need not consider only the models presented in this text.)

3. Estimate the area between the arch and the hill.

4. The artist plans to use strips of mylar 60 inches wide. What is the minimum number of yards of mylar that the artist will need to purchase?

5. Repeat Task 4 for strips 30 inches wide.

6. If the 30-inch strips cost half as much as the 60-inch strips, is there any cost benefit to using one width instead of the other? If so, which width? Explain.

Reporting

Write a memo telling the artist the minimum amount of mylar necessary. Explain how you came to your conclusions. Include your mathematical work as an attachment.

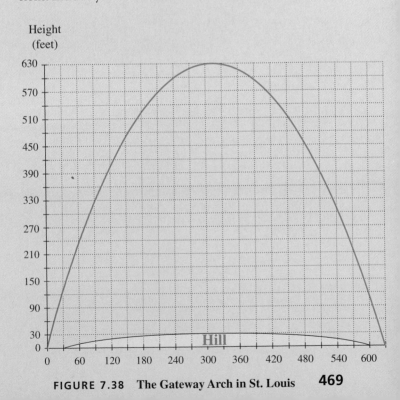

FIGURE 7.38   The Gateway Arch in St. Louis   469

ANSWERS TO SELECTED ACTIVITES
CHAPTER 1

Section 1.1

1. (a) $60
   (b) 3 CDs
   (c) 7 CDs
   (d) The average price of 3 CDs is $12 each.
       The average price of 6 CDs is $10 each.

3. (a) Approximately $9400
   (b) Approximately $340
   (c) Approximately $205
   (d) The graph would pass through (0,0) but would lie below the graph in Figure 1.1.2 because the same monthly payment would pay for a smaller loan amount.

5. (a) 3.7%
   (b) The cost-of-living increase was greatest in 1991 at 5.4%.
   (c) 1994 and 1996
   (d) Benefits increased, but the percentage by which they increased decreased.

7. (a) $6.88
   (b) 9.6 gallons
   (c) $C(g) = 1.25g$ dollars for g gallons of gas

9. $V = 40a + 1.25w + 2.85d$ dollars, where a is the area of the home in square feet, w is the area of the driveway in square feet, and d is the area of the deck in square feet
   $V = 40(1850) + 1.25(400) + 2.85(350) = \$75,497.50$

11. Weight
   (pounds)

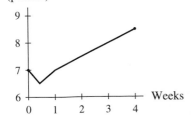

13. (a) Inches
   (b) About 16 inches long
   (c) About 11 inches
   (d) At about 16 years of age, she reached her full height of approximately 60 inches (5 feet).
   (e) During the first 3 years
   (f) Height should not increase or decrease after age 20 (until in her later years, when her height may decline).

15. (a) About 5 inches
   (b) For approximately 5 days

(c) Snow fell.
   (d) The snow settled.
   (e) The snow was deepest (4 feet 4 inches) around February 21.
   (f) Warm temperatures probably caused the decline in late February.
   (g) The most snow fell around February 18.

17. (a)

x	0	1	2	3	4	5
y	500	400	300	200	100	0

   (b) $y = 500 - 100x$ milligrams after x days
   (c) x is between 0 and 5 days.
       y is between 0 and 500 milligrams.

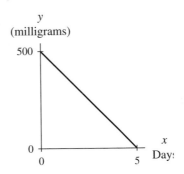

   (d) The x-intercept is 5 days.
       The y-intercept is 500 milligrams.
       The y-intercept is the amount of the drug initially in the patient's body (500 milligrams). The x-intercept is the time when none of the drug remains (after 5 days).
   (e) y is always decreasing (for $0 \le x \le 5$).
   (f) After 3.5 days, 150 milligrams of the drug remain.
   (g) There will be 60 milligrams of the drug after 4.4 days.

Section 1.2

1.

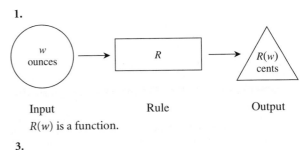

Input   Rule   Output
   $R(w)$ is a function.

3.

Input   Rule   Output
   $B(x)$ is a function.

5.

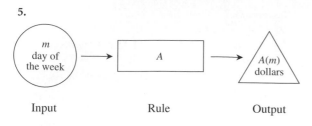

Input   Rule   Output

$A(m)$ is not a function unless you always spend the same amount on lunch every Monday, the same amount every Tuesday, and so on, or unless the input is the days in only 1 week.

7. (a)

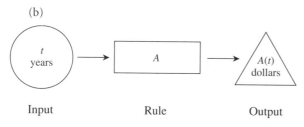

Input   Rule   Output

$B(t)$ is a function.

(b)

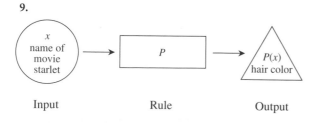

Input   Rule   Output

$A(t)$ is a function.

9.

Input   Rule   Output

$P(x)$ is not a function.

11. Inputs: $w > 0$
Outputs: $R(w) > 0$ (For 1997, these were 32, 55, 78, 101, 124, 147, ...)

13. Inputs: $1 \leq x \leq 365$ and x is an integer
Outputs: $B(x)$ is an integer between 0 and the number of students in the class

15. $A(m)$ is not a function.

17. Inputs: $t \geq 0$
Outputs: $A(t) \geq 0$ and $B(t) \geq 0$

19. $P(x)$ is not a function.

21. This is a function.

23. This is not a function.

25. $P(\text{Honolulu}) = 358.5$
$P(\text{San Antonio}) = 77$
$P(\text{Portland, OR}) = 106$

27. Graphs b and c are functions.

29. (a) Continuous
(b) Discrete
(c) Continuous
(d) Discrete

31. (a) Continuous
(b) Discrete
(c) Continuous
(d) Continuous with discrete interpretation

33. $R(3) \approx 314.255$
$R(0) = 39.4$

35. $Q(12) = 600.36$
$Q(3) = 222.54$

37. $R(w) = 78.8$ when $w \approx 1.001$
$R(w) = 394$ when $w \approx 3.327$

39. $Q(x) = 515$ when $x = 10$
$Q(x) = 33.045$ when $x \approx 0.500$

41. Input is given.
$A(15) = 5,131,487.257$

43. Output is given.
When $y = 2.97$, $x \approx -3.203$.

Section 1.3

1.

Input   Rule   Output

$W(r)$ is the weight in ounces of a first class letter or parcel that costs r cents to mail.

$W(r)$ is not an inverse function.

3.

Input   Rule   Output

$X(b)$ is the day of the year corresponding to the birthday of b students.
$X(b)$ is not an inverse function.

5.

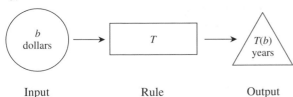

| Input | Rule | Output |

$T(b)$ is the time in years when the value of the investment was b dollars.

$T(b)$ is not an inverse function unless interest is compounded continuously.

7. Reversing the inputs and outputs does not result in an inverse function.

9. Reversing the inputs and outputs does not result in an inverse function because the original table does not represent a function.

11. (a)

Credit hours	Tuition	Credit hours	Tuition
1	$112	10	$1120
2	$224	11	$1232
3	$336	12	$1381
4	$448	13	$1381
5	$560	14	$1381
6	$672	15	$1381
7	$784	16	$1381
8	$896	17	$1381
9	$1008	18	$1381

(b) $T(c) = \begin{cases} 112c \text{ dollars} & \text{when } 0 \leq c \leq 11 \\ 1381 \text{ dollars} & \text{when } c \geq 12 \end{cases}$

where c is the number of credit hours

(c)

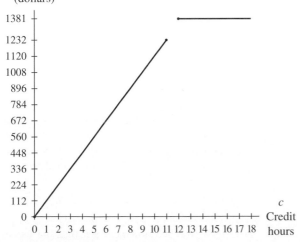

(d) $T(c)$ is a function.

(e) Inputs: integers 0 through 18

Outputs: {0, 112, 224, 336, 448, 560, 672, 784, 896, 1008, 1120, 1232, 1381}

(f)

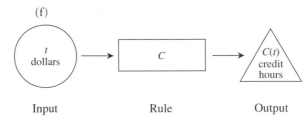

| Input | Rule | Output |

(g) The rule in part f is not an inverse function, because an input of $1381 has seven different outputs.

13. $n(t) = h(t) - p(t) = \dfrac{100}{1 + 128.0427e^{-0.7211264t}} -$

$\dfrac{100}{1 + 913.7241e^{-0.607482t}}$ percent t years after 1924

15. $c(x) = n(x)p(x) = (-0.034x^3 + 1.331x^2 + 9.913x + 164.447)(-0.183x^2 + 2.891x + 20.215)$ cesarean-section deliveries x years after 1980

17. $P(C(t)) =$ profit after t hours of production

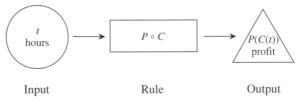

| Input | Rule | Output |

19. $P(C(t)) =$ average amount in tips t hours after 4 p.m.

| Input | Rule | Output |

21. $f(t(p)) = 3e^{4p^2}$

23. $g(x(w)) = \sqrt{7(4e^w)^2 + 5(4e^w) - 2}$

25. (a) $S(85) = \$123.1$ million

$S(88) = \$159.4$ million

$S(89) = \$97.7$ million

$S(92) = \$53.3$ million

(b)

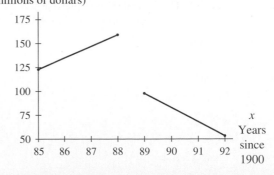

(c) $S(x)$ is a function because each year corresponds to only one amount of sales.

27. (a) The shipping charges encourage larger orders.

(b)
Order amount	Shipping charge
$17.50	$3.50
$37.95	$6.83
$75.00	$11.25
$75.01	$9.00

(c)
$$S(x) = \begin{cases} 0.2x \text{ dollars} & \text{when } 0 \leq x \leq 20 \\ 0.18x \text{ dollars} & \text{when } 20 < x \leq 40 \\ 0.15x \text{ dollars} & \text{when } 40 < x \leq 75 \\ 0.12x \text{ dollars} & \text{when } x > 75 \end{cases}$$

where x is the order amount in dollars

(d)

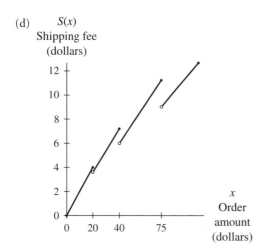

$S(x)$
Shipping fee
(dollars)

(e) Answers will vary.

Section 1.4

1. (a) Slope $\approx \dfrac{-\$2.5 \text{ million}}{5 \text{ years}} = -\$0.5$ million per year

The corporation's profit was declining by approximately a half a million dollars per year during the 5-year period.

(b) The rate of change is approximately -$0.5 million per year.

(c) The vertical axis intercept is approximately $2.5 million. This is the value of the corporation's profit in year zero. The horizontal axis intercept is 5 years. This is the time when the corporation's profit is zero.

3. (a) $321.5 million per year

(b)

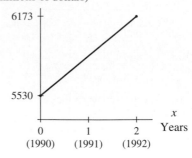

City government payroll
(millions of dollars)

(c) The vertical axis intercept is $5530.167 million. This is the amount of the payroll in 1990 when $x = 0$.

5. (a) Rate of change of revenue =
$$\dfrac{\$99.1 - \$114.1 \text{ million}}{1995 - 1994} = -\$15 \text{ million per year}$$

(b) Each quarter, International Game Technology's revenue declined by $3.75 million.

(c)
Year	1994	1995	1996	1997
Revenue	114.1	99.1	84.1	69.1

(millions of dollars)

(d) Revenue $= -15x + 30{,}024.1$ million dollars in year x

7. (a) The rate of change is -400 gallons per day.

(b) 2800 gallons

(c) 18,400 gallons

The assumptions are that the usage rate remains constant and that the tank is not filled more than once in January.

(d) $A(t) = 30{,}000 - 400t$ gallons t days after January 1

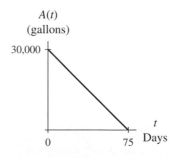

$A(t)$
(gallons)

(e) Solving $A(t) = 0$ gives $t = 75$. Assuming that the usage rate remains constant, the tank will be empty 75 days after January 1, which corresponds to the middle of March.

9. (a) $S(t) = 499.3t + 5036.2$ students t years after 1965

(b) $S(5) \approx 7533$ students

(c) The estimate was 505 students lower than the actual amount; answers vary.

(d) It would not be wise to use this model to extrapolate 31 years outside of the last data point.

11. (a) $P(x) = 0.720x - 47.875$ dollars x years after 1900

  (b) $P(84) \approx \$13$

   $P(86) \approx \$14$

   $P(90) \approx \$17$

   These are interpolations.

  (c) $P(94) \approx \$20$

   This is extrapolation.

  (d) The prediction was accurate.

 (e, f) Answers will vary.

  (g) We assume that the rate of change remains essentially constant. This may or may not be true over a long period of time.

  (h) Short-term extrapolations are sometimes accurate; long-term extrapolations usually are not.

13. (a) $F(y) = -0.152y + 19.514$ percent

   where y is 81 for the 1981–82 school year, 82 for the 1982–83 school year, and so on.

  (b) -0.152 percentage points per year

  (c) $F(93) \approx 5.4\%$

  (d) Answers will vary. One possible answer is a recession, causing the model to overestimate.

  (e) Solving $F(y) = 5$ gives $y \approx 95.7$. Thus the 1996–97 school year is the first year below the 5% level.

15. (a) Rate of change $= \dfrac{\$97{,}500 - \$73{,}000}{1995 - 1983} \approx \$2042$ per year

  (b) $\$97{,}500 + 3(\$2042) \approx \$103{,}600$

  (c) The value was $\$75{,}000$ in 1984 and $\$100{,}000$ in 1996.

  (d) $V(t) = 2041.667t + 73{,}000$ dollars t years after 1983

   $V(9) \approx \$91{,}400$

   The model assumes that the rate of increase of the market value remains constant.

17. (a)

Postage
(dollars)

2.62

0.32

1   11

x
(ounces)

  (b) The first differences are all $0.23.

  (c) $P(w) = 0.09 + 0.23w$ dollars for weight not exceeding w ounces

19. Extrapolating from a model is predicting values beyond the range of the data, whereas interpolating is using a model to estimate what happened at some point within the range of the data. Interpolation often yields a good approximation of what occurred. Extrapolation must be used with caution because many things that are not accounted for by the model may affect future events.

21. (a) 1991

  (b)
$$P(t) = \begin{cases} -7.393t + 1304.857 & \text{when } 85 \le t < 91 \\ \quad \text{thousand people} \\ t + 542 & \text{when } 91 \le t \le 93 \\ \quad \text{thousand people} \end{cases}$$

   where t is the number of years after 1900.

23. (a) The scatter plot does reflect the statements about atmospheric release of CFCs.

  (b) Answers may vary. One possible model is

$$R(x) = \begin{cases} -15.37x + 1554.19 & \text{when } x \le 80 \\ \quad \text{million kilograms} \\ 9.165x - 412.5 & \text{when } 80 < x \le 88 \\ \quad \text{million kilograms} \\ -34.375x + 3413.283 & \text{when } x > 88 \\ \quad \text{million kilograms} \end{cases}$$

   where x is the number of years after 1900.

   (Note that in computing the middle model, the 1980 data point was not included, but in computing the bottom model, the 1988 data point was included. This was done to improve the fit of the model. You will have to use your own discretion to determine whether or not to include the points at which the data is divided when finding a piecewise model.)

  (c)

$R(x)$
CFC-12 release
(millions of kilograms)

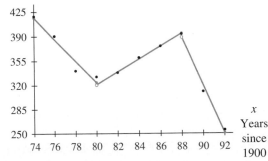

(d) According to the model, by the year 2000 there will be no more atmospheric release of CFC-12.

Chapter 1 Review Test

1. (a) The number of acres increased, but by a smaller amount in 1992 than in 1991.

  (b) Between 1989 and 1991, the yearly gain of wetlands was increasing by approximately 30,000 acres per year.

2. (a)

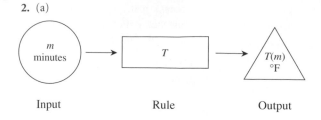

Input   Rule   Output

(b) $T(m)$ is a function because for every possible time, there is only one associated temperature.

(c) If the inputs and outputs are reversed, the result is not an inverse function because one temperature could have more than one time associated with it.

3. (a) Rate of increase =

$$\frac{\$167 - \$114 \text{ million}}{1993 - 1986} \approx \$7.6 \text{ million per year}$$

(b) Sales in 1998 ≈ $205 million

This estimate would be valid if sales rose by approximately $7.6 million per year between 1993 and 1998.

4. (a) $C(t) = 0.0342t + 11.39$ million square kilometers t years after 1900

(b) $C(t)$ is continuous.

(c) The amount of cropland increased by approximately 0.034 million square kilometers, or 34,000 square kilometers per year, between 1970 and 1990.

(d) Answers will vary.

(e) $C(95) = 14.64$ million square kilometers

CHAPTER 2

Section 2.1

1. Exponential

3. The scatter plot is curved, so it cannot be linear. It does not approach the horizontal axis, so it cannot be exponential. While it could be the right half of an increasing logistic, it does not display an inflection point, so we do not choose that model.

5. The scatter plot exhibits a change in concavity, so it cannot be linear or exponential. However, it does not level off like a logistic. It is possible to model this data set with a piecewise continuous model using three linear pieces.

7. $f(x)$ is increasing with a 5% change in output for every unit of input.

9. $y(x)$ is decreasing with a 13% change in output for every unit of input.

11. The number of bacteria declines by 39% each hour.

13. $f(x)$ is increasing with limiting value 100.

15. $h(g)$ is decreasing with limiting value 39.2.

17. (a) $y = 3.003(1.019110)^t$ billion people t years after 1960

(b) Assuming that the model gives end-of-year population values, we have $y(32.5) \approx 5.56$ billion.

$y(37) \approx 6.05$ billion

The article predicts 5.5 billion in mid-1993 and 6 billion in 1997.

(c) $y(65) \approx 10.28$ billion

We assume that the percentage growth will remain constant from 1960 through 2025. However, the article indicates that growth rates are higher now than in the 1960s. The article also states that birth rates are expected to decline, indicating that the percentage growth will not remain constant.

(d) Population in sub-Saharan Africa can be modeled by $P(t) = C(1.03)^t$ people t years from when the article was written, where C was the population at the time the article was written. When we solve for t in $2C = C(1.03)^t$, we see that population will double when $t \approx 23.4$. That is slightly longer than the 20-year statement in the article.

(e) Latin America's population will double in approximately 37 years.

19. (a) $A(t) = 331.693(1.250536)^t$ million CDs sold t years after 1991

(b) 25% per year

21. (a) $P(t) = 12.582(1.026935)^t$ million people t years after 1920

(b) $P(73) \approx 87.573$ million people

(c) The model underestimates the figure reported in the *Almanac* by 1.025 million.

(d) The model represents 2.7% growth each year. This means that the model may overestimate future populations.

23. (a) $R(x) = 0.669(1.039629)^x$ cents x years after 1900

$R(95) \approx 27$ cents

(b) The model lies above the data from 1958 to 1974 and below the data from 1975 on. It doesn't appear to be a very good fit.

(c) $P(x) = 0.107(1.063874)^x$ cents x years after 1900

$P(95) \approx 38$ cents

(d) The model does not seem to follow the trend of the data from 1974 on.

(e) $R(101) \approx 34$ cents

$P(101) \approx 56$ cents

It is probable that neither of these predictions is a good indicator of first-class postage in 2001. $R(x)$ is probably an underestimate, and $P(x)$ is almost certainly an overestimate.

25. (a) $C(t) = -3.048t + 114.036$ million metric tons of CO t years after 1980

$C(t) = 115.033(0.968717)^t$ million metric tons of CO t years after 1980

(b) The exponential model indicates a 3.13% decline in emissions each year. The linear model indicates a decline of 3.05 million metric tons each year.

(c)

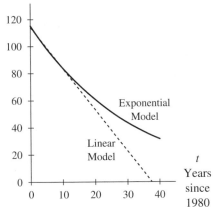

$C(t)$
Emission of CO
(millions of metric tons)

The linear model reaches zero around $t \approx 37.4$ (mid-2018). The exponential model shows a leveling off. The exponential model will continue to get closer and closer to the horizontal axis.

(d, e) Answers will vary.

27. (a) $C(t) = \dfrac{37.195}{1 + 21.374e^{-0.182968t}}$ countries t years after 1840

The model is a good fit.

(b)

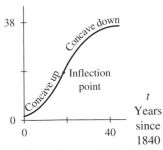

$C(t)$
Countries

29. (a) A plow sulky is a horse-drawn plow with a seat so that the person plowing can ride instead of walk. It was a precursor to the tractor.

(b) $P(t) = \dfrac{2591.299}{1 + 16.848e^{-0.194833t}}$ patents t years after 1871

(c) Patents begin with one innovative idea and grow almost exponentially as more and more improvements and variations are patented. Eventually, however, the patent market becomes saturated, and the number of new patents dwindles to none as newer ideas and products are invented.

31. (a) $N(t) = \dfrac{3015.991}{1 + 119.250e^{-1.023564t}}$

Navy deaths t weeks after August 31, 1918

$A(t) = \dfrac{20{,}492.567}{1 + 518.860e^{-1.212390t}}$

Army deaths t weeks after Sept. 7, 1918

$C(t) = \dfrac{91{,}317.712}{1 + 176.272e^{-0.951129t}}$

civilian deaths t weeks after Sept. 14, 1918

(b) The models given in part a have limiting values less than the number of deaths given in the table for November 30. The models are not good indicators of the ultimate number of deaths.

33. (a) The data are concave down from January through April and concave up from April through June. This is not the concavity exhibited by a logistic model.

(b) The entire data set does appear to be logistic.

(c) $P(t) = \dfrac{42{,}183.911}{1 + 21{,}484.253e^{-1.248911t}}$ polio cases

t months after December, 1948
The model appears to be a good fit.

(d) The model is a poor fit for the January through June data.

35. (a) The scatter plot is decreasing, and it does exhibit an inflection point.

(b) $y = \dfrac{79.294}{1 + 0.122e^{0.210773t}}$ dollars t years after 1970

(c) The inflection point appears to be between 1980 and 1982.

Section 2.2

1. The function is increasing, with a 166.4% change in output for each unit of change in input.

$$f(x) = 100(2.664456)^x$$

3. The function is decreasing, with an 18.9% change in output for each unit of change in input.

$$A(t) = 1000(0.810584)^t$$

5. The function is increasing.

$$f(x) = 39.2e^{0.254642x}$$

7. The function is decreasing.

$$h(t) = 1.02e^{-0.478036t}$$

9. (a) $A = 2000\left(1 + \dfrac{0.045}{4}\right)^{4t} \approx 2000(1.045765)^t$ dollars after t years

(b) $A(2) \approx \$2187.25$

(c) $A(3.75) \approx \$2365.42$

(d) The amount after 2 years and 2 months is the same as the amount after 2 years: $A(2) \approx \$2187.25$

11. (a) APR $= (0.015)(12) \times 100\% = 18\%$

   (b) APY $= \left[\left(1 + \dfrac{0.18}{12}\right)^{12} - 1\right] \times 100\% \approx 19.6\%$

13. (a) The percentage change is approximately 6.4% per year. This is the APY. Balance after 6 years $=$ $(1908.80)(1.064) = \$2030.96$

   (b) $A = 1400(1.063962)^t = 1400e^{0.062t}$ dollars after t years
   The value r is the nominal rate, or APR.

   (c) $A(6) \approx \$2030.89$
   This amount differs from that in part a because it assumes a percentage change of approximately 6.3962% rather than 6.4%.

15. (a) 4.725% compounded semiannually:

   $$\text{APY} = \left[\left(1 + \dfrac{0.04725}{2}\right)^2 - 1\right] \times 100\% \approx 4.78\%$$

   4.675% continuously compounded:

   $$\text{APY} = [e^{0.04675} - 1] \times 100\% \approx 4.79\%$$

   This is the better choice.

   (b) The account with the higher APY will always result in a greater account balance.

   (c) After 10 years, the two options differ by only 79 cents.

17. (a) *Option A:* $\dfrac{0.109}{12}(12{,}300) = m\left[1 - \left(1 + \dfrac{0.109}{12}\right)^{-42}\right]$

   $$m \approx \$353.58$$

   Option B: $\dfrac{0.105}{12}(12{,}300) = m\left[1 - \left(1 + \dfrac{0.105}{12}\right)^{-48}\right]$

   $$m \approx \$314.92$$

   (b) *Option A:* (\$353.58)(42) $= \$14{,}850.36$
   Option B: (\$314.92)(48) $= \$15{,}116.16$

   (c) *Option A:*   \$2550.36
   Option B: \$2816.16

   (d) *Option A* has a higher payment, but the car will be paid off sooner and will cost \$265.80 less than with *Option B*. *Option B* has a lower monthly payment, but it costs more and takes longer to pay off.

19. $A = 2500\left(1 + \dfrac{0.066}{4}\right)^{4(3)} \approx \$3042.49$

21. $2000 = P\left(1 + \dfrac{0.062}{12}\right)^{12(2)}$

   $$P \approx \$1767.32$$

23. (a) $250{,}000 = P\left(1 + \dfrac{0.10}{12}\right)^{12(45)}$

   $$P \approx \$2829.52$$

   (b) $2829.52\left(1 + \dfrac{0.10}{12}\right)^{12(45)} \approx \$249{,}999.92$

   This answer is not \$250,000 because the principal amount was rounded in part a.

Section 2.3

1. Concave up, decreasing from 0.75 to 3, increasing from 3 to 4

3. Concave up, decreasing from 13.5 to 18, increasing from 18 to 22.5

5. Concave down, always decreasing

7. (a)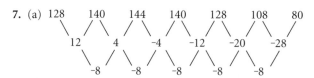

   Second differences are constant, so the data are quadratic.

   (b) After 3.5 seconds the height is 44 feet. After 4 seconds the height is 0 feet.

   (c) $H(s) = -16s^2 + 32s + 128$ feet after s seconds

   (d) $H(s) = 0$ when $s = -2$ and $s = 4$
   The missile hits the water after 4 seconds.

9. (a) $J(12) \approx 374$ jobs
   (b, c) Answers will vary.

11. (a) Because the data are evenly spaced and the second differences are constant, the data are perfectly quadratic.

   (b) 26.5 years of age

   (c) $A(x) = 0.0035x^2 - 0.405x + 32$ years of age x years after 1900

   (d) $A(100) = 26.5$ years of age

   (e, f) Answers will vary.

13. (a) The data do not appear to be concave up or concave down. A linear model is

   $B(x) = 0.002x + 1.880$ dollars to make x ball bearings

   (b) Overhead is \$1.88.

   (c) $B(5000) \approx \$13.64$

   (d) $B(5100) - B(5000) \approx \$0.24$
   This answer could also be found by multiplying the slope of the model by 100. This value is called marginal cost.

   (e) $C(u) = 0.002(500u) + 1.880$ dollars to make u cases of ball bearings

15. (a) $C(t) = -1049.5t^2 + 5988.7t + 33{,}770.7$ AIDS cases t years after 1988

   (b) Answers will vary.

   (c) $C(4) \approx 40{,}934$ cases

(d) The model peaks at approximately 42,314 cases. It never reaches 50,000.

(e) Answers will vary.

17. (a) $y = 0.011x^2 - 0.883x + 16.764$ cents
   x years after 1900
   The model is not a good fit between 1919 and 1958.

(b) $y = 0.014x^2 - 1.242x + 29.427$ cents
   x years after 1900
   There is very little difference between the two models from 1958 on. Neither model is a good description of what happened between 1919 and 1958.

(c) The model in part a predicts 40 cents. The model in part b predicts 42 cents.

19.

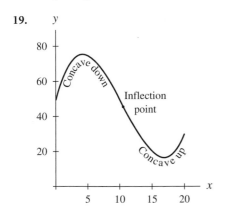

21.

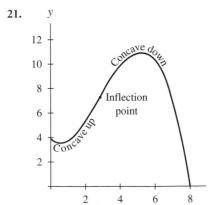

23.

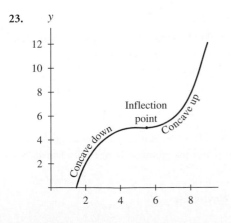

25. (a) The data appear to be concave down until 1987 and then concave up. The implied inflection point indicates that a cubic model may be appropriate.

(b) $A(t) = 0.019t^3 - 23.708t^2 + 2058.423t - 59,587.967$ billion dollars t years after 1900

(c) $A(99) \approx \$285.6$ billion

27. (a) $D(x) = -10.247x^3 + 208.114x^2 - 168.805x + 9775.035$ donors x years after 1975

(b) The model indicates increasingly rapid decline in the number of donors. The organization would currently have no donors if that trend were accurate.

(c) Answers will vary.

29. (a) $P(t) = -0.001t^3 + 0.023t^2 - 0.183t + 1.431$ dollars
   t years after 1978
   $P(15) \approx \$0.55$

(b) The cubic model declines rapidly. Although the purchasing power of the dollar will almost certainly decline, its decline will probably not be as rapid as that indicated by the model unless the economy sees high inflation.

(c) Solving $P(t) = 0.90$ gives $t \approx 11.214$; that is, in 1990. From the data, we see that this actually occurred sometime in 1989.

(d) The data is concave up, and the model is concave down.

(e)
$$A(t) = \begin{cases} -0.001t^3 + 0.028t^2 - \\ \quad 0.202t + 1.442 \text{ dollars} \quad \text{when } t < 10 \\ 0.0075t^2 - 0.2095t + \\ \quad 2.2755 \text{ dollars} \quad \text{when } t \geq 10 \end{cases}$$
where t is the number of years since 1978.

(f) $A(15) \approx \$0.82$

(g) The piecewise continuous model is a better fit, but it shows the purchasing power of the dollar rising after 1992.

31. (a) The behavior of the data abruptly changes in 1989, resulting in a sharp point. None of the five models we have studied will reflect that behavior.

(b)
$$V(t) = \begin{cases} 0.106t^2 - 15.497t + 652.169 \\ \quad \text{pounds per person} \quad \text{when } t < 89 \\ -2.15t + 306.7 \\ \quad \text{pounds per person} \quad \text{when } t \geq 89 \end{cases}$$
where t is the number of years since 1900.

(c) $V(73) \approx 87.5$ pounds per person
   $V(83) \approx 98.4$ pounds per person
   $V(93) \approx 106.8$ pounds per person

(d) Answers will vary.

(e) The model indicates a constant decline of approximately 2.2 pounds per person per year.

(f) Answers will vary.

Section 2.4

1. Because the data are concave up and display a minimum, a quadratic is the only appropriate model.

3. The data appear to be essentially linear. Any concavity is probably not obvious enough to warrant the use of a more complex model.

5. Because of the inflection point, a cubic or logistic model would be an appropriate choice. The choice would depend on the desired behavior of the model outside the range of the data.

7. (a) Linear data have constant first differences and lie in a line.
   (b) Quadratic data have constant second differences and are either concave up or concave down.
   (c) Exponential data have constant percentage differences and are concave up.

9. $T(x) = 0.181x^2 - 8.463x + 147.376$ seconds at age x years

11. $S(t) = 1505.030t + 17,582.764$ dollars t years after 1981

13. Answers will vary.

15. (a) Arguments can be made for using linear, exponential, or quadratic models for the LP data set, but the CD data is best modeled by an exponential model:

   $S(x) = 1596.946(1.288315)^x$ million dollars worth of CDs shipped x years after 1987

   (b) Estimates of LP shipments depend on the model used.
   Estimates of CDs are

   $S(4) \approx \$4399.2$ million in 1991
   $S(5) \approx \$5667.6$ million in 1992
   $S(6) \approx \$7301.7$ million in 1993

17. (a) 10 components
   (b) 25 components
   (c) $N(h) = -0.230h^3 + 2.703h^2 - 0.591h + 7.071$ components after h hours
   (d) 4 hours 41 minutes
   (e) The cubic model shows the total number of components declining, which is impossible (unless the worker begins disassembling components). This model is not a good indicator of what occurs after 8 hours.
   (f) $C(h) = \dfrac{61.859}{1 + 12.299e^{-0.672227h}}$ components after h hours

(g) The model indicates that the number of components assembled is never greater than 62. It is more reasonable than the cubic model, but it still is probably not accurate.

(h) According to the logistic model, the student will assemble only 3 or 4 components after 8 hours.

(i, j) Answers will vary.

19. (a) One choice is a quadratic:
   $S(x) = -72.929x^2 + 429.405x + 2506.083$ thousand cars x years after 1984 although the fit is not very good. Another option is

   $$C(x) = \begin{cases} 403x + 2437.667 \text{ thousand cars} & \text{when } x \le 2 \\ -35.643x^2 + 71.671x + \\ \quad 3267.371 \text{ thousand cars} & \text{when } x > 2 \end{cases}$$

   where x is the number of years since 1984.

   (b, c) $S(8) \approx 1274$ thousand cars
   $C(8) \approx 1560$ thousand cars
   Both predictions are too low, but the piecewise model is closer.

   (d) The additional point indicates a concavity change. A cubic model (which is a much better fit than the quadratic) is

   $S(x) = 11.789x^3 - 194.269x^2 + 739.435x + 2392.556$ thousand cars x years after 1984
   $S(8) \approx 1911$ thousand cars

   A piecewise continuous model is

   $$C(x) = \begin{cases} 403x + 2437.667 \text{ thousand cars} & \text{when } 0 \le x \le 2 \\ 12.556x^3 - 201.452x^2 + \\ \quad 740.754x + 2461.714 \\ \quad \text{thousand cars} & \text{when } 2 < x \le 8 \end{cases}$$

   where x is the number of years since 1984.
   $C(8) \approx 1923$ thousand cars

21. (a) It appears that the ban took effect between 1974 and 1976.
   (b) One possible model is

   $$R(t) = \begin{cases} 0.708t^2 - 71.401t + 1823.427 \\ \quad \text{million kilograms} & \text{when } t \le 74 \\ -19.325t + 1852.133 \\ \quad \text{million kilograms} & \text{when } t > 74 \end{cases}$$

   where t is the number of years since 1900.
   (c, d) Answers will vary.

23. (a) The scatter plot is concave down from 1980 through 1985. From 1986 through 1993, the scatter plot appears to be concave up.
   (b) Although the scatter plot exhibits a change in concavity, it is not the smooth change represented by a

cubic or logistic model. Instead, the concavity change creates a sharp point, which indicates that a piecewise model may be more appropriate.

(c) $P(t) = 0.137t^3 - 33.656t^2 + 2734.950t - 70,611.924$ thousand people t years after 1900

(d) One possible model is

$$I(t) = \begin{cases} -1.932t^2 + 301.25t - 8822.205 \\ \quad \text{thousand people} \qquad \text{when } 80 \leq t < 87 \\ 1.357t^2 - 236.071t + \\ \quad 13,032.429 \text{ thousand people} \quad \text{when } 87 \leq t \leq 93 \end{cases}$$

where t is the number of years since 1900.

(e) Answers will vary.

Chapter 2 Review Test

1. (a) $C(x) = 91.6(1.693000)^x$ thousand subscribers x years after 1984

   (b) 69.3% per year

   (c) Solving $C(x) = 270,000$ gives $x \approx 15.2$, which corresponds to early 2000.

   (d) Answers will vary.

2. (a) $\text{APY} = \left[\left(1 + \dfrac{0.073}{12} \right)^{12} - 1 \right] \times 100\% \approx 7.55\%$

   (b) The largest possible APY occurs when interest is compounded continuously and is $(e^{0.073} - 1) \times 100\% \approx 7.57\%$.

3. (a) The data are increasing and concave up, indicating that either an exponential or a quadratic model is appropriate.

   (b) Possible models include

Quadratic:	$B(t) = 0.455t^2 + 1.114t + 23.116$ thousand births t years after 1975
Exponential:	$B(t) = 22.127(1.131524)^t$ thousand births t years after 1975

   (c) Answers will vary.

4. (a) The scatter plot is essentially concave up and then concave down. A cubic model appears to be appropriate.

   (b) $F(t) = -0.049t^3 + 1.560t^2 - 13.485t + 110.175$ degrees Fahrenheit t hours after midnight August 27.

   (c) $F(17.5) \approx 91°\text{F}$

   (d) Solving $F(t) = 90$ gives three solutions: $t \approx 1.88$, 12.37, and 17.82. The first solution lies outside the time frame of the data. The other two solutions correspond to 12:22 p.m. and 5:49 p.m.

5. (a) Possible answers: (i) The data show no indication of leveling off at both ends, so a logistic model is not appropriate. (ii) Because the data seem to level off in the middle (1991–1992), it is possible that a cubic model may be used. However, there is no obvious concavity change to justify making this the first choice. (iii) Because the data are essentially concave down, a quadratic model is an appropriate choice. (iv) The data are not concave up, so an exponential model is not appropriate. (v) It is possible to argue that the concavity of the data is subtle enough that a linear model is appropriate, but a quadratic or cubic model is probably a better choice.

   (b) Answers will vary.

CHAPTER 3

Section 3.1

1. The stock price rose an average of 46 cents per day during the 5-day period.

3. The company lost an average of $8333.33 per month during the past 3 months.

5. Unemployment has risen an average of 1.3 percentage points per year in the past 3 years. *Note*: Whenever you are writing a change for a function whose output is a percentage, the correct label (unit of measure) is "*percentage points.*" The same is true for a rate of change. The phrase *percentage points per year* indicates that the effect is additive, whereas the phrase *percent per year* indicates a multiplicative effect.

7. Average rate of change $= \dfrac{25,057 - 31,517}{1995 - 1991}$

   $= \dfrac{-6460 \text{ licenses}}{4 \text{ years}} = -1615$ licenses per year

9. (a) Slope of secant line $= \dfrac{6229.09 - 6228.98}{1996 - 1982}$

   $= \dfrac{0.11 \text{ feet}}{14 \text{ years}} \approx 0.008$ foot per year

   (b) In the 14-year period from 1982 through 1996, the lake level rose an average of 0.008 foot per year.

   (c) The lake level dropped below the natural rim because of drought conditions in the early 1990s but rose again to normal elevation by 1996. The average rate of change tells us that the level of the lake in 1996 was close to the 1982 level. Although the average rate of change is nearly zero, the graph shows that the lake level changed dramatically during the 14-year period.

11. (a) The balance increased by $1908.80 − $1489.55 = $419.25.

(b) Average rate of change $= \dfrac{\$419.25}{4 \text{ years}} \approx$

$104.81 per year

(c) You could estimate the amount in the middle of the fourth year, but doing so might not be as accurate as using a model to find the amount.

(d) $A(t) = 1400(1.063962)^t$ dollars after t years

Amount in middle of fourth year $= A(3.5) \approx$ $1739.28

Amount at end of fourth year $= A(4) \approx \$1794.04$

Average rate of change $=$

$\dfrac{\$1794.04 - \$1739.28}{\frac{1}{2} \text{ year}} = \$109.52$ per year

13. (a) Average rate of change $= \dfrac{2338 \text{ deaths} - 2 \text{ deaths}}{6 \text{ weeks}} \approx$

389 deaths per week

(b) Average rate of change $=$

$\dfrac{73{,}477 \text{ deaths} - 6528 \text{ deaths}}{4 \text{ weeks}} \approx$

16,737 deaths per week

(c) Navy, 332 deaths per week; Army, 5559 deaths per week

15. (a) $p(65) - p(59) \approx 72.71 - 61.99 =$
10.72 million people

(b) $\dfrac{p(64) - p(62)}{64 - 62} \approx \dfrac{3.665 \text{ million people}}{2 \text{ years}} =$

1.8 million people per year

17. (a) Between 0 and 2 seconds, the average rate of change is 0 feet per second because the secant line is horizontal.

(b)
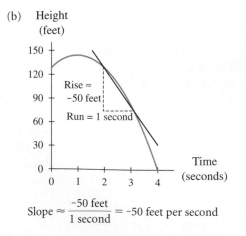

Height
(feet)

Rise ≈
−50 feet

Run = 1 second

Time
(seconds)

$\text{Slope} \approx \dfrac{-50 \text{ feet}}{1 \text{ second}} = -50$ feet per second

(c) $50 \dfrac{\text{feet}}{\text{second}} \times \dfrac{3600 \text{ seconds}}{\text{hour}} \times \dfrac{1 \text{ mile}}{5280 \text{ feet}} \approx 34$ mph

19.

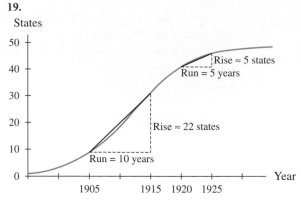

States

Rise ≈ 5 states
Run = 5 years

Rise ≈ 22 states

Run = 10 years

Year

1905　　1915 1920 1925

(a) Slope of secant line $\approx \dfrac{22 \text{ states}}{10 \text{ years}} = 2.2$ states per year

(b) Slope of secant line $\approx \dfrac{5 \text{ states}}{5 \text{ years}} = 1$ state per year

Section 3.2

1. (a) A continuous graph or model is defined for all possible input values on an interval. A continuous model with discrete interpretation has meaning for only certain input values on an interval. A continuous graph can be drawn without lifting the pencil from the paper. A discrete graph is a scatter plot. A continuous model or graph can be used to find average or instantaneous rates of change. Discrete data or a scatter plot can be used to find average rates of change.

(b) An average rate of change is a slope between two points. An instantaneous rate of change is the slope at a single point on a graph.

(c) A secant line connects two points on a graph. A tangent line touches the graph at a point and is tilted the same way the graph is tilted at that point.

3. Average speed $= \dfrac{19 - 0 \text{ miles}}{17 \text{ minutes}} \times \dfrac{60 \text{ minutes}}{\text{hour}} \approx 67.1$ mph

5. Average rates of change are slopes of secant lines. Instantaneous rates of change are slopes of tangent lines.

7. (a) The slope is positive at A, negative at B and E, and zero at C and D.

(b) The graph is steeper at point B than at point A.

9. (a)

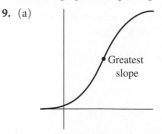

Greatest
slope

(b)

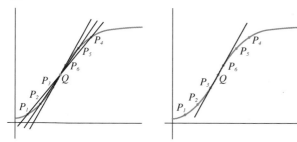

Point of most rapid decline

11. (a, b) *A*: 1.3 mm per day per °C; *B*: 5.9 mm per day per °C; *C*: -4.2 mm per day per °C
    (c) The growth rate is increasing by 5.9 mm per day per °C.
    (d) The slope of the tangent line at 32 °C is -4.2 mm per day per °C.
    (e) At 17 °C, the instantaneous rate of change is 1.3 mm per day per °C.

Section 3.3

1. Tangent lines enable us to measure the slope at a single point on a graph, which is a measure of the instantaneous rate of change at that point.

3. The lines at *A* and *C* are not tangent lines.

5.

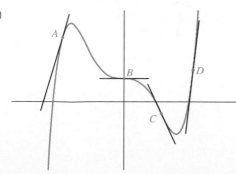

7. (a, b) *A*: concave down, tangent line lies above the curve.
    B: inflection point, tangent line lies below the curve on the left, above the curve on the right.
    C: inflection point, tangent line lies above the curve on the left, below the curve on the right.
    D: concave up, tangent line lies below the curve.

(c)

(d) *A*, *D*: positive slope; *C*: negative slope (inflection point); *B*: zero slope (inflection point)

9.

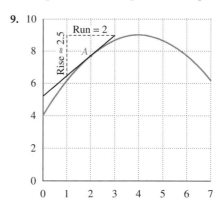

$$\text{Slope} \approx \frac{2.5}{2} = 1.25$$

(Answers may vary.)

11.

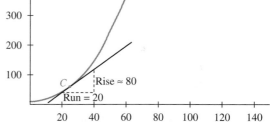

$$\text{Slope at } C \approx \frac{80}{20} = 4$$

$$\text{Slope at } D \approx \frac{90}{60} = 1.5$$

(Answers may vary.)

13. (a) The concave-down graph is lowfat milk consumption. The concave-up graph is whole milk consumption.
    (b)

Per capita consumption
(gallons per person)

Slope for whole milk $\approx \dfrac{-1.5 \text{ gallons per person}}{2 \text{ years}} =$

-0.75 gallon per person per year

Slope for lowfat milk $\approx \dfrac{1.1 \text{ gallons per person}}{2 \text{ years}} =$

0.55 gallon per person per year

15. (a, b, c) Any tangent line to $p(t)$ is the model itself. The slope of any tangent line will be approximately 2370 thousand people per year.

    (d) The slope of the graph at every point will be 2370 thousand people per year.

    (e) The instantaneous rate of change is 2370 thousand people per year.

17.

Nursing students
(thousands)

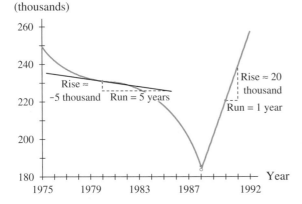

(a) Slope in 1980 $\approx \dfrac{-5 \text{ thousand students}}{5 \text{ years}} =$

-1000 students per year

(b) The slope in 1988 does not exist because the graph is not continuous at 1988.

(c) Slope in 1990 $\approx \dfrac{20 \text{ thousand students}}{1 \text{ year}} =$

20,000 students per year

19. (a) Slope $\approx \dfrac{464.8 - 363.8}{1993 - 1989} = \dfrac{101 \text{ billion dollars}}{4 \text{ years}} =$

$25.25 billion per year

(b) Exports
(billions
of dollars)

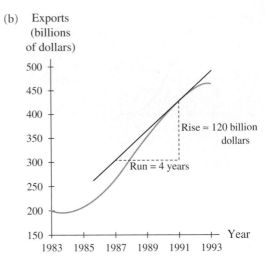

Rate of change in 1991 $\approx \dfrac{\$120 \text{ billion}}{4 \text{ years}} =$

$30 billion per year

Section 3.4

1. (a) Miles per hour

    (b) Speed or velocity

3. (a) The number of words per minute cannot be negative.

    (b) Words per minute per week

    (c) The student's typing speed could actually be getting worse, which would mean that $W'(t)$ is negative.

5. (a) When the ticket price is $65, the airline's weekly profit is $15,000.

    (b) When the ticket price is $65, the airline's weekly profit is increasing by $1500 per dollar of ticket price. Raising the ticket price a little will increase profit.

    (c) When the ticket price is $90, the profit is declining by $2000 per dollar of ticket price. An increase in price will decrease profit.

7. $t(x)$

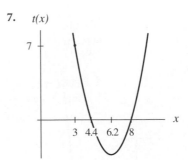

9. (a) At the beginning of the diet, you weigh 167 pounds.
   (b) After 12 weeks of dieting, your weight is 142 pounds.
   (c) After 1 week of dieting, your weight is decreasing by 2 pounds per week.
   (d) After 9 weeks of dieting, your weight is decreasing by 1 pound per week.
   (e) After 12 weeks of dieting, your weight is neither increasing nor decreasing.
   (f) After 15 weeks of dieting you are gaining weight at a rate of a fourth of a pound per week.

   (g)

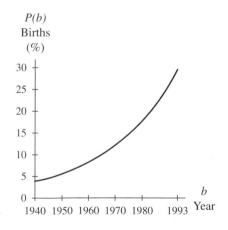

11. We know that the following points are on the graph: (1940, 4), (1970, 12), (1993, 29), and (1980, 18).

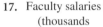

13. (a) It is possible for profit to be negative if costs are more than revenue.
   (b) It is possible for the derivative to be negative if profit declines as more shirts are sold (because the price is so low, the revenue is less than the cost associated with the shirt.)
   (c) If $P'(200) = -1.5$, the fraternity's profit is declining. Profit may still be positive (which means that the fraternity is making money), but the negative rate of change indicates that it is not making the most profit possible.

15. (a) Years per percentage point
   (b) As the rate of return increases, the time it takes the investment to double decreases.
   (c) i. When the interest rate is 9%, it takes 7.7 years for the investment to double.
       ii. When the interest rate is 5%, the doubling time is decreasing by 2.77 years per percentage point. A one-percentage-point increase in the rate will decrease the doubling time by approximately 2.8 years.
       iii. When the interest rate is 12%, the doubling time is decreasing by 0.48 year per percentage point. A one-percentage-point increase in the rate of return will result in a decrease in doubling time of approximately half a year.

17. Faculty salaries (thousands of dollars)

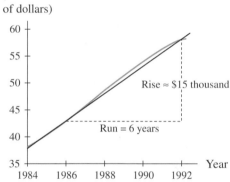

Faculty salaries (thousands of dollars)

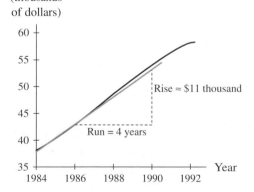

   (a) The slope of the secant line gives the average rate of change. Between 1986 and 1992, faculty salaries increased an average of about $2.5 thousand per year.
   (b) The slope of the tangent line gives the instantaneous rate of change in 1986.

(c) Slope $\approx \dfrac{\$11 \text{ thousand}}{4 \text{ years}} = \$2.75$ thousand per year

In 1986 faculty salaries were increasing by approximately $2750 per year.

(d) By sketching a tangent line in 1992 and estimating its slope, we find that salaries were increasing by approximately $1.5 thousand per year.

19. *G(t)*
Points

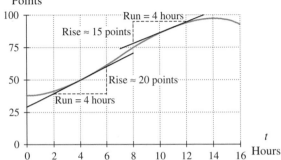

(a, b) Slope at 4 hours $\approx \dfrac{20 \text{ points}}{4 \text{ hours}} = 5$ points per hour

Slope at 11 hours $\approx \dfrac{15 \text{ points}}{4 \text{ hours}} = 3.75$ points per hour

(Answers will vary.)

(c) Average rate of change $\approx \dfrac{36 \text{ points}}{6 \text{ hours}} =$

6 points per hour

21. (a) Rate of change in 1970 $\approx$ 0.8 death per year
Rate of change in 1980 $\approx$ 1 death per year

(b) Answers will vary.

23. Rate of change in 1992 $\approx$

$\dfrac{107 \text{ thousand acres } - \ 116 \text{ thousand acres}}{2 \text{ years}} =$

-4.5 thousand acres per year

In 1992, the annual number of acres of wetlands lost was declining by 4.5 thousand acres per year.

Section 3.5

1. Change = \$463.0 − \$184.2 = \$278.8 million

Average rate of change $= \dfrac{\$278.8 \text{ million}}{10 \text{ years}} =$

$27.88 million per year

Percentage change $= \dfrac{\$278.8 \text{ million}}{\$184.2 \text{ million}} \times 100\% \approx 151.4\%$

3. (a) $12.1 million per quarter

(b) $\dfrac{\$12.1 \text{ million per quarter}}{\$135 \text{ million}} \times 100\% \approx$

8.96% per quarter

5. (a) i. -71 billion cubic feet
ii. 943 billion cubic feet
iii. 311 billion cubic feet

(b) i. -8.65%
ii. 125.7%
iii. 22.5%

(c) i. -4.4 billion cubic feet per year
ii. 188.6 billion cubic feet per year
iii. 155.5 billion cubic feet per year
Natural gas imports fluctuated dramatically between 1970 and 1986. The average rate of change does not describe the fluctuations.

(d) A model that is continuous in 1990 would be needed to calculate the rate of change in that year.

(e) 155.5 billion cubic feet per year (Note that this is the same as the answer to part *c*. iii.)

7. (a) -20.5%

(b) -62.6 thousand cars per year

(c) $C(x) = 11.671x^3 - 192.703x^2 + 734.956x + 2394.172$ thousand cars *x* years after 1984

Percentage change $= \dfrac{C(8) - C(0)}{C(0)} \times 100\% \approx -20\%$

Average rate of change $= \dfrac{C(8) - C(0)}{8 - 0} \approx$

-59.7 thousand cars per year
These answers are close to those obtained using the data.

(d) The percentage change and average rate of change do not describe the rise and fall of imported-car sales during the 8-year period.

9. (a) Change = -1980 accidents
Percentage change = -50.3%

(b) The number of accidents in 1992 was 1980 less than the number in 1976. That decrease represents a 50.3% decline in accidents.

(c) Between 1976 and 1992, accidents declined by an average of 123.8 accidents per year.

11. (a) Change = 0.40 million PCs
Percentage change = 199.5%

(b) Average rate of change = 0.1 million PCs per year

13. (a)

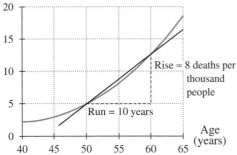

Death rate
(deaths per thousand people)

Rise ≈ 8 deaths per thousand people

Run = 10 years

Age (years)

The slope of the secant line (≈ 0.8 death per thousand people per year of age) is the average rate of change of the death rate between ages 50 and 60.

(b)

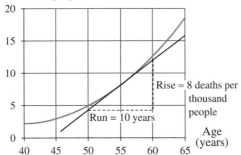

Death rate
(deaths per thousand people)

Rise ≈ 8 deaths per thousand people

Run = 10 years

Age (years)

The slope of the tangent line gives the instantaneous rate of change of the death rate at age 55.

(c) Rate of change ≈ 0.8 death per thousand people per year of age

Percentage rate of change ≈

$$\frac{0.8 \text{ death/thousand people/year of age}}{8 \text{ deaths/thousand people}} \times$$

$$100\% = 10\% \text{ per year}$$

15. (a) From 1980 through 1990, the number of births to women under 15 increased by approximately 1.7 thousand births, representing an increase of approximately 18%.

(b) The rate of change in 1985 was approximately 0.2 thousand births per year, representing an increase per year of approximately 2%.

Chapter 3 Review Test

1. (a) i. *A*, *B*, *C*; ii. *E*; iii. *D*
(b) The graph is steeper at *B* than it is at *A*, *C*, or *D*.
(c) Below: *C*, *D*, *E*; above: *A*; at *B*: above to the left of *B*, below to the right of *B*

(d)

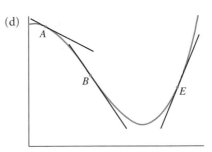

(e)   i. Feet per second per second. This is acceleration.
ii, iii. The roller coaster was slowing down until *D*, when it began speeding up.
iv. The roller coaster's speed was slowest at *D*.
v. The roller coaster was slowing down most rapidly at *B*.

2. (a) Average rate of change = 92.7 students per year
Between the 1991–92 school year and the 1994–95 school year, enrollment increased on average by about 93 students per year.
(b) Percentage change = 14%
Between the 1993–94 school year and the 1994–95 school year, enrollment increased by 14%.
(c) Increased enrollment probably reflects increased enrollment in the university or changes in language requirements, rather than an increased interest in Spanish.

3. (a) Average weekly earnings (dollars)

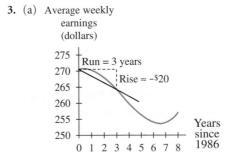

Run = 3 years

Rise ≈ −$20

Years since 1986

The slope of the secant line gives the average rate of change in weekly earnings between 1986 and 1989.

(b) Average weekly
earnings
(dollars)

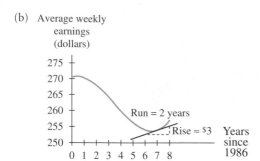

The slope of the tangent line gives the instantaneous rate of change of weekly earnings in 1993.

(c) Rate of change in 1993 $\approx \dfrac{\$3}{2 \text{ years}} = \$1.50$ (1982

dollars) per year

In 1993, average weekly earnings were increasing by $1.50 (1982 dollars) per year.

(d) Average rate of change $\approx \dfrac{-\$26}{3 \text{ years}} = -\$2.00$ per year

Between 1986 and 1989, average weekly earnings fell by an average of $2.00 (1982 dollars) per year.

4. (a)   i. An average 22-year-old athlete can swim 100 meters in 49 seconds.
          ii. The swim time for a 22-year-old is declining by a half of a second per year of age.
    (b) A negative rate of change indicates that an average swimmer's time is improving as age increases.

5. (a) $R'(t) = 6.062$ million dollars per year
    (b) Change = $30.31 million
        Average rate of change = $6.062 million per year
        Percentage change = 75.9%
    (c)   i. A linear model is defined as one that has a constant rate of change.
          ii. Although the rate of change is constant, it is not constant when expressed as a percentage of the output, because output is not constant.
         iii. The average rate of change of a linear model will always be constant (and the same as the slope of the model), because it is the slope of a secant line, and any secant line between two points on a line will be the line itself, having the same slope as the line.

CHAPTER 4

Section 4.1

1. (a)   Cases

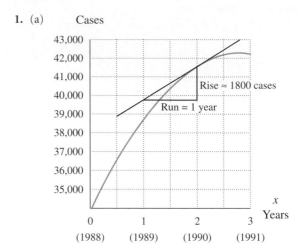

$$\text{Slope} \approx \frac{1800 \text{ cases}}{1 \text{ year}} = 1800 \text{ cases per year}$$

(Answers will vary.)

(b) Input of
close point   Slope
2.1   1685.75
2.01   1780.205
2.001   1789.6505
2.0001   1790.59505
2.00001   1790.6895
2.000001   1790.699
        Trend $\approx$ 1791 cases per year

(c) In 1990, the number of diagnosed AIDS cases was increasing by approximately 1791 cases per year.

3. (a)
Account balance
(dollars)

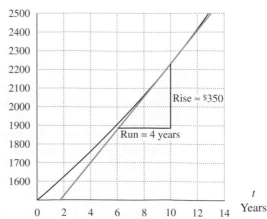

$$\text{Slope} \approx \frac{\$350}{4 \text{ years}} = \$87.50 \text{ per year}$$

(Answers will vary.)

(b) Input of

close point	Slope
10.1	89.35445
10.01	89.19413
10.001	89.17812
10.0001	89.17652
10.00001	89.17637

Trend ≈ $89.18 per year

(c) Part b is more accurate. In part a, the tangent line was estimated, and the rise and run were estimated. In part b, only the trend is not exact.

5. (a) $T(13) = 67.946$ seconds

(b) $T(13 + h) = 0.181(13 + h)^2 - 8.463(13 + h) + 147.376 = 0.181h^2 - 3.757h + 67.946$

(c) $\dfrac{T(13 + h) - T(13)}{13 + h - 13} = \dfrac{0.181h^2 - 3.757h}{h}$
$= 0.181h - 3.757$

(d) $\lim\limits_{h \to 0} (0.181h - 3.757) = -3.757$ seconds per year of age

(e) This method is more accurate as long as no algebraic mistakes are made in the process.

7. (a) $P(t)$
    (%)

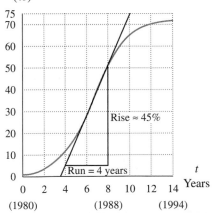

0   2   4   6   8   10 12 14 Years
(1980)   (1988)   (1994)

Rise ≈ 45%
Run = 4 years

$$\text{Slope} \approx \frac{45\%}{4 \text{ years}} = 11.25 \text{ percentage points per year}$$

(b) Input of

close point	Slope
7.1	11.5748
7.01	11.6152
7.001	11.6189
7.0001	11.6192
7.00001	11.6193
7.000001	11.6193

Trend ≈ 11.6 percentage points per year

(c) In 1987, the percentage of households with VCRs was increasing by 11.6 percentage points per year.

(d) The method in part a is fast but can be inaccurate. The method in part b is somewhat tedious but more accurate.

9. (a) $\dfrac{3289.6 - 3162.1 \text{ billion 1987 dollars}}{1987 - 1985}$
$= 63.75$ billion 1987 dollars per year

(b) $I(t) = 75.078t + 2371.521$ billion 1987 dollars t years after 1975

The slope of the model is 75.078 billion dollars per year. This is the rate of change that would have occurred if the data had been perfectly linear. The answer to part a is a better picture of the rate of change around 1987.

11. (a) $\dfrac{198.0 - 168.0}{1991 - 1989} = \dfrac{30}{2} = 15$ index points per year

(b) $CPI(x) = 0.395x^2 + 10.634x + 119.359$ x years after 1985

(c) Input of

close point	Slope
5.1	14.6214
5.01	14.5859
5.001	14.5823
5.0001	14.5820
5.00001	14.5819
5.000001	14.5819

Trend ≈ 14.58 index points per year

(d) The method in part a is simple and fairly accurate. The method in part b is more tedious may or may not be more accurate. Both methods use a slope between two points.

13. (a) $CPI(5) = 182.404$

(b) $CPI(5 + h) = 0.395(25 + 10h + h^2) + 10.634(5 + h) + 119.359 = 182.404 + 14.584h + 0.395h^2$

(c) $\dfrac{CPI(5 + h) - CPI(5)}{5 + h - 5} = 14.584 + 0.395h$

(d) $\lim\limits_{h \to 0} (14.584 + 0.395h) = 14.584$ index points per year

(e) In 1990, the CPI for college tuition was increasing at a rate of 14.58 index points per year.

(f) i. Because the model was rounded in the four-step method, the numerical estimation that uses the full model is more accurate.

ii and iii. We believe that a symmetric difference quotient is appropriate in both of these cases.

15. (a) Input of

close point	Slope
2.9	0.01575
2.99	0.015525
2.999	0.0155025
2.9999	0.01550025

Trend ≈ 0.016 dollars per pound per month

(b) Because the model to the right of $m = 3$ is a line, the slope is -0.028 dollars per pound per month.

(c) Because the slope from the left and that from the right are not the same, the derivative of $p(m)$ at $m = 3$ does not exist.

(d) One possible estimation is a symmetric difference quotient using $m = 3.1$ and $m = 2.9$. This estimation gives a rate of change of approximately \$0.006 per pound per month. From the graph of $p(m)$, you can observe that the maximum price occurred in January 1995, so the rate of change was approximately zero dollars per pound per month. There are many methods that can be used to estimate the rate of change. We have given only two examples.

17. (a) From the left of $m = 3$, we use the formula

$p(m) = -0.0025m^2 + 0.0305m + 0.8405$ dollars per pound

i. $p(3) = 0.9095$

ii. $p(3 - h) = -0.0025(3 - h)^2 + 0.0305(3 - h) + 0.8405 = -0.0025h^2 - 0.0155h + 0.9095$

iii. $\dfrac{p(3 - h) - p(3)}{3 - h - 3} =$

$\dfrac{-0.0025h^2 - 0.0155h + 0.9095 - 0.9095}{-h}$

$= 0.0025h + 0.0155$

iv. $\lim\limits_{h \to 0} (0.0025h + 0.0155) = 0.0155$ dollar per pound per month

(b) From the right of $m = 3$, we use the formula $p(m) = -0.028m + 0.996$ dollars per pound

i. $p(3) = 0.912$

ii. $p(3 + h) = -0.028(3 + h) + 0.996 = -0.028h + 0.912$

iii. $\dfrac{p(3 + h) - p(3)}{3 + h - 3} =$

$\dfrac{-0.028h + 0.912 - 0.912}{h} = -0.028$

iv. $\lim\limits_{h \to 0} (-0.028) = -0.028$ dollar per pound per month

19. (a) $D(R(x)) = \dfrac{1.02^x}{1.4807}$ dollars when x mountain bikes are sold

(b) $R(400) \approx 2754.66$ deutschmarks
$D[R(400)] \approx \$1860.38$

(c) To find $\frac{dD}{dx}$ when $x = 400$, we numerically investigate the following data:

Input of

close point	Slope
400.1	36.8769
400.01	36.8441
400.001	36.8408
400.0001	36.8404
400.00001	36.8404

Trend ≈ \$36.84 per mountain bike

Section 4.2

1. The slopes are negative to the left of A and positive to the right of A. The slope is zero at A.

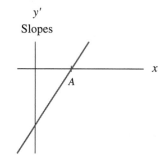

3. The slopes are positive everywhere, near zero to the left of zero, and increasingly positive to the right of zero.

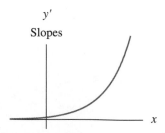

5. The slope is zero everywhere.

$y' = $ slope $= 0$

7. The slopes are negative everywhere. The magnitude is large close to zero and is near zero to the far right.

Slopes

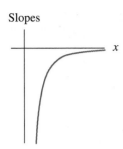

9. The slopes are negative to the left and right of A. The slope appears to be zero at A.

y'

Slopes

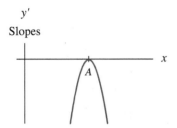

11. (a) The average rate of change during the year (found by estimating the slope of the secant line drawn from September to May) is approximately 14 members per month. (Answers will vary.)

(b) By estimating the slopes of tangent lines, we obtain the following. (Answers will vary.)

Month	Slope (members per month)
Sept	98
Nov	−9
Feb	30
Apr	11

(c) (members per month)

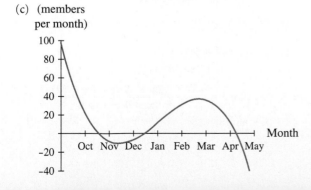

(d) Membership was growing most rapidly around March. This point on the membership graph is an inflection point.

(e) The average rate of change is not useful in sketching an instantaneous-rate-of-change graph.

13. $j'(t)$ (inmates per year)

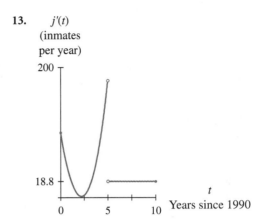

Years since 1990

15. (a) Profit is increasing on average by approximately $600 per car.

(b)

Number of cars	Slope (dollars per car)
20	0
40	160
60	770
80	10
100	−1185

(Answers will vary.)

(c)

Average monthly profit (dollars per car)

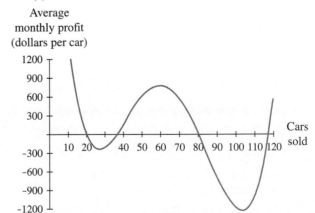

(d) The average monthly profit is increasing most rapidly for about 60 cars sold and is decreasing most rapidly when about 100 cars are sold. The corresponding points on the graph are inflection points.

(e) Average rates of change are not useful in graphing instantaneous rates of change.

17. The derivative does not exist at $x = 0$, $x = 3$, and $x = 4$ because the graph is not continuous at those points.

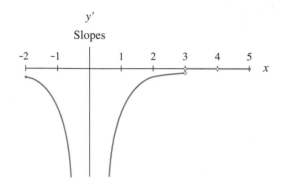

19. The derivative does not exist at $x = 2$ and $x = 3$ because the slopes from the right and left are different.

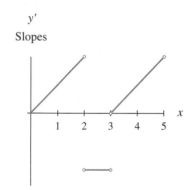

Section 4.3

1.

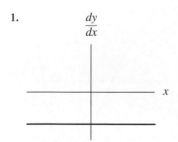

$$\frac{dy}{dx} = -7$$

3.

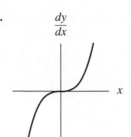

$$\frac{dy}{dx} = 4x^3$$

5.

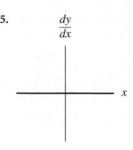

$$\frac{dy}{dx} = 0$$

7. (a) Solving $2.5 = e^h$ gives $h \approx 0.916$ hours ≈ 55 minutes.

(b) $\dfrac{dV}{dh} = e^h$ quarts per hour $= \dfrac{e^h}{60}$ quarts per minute where h is the number of hours the dough has been allowed to rise

(c) $V'(0.4) \approx 0.02$ quarts per minute after 24 minutes
$V'(0.7) \approx 0.03$ quarts per minute after 42 minutes
$V'\left(\dfrac{55}{60}\right) \approx 0.04$ quarts per minute after 55 minutes

9. (a) $P'(t) = 165\,(\ln 1.02228)(1.02228)^t$ thousand people per year t years after 1900

(b) $P(70) \approx 771.6$ thousand people

(c) $P'(90) \approx 26.4$ thousand people per year

11. (a) If $y = e^x$, then $\frac{dy}{dx} = (\ln e)(e^x)$.
Because $\ln e = 1$, $\frac{dy}{dx} = e^x$.

(b) If $y = e^{kx}$, then $\frac{dy}{dx} = (\ln e^k)(e^{kx})$.
Because $\ln e^k = k$, $\frac{dy}{dx} = ke^{kx}$.

13. (a)

x:	-3	-2	-1	0	1	2	3
$\dfrac{dy}{dx}$:	-42	-28	-14	0	14	28	42

(b)

x:	-2	-1	0	1	2
$\dfrac{dy}{dx}$:	0.54134	1.47152	4	10.87313	29.55622

(c)

x:	-3	-2	-1	0	1	2	3
$\dfrac{dy}{dx}$:	-13.5	-6	-1.5	0	-1.5	-6	-13.5

(d) If $y = kx^n$, then $\frac{dy}{dx} = k(nx^{n-1})$.
If $y = ke^x$, then $\frac{dy}{dx} = ke^x$.

15. $\dfrac{dy}{dx} = \dfrac{-k}{x^2}$

17. (i) $(x, 15x + 32)$

(ii) $(x + h, 15(x + h) + 32)$

(iii) Slope $= \dfrac{15x + 15h + 32 - (15x + 32)}{x + h - x} = 15$

(iv) $\lim\limits_{h \to 0} 15 = 15$

19. (i) $(x, 3x^2)$

(ii) $(x + h, 3(x + h)^2)$

(iii) Slope $= \dfrac{3x^2 + 6xh + 3h^2 - 3x^2}{x + h - x} = 6x + 3h$

(iv) $\lim\limits_{h \to 0} (6x + 3h) = 6x$

21. (i) $(x, -3x^2 - 5x)$

(ii) $(x + h, -3(x + h)^2 - 5(x + h))$

(iii) Slope

$= \dfrac{-3x^2 - 6xh - 3h^2 - 5x - 5h - (-3x^2 - 5x)}{x + h - x}$

$= -6x - 3h - 5$

(iv) $\lim\limits_{h \to 0} (-6x - 3h - 5) = -6x - 5$

23. (a) Input of

close point	Slope
1.1	-33.6
1.01	-32.16
1.001	-32.016
1.0001	-32.0016

Trend = -32 feet per second

(b) i. $(t, -16t^2 + 100)$

ii. $(t + h, -16(t + h)^2 + 100)$

iii. Slope

$= \dfrac{-16t^2 - 32th - 16h^2 + 100 - (-16t^2 + 100)}{t + h - t}$

$= -32t - 16h$

iv. $\lim\limits_{h \to 0} (-32t - 16h) = -32t$

When $t = 1$, slope $= -32(1) = -32$ feet per second

25. (a) $D(x) = -0.019x^2 + 1.017x - 9.874$ million drivers age x years

(b) i. $(x, -0.019x^2 + 1.017x - 9.874)$

ii. $(x, -0.019(x + h)^2 + 1.017(x + h) - 9.874)$

iii. Slope $=$
$[-0.019(x^2 + 2xh + h^2) + 1.017(x + h) - 9.874 - (-0.019x^2 + 1.017x - 9.874)] \div (x + h - x)$
$= -0.038x - 0.019h + 1.017$

iv. $\lim\limits_{h \to 0} (-0.038x - 0.019h + 1.017) = -0.038x + 1.017$ million drivers per year where x is the age in years.

Section 4.4

1. $y' = 0$

3. $g'(x) = 3e^x$

5. $h'(x) = 14x - 12$

7. $g'(x) = 17 (\ln 4.962)(4.962)^x$

9. $f'(x) \approx 100{,}000 (\ln 1.051162)(1.051162)^x$

11. $T'(x) = -1.6t + 2$ °F per hour t hours after noon

(a) $T'(1.5) = -0.4$ °F per hour

$T'(-5) = 10$ °F per hour

(b) $T'(-5) = 10$ °F per hour

(c) $T'(0) = 2$ °F per hour

(d) $T'(4) = -4.4$ °F per hour

13. (a) $P(9) \approx 0.88$ million PCs

(b) $P'(7) \approx 0.12$ million PCs per year

(c) $\dfrac{P'(7)}{P(7)} \times 100\% \approx 20.6$ percent per year

15. In 1955, the number of births was rising (barely) at a rate of 0.7 birth per year. In 1980, the number of births was falling at a rate of 80.6 births per year.

17. (a) $P(x) = 175 - 0.0146x^2 + 0.7823x - 46.9125 - \frac{49.6032}{x}$ dollars when x windows are produced each hour

(b) $P'(x) = -0.0292x + 0.7823 + \frac{49.6032}{x^2}$ dollars per window when x windows are produced each hour

(c) $P(80) \approx \$96.61$

(d) $P'(80) \approx -\$1.55$ per window

19. (a) $P(x) = 0.0426x^2 + 2.049x + 251.796 + 68.738 \cdot (1.0213)^x$ million people x years after 1980

(b) $P'(x) = 0.0852x + 2.049 + 68.738 (\ln 1.0213) \cdot (1.0213)^x$ million people per year x years after 1980

(c) $P(14) \approx 381.16$ million people

(d) $P'(14) \approx 5.19$ million people per year

21. (a) We choose a piecewise continuous model:

$$P(t) = \begin{cases} 0.013t^2 - 2.150t + 91.384 \text{ dollars} & \text{when } 78 \le t < 86 \\ 0.002t^3 - 0.542t^2 + 48.074t - 1418.499 \text{ dollars} & \text{when } 86 \le t \le 91 \end{cases}$$

where t is the number of years since 1900.

(b) $\dfrac{1.14 - 1.43 \text{ dollars}}{2 \text{ years}} = -\$0.145$ per year

(c) $$P'(t) = \begin{cases} 0.026t - 2.150 \text{ dollars per year} & \text{when } 78 \le t < 86 \\ 0.006t^2 - 1.084t + 48.074 \text{ dollars per year} & \text{when } 86 < t \le 91 \end{cases}$$

where t is the number of years since 1900.

$P'(79) \approx -\$0.131$ per year

(d) $\dfrac{P'(90)}{P(90)} \times 100\% \approx -3.93\%$ per year

23. (a) We choose a piecewise continuous model:

$$C(t) = \begin{cases} 403t + 2437.667 \text{ thousand cars} & \text{when } t \leq 2 \\ 12.556t^3 - 201.452t^2 + \\ \quad 740.754t + 2461.714 & \text{when } t > 2 \\ \quad \text{thousand cars} \end{cases}$$

where t is the number of years since 1984.

(b)
$$C'(t) = \begin{cases} 403 \text{ thousand cars per year} & \text{when } t < 2 \\ 37.667t^2 - 402.905t + \\ \quad 740.754 \text{ thousand cars} & \text{when } t > 2 \\ \quad \text{per year} \end{cases}$$

where t is the number of years since 1984.

(c) $C'(8) \approx -71.8$ thousand cars per year

(d) $C(9) \approx 1964$ thousand cars

(e) $\dfrac{C'(0)}{C(0)} \times 100\% \approx 16.5\%$ per year

25. (a) $A(x) = 155.139x^3 - 1605.762x^2 + 8996.456x + 61{,}677.214$ million dollars x years after 1983

(b) $A'(x) = 465.417x^2 - 3211.524x + 8996.456$ million dollars per year x years after 1983

(c) $A'(3) \approx 3550.6$ million dollars per year

$\dfrac{A'(3)}{A(3)} \times 100\% \approx 4.53\%$ per year

(d) $A(7) \approx 99{,}183$ million dollars

27. (a)
$$I(t) = \begin{cases} 463.89t - 7625.93 \text{ dollars} & \text{when } 47 \leq t < 77 \\ 92.939t^2 - 15{,}096.582t + \\ \quad 639{,}021.182 \text{ dollars} & \text{when } 77 \leq t \leq 92 \end{cases}$$

where t is the number of years since 1900.

(b)
$$I'(t) = \begin{cases} 463.89 \text{ dollars per year} & \text{when } 47 \leq t < 77 \\ 185.878t - 15{,}096.582 \\ \quad \text{dollars per year} & \text{when } 77 < t \leq 92 \end{cases}$$

where t is the number of years since 1900.

(c)

Year	Rate of change (dollars per year)	Percentage rate of change (percent per year)
1972	463.89	1.80
1980	-226.32	-0.87
1984	517.19	1.94
1992	2004.22	5.45

(d) (Answers will vary.) The negative rate of change in 1980 almost certainly affected Carter's re-election, because it is indicative of the state of the economy as a whole. However, we probably cannot make such statements about the other rates of change.

29. (a) $\text{CPI}(x) = -0.081x^3 + 1.118x^2 + 0.975x + 107.619$, where x is the number of years after 1985

$\text{CPI}(x) = 0.393x^2 + 2.586x + 107.136$, where x is the number of years after 1985

The cubic model is a better fit because it is of a higher order than the quadratic. The data do seem to be concave down on the right; however, the concavity change is very subtle. The cubic model almost certainly does not reflect future behavior of the CPI as well as the quadratic model does.

(b) Quadratic model: 8.1 index points per year
Cubic model: 4.8 index points per year

31. (a) $A'(r) = 10{,}000e^{10r}$ dollars per 100 percentage points or $A'(r) = 100e^{10r}$ dollars per percentage point

(b) $20{,}137.53 per 100 percentage points or $201.38 per percentage point. The rate of change appears to be 100 times as large as it should because the rate of change approximates how the value will increase when r changes by 1. Because the interest rate is input in decimals, an increase of 1 in r corresponds to a change in the interest rate of 100 percentage points.

(c) $A'(r) = 100e^{0.1r}$ dollars per percentage point

(d) $201.38 per percentage point

Section 4.5

1. (a) $f(x(2)) = f(6) = 140$

(b) $\dfrac{df}{dx} = -27$

(c) $\dfrac{dx}{dt} = 1.3$

(d) $\dfrac{df}{dt} = (-27)(1.3) = -35.1$

3. The value of the investor's gold is increasing at a rate of (0.2 troy ounces per day)($395.70 per troy ounce) = $79.14 per day.

5. (a) $R(476) = 10{,}000$ deutschemarks (dm)

On March 12, 1996, sales were 476 units, producing revenue of 10,000 dm.

(b) $D(10{,}000) = 6750$

On March 12, 1996, 10,000 dm were worth $6750.

(c) $\dfrac{dR}{dx} = 2.6$ dm per unit

Revenue was increasing by 2.6 dm per unit sold.

(d) $\dfrac{dD}{dr} = 0.675$ per dm

The exchange rate was $0.675 per dm.

(e) $\dfrac{dD}{dx} = (2.6$ dm per unit$)(\$0.675$ per dm$) \approx \$1.76$

per unit

On March 12, 1996, revenue was increasing at a rate of \$1.76 per unit sold.

7. (a) $p(10) \approx 12.009$ thousand people

In 1995, the city had a population of approximately 12,000 people.

(b) $g(p(10)) \approx g(12.009) \approx 22$ garbage trucks

In 1995, the city owned 22 garbage trucks.

(c) $p'(t) = \dfrac{31.2e^{-0.02t}}{(1 + 12e^{-0.02t})^2}$ thousand people per year

$p'(10) \approx 0.22$ thousand people per year

In 1995, the population was increasing at a rate of approximately 220 people per year.

(d) $g'(p) = 2 - 0.003p^2$ trucks per thousand people

$g'(12.009) \approx 1.6$ trucks per thousand people

In 1995, when the population was about 12,000, the number of garbage trucks needed by the city was increasing by 1.6 trucks per thousand people.

(e) $\dfrac{dg}{dt} = (2 - 0.003p^2)\left[\dfrac{31.2e^{-0.02t}}{(1 + 12e^{-0.02t})^2}\right]$

$= \left(2 - 0.003\left[\dfrac{130}{(1 + 12e^{-0.02t})}\right]^2\right) \cdot$

$\left[\dfrac{31.2e^{-0.02t}}{(1 + 12e^{-0.02t})^2}\right]$

trucks per year t years after 1985

When $t = 10$, $\dfrac{dg}{dt} = (1.6$ trucks per thousand people$) \cdot$

(0.22 thousand people per year)

$= 0.34$ trucks per year

In 1995, the number of trucks needed by the city was increasing at a rate of 0.34 truck per year, or 1 truck every 3 years.

(f) See interpretations in parts a through e.

9. $c(x(t)) = 3(4 - 6t)^2 - 2$

$\dfrac{dc}{dt} = 6(4 - 6t)(-6) = -144 + 216t$

11. $h(p(t)) = \dfrac{4}{1 + 3e^{-0.5t}}$

$\dfrac{dh}{dt} = \dfrac{4(3)(0.5)e^{-0.5t}}{(1 + 3e^{-0.5t})^2}$

13. Inside function: $3.2x + 5.7$

Outside function: u^5

$f'(x) = 5(3.2x + 5.7)^4(3.2)$

15. Inside function: $x^2 - 3x$

Outside function: $u^{\frac{1}{2}}$

$f'(x) = \dfrac{1}{2}(x^2 - 3x)^{\frac{1}{2}}(2x - 3)$

17. Inside function: $35x$

Outside function: $\ln u$

$f'(x) = \dfrac{1}{35x}(35) = \dfrac{1}{x}$

19. Inside function: $16x^2 + 37x$

Outside function: $\ln u$

$f'(x) = \dfrac{1}{16x^2 + 37x}(32x + 37)$

21. Inside function: $0.695x$

Outside function: $72.378e^u$

$f'(x) = 72.378e^{0.695x}(0.695)$

23. Inside function: $0.0856x$

Outside function: $1 + 58.32e^u$

$f'(x) = 58.32e^{0.0856x}(0.0856)$

25. Inside function: $4x + 7$

Outside function: $350u^{-1}$

$f'(x) = -350(4x + 7)^{-2}(4)$

27. Inside function:

$1 + 8.976e^{-1.243x} \begin{cases} \text{inside:} & -1.243x \\ \text{outside:} & 1 + 8.976e^w \end{cases}$

Outside function: $3706.5u^{-1} + 89070$

$f'(x) = -3706.5(1 + 8.976e^{-1.243x})^{-2}(8.976e^{-1.243x})(-1.243)$

$= \dfrac{(3706.5)(8.976)(1.243)e^{-1.243x}}{(1 + 8.976e^{-1.243x})^2}$

29. (a) $A(t) = 1500e^{0.04t}$ dollars after t years

(b) $A'(t) = 1500(0.04)e^{0.04t} = 60e^{0.04t}$ dollars per year after t years

(c) $A'(1) \approx \$62.45$ per year

$A'(2) \approx \$65.00$ per year

(d) The rates of change in part c tell you approximately how much interest your account will earn during the second and third years. The actual amounts are $A(2) - A(1) \approx \$63.71$ during the second year and $A(3) - A(2) \approx \$66.31$ during the third year.

31. (a) $P'(x) = 5609.82(0.1133)e^{0.1133x}$ dollars per year x years after 1980

(b)

Year	1980	1985	1990
Per capita debt ($)	5609.82	9885.00	17,418.25
Rate of change ($/year)	635.59	1119.97	1973.49
Percentage rate of change (%/year)	11.33	11.33	11.33

(c) Answers will vary.

(d) An exponential model always has a constant percentage rate of change.

33. (a) A logistic model is probably a better model because of the leveling-off behavior, although neither model should be used to extrapolate.

(b) $C(t) = \dfrac{1342.077}{1 + 36.797e^{-0.258856t}}$ calls t hours after 5 a.m.

(c) $C'(t) = \dfrac{(1342.077)(36.797)(0.258856)e^{-0.258856t}}{(1 + 36.797e^{-0.258856t})^2}$ calls per hour t hours after 5 a.m.

(d) noon: $C'(7) \approx 42$ calls per hour
    10 p.m.: $C'(17) \approx 74$ calls per hour
    midnight: $C'(19) \approx 58$ calls per hour
    4 a.m.: $C'(23) \approx 28$ calls per hour

(e) The rates of change give approximate hourly calls. This information could be used to determine how many dispatchers would be needed each hour.

35. (a) $c'(4) \approx 1575$ deaths per week

(b) $\dfrac{c'(4)}{c(4)} \times 100\% \approx$ increase of 108% per week

(c) $c'(8) \approx 25{,}331$ deaths per week

$\dfrac{c'(8)}{c(8)} \times 100\% \approx$ increase of 48% per week

(d) Although the rate of change is larger, it represents a smaller proportion of the total number of deaths that had occurred at that time.

37. (a) The data are essentially concave up, which indicates that a quadratic or exponential model may be appropriate. Looking at the second differences and percentage differences indicates that a quadratic model is the better choice.

(b) $u(x) = 177.356x^2 - 342.240x + 5914.964$ units per week x years after 1980

(c) $C(u(x)) = 3250.23 + 74.95 \ln(177.356x^2 - 342.240x + 5914.964)$ dollars per week x years after 1980

(d) $\dfrac{dC}{dx} = \left(\dfrac{74.95}{177.356x^2 - 342.240x + 5914.964}\right) \cdot$
$(354.712x - 342.240)$ dollars per week per year x years after 1980

(e)

Year	1992	1993	1994	1995
x	12	13	14	15
$C(x)$ ($/week)	4015.95	4026.40	4036.31	4045.72
$C'(x)$ ($/week/year)	10.73	10.18	9.66	9.17

(f, g) Although a graph of $C(x)$ may not appear ever to decrease, a graph of $C'(x)$ is negative between $x = 0$ and $x = 0.965$, indicating that cost was decreasing from the end of 1980 to (almost) the end of 1981. A close-up view of $C(x)$ between $x = 0$ and $x = 2$ confirms this.

39. Composite functions are formed by making the output of one function (the inside) the input of another function (the outside). It is imperative that the output of the inside and the input of the outside agree in the quantity that they measure as well as in the units of measurement.

Section 4.6

1. $h'(2) = f(2)g'(2) + f'(2)g(2) = 6(3) + (-1.5)(4) = 12$

3. (a) i. In 1997, there were 75,000 households in the city.

ii. In 1997, the number of households was declining at a rate of 1200 per year.

iii. In 1997, 52% of households owned a computer.

iv. In 1997, the percentage of households with a computer was increasing by 5 percentage points per year.

(b) Input: years since 1995
Output: number of households with computers

(c) $N(2) = h(2)c(2) = (75{,}000)(0.52) = 39{,}000$ households with computers
$N'(2) = h(2)c'(2) + h'(2)c(2) = (75{,}000)(0.05) + (-1200)(0.52) = 3126$ households per year
In 1997, there were 39,000 households with computers, and that number was increasing at a rate of 3126 households per year.

5. (a) i. $S(10) \approx \$15.24$; $S'(10) \approx -\$0.02$ per week
After 10 weeks, 1 share is worth $15.24, and the value is declining by $0.02 per week.

ii. $N(10) = 125$ shares; $N'(10) = 5$ shares per week
After 10 weeks, the investor owns 125 shares and is buying 5 shares per week.

iii. $V(10) = S(10)N(10) \approx \$1905$; $V'(10) = S(10)N'(10) + S'(10)N(10) \approx \$73.50$ per week
After 10 weeks, the investor's stock is worth approximately $1905, and the value is increasing by $73.50 per week.

(b) $V'(x) = \left(15 + \dfrac{2.6}{x + 1}\right)(0.5x) + (100 + 0.25x^2)$

$\left(\dfrac{-2.6}{(x + 1)^2}\right)$ dollars per week after x weeks

7. (500 acres)(5 bushels/acre/year) + (130 bushels/acre)
   (50 acres/year) = 9000 bushels per year

9. (a) $(17,000)(0.48) = 8160$ voters

   (b) $(8160)(0.57) \approx 4651$ votes for candidate A

   (c) $17,000\ (0.48)(-0.03) + 17,000(0.57)(0.07) \approx$
       434 votes for candidate A per week

11. $f'(x) = (3x^2 + 15x + 7)(96x^2) + (6x + 15)(32x^3 + 49)$

13. $f'(x) = (12.8893x^2 + 3.7885x + 1.2548)\cdot$
    $[29.685\,(\ln 1.7584)(1.7584)^x] + (25.7786x + 3.7885)\cdot$
    $[29.685(1.7584)^x]$

15. $f'(x) = (5.7x^2 + 3.5x + 2.9)^3 \cdot$
    $[-2(3.8x^2 + 5.2x + 7)^{-3}(7.6x + 5.2)] +$
    $[3(5.7x^2 + 3.5x + 2.9)^2(11.4x + 3.5)](3.8x^2 + 5.2x + 7)^{-2}$

17. $f'(x) = 12.624(14.831)^x(-2x^{-3}) + 12.624\,(\ln 14.831)\cdot$
    $(14.831)^x(x^{-2})$

19. $f'(x) = (79.32x)\left(\dfrac{1984.32(7.68)(0.859347)e^{-0.859347x}}{(1 + 7.68e^{-0.859347x})^2}\right) +$
    $79.32\left(\dfrac{1984.32}{1 + 7.68e^{-0.859347x}} + 1568\right)$

21. (a) Price $= 0.049m + 1.144$ dollars m months after
        December
        Quantity sold $= -0.946m^2 + 0.244m + 279.911$
        units sold m months after December

    (b) $R(m) = (0.049m + 1.144)(-0.946m^2 + 0.244m +$
        $279.911)$ dollars of revenue m months after Decem-
        ber

    (c) $R(8) \approx \$340.05;\ R(9) \approx \$325.78$

    (d) Because the revenue in September is less than that in
        August, the rate of change in August is probably neg-
        ative.

    (e) $R'(m) = (0.049m + 1.144)(-1.893m + 0.244) +$
        $0.049(-0.946m^2 + 0.244m + 279.911)$ dollars of
        revenue per month m months after December

    (f) $R'(2) \approx \$9.17$ per month
        $R'(8) \approx -\$12.04$ per month
        $R'(9) \approx -\$16.55$ per month

23. (a) $P(t) = 2.351t + 179.421$ million people t years after
        1960

    (b) $N(t) = (2.351t + 179.421)\left[(0.01)\cdot\right.$
        $\left.\left(\dfrac{6.2}{1 + 0.0678e^{0.144t}} + 23\right)\right]$ million people t years
        after 1960

    (c) $N'(t) = (2.351t + 179.421)\cdot$
        $\left[\dfrac{-6.2(0.01)(0.0678)(0.144)e^{0.144t}}{(1 + 0.0678e^{0.144t})^2}\right.$
        $\left. + (2.351)\left[(0.01)\left(\dfrac{6.2}{1 + 0.0678e^{0.144t}} + 23\right)\right]\right]$
        million people per year t years after 1960

    (d) $N'(20) \approx 0.11$ million people per year
        $N'(25) \approx 0.15$ million people per year
        $N'(30) \approx 0.26$ million people per year

25. (a) $m(x) = (-3.029 \cdot 10^{-4})\,x^3 + 0.077x^2 - 6.226x +$
        172.316 million men 65 or older x years after 1900
        $p(x) = (3.304 \cdot 10^{-4})\,x^2 - 0.059x + 2.706$ (percent
        expressed as a decimal) x years after 1900

    (b) $n(x) = m(x)p(x) = [(-3.029 \cdot 10^{-4})x^3 + 0.077x^2 -$
        $6.226x + 172.316][(3.304 \cdot 10^{-4})x^2 - 0.059x +$
        $2.706]$ million men 65 or older living below poverty
        level x years after 1900

    (c) $n'(90) \approx 0.025$ million men per year
        $n'(93) \approx 0.048$ million men per year

        In 1990, the number of men 65 or older who lived
        below the poverty level was increasing by approxi-
        mately 25,000 per year. In 1993, the rate of increase
        was approximately 48,000 per year.

27. (a) $d(x) = (-0.03406x^3 + 1.33145x^2 + 9.91340x +$
        $164.44689)[10(-0.18333x^2 + 2.89121x + 20.21545)]$
        cesarean section deliveries to women 35 and older x
        years after 1980.

    (b)
Year	C-sections per year
1980	6760
1985	8790
1989	6660

    (c) $I(x) = d(x)c(x)$ and $I(7) \approx \$225,024,590$. Obstetri-
        cians made about 225 million dollars from cesarean
        sections in 1987.

    (d) $I'(x) = d'(x)c(x) + d(x)c'(x)$ dollars per year x years
        after 1980

    (e) Answers will vary.

29. (a) $E(x) = -151.516x^3 + 2060.988x^2 - 8819.062x +$
        $195,291.201$ students enrolled x years after the
        1980–81 school year
        $D(x) = -14.271x^3 + 213.882x^2 - 1393.655x +$
        $11,697.292$ students dropping out x years after the
        1980–81 school year

    (b) $P(x) = \dfrac{D(x)}{E(x)} \times 100$ percent x years after the 1980–81
        school year

(c) $P'(x) = D(x)(-1[E(x)]^{-2}E'(x)) + D'(x)[E(x)]^{-1}$ percentage points per year x years after the 1980–81 school year

(d)

x	$P'(x)$ (percentage points per year)	x	$P'(x)$ (percentage points per year)
0	-0.44	5	-0.19
1	-0.38	6	-0.19
2	-0.32	7	-0.22
3	-0.26	8	-0.29
4	-0.21	9	-0.41

In the 1980–81 school year, the rate of change was most negative with a value of -0.44 percentage points per year. This is the most rapid decline during this time period. The rate of change was least negative in the 1985–86 school year with a value of -0.187 percentage points per year.

(e) Negative rates of change indicate that high school attrition in South Carolina was improving during the 1980s.

31. The inputs must correspond in order for the result of the multiplication to be meaningful.

Chapter 4 Review Test

1. (a) $B(t) = 22.127(1.131524)^t$ thousand births t years after 1975

(b) Input of

close point	Slope
15.1	17.5573
15.01	17.4598
15.001	17.4501
15.0001	17.4492
15.00001	17.4491

Limiting value ≈ 17.4 thousand births per year

(c) In 1990, the number of in-hospital, midwife-assisted births was increasing by 17.4 thousand births per year.

(d) $B'(t) = 22.127 \, (\ln 1.131524)(1.131524)^t$ thousand births per year t years after 1975

$B'(15) \approx 17.4$ thousand births per year

2. (a) Slope in 1992 $\approx \dfrac{1019.9 - 526.1}{1993 - 1991} =$

$\dfrac{493.8 \text{ billion dollars}}{2 \text{ years}} = \$246.9$ billion per year

(b)

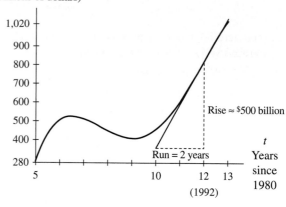

$A(t)$ (billions of dollars)

Slope $\approx \dfrac{\$500 \text{ billion}}{2 \text{ years}} = \$250$ billion per year

(Answers will vary.)

(c) $A'(t) = -6.98264t^3 + 206.4774t^2 - 1932.76t + 5759.1455$ billion dollars per year t years after 1980

$A'(12) \approx 232.8$

In 1992, the total amount of long-term new mortgages was increasing by 232.8 billion dollars per year.

3. The slopes are positive to the left of A and negative from A to B. At A, the slope is zero. From B to C, the slope is a positive constant. To the right of C, the slope is negative and is approaching zero. The slope does not exist at B and C.

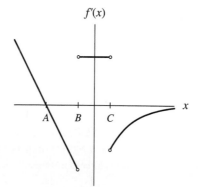

$f'(x)$

4. (a) $\dfrac{dD}{dt} = \dfrac{3175(12.38)(0.3902)e^{-0.3902t}}{(1 + 12.38e^{-0.3902t})^2}$ billion dollars per year t years after 1980

(b) $D'(10) \approx 198.3$

In 1990, the total outstanding mortgage debt was increasing at a rate of $198.3 billion per year.

(c) $D'(13) \approx \$82.8$ billion per year

Total outstanding mortgage debt was growing by $82.8 billion per year in 1993.

5. (a) Rewrite $P(t)$ as $P(t) = 100A(t)[D(t)]^{-1}$.
Use the Product Rule to find the derivative.
$P'(t) = 100A(t)\{-1[(D(t)]^{-2}D'(t)\} + 100A'(t)[D(t)]^{-1}$
percentage points per year t years after 1980

(b)

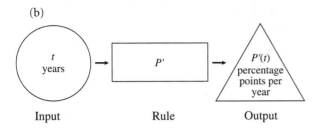

Input   Rule   Output

6. 1. Begin with a point.

2. Choose a close point.

3. Find a formula for the slope between the two points. Simplify completely.

4. Find the limiting value of the slope as $h \to 0$.

1. $(x, 7x + 3)$

2. $(x + h, 7(x + h) + 3)$

3. Slope $= \dfrac{7x + 7h + 3 - (7x + 3)}{x + h - x} = 7$

4. $\lim\limits_{h \to 0} 7 = 7$

CHAPTER 5

Section 5.1

1. Quadratic, cubic, and many product, quotient, and composite functions could have relative maxima or minima.

3.

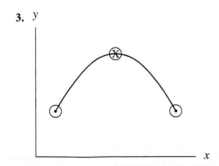

The derivative is zero at the absolute maximum point.

5.

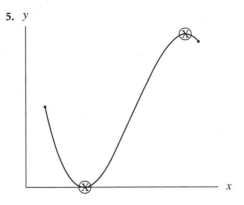

The derivative is zero at both optimal points.

7.

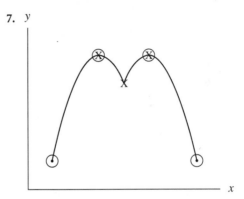

The derivative is zero at both absolute maximum points. The derivative does not exist at the relative minimum point.

9.

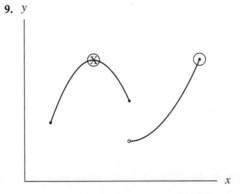

The derivative is zero at the absolute maximum point marked with an X.

11. Answers will vary. One such graph is $y = x^3$.

13. (a) The relative maximum value is approximately 19.888, which occurs at $x \approx 3.633$. The relative minimum value is approximately 11.779, which occurs at $x \approx 11.034$.

(b) The absolute maximum and minimum are the relative maximum and minimum found in part a.

(c)

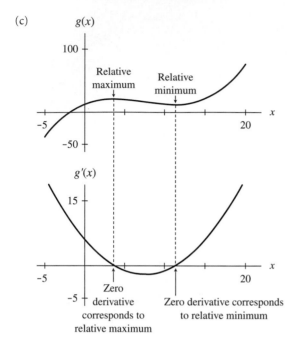

15. (a) The greatest percentage of eggs hatching (95.6%) occurs at 9.4°C.
    (b) 9.4°C ≈ 49°F

17. (a) (3.67, 73.42) is a relative maximum point. (9.20, 71.68) is a relative minimum point.
    (b) The number of medical students was greatest (73.5 thousand) in 1992 and least (70.1 thousand) in 1980.

19. (a) $C(0) \approx 123$ cfs and $C(24) \approx 140$ cfs
    (b) The highest flow rate is 373 cfs; it occurs when $h = 10$ hours. The lowest flow rate is 123 cfs; it occurs when $h = 0$ hours.

21. (a) $S(x) = 0.181x^2 - 8.463x + 147.376$ seconds at age x years.
    (b) The model gives a minimum time of 48.5 seconds occurring at 23.4 years.
    (c) The minimum time in the table is 49 seconds, which occurs at 24 years of age.

23. (a) An exponential model for the data is $R(p) = 316.765(0.949)^p$ dozen roses when the price per dozen is p dollars.
    (b) $E(p) = 316.765p(0.949)^p$ dollars spent on roses each week when the price per dozen is p dollars
    (c) A price of $19.16 maximizes consumer expenditure.
    (d) A price of $25.16 maximizes profit.

25. (a) $G(t) = 0.008t^3 - 0.347t^2 + 6.108t + 79.690$ million tons of garbage taken to a landfill t years after 1960
    (b) $G'(t) = 0.025t^2 - 0.693t + 6.108$ million tons of garbage per year t years after 1960

(c) In 1990, the amount of garbage was increasing by 8.1 million tons per year.

(d)

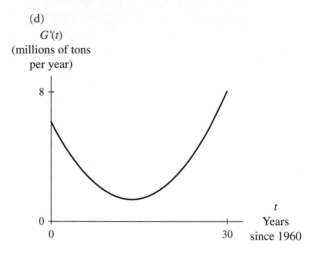

Because the derivative graph exists for all input and never crosses the horizontal axis, $G(t)$ has no relative maxima.

27. (a) $D(x) = -0.100x^3 + 24.808x^2 - 2035.274x + 55,542.450$ billion 1987 dollars x years after 1900
    (b) $D'(x) = -0.301x^2 + 49.616x - 2035.274$ billion 1987 dollars per year x years after 1900
    (c) (76.13, 173.19) is a local minimum point on the model. (88.95, 278.64) is a local maximum point on the model.
    (d) According to the model, defense spending was least in 1976 (173.2 billion 1987 dollars) and was greatest in 1989 (278.6 billion 1987 dollars).
    (e) Relative extrema can be found by locating the two places where $D'(x)$ crosses the x-axis. That is, solve for x in the equation $D'(x) = 0$.

29. (a) $A(p) = 568.074(0.965582)^p$ tickets sold on average when the price is p dollars
    $A(p) = 0.15p^2 - 16.007p + 543.286$ tickets sold on average when the price is p dollars
    The exponential model probably better reflects the probable attendance if the price is raised beyond $35 because attendance is likely to continue to decline. (The quadratic model will begin to increase around $53.)
    (b) $R(p) = 568.074p(0.965582)^p$ dollars of revenue when the ticket price is p dollars
    (c) (28.55, 5966.86) is the maximum point on the revenue graph. This corresponds to a ticket price of $28.55, which results in revenue of approximately $5967. The resulting average attendance is approximately 209.

Section 5.2

1. Cubic and logistic models have inflection points, as do some product, quotient, and composite functions.

3. (a) One visual estimate of the inflection points is (1982, 25) and (2018, 25).

   (b) The input values of the inflection points are the years in which the rate of crude oil production is estimated to be increasing and decreasing most rapidly. We estimate that the rate of production was increasing most rapidly in 1982, when production was approximately 25 billion barrels per year, and that it will be decreasing most rapidly in 2018, when production is estimated to be approximately 25 billion barrels per year.

5. (a)

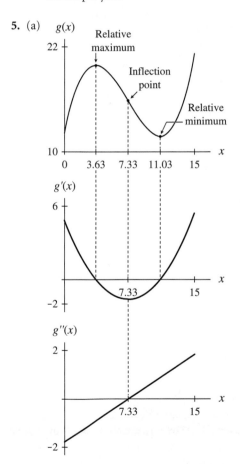

   (b) The inflection point on $g(x)$ is (7.333, 15.834). This is a point of most rapid decline.

7. The inflection point is (1.838, 22.5). After approximately 1.8 hours of study (1 hour and 50 minutes), the rate at which new material is being retained is increasing most rapidly. At that time, approximately 22.5% of the material has been retained.

9. (a)

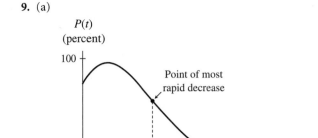

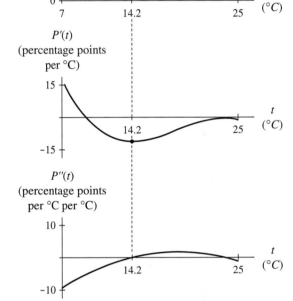

   (b) Because $P''(t)$ crosses the t-axis twice, there are two inflection points. These are (14.2, 59.4) and (23.6, 5.8). The point of most rapid decrease on $P(t)$ is (14.2, 59.4). (The other inflection point is a point of least rapid decrease.) The most rapid decrease occurs at 14.2°C, when 59.4% of eggs hatch. At this temperature, the percentage of eggs hatching is declining by 11.1 percentage points per °C. A small increase in temperature will result in a relatively large increase in the percentage of eggs not hatching.

11. (a)

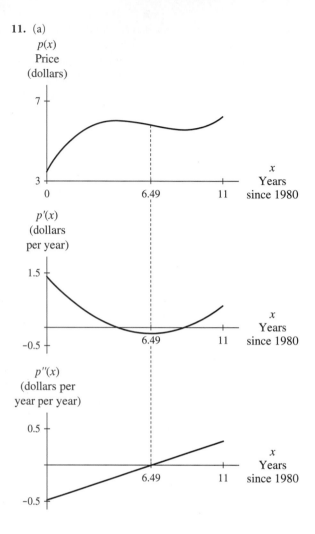

    The minimum point on $p'(x)$ and the x-intercept on $p''(x)$ correspond to the inflection point on $p(x)$.

  (b) The x-intercept of $p''(x)$ is $x \approx 6.487$. This is the input of the inflection point of $p(x)$.

  (c) In 1986, the price of natural gas was declining most rapidly, at a rate of -$0.15 per year.

13. (a) $(0.418, 9740.089)$ is a relative minimum point, and $(13.121, 20{,}242.033)$ is a relative maximum point on the cubic model.

  (b) The inflection point is $(6.770, 14{,}991.064)$.

  (c)   i. The inflection point occurs between 1981 and 1982, shortly after the team won the National Championship. This is when the number of donors was increasing most rapidly.

       ii. The relative maximum occurred around the same time that a new coach was hired. After this time, the number of donors declined.

15. (a) The greatest rate occurs at $h = 3.733$, or approximately 3 hours and 44 minutes after she began working.

  (b) Her employer may wish to give her a break after 4 hours to prevent a decline in her productivity.

17. (a) Using symmetric difference quotients to estimate rates of change, we see that the number of states is increasing most rapidly for a life expectancy of 71 years of age.

  (b) $N(x) = \dfrac{50.498}{1 + 52.689e^{-1.363329x}}$ states, where $x + 68$ is the life expectancy in years
The inflection point is $(2.908, 25.249)$.

  (c) At an age of 71 years $(x = 3)$, the number of states (according to the model) is 27, and the rate of change is approximately 17 states per year of age.

19. (a) Between 1970 and 1975, the average rate of change was smallest at 1 million tons per year.

  (b) $g(t) = 0.008t^3 - 0.347t^2 + 6.108t + 79.690$ million tons t years after 1960.

  (c) $g''(t) = 0.051t - 0.693$ million tons per year per year t years after 1960

  (d) Solving $g''(t) = 0$ gives $t \approx 13.684$ and $g(13.684) \approx 120$ million tons of garbage.

  (e)

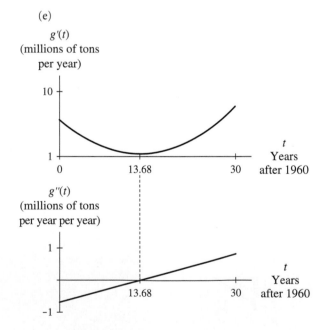

Because $g''(t)$ crosses the t-axis at 13.68, we know that input corresponds to an inflection point of $g(t)$. Because $g'(t)$ has a minimum at that same value, we know that it corresponds to a point of slowest increase on $g(t)$.

  (f) The year with the smallest rate of change is 1974, with $g(14) \approx 120$ million tons of garbage, increasing at a rate of $g'(14) \approx 1.4$ million tons per year.

21. (a) $A(x) = -0.100x^3 + 3.768x^2 - 34.906x + 268.635$ billion 1987 dollars x years after 1970

(b) $A''(x) = -0.601x + 7.537$ billion 1987 dollars per year per year x years after 1970

(c) The inflection point of $A(x)$ is (12.538, 225.916). The amount was increasing most rapidly in 1983 ($x = 13$). In 1983, the amount spent was $A(13) \approx 231.6$ billion 1987 dollars and was increasing at a rate of $A'(13) \approx 12.28$ billion 1987 dollars per year.

(d) Solving for x in the equation $A''(x) = 0$ will give the input value of the inflection point.

23. (a) $L'(t) = \dfrac{22{,}000(5.951)(0.3969)e^{-0.3969t}}{(1 + 5.951e^{-0.3969t})^2}$

thousand workers per year t years after the end of 1981.

(b) $L'(5) \approx 2161$ thousand workers per year

(c) $L(5.5) - L(5) \approx 1066$ thousand workers

(d) (4.494, 118,299.998) is the inflection point of the model. The labor force was growing most rapidly in 1986, at a rate of approximately 2183 thousand workers per year. The size of the labor force at that time was approximately 118,300 thousand workers.

25. (a) $H(w) = \dfrac{10{,}111.102}{1 + 1153.222e^{-0.727966w}}$

total labor-hours after w weeks

(b) $H'(w) = \dfrac{10{,}111.102(1153.222)(0.727966)e^{-0.727966w}}{(1 + 1153.222e^{-0.727966w})^2}$

labor-hours per week after w weeks

(c)

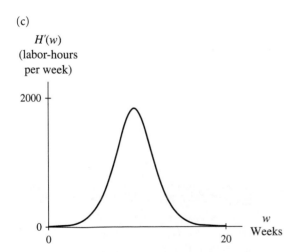

The derivative gives the manager information about the number of labor-hours spent each week.

(d) The maximum point on $H'(w)$ is (9.685, 1840.134). In the tenth week the most labor-hours are needed. That number is $H'(10) \approx 1816$ labor-hours.

(e) The point of most rapid increase on $H'(w)$ is (7.876, 1226.756). This occurs approximately 8 weeks into the job, and the number of labor-hours per week is increasing by approximately $H''(8) \approx 513$ labor-hours per week per week.

(f) The point of most rapid decrease on $H'(w)$ is (11.494, 1226.756). This occurs approximately 12 weeks into the job, and the number of labor-hours per week is changing by about $H''(12) = -486$ labor-hours per week per week.

(g) By solving the equation $H'''(w) = 0$, we can find the input values that correspond to a maximum or minimum point on $H''(w)$, which corresponds to inflection points on $H'(w)$, the weekly labor-hour curve.

(h) The second job should begin about 4 weeks into the first job.

Section 5.3

1. $32\% + (4 \text{ percentage points per hour})\left(\dfrac{1}{3}\text{ hour}\right) \approx 33.3\%$

3. $f(3.5) \approx f(3) + f'(3)(0.5)$
$= 17 + (4.6)(0.5) = 19.3$

5. (a) Increasing production from 500 to 501 units will increase cost by approximately $17.

(b) If sales increase from 150 to 151 units, then profit will increase by approximately $4.75.

7.

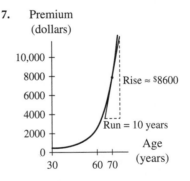

Slope of tangent line ≈ $860 per year of age
Annual premium for 70-year-old ≈ $7850
Premium for 72-year-old ≈ $7850 + 2($860) = $9570
(Answers will vary.)

9. (a)

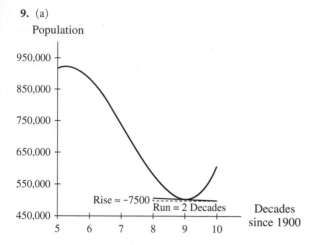

Population

Slope of tangent line ≈ -3750 people per decade
Population in 1990 ≈ 500,000 people
Population in 2000 ≈ 500,000 − 3750 = 496,250 people
(Answers will vary.)

(b) $P(10) = 606,030$ people

(c) Answers will vary.

11. (a) In 1990, the population of South Carolina was increasing by 81.3 thousand people per year.

(b) Between 1990 and 1992, the population increased by approximately 162.6 thousand people.

(c) By finding the slope of the tangent line at 1990 and multiplying by 2, we determine the change in the tangent line from 1990 through 1992 and use that change to estimate the change in the population function.

13. (a) The population was growing at a rate of 1.88 million people per year in 1985.

(b) Between 1985 and 1986, the population of Mexico increased by approximately 1.88 million people.

15. (a) In 1986, the percentage was increasing by 13.02 percentage points per year.

(b) We would expect an increase of approximately 13.02 percentage points between 1986 and 1987.

(c) $p(9) - p(8) \approx 12.87$ percentage points

(d) $48.7 - 36 = 12.7$ percentage points

(e) As long as the data in part *d* were correctly reported, the answer to part *d* is the most accurate one.

17. (a) $R(x) = (-7.032 \cdot 10^{-4})x^2 + 1.666x + 47.130$ dollars when x hot dogs are sold

(b)

x	$R'(x)$ (dollars per hot dog)
200	1.38
800	0.54
1100	0.12
1400	-0.30

If the number of hot dogs sold increases from 200 to 201, revenue will increase by approximately $1.38. If the number increases from 800 to 801, the increase in revenue will be approximately $0.54. For an increase in the number of hot dogs sold from 1100 to 1101, the corresponding increase in revenue is approximately $0.12. And if the number sold increases from 1400 to 1401, revenue will actually decline by approximately $0.30.

(c) The value of x for which $R'(x) = \$1.10$ is approximately 402.59, or 403, hot dogs. The value of x for which $R'(x) = \$0.25$ is 1006.93, or approximately 1007, hot dogs.

19. (a) United States: $A(t) = 0.109t^3 - 1.555t^2 + 10.927t + 100.320$; Canada: $C(t) = 0.150t^3 - 2.171t^2 + 15.814t + 99.650$; Peru: $P(t) = 85.112(2.01325)^t$; Brazil: $B(t) = 73.430(2.61594)^t$; where t is the number of years since 1980.

(b, c)

	U.S.	Canada	Peru	Brazil
Rate of change in 1987 $\left(\begin{array}{c}\text{CPI points} \\ \text{per year}\end{array}\right)$	5.2	7.5	7984	59,193
1988 CPI estimate	143	163	19,134	136,451

21. (a) $A(t) = 300\left(1 + \dfrac{0.065}{12}\right)^{12t}$ dollars after t years

(b) $A(t) \approx 300(1.066972)^t$ dollars after t years

(c) $A(2) \approx \$341.53$

(d) $A'(2) \approx \$22.14$

(e) $\dfrac{1}{4}(\$22.14) = \$5.54$

23. (a) $R(A) = -0.158A^3 + 5.235A^2 - 23.056A + 154.884$ thousand dollars of revenue when A thousand dollars is spent on advertising

(b) When $10,000 is spent on advertising, revenue is increasing by $34.3 thousand per thousand advertising dollars. If advertising is increased from $10,000 to $11,000, the car dealership can expect an approximate increase in revenue of $34,300.

(c) When $18,000 is spent on advertising, revenue is increasing by $12.0 thousand per thousand advertising dollars. If advertising is increased from $18,000 to $19,000, the car dealership can expect an approximate increase in revenue of $12,000.

Chapter 5 Review Test

1. (a) $T'(x) = -1.9216x^3 + 19.905x^2 - 52.252x + 26.981$
   thousand tourists per year x years after 1988
   (b) $T(3) \approx 120.9$ In 1991, approximately 120.9 thousand tourists visited Tahiti.
   $T(5) \approx 145.7$ In 1993, approximately 145.7 thousand tourists visited Tahiti.

2. $T(x)$ has a relative maximum point at (0.682, 143.098) and a relative minimum point at (3.160, 120.687). These points can be determined by finding the values of x between 0 and 6 at which $T'(x)$ crosses the x-axis. (There is also a relative maximum to the right of $x = 6$.)

3. $T(x)$ has two inflection points: (1.762, 132.939) and (5.143, 149.067). These points can be determined by finding the values of x between 0 and 6 at which $T''(x)$ crosses the x-axis. These are also the points at which $T'(x)$ has a relative maximum and relative minimum.

4.

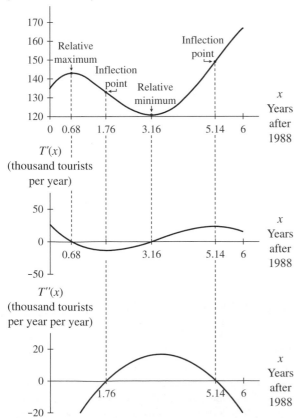

5. (a) The number of tourists was greatest in 1994 at 166.8 thousand tourists. The number was least in 1991 at 120.9 thousand.

(b) The number of tourists was increasing most rapidly in 1993 at a rate of 23.1 thousand tourists per year. The number of tourists was decreasing most rapidly in 1990 at a rate of 13.3 thousand tourists per year.

6. (a) i. (4.7 thousand people per year)$\left(\frac{1}{4}\text{ year}\right) = 1.175$ thousand people

   ii. $225.2 + \frac{1}{2}(4.7) = 227.55$ thousand people

   (b) The answer to part ii is the output of the line tangent to $m(t)$ at $t = 1996$ for an input of 1996.5.

CHAPTER 6

Section 6.1

1. (a) thousand bacteria per hour
   (b) hours
   (c) thousand bacteria
   (d) thousand bacteria
   (e) thousand bacteria

3. (a) The area would represent how much farther a car going 60 mph would require to stop than a car going 40 mph.
   (b) i. The heights are in feet per mile per hour, and the widths are in miles per hour.
   ii. The area is in feet.

5. (a) $N(t) = 160 - 24t$ lots not sold after t years
   (b) In 5 years there will be only 40 lots not sold.
   (c)

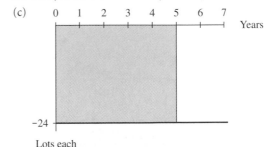

   (d)

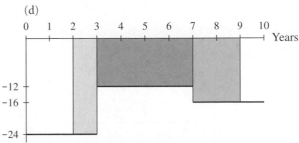

   From the beginning of year 3 through the end of year 9, 104 lots are sold.

7. (a) $V(x) = 200x$ dollars for x shares of stock

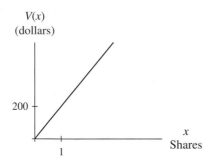

(b) $V'(x) = 200$ dollars per share

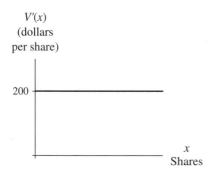

(c) The value increases by $10,000.

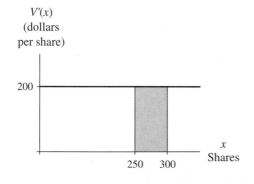

9. (a)

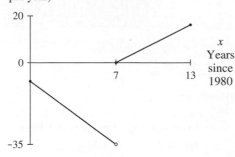

(b) The region is a trapezoid with base 7 and heights 7.841 and 34.889. The area is 149.6 thousand people. Between 1980 and 1987, the population of Iowa decreased by approximately 149.6 thousand people.

(c) This region is also a trapezoid with base 6 and heights 0.069 and 16.353. The area is 49.3 thousand people. Between 1987 and 1993, the population of Iowa increased by approximately 49.3 thousand people.

(d) In 1993, the population was approximately 100.3 thousand people less than it was in 1980.

11. (a)

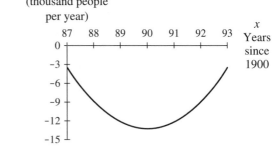

(b) Population was declining most rapidly when $P'(x)$ is a minimum at $x \approx 90$, corresponding to the end of 1990.

(c) Area ≈ 58.9. From the end of 1987 through the end of 1993, the population of the District of Columbia declined by approximately 58,900 people.

13. (a)

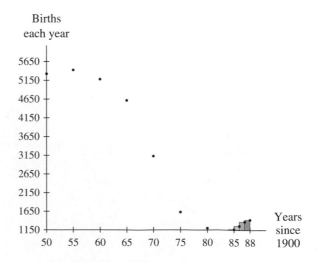

From the beginning of 1985 through the end of 1988, there were 5221 live births to U.S. women 45 years of age and older. Disregarding reporting error, this answer is exact.

(b) $B(x) = 0.417x^3 - 23.651x^2 + 199.682x + 5206.458$ births in the xth year after 1950

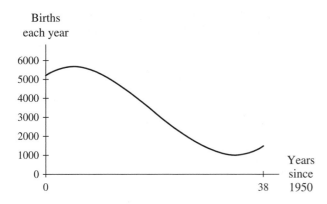

(c) Using 20 rectangles, we estimate the number of births to be 45,329.

(d) To find the exact change, we would need exact data for all years from 1965 through 1984.

15. (a)

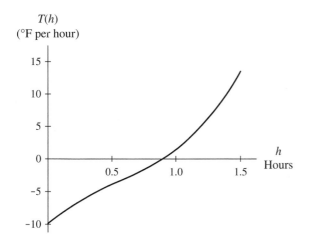

The graph crosses the horizontal axis at $h \approx 0.89$ hour.

(b) The area below the axis represents the amount the temperature declined from the time the storm began.

(c) The area above the axis represents the amount by which the temperature rose from the time it stopped declining until 1.5 hours after the storm began.

(d) Area $\approx 3.4°$F

(e) Area $\approx 4.0°$F

(f) The temperature was higher by approximately 0.6°F.

17. (a)

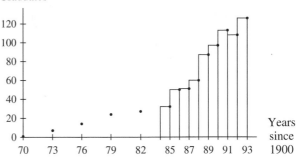

There were 724 graduates from the beginning of 1985 through the end of 1993. This is exact as long as the data were correctly reported.

(b) $g(t) = 0.016t^3 - 0.233t^2 + 2.669t + 1.633$ graduates t years after 1970

(c, d) Summing values of $g(t)$ for $t = 0$ through $t = 23$ gives a total of 971 graduates which is close to the actual number of 987.

19. (a)

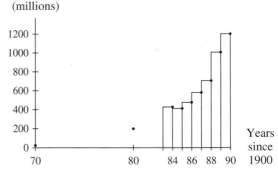

The total number of calls from the beginning of 1984 through 1990 was 4,810,700,000. This number is as exact as the data given.

(b) $C(x) = 25.134(1.210)^x$ million overseas calls x years after the end of 1970

(c) If 21 right rectangles are used, then summing values of $C(x)$ from $x = 0$ through $x = 20$, we estimate that there were 6,447,200,000 overseas phone calls.

Section 6.2

1. (a) Divide the interval from a to b into four equal subintervals. Substitute the endpoints of the subintervals into the function to determine the heights of the trapezoids. Determine the area of each trapezoid by averaging the two heights and multiplying by the width. Add the four areas to obtain the trapezoid estimate.

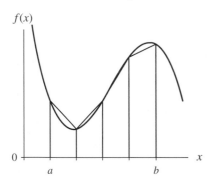

(b) Divide the interval into four equal subintervals. Determine the midpoint of each subinterval, and substitute into the function to find the heights of the rectangles. Multiply each height by the width of the subintervals, and add the four resulting areas.

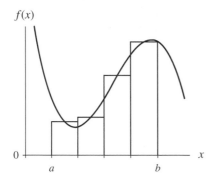

3. Answers will vary.

5. Left rectangles:
   Area ≈ 0.83495
   Error ≈ 0.04956

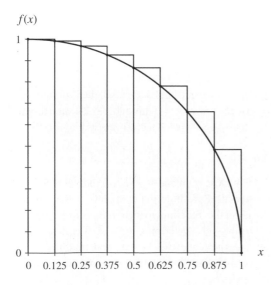

Right rectangles:
Area ≈ 0.70995
Error ≈ 0.07544

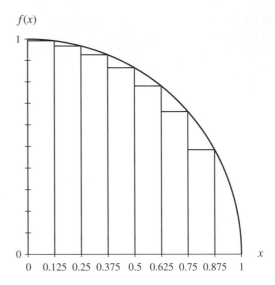

Midpoint rectangles:
Area ≈ 0.78917
Error ≈ 0.00377

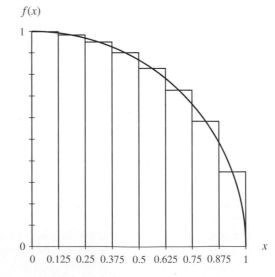

Trapezoids:
Area ≈ 0.77245
Error ≈ 0.01294

$f(x)$

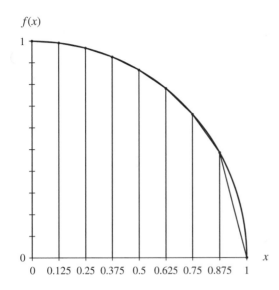

7. (a) To convert the air speeds to miles per second, multiply by 1.15 mph per knot and divide by 3600 seconds per hour.
   (b) $S(t) = -(2.108 \cdot 10^{-5})t^2 + 0.002t - 9.880 \cdot 10^{-5}$ miles per second where t = seconds since taxi began
   (c) Using 18 midpoint rectangles, the area is 0.209 miles. Answers may vary.
   (d) It took approximately 0.2 mile of runway for the Cessna to taxi for takeoff (assuming no headwind).

9. (a)

$P(t)$
(percentage points
per year)

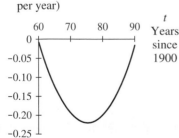

   (b) The percentage of the population living around the Great Lakes was declining from 1960 through 1990.
   (c) The trapezoid area is 2.124. From 1970 through 1980, the percentage of people living around the Great Lakes decreased by approximately 2.1 percentage points.
   (d) The area approximation using midpoint rectangles is 2.131 percentage points.

11. (a) Declining positive rate-of-change data indicate that life expectancies were increasing at a slower and slower rate.
    (b) $E(t) = 0.002t^2 - 0.055t + 0.510$ years per year t years after 1970
    (c) Life expectancy for women increased approximately 4.35 years from 1970 through 1992.
    (d) Using midpoint rectangles yields a change in life expectancy of approximately 4.27 years.

13. (a) $A \approx 43.799$ days after Sept. 30, 1995
    $B \approx 273.382$ days after Sept. 30, 1995
    (b) In about 44 days after Sept. 30, 1995, the level of the lake fell by approximately 0.398 foot.
    (c) Between about 44 and 273 days after Sept. 30, 1995, the lake level rose by about 3.250 feet.
    (d) The lake level was approximately $3.250 - 0.398 = 2.852$ feet higher 273 days after Sept. 30, 1995.

15. (a) Trapezoid area = 10
    (b) Midpoint-rectangle area = 11
    (c) Simpson's Rule area = 10.667

Section 6.3

1. (a) thousand people
   (b) thousand people
   (c) thousand people

3. (a) This is the change in the number of organisms when the temperature increases from 25°C to 35°C.
   (b) This is the change in the number of organisms when the temperature increases from 30°C to 40°C.

5. (a, b) Between 0 and 300 boxes and between 400 and 600 boxes
   (c) NA
   (d) 300
   (e) 400
   (f) 350
   (g) dollars
   (h) less

7. (a) 17.91
   (b) Between 3 and 11 weeks of age, the mouse gained 17.91 grams.

9. (a) The yield from the oil field during the first 5 years is approximately 10.65 thousand barrels.
   (b) The yield from the oil field during the first 10 years is approximately 12.45 thousand barrels.
   (c) $\int_0^5 r(t)\,dt$ and $\int_0^{10} r(t)\,dt$
   (d) The first 5 years account for approximately 85.6% of the first 10 years' production.

11. (a) $A \approx 17.3$ seconds

(b) From 0 to 17.3 seconds, the car's speed increased by approximately 174.7 feet per second (or 119.1 mph).

(c) From 17.3 to 35 seconds, the car's speed decreased by approximately 94.9 feet per second (or 64.7 mph).

(d) The car's speed after 35 seconds was approximately 79.8 feet per second (or 54.4 mph) faster than it was at 0 seconds.

13. (a) The heights will be in meters per second, and the widths will be in microseconds.

(b) The area units will be millimeters.

(c) $V(m) = -(1.589 \cdot 10^{-6})m^2 + 0.001m + 0.137$ millimeters per microsecond after m microseconds

(d) The crack traveled approximately 10.2 millimeters.

(e) $\int_0^{60} V(m)\,dm$

15. (a) For $n = 5$, labor-hours ≈ 9859.
For $n = 10$, labor-hours $\approx 10{,}097$.
For $n = 20$, labor-hours $\approx 10{,}100$.

(b) i

17. (a) $\int_0^{720} c(m)\,dm \approx 2602$ customers

(b) During the 12-hour sale, approximately 2602 customers entered the store.

19. (a) Blood pressure rises when the rate of change is positive. In the table, this is from 2 a.m. to 12 p.m. Blood pressure falls when the rate of change is negative, from 2 p.m. to 12 a.m.

(b) In the table the greatest rate of change occurs at 8 a.m. and the most negative rate of change occurs at 8 p.m.

(c) $B(t) = 0.030t^2 - 0.718t + 3.067$ mm Hg per hour, where t is the number of hours since 8 a.m.

(d) The model is zero at $t \approx 5.59$ hours and at $t \approx 18.13$ hours. These are the times when the blood pressure indicated by the model is highest and lowest, respectively.

(e) From 8 a.m. to 8 p.m., diastolic blood pressure rose by about 2.54 mm Hg.

(f) $\int_0^{12} B(t)\,dt$

Section 6.4

1. derivative graph: b; accumulation graph: f

3. derivative graph: f; accumulation graph: e

5. derivative: a; accumulation function: b

7. (a) $f(t) = 6$ 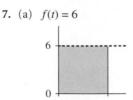 $\int_0^x 6\,dt = 6x$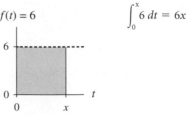

(b) $f(t) = -2$ 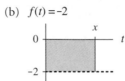 $\int_0^x -2\,dt = -2x$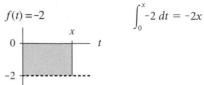

(c) $f(t) = \frac{1}{3}$ $\int_0^x \frac{1}{3}\,dt = \frac{1}{3}x$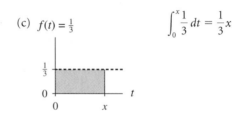

9. (a) $f(t) = -2$ $\int_{-4}^x -2\,dt = -2x - 8$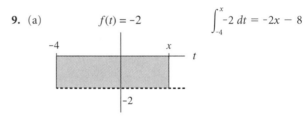

(b) $f(t) = -2$ $\int_2^x -2\,dt = -2x + 4$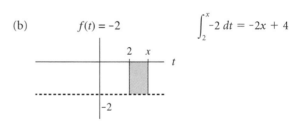

(c) $f(t) = -2$ $\int_{2000}^x -2\,dt = -2x + 4000$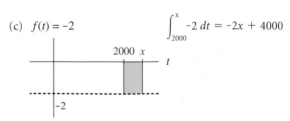

11. (a) $f(t) = 3t$ $\int_5^x 3t\,dt = \frac{3x^2}{2} - \frac{75}{2}$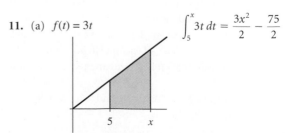

(b) $f(t) = -\frac{1}{2}t$

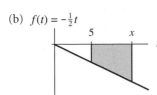

$\int_5^x \frac{-1}{2}t \, dt = \frac{-1}{4}x^2 + \frac{25}{4}$

(c) $f(t) = 100t$

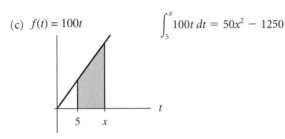

$\int_5^x 100t \, dt = 50x^2 - 1250$

13. $\int_0^x k \, dt = kx$ **15.** $\int_a^x k \, dt = kx - ka$

17. $\int_5^x kt \, dt = \frac{kx^2}{2} - \frac{25k}{2}$

19. (a) $\frac{3(19)^2}{2} - \frac{3(5)^2}{2} = 504$

(b) Between January 5 and January 19 last year, the airline's revenue increased by $50,400.

21. (a) $\int_A^x f(t) \, dt$

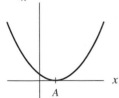

(b) $\int_B^x f(t) \, dt$

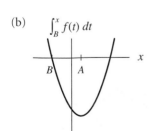

23. (a) $\int_0^x f(t) \, dt$

(b) $\int_A^x f(t) \, dt$

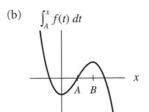

25. $\int_0^x f(t) \, dt$

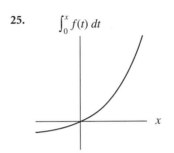

27. (a) Each box represents $4000.

(b) $\int_0^x p(t) \, dt$ represents the accumulation of profit (in thousands of dollars) in the x weeks after the business opened.

(c) Answers will vary.

x	Accumulation function	x	Accumulation function
0	0	28	50.4
4	−3.2	32	48.3
8	3.5	36	41.9
12	15.3	40	32.8
16	28.4	44	23.5
20	39.8	48	17.6
24	47.5	52	19.3

Profit
(thousand
dollars)

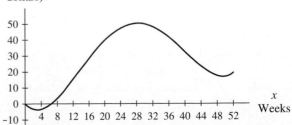

29. (a)

x	$\int_0^x t^2\, dt$
0	0
1	0.33333
2	2.66667
3	9
4	21.33333
5	41.66667

(b) $\int_0^x t^2\, dt = \dfrac{1}{3}x^3$

31. (a)

x	e^x	$\int_0^x e^t\, dt$
0	1	0
1	2.71828	1.71828
2	7.38906	6.38906
3	20.08554	19.08554
4	54.59815	53.59815
5	148.41316	147.41316

(b) $\int_0^x e^t\, dt = e^x - 1$

33. Changing the starting point shifts the accumulation function graph up or down but does not change its shape. Changing the starting point changes the constant term in an accumulation function formula.

Section 6.5

1. b   **3.** c   **5.** a   **7.** c   **9.** b   **11.** b

13. (a) $\dfrac{\text{million dollars of revenue}}{\text{thousand advertising dollars}}$

(b)

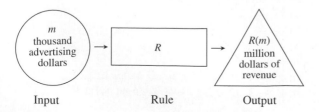

(c) When m thousand dollars are being spent on advertising, the annual revenue is $R(m)$ million dollars.

15. (a) $\dfrac{\text{milligrams per liter}}{\text{hour}}$

(b)

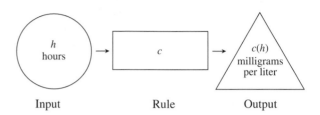

(c) The concentration of a drug in the blood stream is $c(h)$ milligrams per liter h hours after the drug is given.

17. $\displaystyle\int 19.436(1.07)^x\, dx = \dfrac{19.436(1.07)^x}{\ln 1.07} + C$

19. $\displaystyle\int [6e^x + 4(2^x)]\, dx = 6e^x + \dfrac{4(2^x)}{\ln 2} + C$

21. $\displaystyle\int \left(10^x + \sqrt[4]{x} + 8\right) dx = \dfrac{10^x}{\ln 10} + 4\left(\dfrac{2}{3}\right)x^{\frac{3}{2}} + 8x + C$

23. $S(m) = \dfrac{6250(0.92985)^m}{\ln 0.92985} + C$ CDs m months after the beginning of the year

25. $J(x) = 15.29\ln|x| + 7.95x + C$ units, where x is the price in dollars

27. $P(t) = \dfrac{1.724928e^{0.0256t}}{0.0256} + C$ million people t years after 1990

29. $F(t) = \dfrac{1}{3}t^3 + t^2 - 20$

31. $F(z) = \dfrac{-1}{z} + e^z + \left(\dfrac{3}{2} - e^2\right)$

33. (a) $g(d) = -0.00018d^2 - 0.018d + 1.95$ inches per day d days after sprouting

(b) $h(d) = -0.00006d^3 - 0.009d^2 + 1.95d$ inches d days after sprouting

(c) The sweet corn is $h(60) = 71.64$ inches tall.

35. (a) $P(x) = \dfrac{6.178983e^{0.047509x}}{0.047509} + C$ million pounds of poultry x years after 1960

(b) $P(33) - P(0) \approx 493.7$ million pounds of poultry produced between 1960 and 1993.

(c) $P'(x) = 6.178983(0.047509)e^{0.0475509x}$ million pounds per year per year x years after 1960

(d) The answer to part a is the general antiderivative of $P(x)$, and the answer to part c is the derivative of $P(x)$.

37. (a, b) Velocity: $v(t) = -32t$ ft/sec

Distance: $s(t) = -16t^2 + 540$ ft

where t is the number of seconds after the penny was dropped

(c) Solving for t in $s(t) = 0$, we obtain $t \approx \pm 5.8$ seconds. The penny will hit the ground approximately 5.8 seconds after it was dropped.

(d) $v(5.809475019) = -32(5.809475019) \approx -185.9$ feet per second

185.9 feet per second =

$$\left(\frac{185.9 \text{ feet}}{1 \text{ second}}\right)\left(\frac{3600 \text{ seconds}}{1 \text{ hour}}\right)\left(\frac{1 \text{ mile}}{5280 \text{ feet}}\right)$$

$$= \frac{126.75 \text{ miles}}{1 \text{ hour}} \text{ or } 126.75 \text{ mph}$$

39. (a) The impact velocity is -64.99 feet per second or -44.31 mph.

(b) Air resistance probably accounts for the difference.

41. (a) $N(x) = 593\ln|x| + 138x - 478.073$ employees x years after 1978

(b) The function in part a applies from 1982 ($x = 4$) through 1992 ($x = 14$).

(c) $N(14) \approx 3019$ employees. If any employees were fired or quit between 1978 and 1992, this number would not represent the number of employees at the end of 1992.

Section 6.6

1. b **3.** b **5.** a **7.** c

9. (a) $\displaystyle\int_1^x (t^3 - 2t + 3)\, dt = \frac{1}{4}x^4 - x^2 + 3x - \frac{9}{4}$

(b) $\displaystyle\int_1^2 (t^3 - 2t + 3)\, dt = 3.75$

11. (a) $\displaystyle\int_0^x [18{,}000(0.974)^t + 1500]\, dt \approx$

$$\frac{18{,}000(0.974)^x}{\ln 0.974} + 1500x + 683{,}268.1768$$

(b) Area $\approx 173{,}242.2349$

(c) In the 10 years after the owner purchased the property, its value increased by approximately \$173,242.

13. $\displaystyle\int_1^2 (3x - 7)\, dx = \left.\left(\frac{3}{2}x^2 - 7x\right)\right|_1^2 = -2.5$

15. $\displaystyle\int_1^2 \left(1 + \frac{1}{x} + \frac{1}{x^3}\right) dx = \left.\left(x + \ln|x| - \frac{1}{2x^2}\right)\right|_1^2 \approx 2.068$

17. $\displaystyle\int_{10}^{40} [427.705(1.043)^x - 413.226e^{-0.4132x}]\, dx =$

$$\left.\left(\frac{427.705(1.043)^x}{\ln 1.043} - \frac{413.226e^{-0.4132x}}{-0.4132}\right)\right|_{10}^{40} \approx 39{,}236.34072$$

19. (a)

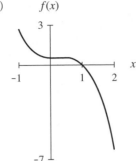

$f(x)$

(b) Area $= \displaystyle\int_{-1}^{1.0544} f(x)\, dx - \int_{1.0544}^{2} f(x)\, dx \approx 3.822$

21. (a)

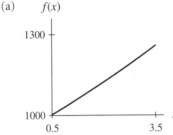

$f(x)$

(b) Area $= -\displaystyle\int_{0.5}^{3.5} f(x)\, dx \approx -3378.735$

23. $\displaystyle\int_{80}^{90} P(t)\, dt \approx -1.24$

Between 1980 and 1990, the percentage of the population of the United States living in the Great Lakes region fell by 1.24 percentage points.

25. $\displaystyle\int_0^5 r(x)\, dx \approx 195.639$

The corporation's revenue increased by 195.6 million dollars between 1987 and 1992.

27. $\displaystyle\int_0^{10} r(t)\, dt \approx 12.45$ (found using a limit of sums)

In the 10 years after production began, the oil well produced 12.5 thousand barrels of oil.

29. (a) $\displaystyle\int_0^{70} s(t)\, dt = 7.938$

In the 70 days after April 1, the snow pack increased by 7.938 equivalent cm of water.

(b) $\displaystyle\int_{72}^{76} s(t)\, dt = -22.768$

Between 72 and 76 days after April 1, the snow pack decreased by 22.768 equivalent centimeters of water.

(c) It is not possible to find $\displaystyle\int_0^{76} s(t)\, dt$ because $s(t)$ is not defined between $t = 70$ and $t = 72$.

31. (a) $\displaystyle\int_0^{0.8955} T(h)\,dh \approx 1.48$

The temperature rose 1.48°F during the field trip.

(b) $\displaystyle\int_{0.8955}^{1.75} T(h)\,dh \approx -1.61$

After rising 1.48°F, the temperature then fell 1.61°F during the field trip.

(c) No, the highest temperature reached was $71 + 1.48 = 72.48$°F.

33. (a) An exponential model for the data is $f(x) = 0.161(1.076186)^x$ trillion cubic feet per year x years after 1900.

(b) From 1940 through 1960, 138.3 trillion cubic feet of natural gas was produced.

(c) $\displaystyle\int_{40}^{60} f(x)\,dx$

35. (a) A quadratic model for the data is $C'(x) = (7.714 \cdot 10^{-5})x^2 - 0.047x + 8.940$ dollars per CD, where x is the number of CDs produced per hour.

(b) $C(x) = \dfrac{7.714 \cdot 10^{-5}}{3}x^3 - \dfrac{0.047}{2}x^2 + 8.940x - 143.893$ dollars when x CDs are produced each hour.

(c) $\displaystyle\int_{200}^{300} C'(x)\,dx = 196.14$ dollars

When production is increased from 200 to 300 CDs per hour, cost increases by $196.14.

Chapter 6 Review Test

1. (a)

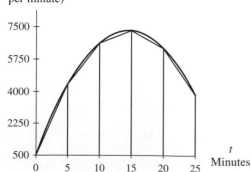

r(t)
(cubic feet per minute)

(b) 134,237.5 cubic feet

(c) In the first 25 minutes that oil was flowing into the tank, approximately 134,238 cubic feet of oil flowed in.

2. (a) A quadratic model for the data is

$S(t) = -1.643t^2 + 16.157t + 0.2$ miles per hour

t hours after midnight

(b)

n	Sum
5	127.131
10	126.869
20	126.803
40	126.786
80	126.782
160	126.781
Trend ≈ 126.8 miles	

3. (a) The area beneath the horizontal axis represents the amount of weight that the person lost during the diet.

(b) The area above the axis represents the amount of weight that the person regained between weeks 20 and 30.

(c) The person's weight was 11.3 pounds less at 30 weeks than it was at 0 weeks.

(d)

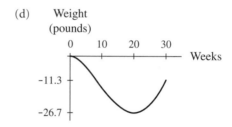

Weight (pounds)

(e) The graph in part d is the change in the person's weight as a function of the number of weeks after the beginning of dieting.

4. (a) $R(t) = 10\left(\dfrac{-3.2}{3}t^3 + \dfrac{93.3}{2}t^2 + 50.7t\right) + 5000$ cubic feet after t minutes

(b) $R(10) - R(0) \approx 41{,}053.3 \text{ ft}^3$

(c) Solving for t in the equation $R(t) = 150{,}000$, we find that the tank will be full after approximately 28 minutes.

5. (a) $2,598.60

(b) At the end of the third quarter of the third year, the $10,000 had increased by $2,598.60 so that the total value of the investment was $12,598.60.

CHAPTER 7

Section 7.1

1. (a, b)

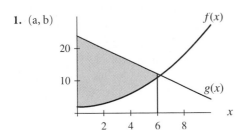

(c) 78

3. (a, b)

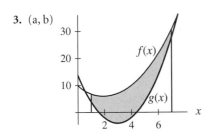

(c) 54

5. (a, c)

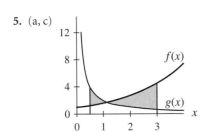

(b) $x \approx 1.134$

(d) $\int_{0.5}^{3} [f(x) - g(x)]\, dx \approx 2.812$

(e) Area ≈ 4.172

7. (a, c)

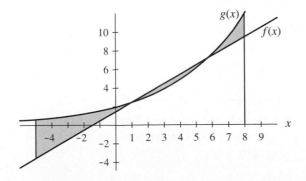

(b) $x = 1$ and $x \approx 5.806$

(d) $\int_{-5}^{8} [f(x) - g(x)]\, dx \approx -11.486$

(e) Area ≈ 15.536

9. (a) When the amount invested in capital increases from $1500 to $5500, profit increases by approximately $1.33 million.

(b) Area $= \int_{1.5}^{5.5} [r'(x) - c'(x)]\, dx = 1329$

11. (a) The company posted a loss of $369,000 during January and February.

(b) The company posted a $972,000 profit between February and May.

(c) $603,000

(d) Because the area of R_1 represents a loss rather than a gain, it must be subtracted from (not added to) the profit represented by the area of region R_2.

13. (a) Before fitting models to the data, add the point $(0,0)$, and convert the data from miles per hour to feet per second by multiplying each speed by $\left(\frac{5280 \text{ feet}}{1 \text{ mile}}\right)\left(\frac{1 \text{ hour}}{3600 \text{ seconds}}\right)$. The speed of the Supra after t seconds can be modeled as

$$s(t) = -0.702t^2 + 20.278t + 2.440 \text{ feet per second}$$

The speed of the Carrera after t seconds can be modeled as

$$c(t) = -0.643t^2 + 18.963t + 5.252 \text{ feet per second}$$

(b) Approximately 17.96 feet

(c) Approximately 18.04 feet

Section 7.2

1. 7

3. ∞

5. 0

7. 0

9. 15

11. 0.3

13. -0.02

15. (a) Approximately 0.0002 milligram; Approximately 0.0015 milligram

(b) $\int_{0}^{\infty} r(t)\, dt = \lim_{N \to \infty} \left. \frac{-1.55(0.9999999845)^t \cdot 10^{-6}}{\ln(0.9999999845)} \right|_{0}^{N} =$

$-99.99999923 \approx -100$

17. (a) $P(10) \approx 3280$ penguins; $P(20) \approx 3465$ penguins

(b) $\lim_{b \to \infty} P(b) \approx 3476$ penguins

Section 7.3

1. (a) $1,950,106
   (b) $1,940,300
   (c) Answers will vary.

3. (a) $r(q) = 350(1.05)^q(0.15)$ million dollars per quarter q quarters after the third quarter of 1996.
   (b) $r(q) = 350(1.05)^q(0.15)(1.13)^{16-q}$ million dollars per quarter for money invested q quarters after the third quarter of 1996.
   (c) If the investment begins with the 4th quarter 1996 profits, then the initial investment is based on a profit of $(350)(1.05) = \$367.5$ million. Thus we sum $367.5\,(1.05)^q(0.15)(1.13)^{16-q}$ for $q = 0, 1, \ldots, 15$ to obtain a value of $3803.23 million.

5. (a) $79.87 thousand
   (b) $58.29 thousand

7. $78,333.08

9. (a) $5.4 billion
   (b) $6.1 billion
   (c) $11.2 billion

11. (a) $7.3 billion
    (b) $5.6 billion
    (c) Answers will vary.

13. $5.2 million

15. Answers given are based on the end of 1998.
    (a) 0.35 million terns
    (b) $r(t) = 2.04(0.83)^{19-t}$ million terns born t years after 1979
    (c) 10.98 million terns

17. (a) None
    (b) $r(t) = (60 - 0.5t)(0.67)^{50-t}$ thousand seals born t years from now
    (c) 90.5 thousand seals

Section 7.4

1. (a) The demand function
   (b) The supply function
   (c) The producers' surplus
   (d) The consumers' surplus

3. (a) To find the price P above which consumers will purchase none of the goods or services, either find the smallest positive value for which the demand function is zero, $D(p) = 0$, or, if $D(p)$ is never exactly zero but approaches zero as p increases without bound, then let $P \to \infty$.
   (b) The supply function, $S(p)$, is a piecewise continuous function with the first piece being the 0 function.

The value p at which $S(p)$ is no longer 0 is the shutdown price. The shutdown point is $(p_1, S(p_1))$.
   (c) The market equilibrium price, p_0, can be found as the solution to $S(p) = D(p)$. That is, it is the price at which demand is equal to supply. The equilibrium point is the point $(p_0, D(p_0)) = (p_0, S(p_0))$.

5. (a) The model $D(p)$ of demand is exponential and so does not have a finite value p at which $D(p) = 0$. Thus the model does not indicate a price above which consumers will purchase none of the goods or services.
   (b) $8815.0 thousand
   (c) 17.0 thousand fans
   (d) $4234.8 thousand

7. (a) $D(p) = 0.025p^2 - 1.421p + 19.983$ lanterns when the market price is p per lantern
   (b) $109.18
   (c) $31.89

9. (a) 18.4 thousand answering machines; 300 thousand answering machines
   (b) $9981 thousand; $3131 thousand

11. (a) $S(p) = \begin{cases} 0 \text{ hundred prints} & \text{when } p < 5 \\ 0.300p^2 - 3.126p + \\ \quad 10.143 \text{ hundred prints} & \text{when } p \geq 5 \end{cases}$
    where p hundred dollars is the price of a print.
    (b) $837.12
    (c) Producers' revenue = $148.5 thousand; Producers' surplus = $27.3 thousand

13. (a) $408.3 hundred
    (b) 297 sculptures; no
    (c) $4542.2 hundred

15. (a) $D(p) = 499.589(0.958086)^p$ thousand books when the market price is p per book
    (b) $S(p) = \begin{cases} 0 \text{ thousand books} & \text{when } p < 08.97 \\ 0.532p^2 - 20.060p + \\ \quad 309.025 \text{ thousand books} & \text{when } p \geq 18.97 \end{cases}$
    where p is the price of a book.
    (c) Approximately $27.15; 156.2 thousand books
    (d) $4728.6 thousand

Section 7.5

1. (a) $V(t) = -1.664t^3 + 5.867t^2 + 1.640t + 60.164$ mph t hours after 4 p.m.
   (b) 68.99 mph
   (c) 72.23 mph

3. (a) $R(t) = 0.004t^3 - 0.908t^2 + 78.048t - 2230.607$ dollars per minute t years after 1900

(b) $0.99 per minute

(c) $0.83 per minute

5. (a) 69.15 million people

(b) Solving Population $\approx$ 69.15 for t gives $t \approx$ 63.1 years since 1921. This corresponds to early-1985.

(c) 1.84 million people per year

7. (a) –123.8 yearly accidents per year

(b) 2943.7 yearly accidents

9. (a) 2.28 feet per second squared

(b) 129.7 feet per second

(c) 4540.7 feet

(d) 4540.7 feet

11. (a) $V(t) = 1.033t + 138.413$ meters per second t microseconds after the experiment began

(b) 174.58 meters per second

13. (a) $B(t) = 0.030t^2 - 0.721t + 3.074$ mm Hg per hour t hours after 8 a.m.

(b) 0.21 mm Hg per hour

(c) 93.4 mm Hg

15. –0.15 percentage points per year

Review Test

1. (a) $R(t) = (0.1)(3000 \times 12 + 500t) = 3600 + 50t$ dollars per year

(b) $1,325,756

(c) $47,930

2. (a) Approximately 1 fox

(b) $f(t) = 500(0.63)^{20-t}$ foxes per year t years after 1990

(c) $\approx$ 1083 foxes

3. (a) No. There is no solution to $D(p) = 0$ because $D(p) > 0$ for all values of p.

(b) $\approx$ $5887.1 thousand

(c) $\approx$ 28.0 thousand televisions

(d) $\approx$ $3980.0 thousand

4. (a) $\approx$ $635.4 hundred

(b) 411 fountains. No; because $D(10) = 594$, supply is smaller than demand at this point.

(c) $\approx$ $6236.0 hundred

5. (a) $\approx$ 5.60%

(b) $\approx$ 0.014 percentage points per month

GLOSSARY

absolute maximum The highest point on a graph in a given input interval. An absolute maximum occurs either at a local maximum or at a point that corresponds to an endpoint of the given input interval. Also see *extreme point.*

absolute minimum The lowest point on a graph in a given input interval. An absolute minimum occurs either at a local minimum or at a point that corresponds to an endpoint of the given input interval. Also see *extreme point.*

accumulated change The accumulated change in a quantity is represented as the area of a region between the rate-of-change function for that quantity and the horizontal axis. In the case where the rate-of-change function is negative, the accumulated change in the quantity is the negative of the area of the region between the rate-of-change function and the horizontal axis.

accumulation function A function of the form $\int_a^x f(t)\,dt$, where the lower limit a is a constant and the upper limit x is a variable. An accumulation function gives a formula for calculating accumulated change in a quantity.

aligning data A renumbering process by which data values are shifted. Large input values are often renumbered to make them smaller before fitting models. Large input values should always be aligned before fitting exponential or logistic models to data.

antiderivative A function $F(x)$ is an antiderivative of another function $f(x)$ if the derivative of $F(x)$ is $f(x)$. If $F(x)$ is an antiderivative of $f(x)$ and C is an arbitrary constant, then $F(x) + C$ is called a general antiderivative of $f(x)$ and is denoted by $\int f(x)\,dx$. The general antiderivative is a group of infinitely many functions, and an accumulation function is one specific function from that group.

APR In an investment context, the Annual Percentage Rate (nominal rate) is the advertised rate of interest. It is denoted by r (in decimals) in the compound interest formulas.

APY In an investment context, the Annual Percentage Yield (effective rate) is the constant percentage change $(b - 1) \cdot 100\%$ for the exponential model $A(t) = ab^t$. The APY for the exponential model $A(t) = Pe^{rt}$ can be found by first converting Pe^{rt} to Pb^t. When comparing compound interest rates, we should use the APY rather than the APR, because nominal rates do not reflect the compounding periods.

average cost The total production cost divided by the number of units produced.

average rate of change The amount that a quantity changes over an interval divided by the length of the interval. The average rate of change is the slope of the secant line. Average rates of change have labels of output units per input unit.

average value of a function A number that equals the accumulated change in the function over an input interval divided by the length of the interval.

biological stream A continuous flow of *individuals* into a population over time in varying quantities. (We call the elements of any population individuals.)

break-even point The number of units (produced or sold) for which revenue equals cost so that profit is zero.

composition A method of combining two functions in which the output of one function (called the inside function) is used as the input of the other function (called the outside function). The symbol for a composite function is $(f \circ g)(x)$ or $f(g(x))$, where g is the inside function and f is the outside function.

compound interest formulas Exponential formulas that are used to determine the amount $A(t)$ accumulated in an account after t years when P dollars are invested. If the interest rate is $r\%$ (in decimals) compounded n times a year, then the accumulated amount is $A(t) = P\left(1 + \dfrac{r}{n}\right)^{nt}$ dollars. If the interest is compounded continuously at a nominal rate of $r\%$ (in decimals), then the accumulated amount is $A(t) = Pe^{rt}$ dollars.

concavity A description of the curvature of a graph. A graph is concave up at a point if the tangent line lies below the graph and concave down if the tangent line lies above the graph near the point of tangency. A point where the concavity of a graph changes is called an inflection point.

consumer price index (CPI) A measure that is 100 times the ratio obtained by comparing the current cost of a specified group of goods and services to the cost of comparable items determined at an earlier date.

consumers' expenditure The actual amount spent by consumers for a certain quantity of goods or services. The consumers' expenditure equals the market price times the quantity in demand.

consumers' surplus The amount that consumers are willing and able to spend but do not actually spend.

continuous Continuous graphs are unbroken curves, and their sets of inputs are assumed to fill up an entire range of values along the horizontal axis. There are no breaks or gaps in a continuous curve, and it can be drawn without lifting the writing instrument from the page. A smooth, continuous graph is one with no sharp points. A continuous function is one whose graph is continuous.

cubic model A function of the form $f(x) = ax^3 + bx^2 + cx + d$. Cubic models have one change in concavity (one inflection point) and no limiting values.

data Real-world information recorded as numerical values.

definite integral The definite integral of a continuous or piecewise continuous function $f(x)$ from a to b is given by the limit

$$\int_a^b f(x)\,dx = \lim_{n \to \infty} [f(x_1) + f(x_2) + \cdots + f(x_n)]\Delta x$$

where $x_1, x_2, \ldots, x_n$ are the midpoints of n subintervals of length $\Delta x = \frac{b - a}{n}$ between a and b. If $f(x)$ lies above the x-axis between a and b, then the definite integral is the area of the region between $f(x)$ and the x-axis from a to b. Whenever $F(x)$, an antiderivative of a continuous function $f(x)$, is known, a more efficient method of evaluating a definite integral of $f(x)$ from a to b is given by the Fundamental Theorem of Calculus:

$$\int_a^b f(x)\,dx = F(b) - F(a)$$

demand curve A graph or equation relating the quantity of goods or services that consumers purchase and the price per unit of those goods or services.

derivative The mathematical term for an instantaneous rate of change. The terms *derivative, instantaneous rate of change, rate of change, slope of the curve,* and *slope of the tangent line* are synonymous. The derivative of a continuous function is given by the limit $f'(x) = \lim_{h \to 0} \frac{f(x + h) - f(x)}{h}$

discrete Discrete information is represented by a scatter plot or a table of data. Discrete graphs are scatter plots of data. In some situations, continuous functions are interpreted discretely; that is, outputs of the function have meaning in the context of a real-life situation only at some, not all, input values.

effective rate of interest *See* APY.

end behavior The behavior of a graph as the input becomes infinitely positive or infinitely negative.

equilibrium point The point at which the demand curve and the supply curve intersect. At this point, there is market equilibrium; that is, the supply of a product is equal to the demand for that product.

exponential model A model of the form $f(x) = ab^x$ or $f(x) = ae^{kx}$. Exponential models are characterized by constant percentage change (percentage differences) in output values when input values are evenly spaced. For the exponential model $f(x) = ab^x$, the constant percentage change is $(b - 1) \cdot 100\%$. In terms of the function $f(x) = ab^x$, exponential growth occurs when b is greater than 1, and exponential decline (decay) takes place when b is between 0 and 1.

extrapolation The process of predicting the output of a model by using an input outside the range of the data. Extrapolation should always be viewed with caution.

extreme point The point at which a maximum or minimum output occurs. At an extreme point on a graph, the slope of the tangent line is zero or does not exist. Extreme points occur at an input value, and the extreme value is an output value. Also see *absolute maximum/minimum* and *relative maximum/minimum.*

fixed costs Also called start-up costs, these costs do not vary with the number of items produced or the number of services performed.

function A function is a rule that assigns to each input exactly one output. Functions can be represented verbally by word descriptions, numerically in tables, graphically with pictures, or algebraically with equations. If x is the symbol for the input and f is the rule, then $f(x)$ symbolizes the output. Input/output diagrams display the input, how the input is measured (input units), the rule that relates the input and output, and the output, including how the output is measured (output units).

Fundamental Theorem of Calculus For any continuous function $f(t)$, the derivative of an accumulation function of $f(t)$ is the function f in terms of x. Symbolically, the Fundamental Theorem of Calculus is expressed as

$$\frac{d}{dx}\left(\int_a^x f(t)\,dt\right) = f(x)$$

This theorem verifies that if we begin with a function $f(t)$, find an accumulation formula or accumulation graph, $\int_a^x f(t)\,dt$, and then write the derivative formula or draw the slope graph, we get $f(x)$, the function with which we began, but in terms of x. This theorem connects two concepts of calculus: rates of change and accumulation of change.

future value The value of an investment at some time in the future. It is calculated using the appropriate compound interest formula. The future value of an income stream is the total accumulated value obtained by depositing the income stream as the money is received plus earned interest.

graph One of the ways to represent a function or a real-life situation. A graph is increasing if it rises, decreasing if it falls, and constant if it neither rises nor falls as you move from left to right along the horizontal axis. Most of the graphs used in this text are discrete, continuous, or piecewise continuous.

improper integral An integral of the form $\int_a^\infty f(x)\,dx$, $\int_{-\infty}^b f(x)\,dx$, or $\int_{-\infty}^\infty f(x)\,dx$. The value of an improper integral, if it exists, is found by replacing the infinite endpoint with a variable, performing the integration, and evaluating the limit of the result as the variable increases (or decreases) without bound.

income stream A flow of money being received continuously over time in varying amounts. When an income stream is invested in an interest-bearing account, the interest is considered to be compounded continuously. A discrete income stream is one into which money flows at regular intervals of time (quarterly, monthly, daily, and so on) with interest compounded at those intervals.

inflection point   A point where the concavity of a graph changes. Cubic and logistic models have inflection points. In real-life applications, this point is interpreted as the point of most rapid change or least rapid change.

input   See *function*.

input/output diagram   See *function*.

instantaneous rate of change   The instantaneous rate of change at a point on a curve is the slope of the curve at that point and the slope of the tangent line at that point. Instantaneous rates of change have labels of output units per input unit.

integration   The accumulation of change (evaluating a definite integral) or the process of recovering a quantity function from a rate-of-change function.

intercept   The input value where the graph crosses or touches the horizontal axis or the output value where the graph touches or crosses the vertical axis.

interpolation   The process of predicting the output of a model using an input inside the range of the data.

inverse function   If a rule obtained by reversing the input and output of a function is also a function, then it is called an inverse function.

left-rectangle approximation   A method of approximating the accumulated change in a quantity between two specified inputs in which the area between the rate-of-change graph of that quantity and the input axis is approximated with the sum of the areas of rectangles whose heights are the values of the rate-of-change graph at the left endpoint of each input subinterval.

limit   A limit of a function is an output value to which the function gets closer and closer as input either increases or decreases without bound or approaches some specified constant.

limit of sums   A method of evaluating a definite integral as the trend in the sum of the areas of approximating midpoint rectangles when the number of approximating rectangles becomes infinitely large.

linear model   A function of the form $f(x) = ax + b$ representing a situation in which change is constant. In the linear model, a is the rate of change (slope) and b is the output corresponding to an input of zero. When data input values are evenly spaced and the first differences of the output values are constant, the data should be modeled by a linear function.

local linearity   The principle that if we graph a smooth, continuous function over a small enough interval around a point, then we see a line that is, indeed, the tangent line at that point. That is, the tangent line and the curve are basically indistinguishable over the interval.

local maximum or minimum   See *relative maximum/ minimum*.

logistic model   A function of the form

$$f(x) = \frac{L}{1 + Ae^{-Bx}}$$

The number L appearing in the numerator is the leveling-off value or horizontal limiting value for the graph of the logistic function.

marginal analysis   A type of approximation of change used in economics. The rates of change of cost, revenue, and profit with respect to the number of units produced or sold are called marginal cost, marginal revenue, and marginal profit. These rates are often used to approximate the actual change in cost, revenue, or profit when the number of units produced or sold is increased by one.

market equilibrium   See *equilibrium point*.

market price   The actual price that a consumer pays for one unit of goods or services.

midpoint-rectangle approximation   A method of approximating the accumulated change in a quantity between two specified inputs in which the area between the rate-of-change graph of that quantity and the input axis is approximated with the sum of the areas of rectangles whose heights are the values of the rate-of-change graph at the midpoint of each input subinterval. A midpoint-rectangle approximation usually gives the least absolute error when compared with the errors involved with corresponding left- or right-rectangle approximations.

model   A mathematical model is an equation that describes a real-life situation. Modeling is the process of translating a scenerio from its real environment to a mathematical setting. There are three important elements to every model: an equation, a label denoting the units on the output, and a description of what the input variable represents.

optimization   The process of finding extreme points.

output   See *function*.

percentage change   When a function is represented by a table, a graph, or an equation, the percentage change in the function over an input interval can be found by dividing the change in the function over the interval by the function value at the beginning of the interval and multiplying the result by 100%.

percentage rate of change   The percentage rate of change can be found by dividing the rate of change at a point by the function value at the same point and multiplying the result by 100%. Percentage rates of change have labels of percent per input unit.

piecewise function   A function formed by combining two or more pieces of other continuous functions. A piecewise continuous function is a piecewise function that has one output value for every input value in some specified interval. A piecewise continuous function is not necessarily a continuous function.

point of diminishing returns   An inflection point beyond which output increases at a decreasing rate.

point of tangency   The point at which a tangent line touches a curve and has the same slope as the curve at that point.

polynomial model   A function of the form $f(x) = a_n x^n + a_{n-1}x^{n-1} + \cdots + a_1 x + a_0$, where $a_0, a_1, \ldots, a_n$ are constants and n is a positive integer called the degree of the polynomial. If $n = 1$, then the polynomial model is a linear model; if $n = 2$, it is a quadratic model; and if $n = 3$, it is a cubic model.

present value The amount of money that would have to be invested in an interest-bearing account now in order for the amount to grow to a given future value.

producers' surplus The amount that producers receive above the minimum amount they are willing and able to accept for a certain quantity of goods or services sold at the market price.

profit Total revenue minus total cost.

quadratic model A function of the form $f(x) = ax^2 + bx + c$. The graph of a quadratic function is called a parabola. When data input values are evenly spaced and the second differences of the output values are constant, the data should be modeled by a quadratic function.

rate of change See *derivative.*

recovering a function The process of beginning with a rate-of-change function for a quantity and finding its anti-derivative to obtain a function for the quantity.

relative maximum A point on a graph that is higher than any point in a small interval around it. A graph increases to the local maximum and decreases after it.

relative minimum A point on a graph that is lower than any point in a small interval around it. The graph decreases to the local minimum and increases after it.

revenue Quantity sold times the price per unit.

right-rectangle approximation A method of approximating the accumulated change in a quantity between two specified inputs in which the area between the rate-of-change graph of that quantity and the input axis is approximated with the sum of the areas of rectangles whose heights are the values of the rate-of-change graph at the right endpoint of each input subinterval.

scatter plot A discrete graph showing points in isolation from one another on a rectangular grid. A graph of data is a scatter plot.

secant line A line through two points on a scatter plot or a function graph. The slope of the secant line through two points is the average rate of change of the quantity between the input values of those two points.

second derivative The derivative of the derivative (provided it exists).

shutdown price An economics term that refers to the price under which producers are not willing or able to supply any quantity of particular goods or services to consumers. The point on the supply curve that corresponds to the shutdown price is called the shutdown point.

slope The rate of change of a linear model. The slope of the line tangent to a curve at a point is the limiting value of the slopes of nearby secant lines. The slope of a curve (graph) at a point is the slope of the tangent line at that point (provided the slope exists). The instantaneous rate of change (derivative) at a point on a curve is the slope of the curve at that point (provided the slope exists).

slope graph Also called the rate-of-change graph or derivative graph, the slope graph depicts the changing nature of the slopes of lines tangent to a graph of a function. Where the function is increasing, the slope graph is positive; where the function is decreasing, the slope graph is negative. Where the function has a maximum, has a minimum, or levels off, the slope graph is zero. Where the function has an inflection point, the slope graph has a maximum or minimum point.

supply curve A graph or equation that expresses the quantity supplied relative to the price per unit.

symmetric difference quotient A method of approximating instantaneous rates of change from data or an equation by using the closest point on either side of the point of interest and the same distance away. The symmetric difference quotient is the difference between the outputs of the two close points divided by the corresponding difference in inputs of the two close points. This technique should not be used if the data are not equally spaced around the point at which the rate of change is estimated.

tangent line A line that touches a graph at a point and is tilted exactly the way the graph is tilted at that point. The tangent line at a point on a continuous graph is the limiting position of the secant lines between nearby points and that point (if the limiting position exists). Provided the slope exists, the slope of the tangent line at a point is a measure of the slope of the graph at that point and gives the instantaneous rate of change of the graph at that point.

total cost The sum of the fixed costs and the variable costs.

total revenue The total amount that producers receive for a certain quantity of goods or services supplied at the market price. The total revenue equals the market price times the quantity supplied.

total social gain The benefit to society whenever consumers and/or producers have surplus funds. When the market price of a product is the equilibrium price for that product, the total social gain is the consumers' surplus plus the producers' surplus.

trapezoid approximation A method of approximating the accumulated change in a quantity between two specified inputs in which the area between the rate-of-change graph of that quantity and the input axis is approximated with the sum of the areas of trapezoids whose sides are formed by the heights of the rate-of-change graph at the endpoints of each input subinterval. The value of a trapezoid approximation equals the average of the corresponding left- and right-rectangle approximations.

variable costs Costs that change according to the number of items produced or the number of services performed.

vertical-line test A method of determining whether a graph is a function. If there is no input at which a vertical line cuts the graph in two or more places, then the graph represents a function.

APPLICATIONS INDEX

General Applications

Applications to Social and Behavioral Sciences

SUBJECT INDEX